FutureFish 2001:

The North Pacific Fisheries Tackle Asian Markets, the Can-Am Salmon Treaty, and Micronesian Seas (1997-2001)

C.D. Bay-Hansen

Illustrations by Charles A. Rondeau

© Copyright 2002 C.D. Bay-Hansen. All rights reserved.

No part of this publication may be reproduced, stored in a retrieval system, or transmitted, in any form or by any means, electronic, mechanical, photocopying, recording, or otherwise, without the written prior permission of the author.

Printed in Victoria, Canada

```
National Library of Canada Cataloguing in Publication

Bay-Hansen, C. D
 FutureFish 2001 : the North Pacific fisheries tackle Asian markets, the
 Can-Am Salmon Treaty, and Micronesian Seas (1997-2001) / C.D. Bay-Hansen ;
 Charles Rondeau, illustrator.
 Includes bibliographical references.
 ISBN 1-55369-293-4
  1. Fisheries--North Pacific Ocean. I. Rondeau, Charles A II. Title. III.
 Title: Future fish 2001.
 SH214.B39 2002       338.3'727'091643       C2002-901330-5
```

TRAFFORD

This book was published *on-demand* in cooperation with Trafford Publishing.
On-demand publishing is a unique process and service of making a book available for retail sale to the public taking advantage of on-demand manufacturing and Internet marketing.
On-demand publishing includes promotions, retail sales, manufacturing, order fulfilment, accounting and collecting royalties on behalf of the author.

Suite 6E, 2333 Government St., Victoria, B.C. V8T 4P4, CANADA
Phone   250-383-6864   Toll-free   1-888-232-4444 (Canada & US)
Fax   250-383-6804   E-mail   sales@trafford.com
Web site   www.trafford.com   TRAFFORD PUBLISHING IS A DIVISION OF TRAFFORD HOLDINGS LTD.
Trafford Catalogue #02-0106   www.trafford.com/robots/02-0106.html

10   9   8   7   6   5   4   3   2   1

Original title:

FutureFish in the Coming Conservative Century

A Complete Social, Political, and Economic Analysis
of the North Pacific Seafood Industry
throughout the 1990s

..."The Coming 'Conservative Century'"
–Irving Kristol, <u>Wall Street Journal</u>
1 February 1993

Cover painting: "Running of the Spring Salmon"
(Chuck Rondeau, 2000)

In Memoriam

Karl Olaf Arntzen
(died ca. 16 March, 1999, Gressvik, Norway)

POST PRODUCTION: 'The Annuals of a Norwegian-American Ex-Seafood Processor, 1991-1999'

"Follow [M]e, and I will make you fishers of men"
- - Matt. 4:19 b (KJV)

During the late 1970s Joe Upton came out with a successful book on the industry, <u>Alaska Blues: A Fisherman's Journal</u> (Anchorage: Alaska Northwest Publishing Co., 1977). I read it during the early 1980s, liked it well enough, but couldn't understand why it had done so well. But ... Joe Upton had been a true-blue <u>salmon</u> fisherman; in <u>Alaska</u> no less. Away down (in the Lower 48) in the industrial pecking order I had, briefly enough, been a lowly shellfish processor and sometime cook. But ... I had also been trained as a Pacific historian (U. Hawai'i at Manoa, 1979 - 1981), and possessed an M. A.; no less. Surely, then, I could write as well as Joe Upton could fish and take photographs? Meanwhile, things weren't working out as I had hoped, and I quit both the University of Victoria and the Province of British Columbia in January 1987, full of self-disgust and in self-disgrace. But ... now settled in Port Angeles, Washington, I did manage to complete an academic monograph on the North Pacific fisheries, eventually splitting the work into two small volumes (<u>Fisheries of the Pacific Northwest Coast</u>, Vols. 1 and 2), and had them put out sequentially (1991 and 1994) by a subsidy publishing company (Vantage Press, Inc.) in New York City.

As I draw near to completing this third book, I realise that if those first two small volumes had <u>too much</u> fish content; <u>FutureFish in the Coming Conservative Century</u> has perhaps too little. As they say, "You can't please 'em all", and <u>Fisheries of the Pacific Northwest Coast</u>, both volumes, did not do well commercially, although the few letters (and fewer reviews) were favourable. I decided by 1994 that <u>Future Fish</u> would have something for everybody -- and it does. My <u>opus magnum</u> started out as one thing, but will end up as something else entirely. For <u>FutureFish</u> has grown far beyond its original parametres, becoming a personal journal (like Joe Upton's), with each chapter (i.e., "The Annuals"; every chapter a year, for 1991 - 1992, 1994 - 1999) remaining unchanged as written at the time. Despite being somewhat a lexicon of popular culture in parts, and always reflecting my traditional Christian world-view, <u>FutureFish</u> is a <u>history</u>. I am, after all, just a historian -- a mere "Moog - synthesizer" of historical information, no matter how apparently low in nature or seemingly insignificant in source. Thus <u>FutureFish</u> stretches a span -- in time and space -- that ranges from Newfoundland to the New Age; from Microsoft to "Macronesia". There are portraits in depth---and brief sketches of personalities, including cyberweasels Bill Gates and Paul Allen; undead counterculture idols Jimi Hendrix and Jerry Garcia; fear-mongers Ted Kaczynski and Jeremy Rifkin; pc and cw scientists Carl Sagan and Arthur C. Clarke; paradigm - shifters Charles Darwin and Albert Einstein; frightful [fellow] Norwegian Knut Hamsun, and fearsome [fellow] American Thorstein Veblen.

Indeed, I have sometimes gotten carried away, and have all too often vented my aching spleen against hypocrisies perceived in Norwegian and American religious, cultural, and political institutions. (Nonetheless, I must have thrown out more than a hundred pages of notes and scribblings.) But the essential subject of discourse has remained constant throughout: The North Pacific fisheries and seafood industry, with its epicentre at North Seattle, Washington, U. S. A., 1991 - 1999. There are solid sections on the Western Pacific Rim, neo-Darwinism and the new physics, Seattle and cyberspace, Norwegian America, the New Age religion of ecologism; "weird science" vs " wise use", and, of course, the future of fish, fishing, and the seafood industry of our Pacific Northwest Coast. (And the reader should note that the root word "conserve", as in "conservative", is the same root word as in "conservation".)

                                                                                      C. D. B - H.
                                                                                      Port Angeles, Wash.
                                                                                      July 1998

Author's Update 2001

This is the companion volume to **FutureFish in Century 21: The North Pacific Fisheries Handle Coming Trends, Radical Environmentalism, and Digital Cyberspace (1991-1992, 1994-1997)**. Like the latter, this volume was (also) gleaned from the original FutureFish--which had by 1999 grown to an unwieldy tome numbering 657 type-written pages! As told in Author's Update 2000 in my first volume of FutureFish in Century 21, the split into two books came naturally and easily (after some soul-searching). But the creation of two smaller offerings was essential in meeting Trafford Publishing's 500 page-limit for "perfect bound". It has ended as a painless Solomonic compromise despite dividing my "baby"!

This companion volume, **FutureFish in Century 21: The North Pacific Fisheries Tackle Asian Markets, the Can-Am Salmon Treaty, and Micronesian Seas (1997-2001)**, also discusses aqua/mariculture, Norwegian Americana, and the significant U.S. elections of 1996 and 1998. Most important for this writer, however, has been the chance to include an addended chapter, 'A Modest Proposal for Micronesia'. As a 1981 U. Hawai'i M.A. graduate in Pacific history, doing a section on the Pacific Islands--after a long absence--has been particularly gratifying. I sincerely hope both Canadian and American readers enjoy it.

Thanks again to Mr. Francis E. Caldwell of Port Angeles, Wash., and to Trafford's Publishing of Victoria, B.C., for the opportunity to publish and publicise both my books. As expressed in the Trafford 'Author's Guide', "Some people see on-demand publishing as a paradigm breakthrough in terms of democratizing book publishing. If it proves to be that significant, it's anyone's guess about all the ramifications." Right on, eh?!

C.D.B-H.
Port Angeles, Wash.
March 2000

Table of Contents

Pages			
	i		Post Production
	iii		Author's Update 2001

	Chapter 1 :		Asia-Pacific Rim Rising
1	Overview Update:		"The Pacific Age"
15	Part I	:	Japan—"The Best and the Brightest"
48	Part II	:	China—"Enter the Dragon"
74	Part III	:	Small Tigers and Little Dragons
98	Flashback :		Once in a Blue Moon

	Chapter 2 :		On the Home (Fishing) Front
104	Entry	:	Anatomy of a Treaty
126	Part I	:	Oh, Canada!
141	Part II	:	Alaska, the Last (Fishing) Frontier
159	Part III	:	The Lower 48, Northwest Coast

	Chapter 3 :		The Summing Up
178	Part I	:	U.S. elections 1996—A Retrospection
191	Part II	:	Future (Fish) Imperfect
210	Part III	:	"Ha det, og takk for all fisken!" ("So long, and thanks for all the fish!")

249	Special Supplementary Bibliography

	Chapter 4 :		"A Modest Proposal" for Micronesia
251	Preface	:	Millennium Madness
255	Foreword	:	Micronesia Notes
264	Part I	:	The United States, the United Nations, and Micronesia
323	Part II	:	UNCLOS III—The Third United Nations Conference on the Law of the Sea
337	Part III	:	Tropical Western and Mid-Pacific Fisheries
358	Part IV	:	"A Modest Proposal" for Micronesia
405	Addendum:		"Just Another Shitty Day in Paradise"? The Writers' View of Post-Pacific War Micronesia, 1945-2001

Supplemental Micronesia/United States Bibliography

Appendices

Acknowledgements

Chapter 1 :	Asia-Pacific Rim Rising

Overview Update: "The Pacific Age"

Part I : Japan -- "The Best and the Brightest"

Part II : China -- "Enter the Dragon"

Part III: Small Tigers and Little Dragons

"We are in a transition to a new era....There will still be faster growth, but it will be more stable with the peaks lopped off and the valleys filled in....It's a very good atmosphere as far as the eye can see."
-- Economist Lawrence B. Krause, U. Cal. San Diego, 1997[1]

Overview Update: "The Pacific Age"

"The Pacific Age ...faster than a speeding train...."
-- Orchestral Manoevres in the Dark, 1986[2]

Much has happened in East and South-East Asia since the writer last tackled the Asia-Pacific economies and fisheries in 1991-1992. The reader would have to turn back to the very beginning of this book to find information on the Western Pacific Rim. The early 1990s were a time of unbounded optimism concerning the entire Asia-Pacific region. An example of this boosterism was reflected in The Pacific Century by Frank Gibney.[3] In it, he wrote: "We will live more and more in a world looking westward, and we stand poised on the verge of what will truly be a 'Pacific Century'."[4] A lieutenant during World War II in U.S. Navy Intelligence, Frank Gibney also served in post-war occupied Japan and returned later as Time magazine bureau chief (1949-1950). Gibney eventually founded the Pacific Basin Institute in 1979 at Santa Barbara, California, and at some point was awarded an honourary doctorate of literature from Kyung Hee University at Seoul, South Korea.[5] So Frank Gibney appeared to be an "old Asia hand" worth heeding, and the extra-large text and lavish photographs within The Pacific Century reflected his enthusiasm for -- and belief in -- Asia-Pacific Rim Rising. This writer, despite his own short experience and limited knowledge of the Asia-Pacific area, shared Gibney's confidence in 1992 and does now half a decade later. But Frank Gibney was merely one example of Asia-Pacific boosterism. Just about all prognostications for the region were glowing. In The Pacific Rim Almanac(1991), an 800-plus-page encyclopaedia of Rim knowledge and facts, economic prophesies were buoyant:

"In the 1990s the Pacific Rim will exceed the rest of the world in economic growth -- as it has for the past several years. Over the next two years [1992, 1993], analysts at the Bank of America predict that the Rim will average 4.8% annual growth, compared with 2.7% growth for Western Europe and 2% for North America. The overall growth figure for the Rim is skewed since Japan accounts for the lion's share of the Rim's economic activity. Japan is expected to grow at a 'modest' 4% per year rate over the next few years. The majority of other Rim markets will grow at a far greater pace than 4.8%."[6]

The Pacific Rim Almanac described both Japan and China as politico-economic subregions of their own: Japan, as an extremely homogeneous and democratic society accounting for 70% of Rim economic activity; heterogeneous China, as a huge potential consumer market with political turmoil (1989) under Communist dictatorship. Both nations, according to The Pacific Almanac, would have profound political and economic effects on the whole Asia-Pacific area.[7] The next major subregion was the aggregate of "newly industrialised countries" (NICs) of East/South-East Asia, the so-called "Four Tigers" of Hong Kong, Singapore, South Korea (R.O.K.), and Taiwan (R.O.C.). The "lesser developed countries" (LDCs) of South-East Asia were Indonesia, Malaysia, Philippines, and Thailand. The common wisdom in 1991-1992 was that the latter LDCs would undergo the more rapid growth throughout the decade.[8] This writer refers to these South-East Asian LCDs as the "Four Little Dragons", due to the predominant political and economic influence of Overseas Chinese. But in 1997 much has changed in both East and South-East Asia. Hong Kong disappeared as a political entity on 1 July, to be absorbed by Mainland China; and Vietnam (with a long coastline on the South China Sea for marine activities) has emerged as a real participant in the greater South-East Asian economy. There are now three small Tigers (instead of four) and five little Dragons (instead of four)...still totaling eight always-improving players on the new Asia Pacific field.

The Asia Pacific Economic Community (APEC) trade group was formed in 1989, with members Australia, Brunei, Canada, China, Hong Kong, Indonesia, Japan, [South] Korea, Malaysia, New Zealand, Philippines, Singapore, Taiwan, Thailand, and the United States.[9] Seattle, Washington, was the host city of the mid-late November 1993 APEC conference in which U.S. President Clinton had a one-on-one series of meetings with Asia-Pacific leaders.[10] This writer, as an adopted Washingtonian, watched and read with great interest the developments of "The Pacific Forum" during November 1993. That year was also the Clintonista first-year frenzy of NAFTA (North America Free Trade Association; i.e.,Canada, U.S.A.,Mexico) and general global free trade fever. Although those days at the Seattle conference were filled with Clintonian double-speak and political posturing, the APEC forum reminded (once again) Pacific Northwesterners -- British Columbians too -- that they are part of the Eastern Pacific Rim; itself the integral half of the vastly important Asia-Pacific Rim. While the omnipresent Bill Clinton schmoozed and smoothed the ruffled feathers of America's disgruntled Asian trading partners, a fully awake Secretary of State Warren Christopher was evidently enjoying his rôle as apex U.S. negotiant. The <u>Seattle Post-Intelligencer</u> cited Warren Christopher as having said for posterity:

"As factories bloom across China, political power will come increasingly from the end of an assembly line....[APEC would help realise the vision of] a new Pacific community based on shared strength, shared prosperity and a shared commitment to democratic values....Trade and investment within the region are weaving a new web of human and commercial relationships.'"[11]

<u>Seattle P-I</u> correspondent Joel Connelly also quoted an observant University of Washington student, one Curt Clark, as having commented, "He [Christopher] wasn't real specific about anything.'"[12]

But -- no matter. It was a time of personal and political triumph for Bill Clinton. For on that very day, 18 November 1993, the U.S. House of Congress voted Aye on the controversial NAFTA free trade pact with neighbours Canada and Mexico -- a crucial victory for the Clinton administration.[13] The American President couldn't afford to look like a loser in the sceptical eyes of Asia-Pacific heads of state. But the ever-politically-alert Bill Clinton remembered not to forget the local [voting] constituents while addressing the APEC-summit organisers: "As we make these efforts, United States business must do more to reach out across the Pacific. I know Seattle's business community understands that potential that lies in the Asia-Pacific region, but millions of our businesses do not.'"[14] In other words, <u>You</u> Seattleites are <u>sophisticated</u> ...unlike some of <u>those</u> other folks....Quick to take the proferred bait was an enthused Bob Gogerty, Seattle corporate and public affairs consultant: "I remember Dan Evans, when he was governor, saying forever, "The Pacific century is about to begin." It has.'"[15]

But this writer remembers wondering, those four years ago, if any of the APEC conference jive <u>cum</u> promises would (or could if ever) benefit the Northwest Coast commercial fishing industry? Or was the entire "Pacific Forum" purely a hyped show at Seattle for the sole attention (and advantage) of Boeing and Microsoft? Time would tell. It is now late October 1997. Asian stock markets are presently in a turmoil, adversely affecting those of Australia, New Zealand, United States and Canada. Whatever the short-range outcome, the long-range outlook for the Asia-Pacific Rim is truly awesome, with the trends of the early 1990s being well-nigh irreversible. This writer still concurs with retired U.S. Pacific Fleet commander, Admiral James A. Lyons Jr., who projected that

"The compass needle of history has swung away from the Mediterranean and Atlantic civilizations and is pointing toward Asia. By 1990, 62% of the world's population will live around the Pacific Rim. The Pacific region already accounts for 60% of the world's GNP (gross national product, the value of all goods and services). Our trade with Asia is nearly 40% greater than with Europe. The next 100 years will be the century of the Pacific.'"[16]

There have been attempts within the Asia-Pacific community to establish localised trading alliances, with varying results. In early 1993 Brunei, Indonesia, Malaysia, Philippines, Singapore, and Thailand formed the ASEAN Free Trade Area (AFTA).[17] AFTA was set up by the six member nations to lower trade barriers among themselves; but, despite a proposed 15-year period to fully implement the agreement, there has been both internal and external discord. Many Asian countries seem to prefer membership under the powerful, global free-trade umbrella of the 15-nation APEC.[18] But, as futurist John Naisbitt has written, Asians are interested in keeping their borders -- and options -- open. Indeed, some envision the eventuality of a Pacific Area Free Trade Agreement, coupling APEC to NAFTA, thereby spawning the inevitable PAFTA (Pacific Area Free Trade Area).[19]

Futurist John Naisbitt of Telluride, Colorado,[20] is fascinated with Asia, especially the Chinese (Mainland and Overseas), and his keen interest shows in his superior 1996 work, <u>Megatrends Asia: Eight Asian Megatrends That Are Reshaping Our World</u>.[21] In it, Naisbitt has updated the impressive growth market numbers for ASEAN since his <u>Global Paradox</u> (1994). In 1996, the six nations comprising ASEAN encompassed a combined population of 359 million people and an import market valued at $226 billion U.S.[22] And since ASEAN admitted Vietnam in 1995, the population has expanded to nearly 450 million (almost half a <u>billion</u> people!). According to Naisbitt, Cambodia (Kampuchea) has applied for observer status and both Burma (Myanmar) and Laos are candidates for future membership. Malaysia invested more than $600 million in next-door Indonesia during 1994; over ten times the amount invested during 1993. And, in Naisbitt's words, "Thai companies led by communications giant Shinawatra, agri-industrial conglomerate Charoen Pokphand and the government petroleum authority are starting various projects in the Indochina states of Laos,

Cambodia and Vietnam."[23] Whew! When this writer was briefly in Thailand during 1971, local entrepreneurs were literally begging visiting Americans, Australians, and Japanese for business investments. Conditions have assuredly changed in the new Asia.

An example of the new Asia success story has been Taiwan, the Republic of China (R.O.C.). Today Taiwan is the world's 13th-largest trading nation, a financial and manufacturing colossus (in-between the sizes of Vancouver Island, B.C. and Switzerland). Less than half a century ago, Taiwan (Formosa) was an Asian agricultural "backwater" land, but by the 1960s the GNP rate averaged more than 9%; 10.2% in the 1970s, and 8.2% in the 1980s. The R.O.C.'s GNP is set to increase on an average of 9 to 10% from 1995 to 2005 A.D.[24] Most of Taiwan's stunning prosperity has been engendered by export, mostly labour-intensive manufactured produce like clothing and footwear. In addition to exporting textiles and gym-shoes, Taiwan is now a leading innovator of high technology. In 1994, Taiwan's exports totaled $91 billion, with imports amounting to $84 billion -- a fair trade. Taiwan's good accounting and smart economics have resulted in the tiny island-nation possessing the globe's largest foreign-exchange reserves. The R.O.C.'s next goal, according to futurist Naisbitt, is..."to become a regional center for international corporations...[and that is] realistic, considering that it lies between northern and southern Asia [i.e., East and South-East Asia], is near [Mainland] China, and has a skilled labor force, immense capital and a long tradition of supporting the manufacturing needs of multinationals...."[25]

Taiwan is also on the cutting edge in Asian cyberspace, and has set aside $114 million to be spent on computers for national schools by 1998.[26] Furthermore, Taiwan's Acer is all Asia's principal computer company, and is determined to create localised versions of itself in all neibouring countries.[27] Getting onto the Internet has become the latest Asian craze. The Internet is, of course, the crucially important entrance-ramp onto the Infobahn, and a rising generation of home-grown cybernauts is already familiar with electronics technology, is receptive to new ideas, and will we hope (especially for Mainland China and North Korea) act as agents of social, cultural, and political change throughout Asia.[28] Futurist Naisbitt has passed on a telling anecdote:

.... "An expatriate executive at Microsoft in Singapore was amazed at how much interest teenagers have in computers compared with their peers in the United States. He said, when he returns home, the common talk among teenage girls is about becoming beautiful, going out on a big date and other stuff about having fun. In Asia, the girls are talking about Pentium, how many megabytes you need to run a certain software and Windows 95. They are ready for cyberspace."[29]

(Earlier in this work, the writer expressed the thought that the politically-incorrect personae of tomorrow might well be fish-eating Asian females rather than meat-eating European males! Perceived superiority in one group breeds resentment in another.)

James Dale Davidson, of the Baltimore-based Strategic Investment Ltd. Partnership, has written that enormous amounts of clerical work are at present being farmed out to places so disparate as Ireland (Éire) or the Philippines -- where labour costs are cheaper, and the work force is both more motivated and more literate.[30] (And this writer would wager that the vast majority of these quick-witted, nimble fingered workers are women.)

There is a downside to digitisation, and two South-East Asian nations have been variously commended and castigated in the Western press for over-zealous government [mis]use of electronic information. A 1991 Time magazine article described the Thai government as amassing personal statistics on private Thai citizens; by 2006 A.D. such vital electronic data will have been stored on 65 million Thais in a single computer network. The Thai system is being seriously considered for use by Indonesia and the Philippines.[31] (Imagine, gentle reader, such a system under the control of a military dictatorship or a unitary world government!) And the city-state of Singapore might soon become the globe's very first cashless society...whether or not the voting populace wants plastic rather than paper. According to Insight magazine,

...."No cash transactions are permitted in major department stores, supermarkets, gas stations, hospitals and government offices. The scheme, under which shoppers use their plastic automated teller machine [ATM] cards to pay for goods and services, began weeks ago....To promote the new system, the government closed all cash payment offices, leaving citizens little choice but to join Singapore's ceaseless drive toward a high-tech -- but cashless -- society."[32]

Religion on the Rim sharply divides the Asian mainland (save Russia) from the Pacific Islands and Eastern Pacific littoral. The Asian mainland is overwhelmingly Buddhist, Taoist, and Confucianist, with substantial numbers of Christians in China and South Korea. In the R.O.K. a full 40% of the people are Christians; 700 thousand of them members of the Reverend David Yonggi Cho's church in Seoul.[33] The

Reverend has claimed his church as the "largest Christian congregation in the world.'"[34] David Barret, editor of the World Christian Encyclopedia, has estimated that 8% of Asia's population is Christian. In China, some 25,000 individuals join a Protestant denomination every single day; three new churches open every other day.[35] Island South-East Asia is home to the world's largest Islamic country, Indonesia, with the Hindu island of Bali as an anomaly. Neighbouring Brunei and Malaysia are also Islamic nations, with Muslim enclaves in the southern Philippine areas of Mindanao and Sulu. The remaining 85% of the Filipino population is Christian, in Asia's only predominantly Roman Catholic country.[36]

It is not the Protestant work ethic, therefore, which drives the red-hot Asian engine; it is, rather, what Canadian writer Eric Downton in 1986 dubbed "The Confucian Connection."[37] It was Eric Downton's sub-thesis, in his excellent Pacific Challenge, that [Asian] societies with Confucian roots were proving better able to cope with industrial changes and technological challenges on the eve of Century 21, than those [Western] cultures with a Judaeo-Christian heritage.[38] Many Western observers, like Eric Downton, have perceived the Confucian work ethic -- spread throughout the Western Pacific Rim by tens of millions of energetic and acquisitive Overseas Chinese -- to have been the prime mover and motivating factor in East and South-East Asia's present material prosperity and financial power. The up-side of Confucianism is a "respect for learning, compulsion to excel, and happy cooperation between amenable subordinates and responsible superiors."[39] There is a dark side to Confucianism; namely official theft and despotic corruption, but it is the up-side Confucian work ethic that has created the atmosphere and conditions for ushering in the Pacific Age.[40]

Confucianism, of course, is derived from the fifth century B.C. thoughts and precepts of self-educated teacher, Master Kung, known to the West as Confucius (K'ung Ch'iu, Kung Fu-tze, or Kong Fuzi). Master Kung (ca.551-478 B.C.) lived, moved, and had his being in the North China state of Lu. The authorship of Analects is attributed to Master Kung, a little book which has had a far greater (and more lasting) impact on the Chinese cultural sphere than that of Chairman Mao.[41] Master Kung's Analects emphasised propriety in personal and social behaviour; filial piety and the correct observance of ancestor "worship"; and bilateral responsibility between master and servant, ruler and subject. By the second century B.C., Confucianism had become enshrined as the state "religion" of China, and the teachings of Master Kung became the bed-rock foundation of the Chinese cultural sphere -- China, Japan, Korea, Singapore, Taiwan, and Vietnam. (And for Overseas Chinese living in the Indian cultural sphere-countries of Cambodia, Laos and Thailand.) As we have seen, the Confucian model is credited for the resounding success story of Japan and the Four Tigers (Hong Kong, Singapore, South Korea, and Taiwan).[42] Once Mainland China -- plus North Korea and already evolving Vietnam -- becomes a free and democratic society, the Confucian model will retransform its ancestral home. That change is already under way.

Regarding Eastern religions about the Rim; it has been a case of what has gone around has come around. Zen (Ch'an) Buddhism and Taoism were eagerly received in North America by faithless Gentiles and secular Jews during the tumultuous 1960s, mostly on the West ("Left") Coast. Since then, the state of California (mainly) has served as the epicentre of the planet's New Age movement, making-over Eastern religions such as Zen Buddhism and Taoism and re-exporting them back in recycled form to East and South-East Asia. In some cases the occultic New Age beliefs have been re-processed, re-packaged, and sent back (yet again) to the West Coast. The Pacific Rim Almanac told the tale of Thai-Chinese master Mantak Chia, who in the late 1980s, promoted Taoist longevity secrets, personal alchemy practices, and esoteric sexual techniques in the West.[43]

And none other than Scottish-born guru-channeler Benjamin Creme visited Japan during the early 1990s. At one New Age convocation in Tokyo -- complete with chanting Buddhist monks and dramatic Nō players dancing -- Creme informed his enraptured audience that his messiah, the Lord Maitreya (living in London's East End), had caused the recent jolt to the (Nikkei) Stock Exchange. This would be a prelude to a world-wide stock market collapse, explained Creme, as the globe is moving away from a monetary system. Benjamin Creme even had the temerity to play a tape of the channeled Lord Maitreya (known in Japan as Miroku Bosatsu), who spoke with a decidedly British accent![44]

Also during the early 1990s the well-known Japanese actor, Tetsuro Tamba, personally organised the "Afterlife Research Association". As self-appointed PR man for the spirit world, Tamba-san would help sell ouija-boards, occultic literature, and conduct "tours of the afterlife" (complete with a guarantee of no physical harm).[45] In today's Japan (the most completely "Californicated" Asian country, with the Philippines being the most "Hawaiianized"), reflecting ever-increasing interest in New Age "spiritualism", are bookstores with whole New Age sections. Like their counterparts in California and the West, these Japanese bookshops feature works heavily "into" reincarnation and channeling.[46] The Cosmo Space New Age Center in Harajuku, Tokyo, sells up to $2 million U.S. worth of New Age titles per annum. The Cosmo Space New Age Center's best-selling gimmicks include "power stones that are said to help people meditate or regain their stability'."[47] (Is the Rising Sun already setting?)

In California itself on the New Age Pacific Rim, hard-driving but laid-back executives, during the early 1990s, discovered a new "home"-grown productivity-tool for the office: A didjeridu (dijeridoo). Yes indeed; the 40,000 year-old Australian Aboriginal woodwind instrument would empower Left Coast [customer] businessmen. Thus conjectured one Fred Gray Tietjen, a San Francisco entrepreneur who had developed (and was marketing) a California high-tech version of the ancient musical instrument.[48] Rationalised Tietjen (who also ran a Bay Area "shamanic counseling" practise): "By triggering the sound of the didjeridu, they [the goofy executives] can learn to alter their state of consciousness and become extremely empowered.'"[49] (It cost $85 U.S. to find out how to get a good drone going in about ten minutes.)

It is Hawaii, though, in this writer's own experience, that reigns as the Pacific Rim's New Age paradise. At least one author, Englishman Simon Winchester, seems to agree, even if arriving at a very different conclusion. Winchester's opus parvum on the Rim, Pacific Rising, was published in 1991 and grandly subtitled: The Emergence of a New World Culture.[50] There has been a long-time trend, for the last half-century in writings on the Rim, to hail the Coming of the Pacific [New] Age; to worship the (new/old neo-Pagan) "Pacific Way". Author Winchester appears to be one of those haole (White) worshipers who await the Coming. Despising Atlantic (i.e., European) civilisation and eschewing Western (i.e., Judaeo-Christian) credoes, these writers on the Pacific Rim have ranged from late (and great) James A. Michener during the 1950s to John Naisbitt and Simon Winchester in the 1990s. This writer is not about to make too much of the above cultural "correctness" as a Pacific man himself; however, when encountered, the new pc (Pacific correctness) would try the patience of a Job. Simon Winchester's knowing wink and patronising smirk were virtually palpable when he leeringly dealt with two wahine (female) New Agers on Kaua'i:

"...[O]ne of the people I was hoping to see was a woman from Pittsburgh who had settled in the old Kauaian sugar town of Hanapépé because she believed in the presence of 'a vortex' nearby. On arrival she had changed her name from Ann Merritt to Roberleigh Hale because, she explained, 'numerologically speaking, Ann Merritt sucks', and she was a woman deeply impressed by the strange 'magick' of numbers. She was also very anxious about health -- hers and that of those around her. So she bought herself a tiny hotel and now offers to her visitors a sombre regime of massages, strenuous bouts of colonic irrigation, meals purged of all free oils and dairy products, also phrenology, iris diagnosis, aromatherapy -- and palm readings."[51]

And <u>wahine</u> number two... "[a] fanciful woman, wife of a teacher at Waimea School, apparently her present task -- undertaken with the kind of studied seriousness found among Hawaiian eccentrics -- was to compile an illustrated sex manual for centaurs, a job made difficult [she said] by her inability to draw <u>good hooves</u>. Centaurs, she assured me... tended to favour the missionary position, somewhat different from the standard equine approach liked by regular horses. She liked Kauai for the same reason as Dr. Hale, in that it was sited on a vortex of celestial power -- she had been up at dawn, praying for the successful conclusion to the Harmonic Convergence of the Planets, due that particular day and visible (or whatever) from Kauai. 'It is that sort of place', she said in a whisper. 'We're into things like that on Kauai. Nowhere else in the States seems quite so tolerant.'"[52]

No indeed! Where else -- save California perhaps -- would countenance such metaphysical nonsense? From the shores of the U.S. Mainland whence once sailed trickles of resolute Christian missionaries, now fly in aeroplane loads of human social flotsam, cultural jetsam, religious froth, and spiritual scum, from the teeming, pestilential cities of the "great harsh continent of Eldollarado" (to quote Ian Fleming). It is escape to <u>California Ultima</u>, flight to <u>Hawai'i Nei</u>. But it is also the magic -- not "magick" -- of the Pacific Basin. It is beautiful, it is vast, and it is a place of starting over, of renewal. (If Siberia and Manchuria are considered parts of the Pacific Rim, so too then are Alaska, "the last frontier", and Montana, "the last best place".) The Pacific Age is imminent; of that there is no doubt. But rather than morphing into a Macronesia or greater California, the entire Asia-Pacific region may well become centre-stage for a religious revival; in addition to the cultural, social, and economic transformation in Century 21. Today, there are most likely millions more born-again Christians in imprisoned China than there are in all the free United States. Who knows? The islands and littoral of the Pacific Basin (this time around) might be re-Christianised, during the next millennium, by today's inspired believers from the Chinese mainland instead of yesterday's traditional missionaries from the American mainland! <u>Pakē</u>s (Chinese) converting <u>haoles</u> instead of the other way around!

But for now, within the United States, there hangs the spectre of oxymoronic same-sex marriage over Hawaii. This pernicious legal carcinoma -- if approved by the Aloha State -- would quickly metastasise throughout the Left Coast, and from there into the American heartland. So at present, with California embodying the epitome/epicentre of terrestrial and celestial wrath, One World and New Age hopes and dreams reside in the Pacific Way. Palaeo-conservative commentator, Philip Jenkins of <u>Chronicles</u> magazine, has written in warning of the coming Pacific century:

....."The model megalopolis of the new era is the utopian city of Los Angeles, with its 'correct' orientation toward the Pacific, rather than to what was once the American heartland. To fulfill our destiny in the Pacific century, it is first necessary to abandon those financial and economic restraints which prevent the ultimate merger into the nationless world federation. And someday, the final frontier will be attained: 'Next year in Tokyo...' is the ideal."[53]

We shall see. This writer, despite the high-flying pretensions of APEC, perceives the Pacific Century as having the exact opposite effect. Confucian pragmatism plus the common sense of the Protestant work-ethic, will combine to ensure natural financial and economic restraints. And the ancient patriotisms of the Western Pacific will work together with the newer nationalisms of the mid-Pacific and Eastern Rim, to stay the globalist hand from enforcing political union.

NOTES

1. 'Pacific future bright', <u>Peninsula Daily News</u>, 8 June 1997, p.E-1.

2. O.M.D., <u>The Pacific Age</u>, Virgin Records, Ltd., 1986.

3. Frank Gibney, <u>The Pacific Century</u> (New York: Charles Scribner's Sons, 1992), inside front cover.

4. <u>Ibid</u>., inside front cover.

5. <u>Ibid</u>., inside back cover.

6. Alexander Besher, ed., <u>The Pacific Rim Almanac</u> (New York: HarperCollins Publishers, 1991), p.78.

7. <u>Ibid</u>., p.79. Cf. Alvin and Heidi Toffler, <u>War and Anti-War: Survival at the Dawn of the 21st Century</u> (Boston: Little, Brown and Company, 1993), pp.214-217.

8. <u>Ibid</u>., p.79. NB: The Association of South-East Asian Nations (ASEAN). The other two ASEAN members were Brunei and Singapore, but even then hardly LDCs.

9. Merrianne Bieler, 'Barriers to International Trade', <u>Studium Generale Program Notes</u>, 18 November 1993, Peninsula College, Port Angeles, Wash.

10. Joel Connelly, 'Clinton arrives today', <u>Seattle Post-Intelligencer</u>, 18 November 1993, p.A-1, NB: The APEC conference was grandiosely referred to as "The Pacific Forum" by Seattle's self-important chattering-class clerisy.

11. Joel Connelly, 'Clinton: Machinists will greet him', <u>Seattle Post-Intelligencer</u>, 18 November 1993, p.A-16.

12. <u>Ibid</u>., p.A-16.

13. 'Leaders, Clinton meet for APEC', <u>Peninsula Daily News</u>, 19 November 1993, p.A-1.

14. 'APEC EXTRA -- Quotebook', <u>Seattle Post-Intelligencer</u>, 20 November 1993, p.A-7.

15. <u>Ibid</u>., p.A-7.

16. <u>The Pacific Rim Almanac</u>, p.287.

17. John Naisbitt, <u>Global Paradox</u> (New York: Avon Books, 1994), pp.308-309. ASEAN: Association of South-East Asian Nations.

18. <u>Ibid</u>., p.309.

19. <u>Ibid</u>., p.309. NB: For dangers inherent in APEC and the inevitable PAFTA, see Grant R. Jeffrey, <u>Final Warning: Economic Collapse and the Coming World Government</u> (Toronto: Frontier Research Publications, Inc., 1995), pp.88-89.

20. Telluride, Colo., has its own bumper-sticker: "T-Hell-U-Ride". The word <u>telluric</u> means "pertaining to the earth", and Telluride appears to be yet one more well-heeled and self-confident Colorado New Age community.

NOTES (cont'd)

21. New York: Simon & Schuster, 1996.

22. Megatrends Asia, p.134.

23. Op.cit., John Naisbitt, p.134.

24. Ibid., p.177.

25. Op.cit., John Naisbitt. This writer's brackets.

26. Ibid., p.76.

27. Ibid., p.120.

28. Ibid., p.77.

29. Op.cit., John Naisbitt, p.197.

30. James Dale Davidson, The Capitalist Manifesto (Baltimore, Md.: Strategic Investment Ltd. Partnership, 1995), p.69.

31. 'Peddling Big Brother', Time, 24 June 1991.

32. Insight, 10 March 1986. NB: For more on Singapore's "benign techno-dictatorship", see Tal Brooke, ed., Virtual Gods (Eugene, Ore.: Harvest House Publishers, 1997), pp.32,87.

33. Megatrends Asia, p.80. Author's note: The Reverend David Yonggi Cho is most probably the son of Pastor Paul Yonggi Cho, discussed in the previous chapter.

34. Ibid., p.80.

35. Ibid., p.79.

36. Ibid., p.79.

37. Eric Downton, Pacific Challenge (Toronto: Stoddart Publishing Co. Ltd., 1986), p.68, ff.

38. Ibid., n.p.

39. Op.cit., Robert Elegant, 'Where the Action Is', National Review, 10 May 1993, pp.36-37.

40. Simon Winchester, Pacific Rising (New York: Prentice Hall Press, 1991), pp.61-62.

41. The Pacific Rim Almanac, p.392.

42. Ibid., p.393. Cf. Lin Yutang, ed., The Wisdom of Confucius (New York: Random House, Inc., 1966), passim.

43. The Pacific Rim Almanac, p.393.

NOTES (cont'd)

44. Ibid., p.394.

45. Ibid., p.394.

46. Megatrends Asia, p.80.

47. Ibid., p.80.

48. The Pacific Rim Almanac, p.393.

49. Ibid., p.394. This writer's brackets and contents.

50. New York: Prentice Hall Press, 1991.

51. Op.cit., Simon Winchester, pp.368-369.

52. Ibid., p.374.

53. Philip Jenkins, 'Friends All Over the World', Chronicles, Vol.19, No.5, May 1995, p.33.

Part I: Japan--"The Best and the Brightest"

"Why are you [Americans] neglecting us to pay so much attention to China? China can never be as important to you as Japan!'" -- Wataru Hiraizumi (Japanese chairman of LDP Foreign Affairs Committee) to Robert Elegant, author of Pacific Destiny, 1990

"Don't forget, education is the stem which winds the watch.'" -- Journalist — historian Frank Gibney to David Halberstam, author of The Next Century, 1991

"America has a high standard of living of low average quality," critic Paul Goodman in Growing Up Absurd, 1966

The United Nations Organization has designated 1998 as the Year of the Oceans; will 1998 also be the Year of the [resurgent] Japanese Tiger? (Probably not, due to current dynamics and circumstances beyond the Tiger's control.) From the late 1980s -- the high-tide of Japanese economic power -- to the late 1990s, there has been a precipitous decline in Japan's fortunes. For twenty-five years and more (starting from ca.1964 -- the Tokyo Olympics -- until the early 1990s), Japan was beheld in growing wonderment and awe by the West. By the late 1980s, with Japan at the pinnacle of her financial prestige, she was perceived as an unstoppable colossus; especially in America. It prompted "old Asia hand" Robert Elegant to declare in 1990:

"East Asia, with Japan in the lead, is the new and most dynamic center of world power. After centuries of backwardness and subjugation, Asians have asserted themselves in realms where Americans and European[s] had been supreme: Technological innovation, industrial production, and mass marketing. Their prowess is a mortal challenge to which the West, with the United States in the lead, has responded falteringly."[54]

The key phrase here is "with Japan in the lead". As Robert Elegant and other journalist-historians have observed over the years, the Japanese-American relationship has always been very special. From the time of the intrusive "black ships" of U.S. Commodore Matthew Perry in Tokyo Bay during the early 1850s, the fates of Japan and the United States have been intertwined. And since the end of the Great Pacific War (1941-1945), Japan and the United States have shared virtually identical political and economic interests. Both nations probably need each other more than any two non-contiguous countries on earth. Hence the "Japan-bashing" in America is matched in equal fervor by "American-bashing" in Japan.[55]

Today, Americans are still generally aware of the post-Pacific War Japanese miracle; of her arising broken and bankrupt from the ashes of defeat to become the globe's greatest creditor nation. This writer mentions the success of the 1964 Tokyo Olympics as the turning point for Japan's rôle on the world stage.

But so late as 1965, the American magazine Look (trying to be Life-like) scornfully characterised the Japanese economy as "like a confectioner desperately eating his own chocolates on a raft, hoping that a passing ship will stop to purchase his surplus.'"[56] But it wasn't long before Look's imaginary passing ship, as Pacific Ager Simon Winchester has written, "started to buy the chocolates -- the cars, television sets, cameras and small...esquisitely designed pieces of electronic wizardry which the Japanese factories began to turn out with such efficiency and zeal -- with glorious and prosperous abandon".[57] Indeed, "Made in Japan" by the 1970s was no longer a pejorative phrase. The defeated, resource-poor nation which had to "export or die", was at last shipping out her goods from a vast technological cornucopia in increasingly gargantuan amounts.

The Japanese, moreover, saved money, paid with cash, and avoided the Western pitfall of hire-purchase or buying on credit. The Japanese virtues of true grit, toughness, simplicity, and spartan frugality were literally starting to pay off. And when the industrial deluge at last translated itself into true prosperity for the Japanese, they collectively determined to not throw it all away in the reckless extravagance of Westerners.[58] And it worked (for a while). The astounded American clerisy guessed -- haphazardly as is their wont -- that governmental policies and industrial strategies were solely responsible for Japan's stunning financial and technological supremacy. But no: The consistent high quality in Japanese plants and factories mostly came from following the post-War advice of the legendary production expert W. Edwards Deming, an American! If the high-handed U.S. companies had listened to Edwards Deming after 1945, they would have learned a whole lot.[59] But that is another story....

Thus, when the Japanese had come so far as to build their auto factories in America during the 1980s, they did so quietly and carefully, avoiding major urban areas in favour of more rural settings (i.e., Kentucky, Tennessee et al.). Naturally, U.S. labour bosses, liberal politicians, and the pc lap-dog press raised a hubbub, accusing the "racist" Japanese of avoiding the inner-city poor (Blacks, Hispanics, et al.).[60] But to Chuck Colson and Jack Eckerd, writing from a conservative-evangelical perspective, the charge was false:

"[T]he Japanese sought the middle-American people because of their farming background, strong family ties, and willingness to learn and perform many different tasks. The Japanese were looking for employees who had something extra to bring to their line jobs -- creativity, a broad range of skills, and the ability to look at a problem from several different perspectives".[61]

Liberal author and editor, James Fallows, has commented on Japanese views of race in his 1989 book,

More Like Us. When Fallows asked the manager of a semi-conductor plant in Kyushu why his chips contained less defects than those made in California or Texas, he replied, "'We are all Japanese working here, one race.' His company was planning to open a plant in rural Wisconsin. 'The people [there] are all Swedish', he [the Kyushu plant manager] said. 'They will understand each other without many words. Like us.'"[62] Fallows conceded that the Japanese word for "race", minzoku, is better translated as "tribe". But the implied questions remains: Are Japanese (and other overseas) corporations investing in America to be subjected to the idiotic official U.S. standards of "affirmative action"?

None other than arch-Democrat 1960s apologist, David Halberstam, has cited a glaring case of American racialism in reverse, employing as exemplar Democrat candidate for president, "an ambitious young congressman from Missouri", Rep. Richard ("E.T.")Gephardt.[63] "Japan-bashing" boiled over during the 1988 U.S. presidential election, and reached critical mass on Gephardt-televised dis-infomercials. These featured American cars and what they might cost in Asian countries (a selected grain-of-truth reality), but were intentionally incendiary; designed to reinforce deeply-held American fears of the Japanese and (to a lesser extent) the Koreans. Halberstam remarked aside that Gephardt personally knew far better than his blatant propaganda allowed; that Gephardt had even admitted in private conversation that opened Japanese markets would do little to reduce the chasmal U.S. trade deficit.[64] The crusty Halberstam reported watching Gephardt's dis-infomercials and envisioning the Dick Gephardt voter:

"[S]omeone who votes [N]o on all school bond issues, doesn't supervise his or her children's schoolwork, and then wonders why the Asians [i.e., the Japanese, Koreans et al.] are doing better than we are."[65]

David Halberstam has concluded that "[m]aking things" has a higher social value in Japanese than in American society. Those who manage plants, and work in them (Gephardt's Big Unionism not withstanding), are regarded with far more respect in Japan than in America. As a nation, the Japanese "are far closer to being the true children of the original Henry Ford than we are."[66]

Japanese and American education reflect the belief system (or lack thereof) of their respective society. In the 5 July 1993 issue of National Review magazine, was an advertisement touting the annual U.S./Japan Teachers Exchange Program (sponsored by Hitachi). Since 1987, the Program has enabled the visits of 28

American teachers to Japan, with 13 Japanese teachers to the United States. The year 1993 saw the largest exchange to date -- four American and four Japanese teachers. For three weeks they toured the sites, taught the children, and lived in the homes of their respective hosts. The Exchange, of course, ostensibly "helps promote mutual understanding, cultural awareness, and friendship between our societies on a personal level."[67] A Michigan high school social studies teacher was quoted as mouthing the usual fluff about bridging-the-gulf-between-our-cultures etc., etc. Another social studies teacher, from California, was struck by the support of Japanese parents for their school system, and the respect accorded the teacher in the classroom. He said: "'The support and love of the family seems to carry out not only in the school system but also into their companies '."[68]

Well, indeed. But a fair exchange? This writer thinks not. The average Japanese public school/student body is probably like the proverbial Greyhound bus driver -- "safe, reliable, courteous". But far too many U.S. public school/student bodies are the exact opposite -- dangerous, unreliable, surly. Naming Brave New Schools in the inner city "Malcolm X Intermediate" or "Pancho Villa High" just ain't goin' to get it; these "schools" are pestilential sink holes of ignorance, drug use, gang violence and rape. Every concerned parent in the U.S.A. -- Black, White, Hispanic, Asian -- does everything he or she is able to avoid sending their child to public school. In the Hitachi-sponsored U.S./Japan Teachers Exchange Program, the American teachers got the fair exchange; the Japanese the foul. For what on earth could a Japanese teacher -- accustomed to disciplined learning in a clean and safe environment -- possibly learn from a U.S. public school?

The proofs of the failure of U.S. public schools are legion. Some of the low marks for American education come from the U.S. Department of Education itself. The giant federal agency helped fund a study in 1989 of the mathematics and science skills of thirteen-year-olds in several countries. The American kids arrived dead last, with lower scores than children from Éire, United Kingdom, Canada, Spain, and the Republic of Korea.[69] (The South Korean teenagers were number one.) In another report released by the National Center for Education Statistics in 1992, thirteen-year-olds from twenty countries were graded on math and science skills. Americans placed thirteenth in science and fifteenth in math. Among countries that ranked higher in both categories than the United States were South Korea (ROK), Taiwan (ROC), Canada, France,

Switzerland, Hungary, Slovenia, Israel, and the former Soviet Union.[70] (Even though the statistics cited in both studies provided no data on Japanese performance, this writer would wager that it was on a par with that of South Korea's.)

Disciplined learning in a clean and safe environment translates into low crime figures, too. Japan (and, for instance, Switzerland) is a very traditional family-values nation in which crime figures are very low. Indeed (despite the yakuza movies), there are more murders committed in New York City during one weekend than in all Japan during one year![71] As a proud United States citizen and former New Yorker, the writer is not elated by this information. In the updated 1992 edition of The Next Century, David Halberstam listed yet one more U.S. institution as having participated in a multi-national examination of math and science aptitudes among thirteen and fourteen-year-olds. The results were neither heartening nor fully unexpected for Americans:

"....Not surprisingly, the Japanese and Korean children recorded the highest scores. Again, not surprisingly, the American children were at the bottom. The children were also asked how well they thought they had done. The results showed that the American children thought they had done the best of all."[72]

Why yes -- U.S. public school pedagogues have been busy teaching American kids the importance of high self-esteem, rather than the realisms of readin', writin', and 'rithmetic.[73]

Low self-esteem was not a problem for upper-echelon students at Deerfield Academy, the posh prep-school attended by this writer from 1959-1963. The writer was far too miserable, both at home and at school, to be a top-drawer student and even flubbed his I.Q. test! At élitiste Deerfield, a perceived less than near-genius intelligence quotient relegated the bearer to a "prole", a gamma status. The writer will never forget the contemptuous sneer on the wizened, wrinkled visage of legendary octogenarian headmaster, Frank L. Boyden, as the Great Man related the middling-low (i.e., average) I.Q. test results. This writer would subsequently wear the middling-low I.Q. tag, throughout the four years, like the Mark of Cain. (But a middling-low I.Q. didn't matter so long as the dummy's parents were rich.) It was not until an entire decade and a half later, after graduating with high marks from Seattle Pacific University, that the writer dared consider himself suitable collegiate material. That was in truth low self-esteem, and it spurred the writer to try harder in the long run. (But this writer has loathed upper-echelon private academies ever since.)[74] Paul Davies, the

Australian scientist, has recently written "[a]fter all, psychologists are divided about how to define and measure human intelligence anyway."[75] But this knowledge was of no comfort to a self-conscious and insecure adolescent at Deerfield Academy during the early 1960s.

But, strangely enough, it was at Deerfield where the writer came to know his first Japanese. After being graduated, Yasuteru ("Yatch") O. and the writer attended Boston University's two-year junior college, and, in the summer of 1965, toured Spain's northern "Costa Verde" together (along with Yatch's widowed mother) by car. Yatch O. had done well enough at B.U. to go on to the Rhode Island School of Design ("Risde"). The writer mucked around New York City for a while before returning, defeated, to Norway. Yatch O., the one and only Japanese at either Deerfield or at B.U.'s junior college, was considered "cool" by his peers. He sported fashionably-long hair, wore a camel-hair coat, played the saxophone, had an obligatory blonde girl-friend, owned and drove a red MG, and favoured the hip jazz (Art Blakey, Errol Garner), and Bossa Nôva (João Gilberto, Luiz Bonfá) of the early mid-1960s. Yatch went on to become a famous photographer, and once did a nude lay-out of Norwegian playgirl, Liv Lindeland. After not having seen Yatch for almost twenty years, the writer finally looked him up in New York in January 1995. This writer, who had lived far away from the East Coast since 1971, was eagerly looking forward to meeting his old friend again. But life had not been kind to Yatch O. Even though he lived [presumably] well on Manhattan's sumptious Upper East Side with his [Japanese] wife and children, Yatch looked wigged-out, unhappy, unfulfilled. He reeked of last night's alcohol (not so cool now), and brunch at the Algonquin Hotel on 45th St. in Midtown turned out to be a laconic and lugubrious affair. For Yatch, financial gain and material success hadn't carried over into physical sobriety and mental serenity. You really can't go home again.

The writer left the Algonquin -- and Yatch O. -- wondering how his old friend's grown daughter and young son were faring in Fun City. Like other Deerfield alumni, Yatch had probably had his children attend private school. The writer's three children are all graduates of Port Angeles High School, having there received a first-class education. But if the writer had remained in metropolitan New York, he would most likely have sent his kids to private/parochial school, too. The difference is that this writer, and other political conservatives, want all American parents to have that option. And unlike those in Japan, U.S. public schools vary widely from city to city, and state to state, with not a few that are hazardous to a pupil's physical and mental health.

1994 was the year of the U.S. congressional elections which promised much rhetorically, but delivered very little politically in the long run. 1994 was also the year of The Bell Curve, a book that caused a furore both inside and outside the Washington, D.C., "beltway".[76] The book -- not read by this writer -- compared educational and economic achievements among various ethnic groups in the United States. Throughout late 1994 and early 1995, network and cable TV talking-heads decried the now-notorious Bell Curve of Richard Herrnstein and Charles Murray. And "meritocracy" became yet another dirty word in the pc newspeak/groupthink Nineties. But the idea of a high I.Q./high achievement meritocracy had been bandied about by academic élites at the universities and think-tanks so early as the 1970s, when Richard Herrnstein (he of the later Bell Curve infamy), came out with I.Q. in the Meritocracy (Boston: Atlantic Monthly Press, 1971). What has bothered left-wing American social scientists so much is the inescapable fact that their favoured minority-group students, Blacks and Hispanics, are [still] out-performed everytime academically by Whites and [especially] Asians. This, in spite of years of "affirmative action" and an official educational spoils system. The great irony continues in that the chattering classes themselves still scream the loudest at the very idea of an intelligence meritocracy.

A prototype of the above mind-set is one James Fallows, former editor of The Atlantic Monthly (from 1979 until 1996) and current editor of U.S. News & World Report. (Fallows' very last editorial for The Atlantic Monthly in October 1996 was to do a "puff piece" on Bill Clinton.) Fallows -- super WASP Harvard graduate and Rhodes scholar -- has edited several periodicals (Washington Monthly, Texas Monthly), and served as speechwirter for candidate Jimmy Carter; dutifully following the ever-grinning Georgian to the White House.[77] Fallows has written three books, most recently Breaking the News: How the Media Undermine Democracy (1996). In Breaking the News, Fallows expressed the outrageous views that Ira Magaziner and Hillary Clinton should have been permitted to develop their stealth-health care plan in secret; that The New York Times reporter, Robert Pear, actually impeded democracy by informing readers what exactly the Magaziner-Clinton health [s]care plan scheme entailed.[78]

But it is James Fallows' More Like Us (1989) which has some interest for Japan watchers; for although containing precious little about Japan (or East Asia), the book at least appeared to be right about Japan some of the time (and wrong about America almost all the time). It must have been precisely because there was so little Japan (or East Asia) content in More Like Us, that this writer remained frustrated throughout the reading of the book...and is consequently biased in its assessment. Despite some good ideas, Fallows couldn't seem to make up his mind as to whether he was a liberal or libertarian on U.S. economics and immigration. On all else, though, Fallows was numbingly predictable. He held up, as Black American political paragons, the Reverend Jesse Jackson (whose dime-a-rhyme couplets have included "From the agony of confrontation to the ecstacy of reconciliation") and former Michael Kinsley Crossfire-clone, Juan Williams.[79]

A White ideal proffered by Fallows was the sonorous conscience of U.S. politics, gusty Senator Paul Simon (D-Ill.), from the Land o' Lincoln; Democrat Jiminy Cricket and presidential aspirant -- along with Richard Gephardt and others -- in 1988. Tragically laughable in light of the South Central L.A. riots of 1992, it was Paul Simon who loftily intoned, "Americans instinctively know that we are one nation, one family."[80] (Yeah, right.)

In More Like Us, Fallows carefully circumvented any mention of the breakdown in discipline and structure for the American decline. Unlike other experts commenting on the Japanese/East Asian success story, Fallows completely discounted Confucianism (i.e., meritocracy) as the key motivating factor.[81] Instead, Fallows opined that "[i]n America, the Confucian idea that society should be more orderly is an unhealthful alien influence."[82] (Well, so much for America borrowing at least something of great value from Japan/East Asia!) Fallows did impart to this reader one Deep Thought by which to ever remember More Life Us: "Japan is strong because of its groups; America is strong because of its individuals."[83]

To his credit, James Fallows wrote candidly about his family history and his avoidance of U.S. military service while attending Harvard. As a Harvard graduate, Fallows falls into the category of "The Best and the Brightest" along with David Halberstam and Zbigniew Brzezinski. (Part I of Chapter 1 is titled 'The Best and the Brightest' because this writer has utilised the opinions of Fallows, Halberstam, and Brzezinski -- Harvard men all -- as his main primary sources. Part I was originally called 'The Best and the Brightest:

Three Harvard Men Comment on the United States and Japan'. But after five full years of R&D on Japan, the writer has come to feel that the Best and Brightest signifies the Japanese themselves.) Fellow Harvard colleague Halberstam aided and abetted Fallows on More Like Us (as did Frank Gibney among others), and it showed in this donnish pronouncement regarding America:

....."Of course America still has some important middle-class institutions, such as network TV and USA Today. And, as these examples remind us, broad-brush mass-cult institutions don't make up a distinguished civilization all by themselves. But they have been valuable in offsetting the centrifugal forces built into American society, which is why their decay is a problem."[84]

From on high atop his Ivory Tower at USN&WR sits a man who has pontificated that horrendous network TV and USA TODAY (news bubblegum) are good for Middle (i.e., proletarian) America. (And U.S. News & World Report was, at one time, a Middle American alternative to Time and Newsweek.) An unperturbed David Halberstam, however, critiqued More Like Us as "[a]n exceptionally original rumination on the dramatic differences between Japan and America. What run through these pages are Fallows' intelligence and practicality -- and, oddly enough, his optimism for the future."[85]

David Halberstam, the second in our Harvard triumvirate, should be familiar to most North American readers. He is the author of numerous books, including the eponymous The Best and The Brightest[86] (U.S.A. vs. North Vietnam), and The Reckoning[87] (Ford vs. Nissan/U.S.A. vs. Japan). Halberstam is a prolific writer within his works as well -- The Reckoning is an intimidating and ponderous tome of over 700 pages ("Homer nods"). But for a short and succinct commentary on Japan and America, this reader could find no better than Halberstam's slender volume, The Next Century...again. (This writer used the original 1991 edition as a primary source for the Introduction to Chapter 1, 'Pacific Shift'.)[88]

In the 1992 edition of The Next Century, Halberstam has added numbered epilogues to update it. Halberstam managed, in one epilogue, to "trash-talk" Ronald and Nancy Reagan, George Bush, and even the aging and admired King Hussein of Jordan. His negative portrait choices were in stark contrast of Fallows' wide-eyed laudations of the Rev. Jesse Jackson and Sen. Paul Simon. But there was nothing derogatory in the style which Halberstam handled Japan and the Japanese. On Japan-bashing he accurately observed:

"For an America less than eager to find fault with itself, Japan, with its trade policies, its legal and cultural protectionism, becomes the perfect foil. If 90 percent of what the Japanese do is easy for us to admire, then it is also comforting to focus on the 10 percent we find so annoying. If there were no Japanese, we would have to invent them. They seem to mock us [Americans]: hardworking, careful, self-absorbed, utterly devoted to their narrowly defined self-interest. The current economic situation [late 1980s, early 1990s] is a dilemma for us both: our economic decline, which our political system has yet to accept; their economic surge, which their political and social systems have not kept pace with."[89]

Halberstam paralleled Japan to America during her two wars in Asia (Korea and Vietnam), with what America was to Europe during her two great wars (World Wars I and II). During the Cold War, Japan never thought in terms of Capitalism or Communism but always what was best for the Japanese polity. Japan's economy reflected her powerful nationalism, and, coupled with non-ideological pragmatism, ultimately represented a form of "state-guided communal capitalism."[90] And from the old days of national polity, (kokutai), Japan still frowns on any individual citizen who uses too much of anything -- food, money, or even personal freedom of speech. The last is seen, through collective Japanese eyes, as one taking at the expense of others. Halberstam perceptively posited that these boundaries not only limit over-use of precious Japanese resources, but the abuse of individual personal freedoms. As Naohiro Amaya has noted, "[s]ardines packed in a tin cannot be too individualistic.'"[91] In other words, Japanese [collective] culture carries inherent standards of individual sacrifice for the good of [Japanese] society as a whole.

David Halberstam first met Naohiro Amaya in 1983 when he was in Japan researching and developing The Reckoning. Amaya then served as a chief official at the Ministry of International Trade and Industries (MITI). Halberstam was at once impressed by Amaya, feeling he had encountered a historian rather than a bureaucrat. While most Japanese were self-congratulatory over the nation's new manufacturing might, Amaya was one of those few warning that the economy should get away from Second Wave smokestack industries and turn to future high technology. The prescient Amaya believed that Japan's superiority in education would be better utilised riding the crest of the Third Wave.[92] Halberstam visited Amaya in Tokyo for a couple of days during 1989. By then, Amaya had become involved in a major attempt to reform the Japanese educational system in toto. For, excellent as Japanese education was, Amaya was convinced that Japan produced drone-like workers instead of fully-rounded citizens. For the next millennium, asserted Amaya, the Japanese educational system must stop turning out narrow and self-isolated individuals who could soon be functionally obsolete. In an aside, Amaya confided to Halberstam that it was easier to make a good [Japanese] car than a good [working] man.[93]

It was also Naohiro Amaya who pointed out that "The American Century" (coined by Henry Luce of Time-Life, Inc.) in reality was an Oil Century or Oil Culture. Henry Ford had invented the Model-T automobile at about the time prospectors hit the great oil-gushers in the U.S. Southwest. So good a date as any to mark the true advent of the American Century was the Battle of Midway (against Japan) from 4-6 June 1942.[94] Thus it <u>was</u> an Oil War (the Great Pacific, 1941-1945) in the Oil Century (the twentieth), but only the Allies had an easy and total access to oil. As Clarence "Bud" Anderson, U.S. flying ace, put it in his World War II memoir To Fly and Fight: "Oil was everything, the lifeblood of war. Nations can't fight without oil.'"[95]

According to Halberstam, the Vietnam War helped bring the American Century to a close. If the American Century was an Oil Century as Amaya had theorised, then all of World War II had been the ideal Oil Century war. For although both antagonists were highly mechanised, America had the oil -- lots of it; Japan did not. Halberstam has drily concluded:

"World War II gave our [American] policy makers an illusion of American military power in Asia that subsequent Pacific wars [Korea (1950-1953) and Vietnam (1954-1975)] did not bear out, since our adversaries, unlike the Japanese, were not Oil Century military powers."[96]

U.S. Senator Robert A. Taft (R-Ohio) argued heatedly during the 1940s against American involvement in a European war; directly contradicting Henry Luce's grand notion of an American Century. Senator Taft warned that were the United States to win, an era of Anglo-American imperialism would follow. The United States, Taft averred, with her domestic instincts and democratic values, was poorly suited to become the global policeman envisioned by Henry Luce. America, then, would be assuming the rôle in the twentieth century which Britain had had during the nineteenth. The American Century, pronounced Taft in the spring of 1943, "is based on the [Henry Luce] theory that we know better what is good for the world than the world itself. It assumes that we are always right and that everyone who disagrees with us is wrong.'"[97]

The American Century has resulted from the economic power, industrial might (plus oil) and military hegemony (plus oil!) of the United States. The U.S.A. has been blessed with a vast and productive heartland, a naturally-rich hinterland, a large and well-educated population, and a binding patriotism; and all-motivating sense of just what it means to be an American. And the U.S.A. was further gifted with a political system in which government actually encouraged free enterprise. Japan, on the other hand, is a nation with no natural resources at all, but has achieved economic superiority through her educated and productive population. Japan, without minerals (and no oil!) and smaller than Montana, has nonetheless..."marked the beginning of a new century by coming forward as a powerful new international player, but even more important, it has given the world a new definition of economic power."[98]

But it has often been a long and bumpy road for Japan. Along the way to modernisation, wrongheaded Japanese borrowings included British-style diplomacy, a French-style military, and a German-type of parliament. (And not all Western historians have been convinced of Japanese effectiveness at adaptation.) Anyway; the third member of the Best and Brightest Harvard triumvirate, which this writer chose for commentary on Japan, was Zbigniew Brzezinski -- Carterite trilateralist and former national security adviser. Brzezinski, in his <u>Out of Control</u> published in 1993, was downright pessimistic about Japan's future prospects.[99] As the 1990s have unfolded, the myth of Japanese infallibility has faded somewhat, and Brzezinski's book was more recent that of Fallows (1989) or Halberstam (1991,1992). In <u>Out of Control</u>, Brzezinski didn't comment at length about the Japanese leadership challenge, however his short summation is worthy of extensive citation:

...."Japan impacts on but does not speak to the world. As a society, it offers the world neither an appealing social model nor a relevant message. It's cultural homogeneity makes Japan -- quite unlike America -- into a less congenial participant in the increasingly open and organic global political process. America is the world in a microcosm; insular Japan is unlike the rest of the world. That makes American leadership seem somewhat more natural and thus more acceptable. In contrast, any attempt at the assertion of Japanese global leadership would be instinctively resisted as remote and organically alien.
"Furthermore, Japan's compressed and crowded cities do not create the image of a life-style that others yearn to imitate and enjoy. Its language -- difficult and spoken by no one else -- reinforces a sense of exclusiveness and distance. Its politics -- a translated adaptation to Japanese culture of American democracy -- have become corrupt and secretive, with traditional loyalties and personal political fiefdoms becoming at least as important as formal constitutional structures. As one Japanese politician put it, 'Our country has a first-rate economy, second-rate standard of living, and third-rate political system' -- a combination not likely to help translate global economic power, and even nascent military power, into effective global authority."[100]

As noted, "the American Century" was a phrase coined by Henry R. Luce, the autocrat of Time-Life, Inc., in an article under that heading in <u>Life</u>, February 1941.[101] Luce was a member of the Council of Foreign Relations and...a confirmed Sinophile.[102] CFR-member Luce, in referring to U.S. World War II aims, spoke bluntly: "[T]yrannies may require a large amount of living space. But Freedom requires and will require far greater living space than Tyranny.'"[103] Luce was alluding to the "Grand Area", the U.S.-led non-German bloc in 1941, an interim measure to deal with the emergency situation of 1940 and early 1941. The Grand Area was required for American "elbow-room"; the preferred ideal for the CFR was greater still -- a one-world economy dominated by the United States. In its final form, the Grand [American] Area would consist of the entire Western Hemisphere, the United Kingdom, the remaining British Empire and Commonwealth, the Dutch East Indies (Indonesia), all China, and Japan herself.[104]

(Conspiracy theorists take note: The Grand Area was proposed by the Council on Foreign Relations some time <u>before</u> the Japanese bombardment of Pearl Harbor, Hawaii, December 1941.)[105]

The <u>Pax Romana</u> was centred around the Mediterranean Sea and the outlying areas containing Rome, Athens, Cairo. The power-shift took place by land, having one millennium of years as a unit. The <u>Pax Britannica</u> encompassed the North Atlantic Ocean, and London, Paris, and New York were the participating power-node cities. The power-shift of <u>Pax Britannica</u> took place by sea and spanned a century.[106] Since the end of the Great Pacific War in 1945, the <u>Pax Americana</u> has expanded the number of city-nodes participating in the power-shift, to encompass the entire Pacific Rim. Power-node cities include Los Angeles, Tokyo, Beijing, Hong Kong, Singapore, and Sydney. The <u>Pax Americana</u> power-shift took place by air, and has spanned half a century.[107]

The idea of a once and future <u>Pax Japonica</u> has been conceived by one Tosiyasu Kunii, professor and chairman of the department of information science at the University of Tokyo. Professor Kunii visualised the <u>Pax Japonica</u> during his country's economic hey-days during the mid-late 1980s and early 1990s. He has projected that the <u>Pax Japonica</u> is a humble instrument of coördinating global activities of an economic and cultural nature, rather than a military preponderance.[108] As the professor expressed: "Pax Japonica in the era of the Pacific Rim means quality-intensive world management....And not, as in the past, simple domination of the world."[109]

Because of Article Nine in the Japanese constitution, imposed by the Supreme Commander Allied Powers (SCAP, ie., General Douglas MacArthur) on a defeated nation, post-World War II Japan has been prohibited from any military/naval presence save for rudimentary self-defense forces. This state of affairs has given rise to extremely odd geo-political situation: A land of nearly 140 million people, with the second highest GNP in the world, is virtually defenseless in the face of powerful (and hostile) neighbours; is totally dependent for armed protection on a distant ally ...which is also a former (bitter) enemy and Number One (Ichiban) current economic rival.[110] So if there is to be a once and future Pax Japonica as envisioned by Tosiyasu Kunii, it could only be of an economic and cultural nature. As for a "quality-intensive world management", the Japanese Tiger in A.D. 1998 looks much the worse for wear; its economic teeth are worn and its cultural claws blunted. The writer remarked early in this chapter, and now shall repeat the statement, that there will not likely be a resurgent Japanese Tiger anytime soon, due to cultural dynamics and economic circumstances beyond its control.

Futurist John Naisbitt flatly declared in 1996 that "Japan was the star performer of the industrial world, but it is now the sick economy of Asia".[111] Since 1990, the big investors in the ASEAN (Association of South-East Asian Nations) countries have not been Japanese but increasingly from the Four Tigers of East Asia -- Hong Kong, Singapore, South Korea, and Taiwan -- plus the Dragon, China. The Dragon, in fact, had a phenomenal growth rate of ca. 12% from 1993-1995.[112] The Small Tigers of East Asia are now financing the Little Dragons of South-East Asia. And by 1995, for the initial time in a decade, Japan fell from top place as the world's most competitive economy. The United States ranked first, with Singapore second, Hong Kong third, and Japan in fourth place.[113] (But all first four countries were on the Pacific Rim.)

There are many reasons for the Japanese decline, four of which [obvious Sinophile] Naisbitt analysed in a long list: (1) Japan is way over-regulated, which hampers economic growth; (2) there is also an over-centralisation of people, money, and information in the primate megalopolis, Tokyo; (3) the self-sacrificial post-Pacific War generation has been replaced by a generation of selfish, pleasure-seeking, "risk-averse" yuppies; (4) because of a very low birthrate, with people living longer and with virtually no in-migration, Japan is aging faster than any other nation on earth. According to Naisbitt, fully 25% of all Japanese will be sixty-five years or more by 2020 A.D.[114]

Of course there's more. In his short section on Japan in Out of Control, Zbigniew Brzezinski made some sound observations on Japanese cultural attitudes. One of these was that, in spite of Japan's internationalisation after World War II, there persists a deeply embedded "Us versus Them" mentality; a mind-set which has adversely affected Japanese business overseas.[115] A Japanese company overseas often counts Japanese only as <u>bona fide</u> members of their work force, and the majority of Japanese companies overseas insist that all power must flow from Tokyo.[116] Even when Japanese executives delegated with real authority are sent overseas, company headquarters is still ambivalent about accepting command decisions made abroad, or micro-managing from home. Japanese multi-nationals with this typical mind-set fail to attract top overseas local (<u>gaijin</u>) talent. An American woman, who has worked in Japan for ten years, remarked to <u>Business Week</u> magazine in May 1996, "There's no glass ceiling [in Japan], [t]here's a cement ceiling. You can see it coming a mile away'"[117]

Indeed, by the mid-1990s, there was much talk and speculation on the Japanese decline. There have been many voices -- beside that of futurist John Naisbitt -- speaking of the coming Millennium of China; a Dragon Century. But not all those anticipating Japan's imminent demise are doing so with such <u>schadenfreude</u>. William Letwin, former member of the U.S. Armed Forces which first defeated and then occupied Japan, wrote in 1995 of China's coming eclipse of Japan:

"...Japan may be overtaken by China well within the next fifty years. For the time being, China and Japan are well-matched trading partners, the raw materials and fabricated products of the one being exchanged for the high-tech and heavy-industry products of the other. But s the Chinese economy develops, in response to privatization and liberalization, Japan may sink into the condition of junior partner. Such demotion would badly dent the long-cherished belief of the Japanese that they are intrinsically superior to all other Asians, whom they ought therefore to lead.
"Japan's next fifty years may be less gratifying than the last fifty. It comforts me that I will not be here to be blamed if my forecast is mistaken."[118]

<u>Fifty</u> years, Mr. Letwin? Heck, Sir, at this rate (1998) it will be less than <u>twenty-five</u> years before China surpasses Japan! This writer is also an admirer of modern Japan, but believes that William Letwin has been too charitable to his old foe. (That was, however, in 1995.)

The recent sub-heading of an Associated Press report (January 1998) on Japan read: 'Economic ills at home pos[e] threat to leadership role in solving region's deepening problems.'[119] Japan is still the world's biggest creditor nation, and has financially paved the way to international markets; has loaned and invested the major portion of funds which fueled Asia's economic engine. Thus Japanese banks, staggering and reeling from their own predicament, could ignite a two-front financial fire-storm when coupled with the concurrent

Asian crisis. The fall-out would have world-wide ramifications.[120]

Years of propping up the home economy by pouring massive amounts of money into public works have saddled the Japanese government with a colossal $4 trillion (U.S.) debt. The world's biggest creditor nation has also one of the globe's greatest public debts. As if all the above were not enough, real estate and stock prices in Japan have flattened following the burst speculative bubble of the 1980s.[121] It will take all the choler, toughness, and resourcefulness of the present prime minster (January 1998), Ryutaro Hashimoto, to avert a major financial panic -- and loss of the will to win -- in today's Japan. For with China's burgeoning economy suffering ever less constraints, the road to victory is paved in the Dragon's favour: A huge population and land mass, natural resources, and the ancestral home of the winning "Confucian Connection" (Downton: 1986). It's only a matter of (short) time before China reaches technological and economic parity with Japan. Then the only advantages remaining to Japan will be her ethnically homogenous (and harmonious) people, superior public education, and determined national spirit.

But does Japan have a determined national spirit anymore? This writer wonders, as he contemplates the wretched Japanese excesses of the 1980s. In Megatrends Asia, John Naisbitt scathingly asked, "$39 Million for a Van Gogh Sunflower, Anyone?"[122] The writer well remembers the overly-dressed Japanese tourists in Honolulu (1978-1982), with their over-priced fashion-designer label clothing, luggage, and accoutrements. Mōdan ("modern") and ima-yō (absolutely now") had been almost tenets of religion with the up-scale visitors. A recent Associated Press article has illustrated that "high camp" has not abated during the 1990s:

"Lucky Kentucky, an American-made shampoo for horses, has become all the rage among young Japanese women.
"Japanese are the most trend-conscious people in the world', said importer Shizue 'Suzie' Shimizu. 'Besides, who doesn't want their hair to shine like a thoroughbred's coat?'
"The latest American fad to hit Japanese shores has sold nearly a million bottles since its debut in the spring last year -- and has spawned several Japanese competitors."
"The scents and formulations vary, but all are marketed in the image of the Star[s] and Stripes and the American wild west...."[123]

Port Angeles, Washington, has been the official "sister city" of Mutsu City, Japan, since August 1995. In November 1997, two faculty members from the local Peninsula College rendered a presentation on their past spring journey to Mutsu City. (The trip had been the second annual delegation of PIRA, the [Olympic] Peninsula International Relations Association.) But this attendant writer -- rather than paying heed to the usual rhetoric on the joys of global interconnectedness -- was distracted by the matching garb of the Japanese exchange students on stage. For, to a [young] man (and young woman), they were all wearing "The North Face" jackets; the "in" hiker/biker upper garment of choice.[124] (It is at times like those that the writer finds it hard to commiserate with a once-again humbled Japan.) Obviously, if the few Japanese exchange students at Peninsula College all felt compelled to wear The North Face togs as uniform, how would they ultimately react to the nonsensical American collegiate newthink of "Porn Theory", "Queer Scholarship", or "Whiteness Studies"? Mayhap the brain-washed groupthink drones of Japanese mass education -- as perceived by Naohiro Amaya -- are to finally morph into brain-washed groupthink drones of a newthink mind-set. These exchange students, certainly, could not be the inheritors of Tosiyasu Kunii's avowed goal of "quality-intensive world management" via ostensibly superior Japanese education.[125]

Yet another portrait of Japan has been explicated by British author, Simon Winchester. In an extravagant concluding chapter of Pacific Rising, Winchester illustrated a melodramatic scenario of metroplexes Los Angeles, California, and Tokyo, Japan.[126] Winchester's picture was overdrawn but intriguing. He patiently informed us that Bei-jing in classic Chinese characters, means "Northern capital"; Nan-jing is "Southern capital", Xi-jing is "Western capital", and Dong-jing is "Eastern capital". Dong-jing is also the Mandarin word for Tokyo, Japan. But in a neat geographical legerdemain, Winchester proposed that Los Angeles, California, is the true Dong-jing of the Pacific Rim.[127] (Conversely, Tokyo is the implied Xi-jing of the Pacific Rim.) Winchester continued, in grandiose fashion, to compare and contrast the "ineluctably entwined...twin capitals of the Ocean, the two ends of the axis upon which tomorrow's world spins."[128] The metroplexes of Tokyo and L.A. are mirror-images geographically and geophyscially -- in the mid-30°s N. latitude at each end of the Pacific Rim; plagued by earthquakes and ravaged by fires. But here all likeness ends. Tokyo is a city that travels by train, while L.A. drives by car. Tokyo manufactures a good number of the autos that have become an integral part of California (car) culture. L.A. produces the movies and music-videos that are recorded on Japanese-made cameras, then reproduced on Japanese-made VC-recorders and CD-players.[129]

This mutually-hostile symbiosis shall soon stop after China reaches technological and economic parity with Japan. All the remaining urban sociology of Simon Winchester, although colourful and entertaining, is already becoming irrelevant vis-à-vis Tokyo. By, or before, the year 2020 A.D., the super metroplex of Shangai, China, shall reign as the commercial "Western Capital" -- Xi-jing -- of the Pacific Rim. 1998 is still the Year of the Japanese Tiger, but it might well be the last; for the coming millennium shall be most assuredly a Dragon Century. Will the best and brightest years of the Japanese Tiger in the future be measured from 1964 (the Summer Olympics at Tokyo) to 1998 (the Winter Olympiade at Nagano)?[130]

1990s Japan Fisheries Update and Seafood Market Overview

It will be a long time, though, before the sun ever sets on the Japanese fishing fleet or Tokyo's renowned Tsukiji (pronounced "skiji") Central Wholesale Market. Japan is an archipelago in the western North Pacific which has always derived sustenance from the seas around her. Fishing and aqua-/mariculture have been a way of Japanese life for millennia.[131] The Japanese are a truly maritime people: Indeed, the late Shōwa emperor, Hirohito (1901-1989), was a serious student of marine biology and wrote several books on the subject. And what local waters could not sustain, or far seas would not yield, the adaptable Japanese supplied themselves through invention and innovation. A recent example of Japanese quick reaction followed the international hue and cry concerning East Asian driftnet fishing (during the late 1980s and early 1990s).[132] The Japanese Fisheries Agency responded by successfully developing bioplastic nets and lines which disintegrate in salt water after a certain amount of time.[133] But Japanese fishermen (along with South Korean and Taiwanese) will be forever remembered -- and damned -- by North Americans as the East Asian driftnetters who ravaged the world's oceans with their "curtains of death".[134]

Also, during the early 1990s, was the resurfacing controversy of the Commercial Vessel Anti-Reflagging Act of 1987. The law was originally enacted to require majority American ownership for corporations to operate vessels fishing within the U.S. 200-MEZ (mile economic zone). In addition, the vessels had to be built or rebuilt in U.S. shipyards. The court ruling was subsequently blamed for millions of dollars in losses by foreign-owned (i.e., mostly Japanese) fishing corporations that operated within the U.S.

200-MEZ.[135]

Another provision of the Commercial Vessel Anti-Reflagging Act had permitted continued licensing of vessels which had been allowed in U.S. waters, or were being built, or contracted to be rebuilt, before 28 July 1987. The purpose of the provision had been to restrict U.S. waters to vessels manufactured and owned/operated by American interests. Enforcement of the provision was challenged by U.S. shipyard owners and others who accused the U.S. Coast Guard of having failed to properly enforce the regulation. For under an early Coast Guard interpretation of the provision, vessels covered by the original "grandfather clause" could remain in American waters after changes of ownership, even if said vessels were purchased by foreign (i.e., mostly Japanese) interests.[136] In other words, the U.S. Coast Guard had created an ambiguous loophole, for any number of prospective foreign buyers (i.e., mostly Japanese), by applying the provision to the vessel -- meaning a boat could be bought and re-sold, any number of times, without losing her license to fish in American waters.[137]

The upshot had finally been a May 1991 decision by U.S. District Judge John Garrett Penn in Washington, D.C., to decide that the Coast Guard's spin on the provision "effectively obliterates the primary purpose of the Anti-Reflagging Act.'"[138] As a result, many foreign-controlled (i.e., mostly Japanese) factory trawlers and other Alaska-bound boats based in Seattle were prohibited from fishing Up North. By late November 1992, however, John Garrett Penn's ruling had been overturned by a unanimous three-judge panel for the U.S. Court of Appeals for the District of Columbia (that *other* Washington!). The verdict of the appeal panel read:

"Whether the (grandfather) clause gives too much protection; whether too many ships retain grandfathered status; whether as a result the phase-out (of foreign-owned ships) will take too long are the judiciary imponderables....
"We can say only that the savings (grandfather) clause, as written, makes the exemption from the citizen-control requirement run with the ship...'"[139]

An immediate example of the November 1992 aftermath had been the release of the 240-foot factory trawler, Resolute, operated by Pacific King Fisheries of Seattle. The U.S. Coastguard had forced The Resolute earlier during the year to leave the Alaska fishing grounds....And to change its Japanese ownership.[140] Whether or not real American justice was truly served, the whole re-flagging show made for an extra round of Japan-bashing. So there have been two blackened eyes for the Japanese in the 1990s -- one for driftnet fishing, and the other for the reflagging controversy. Either way, the American sentiment went, "Hey! Them Japs are takin' our fish!"

Meanwhile, the vilified Japanese have gone quietly about their fishing and fish-farming business throughout the 1990s. Aquaculture has been practised in the Japanese archipelago for millennia. Cultivated at first were Pacific oysters (kaki) and various edible seaweeds, especially nori. More recently, Japan's aquaculture has concentrated on such high-cost species as Kuruma prawn (ebi), yellowtail (hamachi), flounder (ohyō), and red sea bream (madai).[141] Research and development (R&D) in Japanese aquaculture has accelerated greatly during the 1990s. Renowned for their salmon consumption, the fish-loving Japanese in 1991 raised 27,000 tons of farmed salmon; almost as much as Canada and Chile during that year. According to a 1994 report in National Fisherman, there were 27 species of fin-fish being commercially cultivated in Japan, with hatchery techniques so advanced that a further 20 species were being grown to fingerling size and then released into the sea.[142] (Due to limited land space in Japan for pond aquaculture, most fish-rearing beyond fingerling stage is undertaken in floating cages.)

The Japanese have also been successful in commercial shellfish culture. Japanese scallop (hotate-gai) farming ("seeding", really) didn't commence until the 1960s. but by the 1990s has become a chief factor in the aquaculture industry. Scallop spat is still gathered into onion bags stuffed with netting. Until the mid-1970s, all scallop spat 2 inches in size was transferred to lantern nets for final grow-out. But in the 1990s, scallop spat -- after initial grow-out in pearl nets -- is now often sown directly onto a sea-bed cleared of predators. This method, plus suspended-culture spat, has accounted for half the scallop landings in Japan.[143] By the mid-1990s, farming abalone (awabi) and sea urchin (uni) were being effectively raised to match the established commercial success of oysters and scallops. Abalone have traditionally been reared in hatcheries, but some farming coöperatives are growing abalone in suspension culture to market size. Cages are utilised containing plastic piping to increase feeding surface area. Studies have shown that Japanese-farmed abalone can reach market size in less than two years, with a 70% survival rate.[144]

In 1985 a government-subsidised R&D organisation, Marino-Forum 21, was set up to formulate strategies and develop technology to take Japanese aquaculture into Century 21. Plans included, in the words of National Fisherman:

"[T]he expansion of submerged fish farming sites into offshore waters and the polyculture of different species -- including finfish, mollusks and seaweeds -- at the same site to lessen the nutrient and waste problems."[145]

By the mid-1990s, a further number of fish species were being planned for farming in Japan, among them the highly-valued king crab (tarabagani) and the -- getting ever scarcer -- greatly esteemed giant bluefin tuna (kuro-maguro).[146]

Marino-Forum 21, as mentioned, is sponsored by the Japanese government. Indeed, all aquaculture in Japan is heavily subsidised. One John Manzi, president of Atlantic Littlenecks of James Island, S.C., has indicated that the Japanese "spend more [money] per capita than any other country on aquaculture development and research.'"[147]

The Deep Thought bequeathed to us by James Fallows, formerly in this chapter, was that Japan's strength is in her groups.[148] And much of the aquaculture undertaken in Japan's waters is under the auspices of fishing coöperatives working at stock enhancement. The fishing (cum mariculture) coöperatives also share in sea-farm harvests. For the Japanese, the advantages of these coöps are three-fold: 1) Members don't fish competitively (hence money is made and kept and not wasted on investments in power or technology); 2) coöps have set geographical territories in which they retain property rights (there is resulting incentive for wise-use conservation and smart husbandry); 3) and coöps spread the risk around in speculative ventures. Probably no single fisherman in a Japanese region has the personal finances to start-up a mariculture operation on his own, but collectively the coöp can invest the needed capital without risking more than it can afford to lose. Potential fiscal loss is thereby small for the individual (goes the thinking), but the [shared] rewards for the community as a whole are large.[149]

Korean hair crab, anyone? A joke might be made that eating the merus meat of this crustacean will help hirsutically-challenged males grow more hair. But the Korean hair crab is valued by the Japanese for its (body) "butter", and there exists a very small fishery around Alaska's remote Pribilof Islands, producing about 450 tons (annually).[150] Alaska fishermen also provide the Japanese market with Alaska [walleye] pollock, mostly consumed as surimi seafood. Also popular in Japan is hiraki-sukesodara (salted pollock), tokan-hin (freeze-dried pollock), and mamijiko (coloured, salted pollock roe).[151] No matter what the fin-fish or shellfish species, the Japanese are by far the greatest importer-consumers of North American seafood.

The manner in which seafood is sold and distributed in Japan, however, began a sea-change in the early 1990s. North Americans in Japan during 1991-1994 heard of the latter years referred to as "the depression."[152] The Japanese "depression", a profound and prolonged recession certainly, followed the golden years of 1985 to 1991 when the Japanese yen rose in value from ¥225 per U.S. dollar to ¥150. The late 1980s had been frivolous and foolish hey-days when in one instance, newly-rich Japanese paid $39 million U.S. for a Van Gogh sunflower painting (Naisbitt: 1996). Because of the huge demand for foreign goods throughout the late 1980s, Japanese supermarket chains had been permitted limited expansion to cope with consumer demands. Thus came the sea-change in the Japanese distribution system; a system historically composed of a myriad middlemen, and a multitude of little "Mom and Pop" specialty stores.[153] The advantages of the convoluted old Japanese distribution system were much like those of the fishing coöperatives: 1) It reduced the danger of risk by "spreading it around" (i.e., from importer to processor to wholesaler to reprocessor to secondary wholesaler to retailer to consumer); 2) the established system provided a dependable structural framework of cold storage, transportation, and reprocessing plant facilities.[154] But the old, established, and costly Byzantine distribution system of Japan's seafood was about to be washed away in the 1990s.

No industrialised nation has been hit so hard or so fast as Japan during the recession of the 1990s. When the economic bubble burst after the high and heady times of the late 1980s, the effect on seafood sales was immediate and deeply felt by the industry. Corporate Japan reined in their high-riding and high-flying company executives; their free-wheeling and free-spending businessmen. Personal expense accounts were cut, causing expensive orders in Japanese seafood restaurants to plunge. Sales of high-priced seafood like bluefin tuna, blowfish (fugu) and king crab took a dive, by so much as 30% by some estimates.[155] The subsequent Japanese business mantra of the no-growth 1990s has been: Cut costs! In the Japanese fishing and seafood industry, none has been spared contraction pains -- not the fisherman, processor, distributor, restaurateur or retailer.[156] In 1994, Seafood Leader editor Peter Redmayne predicted that:

"There's little doubt that the Japanese seafood market of the 1990s will be a lot different than the Japanese seafood market of the 1980s. The decline in expense-account entertainment, increasing competition and rampant cost cutting will keep the pressure on to keep prices low in spite of the strong yen. That's bad news for a lot of [North American] fishermen and fish farmers that rely mainly on the Japanese market..."[157]

There is also the long-range effect of the new generation in Japan. The Baby Boomers, Busters, and Gen-X young in today's Japan often eat beef -- lots of beef -- and dairy products at McDonald's and Denny's; dine-out American-style at Kentucky Fried Chicken (KFC), et al. Meat consumption in Japan has risen dramatically from 1972 to 1992, due to its affordability plus (American-influenced) market trends.[158] And Baby Busters and Gen-X'ers, despite coming labour shortages in Japan due to down-sizing, will hardly flock for work to ("lousy jobs") fish processing or reprocessing plants. The work is hard, low-paying, unpleasantly smelly, and cheaper labour markets for Japanese processors beckon in China and Korea. Processing new foods for changing tastes in the new millennial Japan will entail value-added seafood-processing done outside Japan -- at the place where the fish was landed; or in fisherispeak, "adding value at the point of primary production."[159] This new trend in seafood processing will have enormous consequences for Japan.

The Tokyo International Seafood Show was held during August 1995 at the Harumi Trade Fairground. The British trade magazine, Seafood International, reported that 230 exhibitors displayed their wares to 11,119 visitors attending. This was a small but significant show for fish exporters doing business with Japan. Most of the exhibitors were firms with established links to Japan and the Far East; from Norway, Canada, and Australia.[160] Although an expensive array of seafood products still caught the eye of browsing onlookers, the unspoken message of Japanese consumers at the Tokyo International Seafood Show was a demand for cheaper and more value-added products. That meant, of course, processing outside Japan and by-passing domestic trading houses.[161] B.C. Salmon Market Council's Christina Burridge put the two-fold case well:

"The Japanese consumer is tired of paying really high prices for things, and if we're shipping a finished product, we're saving money on transport.
"The Japanese are having problems finding staff to do processing, as the work is not very popular, so they are getting it done in other countries and are working more closely with these processors'."[162]

Dalton Hobbs represented the Oregon Department of Agriculture, from a Pacific Northwest state which exports [mostly frozen] crab, black cod, and salmon to Japan. Hobbs made the point that U.S. seafood corporations exporting directly to Japanese retailers, rather than through trading houses or regional wholesale markets, was increasing. Explained Hobbs:

"For many years the Japanese consumer has paid a very high price for seafood, but now they are realising they don't have to.

"You [don't] have numerous middle men in between now, so margins have to be much smaller. But because Japanese companies are still very new to this, they don't have much buying experience.'"[163]

Katsumi Hirai, president of Tokyo-based Hirai & Company, expressed his belief that seafood importers would start opening up to foreign exporters, and that Japanese retailers would begin dealing directly with them. Expanded Hirai:

"I myself am targeting supermarkets in Japan: I want to sell to them rather than to wholesale markets....

"Some of the supermarkets can't handle big sizes of salmon for example, so it's up to us to process it into manageable sizes for them.'"[164]

Dining Dangerously on Fugu: A Strange Japanese Fish Tale

A small side-bar article appeared in an autumn 1996 issue of Seafood Leader magazine, under the heading 'Fugu Fades in Japan.'[165] The article related how fugu (puffer, blowfish) consumption in Japan has, for centuries, been a blend of seafood dining and Russian roulette. A single slice of the tetrodotoxic fish could cause agonising death to the diner within minutes -- moments -- after ingestion. (One fugu, we were told, contains sufficient poison to kill 30 people.)[166] But today's Japanese seafood chefs are so skilled at cleaning and cooking fugu, that they've all but eliminated the joyous [intensely Japanese] risk of eating the dangerous fish. Said a Mr. Kiichi Kitahama, owner/operator of a fugu restaurant in Osaka, "With so few deaths, fugu has become mundane.'"[167]

Despite knowing Yatch O. and his Japanese family since school days, this writer hadn't heard of fugu until reading Ian Fleming's (1908-1964) You Only Live Twice some years later.[168] In doing research for this chapter, the writer came across a brief reference to fugu in Robert Elegant's Pacific Destiny:

"...[T]ourists stopped at a miniscule shop where a portly craftsman was making lanterns. Each the skin of a single fugu...the puffed-up lanterns were menacing with their clenched, malign features and their spiky, gray surface. A demicult of death centers on the fugu, whose gall and ovaries contain tetrodotoxin, which brings almost instantaneous death by paralyzing the respiratory tract.

"The Japanese passion for sashimi...finds macabre expression in the eating shops that specialize in fugu. Since a slip of the knife can bring death, iron lungs used to stand in the corner. Nowadays, state-licensed slicers are certified safe, but the fillip of danger still titillates the daredevil gourmets."[169]

But to find the definitive fish pharmacology of the fugu, this writer had to consult a book on -- of all things -- the ethnobiology of the Haitian zombi. And it had all happened quite coincidentally. During the late 1980s. toad-licking by certain wiccan and Mother Earth cultists to obtain a "natural high" had been all the rage. (The writer lives in the ecotopian Pacific Northwest where some countercultural types use the fly agaric and magic button mushrooms -- "shrooms" -- in lieu of pakalolo, booze, coke, or pills.) Horrified but intrigued, the writer (then a father of three kids 'twixt twelve and twenty) pursued the topic of toxic toads...which led to notions of the toad as Earth Mother in Meso-American religion. In turn, questions arose concerning the influence of psychotropic flora and fauna on specifically Maya beliefs; thereby headed for the idea of Bufo marinus (or Bufo aqua) as possible hallucinogen. Literary exploration of water-friendly toxic toads at last unlocked a tropical vista of frogs, newts, and, ultimately fish.[170] A countercultural "head"/book shop in Downtown Port Angeles finally yielded informational pay-dirt to the writer's probing eyes and fingers. The Serpent and the Rainbow by Wade Davis, a transplanted British Columbian, turned out to be a minor lexicon (but major source) of fauna and fish pharmacology.[171]

As noted, the book by Wade Davis was basically the study of voodoo (vodoun), and an ethniobiology of the Afro-Haitian zombi. Davis described the substance responsible for the suspended state of zombisme as one of the most toxic substances known from nature; derived from two varieties of fish found in West Indian waters. There were two puffers, or blowfish, known locally as fou-fou ("crazy-crazy") and crapaud de mer ("sea toad"). Of the family Tetraodantidae, the puffer, or blowfish, is so called for its ability to imbibe large amounts of water, thus assuming a globular shape, making it hard to be swallowed when threatened by a predator. Scientifically, the two puffers in question are identified respectively as Diodon hystrix and Sphoeroides testudineus (Davis: 1987). Both fish belong to a large pan-tropical order, many of whose members contain tetrodotoxin in their skin, liver, ovaries, and intestines. Their non-protein neuro-toxin is 160,000 times more potent than cocaine and 500 times more powerful than cyanide. The deadly Red Sea puffer, according to Davis, was the reason for the Deutoronomical injunction against eating fish without scales.[172]

The first of the great pharmacopeia was ancient Chinese, the Pentsao Chin, thought to be written during the reign of the legendary Emperor Shun Nung (2838 B.C.-2698 B.C.?), which recognised the puffer's toxicity. But throughout Chinese history there continued a real record reflecting knowledge of blowfish biology and toxicology. By the era of the Han Dynasty (202 B.C.-A.D.220), the puffer liver had been found to be a source of the fish toxin. Four centuries later in the Sui Dynasty, a medical treatise disclosed that toxicity resided in puffer eggs and ovaries, in addition to their livers. In A.D. 1596, the last of the [Chinese] Great Herbal[s], the Pentsao Kang Mu, declared that levels of toxin vary in diverse species of the puffer, and that within any single species levels may fluctuate with the seasons. And this Great Herbal was different -- the Pentsao Kang Mu not only depicted the dangers of ingesting puffer livers and eggs, but related the extraordinary development in Mandarin society of the puffer having become a culinary delicacy![173]

By the 18th century, the Japanese had adopted the Chinese practise of dining on the puffer, but with an intensity that was strictly Japanese. (As mentioned, the Japanese hunger for fugu later became a national obsession.) During the 1980s, just about every seafood resoturan in Japan served fugu, with the government actually licensing specially-trained chefs who alone were (and still are) permitted to prepare the poisonous fish. But for some diners living dangerously in the 1990s, eating plain fugu is not enough -- they call for chiri; a dish of partially cooked fillets taken from a steaming pot full of toxic skins, livers, and intestines. As a result, connoisseurs of chiri invariably place high on the list of annual Japanese fatalities.[174]

Ethnopharmacologist Wade Davis has queried, with some wonderment, Why would any one [people, ethnos] play Russian roulette by eating such fish? Ethnopsychologist Wade Davis has answered his own question extremely well. First, the fugu is one of the few substances in the twilight zone between food and drug. Second, Davis has posited:

"For the Japanese, consuming fugu is the ultimate aesthetic experience. The refined task of the fugu chef is not to eliminate the toxin, it is to reduce its concentration while assuring that the guest still enjoys the exhilarating physiological aftereffects [sic]. These include a mild numbing or tingling of the tongue and lips, sensations of warmth, a flushing of the skin, and a general feeling of euphoria. As in the case of so many stimulants, there are those who can't get enough of a good thing. Though it is expressly prohibited by law, certain chefs prepare for zealous clients a special dish of the particularly toxic livers. The organ is boiled and mashed and boiled again and again until much of the toxin is removed. Unfortunately, many of these chefs succumb to their own cooking....[L]ike all of those who eat the cooked livers...[they] enjoy 'living dangerously'."[175]

As reported in the Seafood Leader article of September/October 1996, fugumania is waning in Japan; concurrent with a waxing gilded youth (a jeunesse dorée) which shuns seafood, both at the dinner table and in the workplace. Despite the probability of the new Millennial Japan becoming a less fish-conscious place, it will be a long time before the sun sets on the Japanese fishing fleets or Tokyo's renowned Tsukiji Central Wholesale Market.

NOTES

54. Robert Elegant, <u>Pacific Destiny: Inside Asia Today</u> (New York: Avon Books, 1990), Chapter 1, title page.

55. <u>Ibid.</u>,p.115. NB: Robert Elegant reflected the views of Wataru Hiraizumi, who, at the time of writing, served as chairman of the Foreign Affairs Committee of the (then) ruling Liberal Democratic Party.

56. Cited by Simon Winchester, <u>Pacific Rising: The Emergence of a New World Culture</u> (New York: Prentice Hall Press, 1991), p.278.

57. <u>Ibid.</u>,p.279. See Appendix **A**.

58. <u>Ibid.</u>,p.279. See Appendix **A**.

59. James Fallows, <u>More Like Us: Making America Great Again</u> (Boston: Houghton Mifflin Company, 1989), p.22. Subtitle on cover: <u>Putting America's Native Strengths and Traditional Values to Work to Overcome The Asian Challenge</u>. NB: Sam Walton (1918-1992) of Wal-Mart praised both Japan and the influence of the American, W. Edwards Deming, in his autobiography, <u>Made in America, Sam Walton: My Story</u> (New York: Doubleday, 1992), pp.227,254-255.
   For more on W. Edwards Deming, see 'The Road to Recovery', <u>The Pacific Almanac</u>, pp.556-557.

60. Chuck Colson and Jack Eckerd, <u>Why America Doesn't Work</u> (Dallas: Word Publishing, 1991), p.139.

61. Op.cit., Colson and Eckerd, p.139.

62. Op.cit., James Fallows, p.33. NB: Brackets and contents this writer's.

63. David Halberstam, <u>The Next Century</u> (New York: Avon Books, 1992 ed.), p.120.

64. <u>Ibid.</u>,p.121. NB: This same Dick Gephardt today touts himself as the [Big Union] labour candidate for the U.S. presidency in 2000 A.D.

65. Op.cit., Halberstam, p.121. NB: Brackets and contents this writer's.

66. <u>Ibid.</u>,p.111.

67. 'Japan and America: Where the Twain Meet', <u>National Review</u>, 5 July 1993, p.3.

68. <u>Ibid.</u>,p.3.

69. John Chancellor, <u>Perils and Promise</u> (New York: Harper & Row, 1990), p.47. Quoted from Dave Breese, <u>Seven Men Who Rule the World from the Grave</u> (Chicago: Moody Press, 1990).

70. Richard Bernstein, <u>Dictatorship of Virtue</u> (New York: Alfred A. Knopf, 1994), p.198.

71. James Q. Wilson and Richard J. Herrnstein, <u>Crime and Human Nature</u> (New York: Simon & Schuster, 1985), pp.439-458. Cited in William D. Gairdner, <u>The War Against the Family</u> (Toronto: Stoddart Publishing Co., Ltd., 1992), pp.348,622.

72. David Halberstam, 'Epilogue Three', <u>The Next Century</u>, p.160. Emphasis Halberstam's.

NOTES (cont'd)

73. See James Fallows, p.175. Fallows has referred to the difference in attitudes of Japanese and American high school seniors. The Japanese listed grades and university admissions as their paramount concern; the Americans listed sexual activity.

74. For an idealised, gushing (and wholly inaccurate) portrait of Frank L. Boyden, see John McPhee, The Headmaster (New York: The Noonday Press, 1992 ed.) NB: Instrumental in teaching this writer about God, history, and scholastics at SPU, was Swedish-born Dr. Roy Swanstrom.

75. Paul Davies, Are We Alone? (New York: Basic Books, 1995), p.89.

76. Herrnstein, Richard and Charles Murray, The Bell Curve (New York: The Free Press, 1996 ed.)

77. '77 North Washington Street', The Atlantic Monthly, October 1996, p.6.

78. Maureen O'Dowd, 'Revisionist views of media not needed', Peninsula Daily News, 1 April 1996, p.A-6.

79. James Fallows, More Like Us, p.121.

80. Ibid.,p.209.

81. Ibid.,p.212 and passim.

82. Ibid.,p.131.

83. Ibid.,p.208. NB: A generalisation to which Naohiro Ayama, former Japanese vice-minister of foreign affairs, has replied: "Sardines packed in a tin cannot be too individualistic'" (Halberstam: 1992). Amaya-san in 1992 directed the Institute for Human Studies.

84. Ibid.,p.178.

85. Ibid., review on front cover.

86. Fawcett Publications, Greenwich, Conn., 1969.

87. Avon Books, New York, 1986.

88. William Morrow and Co., Inc., New York. See Intro., FutureFish 2000.

89. David Halberstam, p.108. NB: All contents of brackets this writer's.

90. Ibid.,p.82.

91. Ibid.,p.83.

92. Ibid.,p.115.

93. Ibid.,pp.116-117.

94. Ibid.,p.58.

NOTES (cont'd)

95. Ibid.,p.59.

96. Ibid.,p.66. NB: All brackets and contents this writer's.

97. Ibid.,p.67. NB: Robert A. Taft, son of President William H. Taft (1909-1913), served in the U.S. Senate from 1939 to 1953. (Brackets and contents this writer's)

98. Ibid.,p.87.

99. Subtitled: Global Turmoil on the Eve of the Twenty-First Century (New York: Charles Scribner's Sons, 1993).

100. Op.cit., Zbigniew Brzezinski, p.126.

101. Henry R. Luce, 'The American Century', Life, 17 February 1941, p.64. Cited in Holly Sklar, ed., Trilateralism: The Trilateral Commission and Elite Planning for World Management (Boston: South End Press, 1980, p.156. NB: Henry Luce's empire has devolved from Time-Life monster to Time-Warner mega-mediocrity.

102. Barbara W. Tuchman, Notes from China (New York: Collier Books, 1972), pp.108,109.

103. H.R. Luce quote in Holly Sklar, ed., Trilateralism, p.151.

104. Ibid.,pp.140-141.

105. (U.S.) War Plan Orange had been promulgated so far back as the end of the Russo-Japanese War, 1904-1905. See Mark R. Peattie, Nan'yō: The Rise and Fall of the Japanese in Micronesia, 1885-1945 (Honolulu: U.H. Press, 1992 ed.), pp.35,48,254.

106. Tosiyasu Kunii, 'Pax Japonica in the Era of the Pacific Rim: Toward Quality Intensive World Management', The Pacific Rim Almanac, p.517.

107. Ibid.,p.518.

108. Ibid.,p.518.

109. Ibid.,p.517.

110. Zbigneiw Brzezinski, Out of Control, p.123.

111. John Naisbitt, Megatrends Asia (New York: Simon & Schuster, 1996), p.39.

112. Ibid.,p.39.

113. Ibid.,p.41.

114. Ibid.,p.41.

115. Brzezinski, Out of Control, passim.

NOTES (cont'd)

116. Edith Hill Updike and William J. Holstein, 'Mitsubishi and the "Cement Ceiling"', Business Week, 13 May 1996, p.62.

117. Op.cit., Denise Meagles, p.62.

118. William Letwin, 'Return to Japan', National Review, Vol.XLVII, No.16, 28 August 1995, p.29.

119. 'Japan's outlook dims', Peninsula Daily News, 20 January 1998, p.B-4.

120. Ibid.,p.B-4.

121. Ibid.,p.B-4.

122. Op.cit., John Naisbitt, Megatrends Asia, p.115.

123. 'Japanese in lather over shampoo', Peninsula Daily News, 7 November 1997, p.C-4.

124. Dr. Fred Thompson et al., 'Bridging the Ocean Between Us', Studium Generale Notes, 6 November 1997. NB: Were not these exchange students some of Japan's "best and brightest"?

125. Tosiyasu Kunii, 'Pax Japonica...', The Pacific Rim Almanac, p.517.

126. Simon Winchester, 'Eastern Capital', Pacific Rising, pp.409-441.

127. Ibid.,p.409.

128. Ibid.,p.438.

129. Ibid.,p.439.

130. The years 1964-1998: Thirty-four years -- a mere third of a century! And just in time to dampen international ardour for the Winter Olympiade at Nagano, has been the publication of The Rape of Nanking: The Forgotten Holocaust of World War II (New York: Basic Books, 1997) by Iris Chang.

131. See C.D. Bay-Hansen, Fisheries of the Pacific Northwest Coast, Vol.1 (New York: Vantage Press, Inc., 1991), pp.201-218.

132. See Ibid., Vol.2, (1994), pp.130-146.

133. The Pacific Rim Almanac, p.407. NB: For Japanese small-meshed nets snaring porpoises (and resultant "Dolphin-Friendly" tuna-fish cans), see Elegant (1990), p.169, and Winchester (1991), pp.152-154.

134. See Ch.1, Pt.II, in `FutureFish 2000`.

135. 'Coast Guard gets OK to enforce reflagging', Peninsula Daily News, 29 November 1992, p.A-5.

136. Ibid.,p.A-5. Cf. Donna Parker, 'Anti-Reflagging Decision Reaffirmed', Pacific Fishing, March 1992, pp.28,29.

NOTES (cont'd)

137. 'Judge bars foreign boats from waters', Peninsula Daily News, 9 February 1992, p.D-3.

138. 'Coast Guard gets O.K. to enforce reflagging', PDN, 29 November 1992, p.A-5.

139. Ibid.,p.A-5.

140. Ibid.,p.A-5.

141. Ken Kelley, 'Japanese fish farming: Offshore pens are home to dozens of species', National Fisherman, Vol.74, No.12, April 1994, p.58.

142. Ibid.,p.58. NB: All parentheses and contents this writer's.

143. Ibid.,p.58. NB: All parentheses and contents this writer's.

144. Ibid.,pp.58-59.

145. Ibid.,p.60.

146. Ibid.,pp.58,60. NB: For an environmental vision of the vanishing northern bluefin tuna (Thunnus thynnus), see Carl Safina, Song for the Blue Ocean (New York: Henry Holt and Company, 1998), pp.7-116. A definitive study although marred by the usual eco-political bias.

147. Op.cit., James Manzi, p.60.

148. James Fallows, More Like Us, p.208.

149. Terry Johnson, 'Journey to Japan: In Search of Seafood's Future', Pacific Fishing, August 1993, p.45.

150. 'From the Grounds', Seafood Leader, November/December 1993, p.146.

151. 'Whole Seafood Catalog', Seafood Leader, September/October 1993, p.138.

152. Buck and Shelly Laukitis, 'Fishing for Opportunity in Japan', Pacific Fishing, Vol.XVI, No.8, August 1995, pp.30-31.

153. Ibid.,p.31.

154. Ibid.,p.32.

155. Peter Redmayne, ed., 'Japan: In the '90s, Selling Seafood Is Not So Simple', Seafood Leader, July/August 1994, p.40.

156. Ibid.,p.40.

157. Op.cit., Peter Redmayne, p.44. NB: Brackets and contents this writer's.

158. Ibid.,p.44. Cf. Buck and Shelly Lukaitis, p.33.

NOTES (cont'd)

159. Ibid.,p.49. Cf. Buck and Shelly Lukaitis, p.33.

160. Helen Gregory, 'Japanese move with the times', Seafood International, Vol.10, Issue 8, August 1995, p.27.

161. Ibid.,p.27.

162. Op.cit., Christina Burridge, p.29.

163. Op.cit., Dalton Hobbs, p.29.

164. Op.cit., Katsumi Hirai, p.29.

165. Vol.16, No.5, September/October 1996,p.116.

166. Ibid.,p.116. NB: According to Seafood Leader, the fugu "industry peaked in 1958, when 176 people died of fugu."

167. Op.cit., Kiichi Kitahama, p.116.

168. Ian Fleming (1908-1964), You Only Live Twice (New York: Signet Books, 1993 ed.,)

169. Op.cit., Robert Elegant, p.111.

170. Tetrodotoxin has been found in atelopid frogs from Costa Rica, the California newt, the goby fish from Taiwan, and the blue-ringed octopus from Australia's Great Barrier Reef. See Wade Davis (below), pp.197, 344-345.

171. A Harvard Scientist Uncovers the Startling Truth About The Secret World of Haitian Voodoo and Zombis (New York: Warner Books, Inc. 1987).

172. Ibid.,pp.134-135. NB: A Brazilian puffer is Tetrodon psittacus; a Baja, Calif., botete is Sphoeroides lobatus. In Davis, pp.196,144.

173. Ibid.,pp.135-136.

174. Ibid.,pp.137-138.

175. Op.cit., Wade Davis, p.139. NB: All brackets and contents this writer's.

Part II: China -- "Enter the Dragon"

"China? There lies a sleeping giant. Let him sleep! For when he wakes he will move the world."
-- Napoleon Bonaparte

"In China the educated believe nothing and the uneducated believe everything."
-- Alasdair Clayre of the BBC, quoting a Christian Missionary in 1984

"What's a 'house church'?"
-- U.S. ambassador to China, James Sasser

Noted historian Barbara W. Tuchman (The Guns of August), in her 1972 Notes from China, compared the Chinese with the Jews throughout the millennia:

....."They are the two oldest peoples with a continuous history and a continuous language and the only two now maintaining sovereignty over the same territory as 3,000 years ago. Both went through a long struggle and a final armed fight to achieve that sovereignty, both came to power at about the same time, 1948-49, both pursued a dominant idea, in one case revival, in the other revolution. Socialism if not Marxism was the early Zionist goal, and the communal system of the kibbutzim antedates the present communes of China. Both nations stress self-sufficiency for similar reasons, and both live in fear of invasion."[176]

But this writer, who worked at Kibbutz Hazorea (near Haifa, Israel) during the summer of 1961, didn't hear much Marxist rhetoric; he did learn, however, that the Israelis are truly a unique nation. Since those all-too-few wondrous weeks in 1961, the writer has discovered the amazing range of characteristics encompassed within the Jewish people. Thus for every [actor-director] Woody Allen there is a [Rabbi] Dan Lapin; for every [comedian-actor] Billy Crystal there is a [rabbi, writer, comedian] Jackie Mason; for every [disc jockey-celebrity] Howard Stern there is a [film critic] Michael Medved. In U.S. politics, for every ultra-liberal Charles Schumer (Democrat, N.Y.) there is an arch-conservative Bruce Herschensohn (Republican, Calif.). The list is virtually endless. Within the highly varied Jewish ethnos there are the Ladino (Latino)-speaking Sephardim and the Yiddish (Jüdisch)-speaking Ashkenazim. In the New York City of this writer's youth, Ashkenazim themselves were split into battling cultural schools of Lithuanian-Polish Litvaks and Polish-Ukrainian Golitzianers. (But both groups told and re-told the old joke: For every four Jews there are five opinions!)

As noted by Barbara Tuchman, the Chinese are also an ancient nation with an unbroken history. The Chinese, like the Jews, are blessed with qualities which are simultaneously philosophical and acquisitive.

Overseas Chinese have thrived culturally and commercially in a way similar to Jews of the Diaspora. The Israelites have given the West Moses and the Torah; the ancient Chinese have given the East Confucius and the Analects. There are also many divisions within the Chinese nation -- the main one being between the Mandarin-speaking north and the Cantonese-speaking south. And again, in a like manner, the Chinese are a complex people of varied personalities: For every Master Kung there has been a Chairman Mao; for every dissident Harry Wu there has been a Premier Li Peng. A famous [now forgotten] author once wrote that every civilised man should love his own country and [Mediaeval] France. A self-avowed <u>illuminatus</u> -- acquainted with this writer -- added [Ancient] Greece to the list of the forgotten <u>literatus</u>. An evangelical friend once said that all Christian believers ought to feel patriotic loyalty to their own nation, and a spiritual reverence for [<u>Eretz</u>] Israel.

But this contrarian writer, although a Christian, a patriot, and in general accord with the above sentiments, is also a cultural determinist and political realist as well as a serious student of history. And in China you get it all -- religious revival, living culture, power politics and continuous history. China-watching by the West should clearly (and closely) follow cultural loyalty (to France or Greece), spiritual devotion (to Israel), or nationalistic patriotism (to Norway or Canada or the U.S.A.). For in out time, tiny Israel is losing the little land (<u>eretz</u>) she has, while huge China is adding to her already vast real estate. (Meanwhile, Israel is selling China advanced weapons systems technology. Go figure.) The Dragon is armed, primed, and ready to rule in the coming Chinese century.

On 1 July 1997 the British crown colony of Hong Kong (not-so "Fragrant Harbour") was literally "handed over" to the People's Republic of China. (The nearby Portuguese possession of Macau will be retroceded to the Mainland in December 1999.) The United States apparently saw nothing amiss with the abandonment of 6.4 million human beings to the tender mercies of Communist rule.[177] Indeed, the attendant U.S. Secretary of State, Madeleine Albright, appeared to almost welcome the Beijing takeover: This was, after all, an Anglo-Chinese problem dating from the 1840s. And even though the American public has overwhelmingly disapproved of granting China "most favoured nation" (MFN) status, in June 1997 the U.S. House of Representatives voted -- once again -- to preserve Beijing's MFN status.[178] In light of China's grisly human rights record, Boston Globe columnist Jeff Jacoby has observed:

"Profits before principles....In the days when Nelson Mandela was still imprisoned on Robben Island, no American retailer dared offer merchandise labeled 'Made in South Africa'. It should be equally unthinkable to market products made in China."[179]

Leftist double-standards, especially regarding China, have been endemic in American geo-political life since the 1960s. (There is the example of Cuba, too, as an example of Rightist hypocrisy.) But China-lobbying has transcended U.S. party lines. "Profits before principles" is the position taken in the Clinton White House -- as it was in the Bush White House before the current Clinton administration. It is also the position held in the plush suites of [Henry] Kissinger Associates, and by [China] lobbyists for the National Association of Manufacturers. Above all, columnist Jeff Jacoby has told us, "it is the position advocated by the CEOs of Fortune 500 corporations eager for China's business."[180]

Also in 1897, Ross H. Munro and Richard Bernstein (Faded Glory, Dictatorship of Virtue) were the co-authors of The Coming Conflict with China.[181] In it, Munro and Bernstein referred to American corporate and ChiCom-friendly special interest groups as the "New China Lobby". This lobby is spear-headed by former U.S. Secretaries of State Henry Kissinger (as noted) and Alexander M. Haig, Jr. According to The New American, Messrs. Kissinger and Haig are

"...the two most conspicuous practitioners of this corrupt trade, and also the most slavishly devoted to the Bejing party line. Each has written op-ed columns praising the late Deng Xiaoping and urging closer U.S.-China cooperation, and each has used his influence on Capitol Hill in defense of Chinese interests."[182]

The New American further reported that General Haig, in fact, was the sole United States citizen to attend the People's Republic's 1989 National Day commemorating the founding of the Communist régime.[183] Thus, a U.S. general appeared in the public eye as an honoured guest of a Communist government -- a government whose army, only four months previously, had slaughtered hundreds of dissenting Chinese civilians at Tiananmen Square. If the reader (or this writer) might consider the above act to be truly degrading to American credibility, (we must) think again. Alvin and Heidi Toffler, former Marxist now Third Wave gurus, have pooh-poohed U.S. concerns about undue ChiCom influence on domestic politics (i.e., the presidential election of 1996). Rationalised the Tofflers:

"[T]he concept that foreign-linked funding is necessarily inappropriate is based on obsolete conceptions of both sovereignty and nationhood. It is based on the idea that states have absolute sovereignty and that nations are closed systems. All that is coming apart as we move into the global economy. Transnational flows of political influence will increase, not decrease, in the years ahead. For good and for ill, it is time to recognize that politics...like economics and information, is going transnational."[184]

The Tofflers -- perhaps due to hold-over thinking from their days of statist haze -- don't seem to comprehend that Chinese issues transcend mere national sovereignty; these issues have to do with human freedom -- the right to speak, assemble, and worship as the private citizen chooses. Harry Wu, the now-famous Chinese-American dissident, spent nineteen long years in a People's Republic political <u>laogai</u> (prison camp) and knows well the two antonymous faces of Communist Chinese "capitalism". In a revealing <u>Human Events</u> interview in early 1997, Harry Wu acknowledged China's current economic boom -- made possible, of course, by Western capitalism -- and mammoth earnings accrued by both China and the West.[185] But the ordinary Chinese civilian, asserted Wu, sees a mere modicum of this prosperity. (The People's Republic possesses a true trickle-down economy.) The People's régime mostly pumps the huge profits into upgrading its extensive weapons systems, and into "maintaining the nation's political stability."[186]

The People's régime, however, has taken full advantage of China's economic prosperity to make itself look good to the outside world. Meanwhile in China herself, the inmates of the 1,100-camp <u>laogai</u> system continue, as slave labour, to grow a significant portion of China's cotton and a third of China's tea; to produce 60% of China's rubber vulcanising chemicals, and hand tools etc.[187] And back in America, President Clinton -- who in 1993 so strongly (with jutting chin) condemned despots and dictators from Baghdad to Beijing -- has, since 1994, delinked the human rights issue from U.S. trade policy.[188] Reversing himself politically has been nothing new for Bill Clinton, but to do so as quickly at such a price for the Chinese people.... But pc Communist China is <u>not</u> Apartheid South Africa, and as the <u>Boston Globe's</u> Jeff Jacoby has aptly written, "Profits before principles."

There is religious as well as political persecution in today's China. A.M. Rosenthal, a columnist for the liberal-élitist <u>New York Times</u>, would appear an unlikely candidate to noise abroad the plight of Chinese Christians -- but he has. The forthright A.M. Rosenthal admitted in a February 1997 op-ed piece that "[i]t was a Jew, Michael J. Horowitz of the Hudson Institute, who screamed me awake, as he has so many Christians."[189] Not only that, but Rosenthal dared voice the hope that prominent Jews of faith would unite in opposition to the persecution of Christians in China and Buddhists in Tibet. Furthermore, Israel should show the United States the path to righteousness by ending her (Israel's) arms trade with China.[190] And that wasn't all. Rosenthal extensively quoted Nina Shea, director of the Puebla Project at Freedom House, and author of In the Lion's Den:

> "Millions of American Christians pray in their churches each week, oblivious to the fact that Christians in many parts of the world suffer brutal torture, arrest, imprisonment and even death -- their homes and communities laid waste -- for no other reason than that they are Christians. The shocking untold story of our time is that more Christians have died this century simply for being Christians than in the first nineteen centuries after the birth of Christ. They have been persecuted and martyred before an unknowing, indifferent world and a largely silent Christian community.'"[191]

Among eleven countries indicated by Nina Shea as persecuting Christians were China, North Korea, Vietnam and Laos. In these Communist nations, thousands of Roman Catholics and Protestant evangelicals have been imprisoned for preaching the gospel, holding worship services, and distributing Bibles. And no political unit on the globe incarcerates more Christians than the People's Republic of China.[192] So when President Clinton met with his Chinese counterpart, during the UNO's October 1995 50th anniversary celebration at New York City, the U.S. president "proclaimed that the greatest threat China now poses to the world is pollution'."[193] U.S. officials at the time also scolded the ChiComs about pirating American software, but never was heard a discouraging word to poop the UN party about the brutal Chinese crackdown on "house-church" members, Roman Catholics, or Tibetan Buddhists.[194] (Although Bill Clinton probably did alot of jaw-clenching and lip-biting in every appearance of deep concern.)

William Norman Grigg of The New American has called the malign neglect of the current U.S.-China policy regarding Christians as arising from the generally identical goals of both mega-powers: "Renewal of most favored nation (MFN) status for China, increased trade between China and the U.S., and joint U.S.-Chinese diplomatic and military ventures."[195]

In conclusion Nina Shea suggested that, as China is subject to economic pressure, concerned Christians must work unceasingly ("agitate tirelessly") to turn back America's slow but sure drift into economic, political, and military interdependence on China (and, to a lesser extent, on Vietnam).[196]

There was a stock joke told throughout the Cold War years (1945-1990) which went: Optimists learn

Russian; pessimists Chinese. The gallows-humour saying enjoyed especial prominence during the Sino-Soviet split, commencing in the early 1960s. But since the official extinguishing of the old Soviet Union on 31 December 1991, will a broke and tentative New Russia snuggle up to the economically rejuvenated and militarily expansionist Red China? In mid-1997, at the time of the Hong Kong Handover, an article in The Economist (of London, U.K.), reprinted in The National Times (of Washington, D.C.), plaintively asked, 'Russia and China... Can a Bear Love a Dragon?' The Economist's sniffy answer to its own question was that the present attraction/alliance between Russia and China "is really only a temporary one....which will not last long."[197] But The Economist based its wishful thinking on three hard geopolitical facts: (1) Russia and China harbour a shared "indignation" (i.e., massive grudge) against the West; (2) the Bear and the Dragon hold common interests for regional stability; (3) and both have a blossoming, mutually-convenient arms trade.[198]

But there are additional hard geopolitical facts regarding the former Soviet, now independent, Islamic republics of Central Asia; Russo-Chinese attraction/alliance or not. Communist China -- along with Hindu India -- will continue to be a fire-wall containing the flames of Islam, which have threatened to spread from the Middle East to Central Asia since the Iranian Revolution of 1979. Despite ChiCom sales of silkworm missiles to Iran and purchases of Israeli weapons technology, insular China has no illusions concerning military hegemony within her own borders. Chronicles magazine contributing editor, Dhimitrios Gheorghiou, has got it just right:

....."The Chinese authorities, no matter who ends up holding power in the Forbidden City [Bejing], will suppress the Uighur-led [minority Chinese Turkic Muslim] jihad with whatever force is necessary. Belgrade may let the Serbs be driven from Sarajevo today, Paris may abandon Marseilles to the Algerians tomorrow, but Beijing will not tolerate the eradication of the Han [ethnic Chinese] from Xinjiang [western Sinkiang province] or the creation of an Islamic superpower in Asia."[199]

Before his sudden death by drowning in the mid-1990s, William E. Colby, former director of the C.I.A., forecast two possible causes of World War III: "A successful military coup which turns Russia back to imperialist ambitions"....[or] (2) "North Korea attacks South Korea and threatens Japan with a nuclear weapon.'"[200] The worldly-wise Bill Colby omitted mentioning The Coming Conflict With China,[201] as the ex-Central Intelligence Agency chief didn't live long enough to either read the Bernstein-Munro book or to heed its contents. If he had, Colby would have probably suggested a third cause of World War III -- the

Dragon rampant.

The Paracel Islands are 130 small coral islands 220 miles south-east of China's big island of Hainan, and 250 miles east of Vietnam's long coastline on the Gulf of Tonkin. The Paracels are claimed by both [Communist] China and [Communist] Vietnam. The Spratly Islands (Chin. <u>Nanshan</u>) lie midway between Vietnam and the Philippines in the South China Sea, and are composed of islets, reefs, cays, and shoals.[202] Sealife around and about the islands includes groupers, red snappers, sharks, and hawksbill turtles, but the marine resource of interest (and increasingly of contention) is off-shore oil deposits. Despite the Spratly Islands being approximately equidistant between Vietnam and the Philippines, it is China which claims the Spratlys as well as the Paracels.[203] Mainland China has become increasingly assertive during the preceding quarter century. In 1974, the ChiComs seized several of the Paracels from [weakened by war] South Vietnam (RVN).[204] The Chinese sank three Vietnamese ships of the Spratlys in 1988; more than 75 Vietnamese sailors were killed or MIA during the engagement.[205] (The last occurring thirteen long years after the "reunification" of all Vietnam by the Communist invaders from the North.) The two Communist powers fought another battle in the South China Sea during 1992. Since then, the Chinese have gradually moved in on ever more islands in the Spratlys. And during 1995 a ChiCom naval force took over Mischief Reef in the Spratlys from the Philippines.[206]

Mischief Reef in the Spratly Islands is located about 800 nautical miles from Hainan Island (China), but circa 135 miles from the Philippine island of Palawan.[207] Not only, then, is Mischief Reef noticeably closer to the Philippines, but the 1992 ASEAN-sponsored agreement in the Manila Declaration promised, moreover, to resolve peacefully the dispute over the sovereignty of the Spratlys. The resource-rich (i.e., oil) archipelago also interests Brunei, Malaysia, and the R.O.C. (Taiwan) as claimants. Tensions have mounted between all six Asian countries, and it has been suggested that the United Nations act as a mediating force.[208] (But most non-partisan observers already know that the whole scenario in the South China Sea is merely a case of the Dragon extending her claws.)

American master of the techno-thriller, Tom Clancy, appears to have no qualms identifying the future perpetrator of World War III -- or where it will commence. On the front page of <u>SSN: Strategies of</u>

Submarine Warfare (1996), Tom Clancy fired straight from the hip with "America is at war....China has invaded the oil-rich Spratly Islands and the Third World War has begun."[209] Another author of techno-thrillers, one Ian Slater, has written an entire series of would-be World War III pulp-fictions. (It's been a 1990s literary genre.) But Ian Slater, an expatriate Australian Joint Intelligence Bureau veteran, teaches (1996) political science at U. British Columbia and serves as managing editor of Pacific Affairs. So Slater, like Clancy, knows his subject matter. The one book of Slater's this writer read, WWIII: South China Sea, was compelling, despite the typical Anglophone ethnic and military stereotypes of good guys (Royal Gurkhas, British SAS, and U.S. Special Forces) versus a villainous Frenchman and howling hordes of blood thirsty Chinamen.[210] (But Slater, to remain acceptably Asian pc, included resuscitated Vietnamese as good guys. But he showed the Frenchman no such mercy.) Beside and despite these colourful and exciting techno-thrillers, there remain the very hard facts of Pax Americana and Dragon rampant. The New York Times, in a November 1994 article, for once expressed geopolitical concern about a Communist power:

"For Vietnam, which fears that China will one day use military power to enforce its claim over the entire South China Sea and its potentially rich oil reserves, the desirability of inviting American forces back into the disputed neighborhood is a form of insurance against what is seen as Chinese hegemonism."[211]

But the Philippines kicked the U.S. Armed Forces out of Subic navy yard and Clark airfield. (Isn't that what The New York Times has always wanted?) As a result, the small and poorly-equipped Philippine naval patrol units failed to stop the 1995 Chinese occupation of Mischief Reef in the Spratlys.[212] Richard Bernstein and Ross H. Munro related the almost-amusing story of the Filipino fisherman stumbling onto Mischief Reef in January 1995, and seeing ChiCom military structures and personnel everywhere. The Chinese government claimed that the occupiers were "fishermen"; of course nobody believed it.[213] In late 1996, an [always anxiously] alert Casper "Cap" Weinberger, Cold War warrior extraordinare and Ronald Reagan's Secretary of Defense for seven years, commented on China's militant expansionism to Human Events:

"Now they [the Chinese] have changed...to a very aggressive military posture, including declaring themselves sovereign over the Spratly [I]slands and some of the other islands in the South China Sea, where there is a good likelihood of finding oil."[214]

At last look, the Chinese had constructed a 2,600 metre-long airfield runway in July 1990 on Woody Island (Yong xing) in the Paracels. The new airbase, according to Bernstein and Munro, regularly accommodates thirty to forty fighters and bombers and possesses storage depots for aircraft, fuel and air-to-air missiles.[215] The airbase not only puts ChiCom warplanes within combat range of the Spratlys, but greatly expands their combat range generally. In March 1994, China constructed an early-warning-radar installation on Fiery Cross Reef in the Spratlys to support and facilitate coming People's Liberation Army (PLA) operations in the South China Seas area.[216] The sleeping Dragon has awakened and is growling with hunger.

Super-Sinophile John Naisbitt exulted in <u>Megatrends Asia</u> (1996) that "Japanese economic power has reached its peak and begun to recede as the Chinese prepare for the year 2000, the Year of the Dragon that will usher in the Dragon Century."[217] And this writer believes that the Colorado futurist is right: China's economy <u>did</u> double from the six years between 1990 and 1995.[218] But like most Sinophiles, and American ones in particular, Naisbitt has overlooked Chinese political action; ignored ChiCom military reaction. On China's intentions in the Paracels and the Spratlys, Naisbitt euphemised in a spin-cycle worthy of the Clinton White House:

> "Unlike Japan of the 1930s, China's interest is not in conquests. The complexities of domestic conditions, the need for capital to fund growth, the widening rural/urban gap and impending social instability, the weakening of Beijing's control and authority on the provinces and a whole range of political and economic concerns will undermine China's potential hegemonistic intentions or ambitions. However, the same reasons could force China to reach out to secure valuable resources, such as control over oil and gas reserves in the South China Seas, to address domestic needs."[219]

In a section on the rôle of the People's Liberation Army (PLA) as China's "privatization project", Naisbitt proudly listed the number and extent of capitalist ventures in which the PLA is engaged. According to Naisbitt, maybe so much as half of all PLA personnel are involved in (or have been) in non-military commercial undertakings. Indeed, since 1984 the PLA has set up more than 20,000 companies, producing profits estimated at $5 billion U.S.[220] The 999 Pharmaceutical Company, for example, is (1996) part of the San Jiu Enterprise Group, which is owned by the PLA. A prime investor is the U.S. mega-firm, Citicorp. Another American co-partner with the PLA is the ostensibly squeaky-clean ice cream manufacturing company, Baskin Robbins. Other successful PLA-operated businesses are Guangzhou's (Canton) three-star hotels, big buck-earning karaoke bars, the best bus routes between Guangzhou and Shenzhen (just north of Hong Kong), and Shenzhen's fastest-growing securities companies. The PLA was even the proprietor of JJ's, Shanghai's biggest nightspot, until it was closed down.[221]

"PLA, Inc.", as the People's Liberation Army is known to the local media, already had its entrepreneurial tentacles in Hong Kong before the Handover of 1997. John Naisbitt admiringly informed his readers that the PLA has "a stake in Nine Queen's Road Central and offices in the old Bank of China building."[222] As if all the above were not sufficiently laudatory, Naisbitt went on to boast that China United Airlines is PLA-owned and operated -- with the singular advantage that China United may fly to places that civilian airlines are not permitted.[223] (Some advantage! "Fly United" -- the only way to fly!) The PLA is the same organisation of people-unfriendly Chinese businessmen who massacred hundreds of dissenting (mostly) young people at Tiananmen Square, Beijing, 4 June 1989. The PLA took care of business there, too. Naisbitt ended his puff-piece on the PLA crowing, "The People's Liberation Army is now the largest privatization project in the world."[224]

The South China cities of Guangzhou and Shenzhen, alluded to by John Naisbitt, are in the prosperous province of Guangdong (Kwantung). This southern Chinese province alone has more than 60 million human beings -- the population size of a major European country.[225] Zbigniew Brzezinski, writing in 1993, informed his readers that the rate of economic growth in Guangdong had reached 13% per annum. At that rate, Brzezinski ventured (including the assimilation of Hong Kong), all China by 2010 A.D. could become the fourth global economic power; preceded only by the United States, the European Union, and Japan.[226] (This writer, as noted, predicts that China will eclipe Japan economically by 2020 A.D. or even earlier.)

By 1994, much of South China's economic development had been fueled by U.S. multinational corporations such as Amway, Avon, Heinz, and Wrigley's.[227] By 1996, Microsoft Corp. (of Redmond, Wash.) together with Compaq Computer Corp. (of Houston, Tex.), were sponsoring a Beijing-produced digitally-oriented TV sitcom that debuted in China. The show, <u>My Computer Family</u>, headed the offerings of a new television network -- the Education Channel -- a joint venture (JV) of Seattle-based RXL Pulitzer and Beijing TV. Microsoft and Compaq were to spend $475,000 U.S. for the first 26 episodes.[228]

Also in 1996, the National Broadcasting Co. (NBC) launched a brand new TV network in Asia. NBC in Asia, a 24-hour English-language (rather than Mandarin or Cantonese) cable service, offered a TV mix of world and all-Asia news, documentaries, dramas, and children's programmes for "the Pan-Asia region."[229] Based in Hong Kong, the new network was slated to feature borrowings from the business-oriented sister channel CNBC Asia, as well as the newspeak NBC Nightly News and Dateline NBC. The admitted viewer-target of NBC/CNBC Asia was to be Asian "business professionals" (read, yuppies) and their families, (mostly Western) expatriates in Asia, and (mostly Japanese) tourists.[230] Time will tell if the gerontocracy in Beijing approves of such American candy-floss broadcast fare as Seinfeld or The Tonight Show With Jay Leno in post-libertarian Hong Kong.

Meanwhile, China's influence on America is growing. An Associated Press article of 29 October 1997 reported that the Chinese imprint on Main Street U.S.A. becomes more evident every day:

"The impact [of the People's Republic] on America goes far beyond 'Made in China' Mickey Mouse magnets and Christmas toys. Baby boomers are adopting Chinese children. College students are signing up for Chinese language classes in record numbers. Young and old are taking martial arts courses."[231]

There is an undeniable reverse flow, both cultural and economic. Without thinking about it, Americans outfit themselves and their children in Chinese-manufactured shoes and clothing; purchase audio, video, and book gift packs -- all made in China -- for themselves and their children. While China restricts many U.S. exports, the reverse flow of goods earned the Chinese a record $5.9 billion by August 1997. Fully half the toys sold in the United States are from Mainland China.[232] The imbalance of the China trade has angered many Americans, including garment union members who have been understandably upset at Chinese apparel manufacturers utilising low-wage workers denied basic human rights (not to speak of the slave-labour in laogai prison camps).[233] But the great bulk of Americans are blissfully unaware of social and political conditions in the People's Republic of China.

In April 1994, head cyberdweeb Bill Gates of Microsoft Corp. posed for the global media with Jiang Zemin, general secretary of the Communist Party of China. Thus, as Boston Globe columnist Jeff Jacoby has observed, the world's foremost capitalist shmoozed with the world's foremost dictator. Said Bill Gates, in reference to de-linking free trade with civil rights in China, "It is basically interference in internal affairs.'"[234] This poses a real dilemma for the people of the United States. Americans of religious faith and humane conscience may choose to invest in such ChiCom-friendly companies as Boeing, GE, Heinz, IBM, KFC, Motorola et al. -- or not. As Jeff Jacoby has reminded us, there are plenty of big bucks to be made on Wall Street without enriching the above corporations, which help shore up the most evil empire since Stalin's Soviet Union.[235]

China-watchers like this writer are convinced that the closing of the 20th century, the American Century, will usher in the 21st century, the Chinese Century. The question remains: Are U.S. corporations like Boeing and Microsoft helping to perpetuate the present cruel régime in Beijing, thereby ensuring a coming Dragon century?

1990s China Fisheries Update and Seafood Market Overview

The rumours started in July 1993, and were confirmed by late summer in Urner Barry's Seafood Price-Current (14 September 1993): The current crop of Mainland China's nascent farmed shrimp industry had crashed. At first, no one really knew how -- or how many of -- the China white shrimp had perished. There had been speculation that pollution or red tide or even a killer virus had decimated the China whites (Penaeus chinensis).[236] By the time of the Urner Barry seafood report, American estimates of China white numbers had become so pessimistic as to engender the industry statement, "Quotations have been removed pending offerings of new season product.'"[237] A year later, in autumn 1994, there were still no quotations for China white shrimp.

It hadn't always been that way. The Mainland Chinese shrimp farming industry, as a commercial enterprise, was started up during the late 1970s near the northern port city Lianyungang, Jiangsu province (south of the Shandong peninsula).[238] During the initial year, the total shrimp harvest was less than 1,300 tons. But within seven years (during the mid-1980s), Mainland China became the globe's greatest producer of farmed shrimp, and would remain Number One for six years (until the early 1990s) until finally overtaken by Thailand.[239] The vast bulk of China whites were produced along North China's Gulf of Bohai (north of the Shandong [Shantung] peninsula). Two other species of shrimp, black tiger (P. monodon) and red-tailed shrimp (P. penicillatus), were mostly cultured in the southern provinces of Guangdong and Fujian. (Warmer temperatures and a longer growing season allow two shrimp harvests per annum.)[240]

It was in the southern province of Fujian (Fukien) where farmers, during the summer of 1993, began noticing white spots, rotting or swollen gills, crooked legs, and broken antennae on their red-tailed shrimp.[241] Within two to four days of shrimp manifesting the above symptoms, mortalities of red-tails were absolute. Spreading south from Fujian, the fatal malady sought out and destroyed black tigers in the farms of Guangdong. Then the disease diffused from the southern provinces upward to North China; all the way taking a deadly toll (of at least 70%) of the white shrimp ponds around the Gulf of Bohai.[242]

So what caused the great China shrimp kill of 1993? When all the facts finally came to light, the answers were simple: As a fish pond ages, its biological chemistry changes; the pond's bottom deteriorates over time; bad pond-cleaning, worse water exchange, plus Chinese reliance on natural feed (cheap and abundant trash-fish, dead shrimp, crushed mussels and clams) producing high organic waste -- all the model conditions for the lethal intrusion of a pathogenic vector.[243] As a former shellfish processor, this writer has worked around live shrimp (and crab) in tanks. Diligent tank cleaning and a constant flow of fresh, aerated seawater were the two tried-and-true methods for keeping live shrimp alive. And these shrimp tanks were located in a cool, well-aired shed on a dock overlooking the bay in Port Angeles (Washington) harbour. The reader (and the writer) might just imagine the conditions at the bottom of an ageing, uncleaned shrimp pond by the Bay of Bohai during a sweltering Chinese summer! After the Chinese shrimp crash of 1993, U.S. fishmongers were buying their penaeid shrimp from Ecuador, India, and Bangladesh -- for a while, anyway.[244]

Could China make a seafood comeback after 1993? In a side-bar to a <u>Seafood Leader</u> article of late 1994, the prognosis was..."not good, at least not any time soon."[245] As 1994 closed, many pragmatic Chinese fish-farmers were already switching to low-risk/low-maintenance species like East Asian carp (family <u>Cyprinidae</u>) and milkfish (<u>Chanos chanos</u>). Meanwhile, <u>Seafood Leader</u> opined, Mainland China's shrimp farmers had <u>not</u> learned their lessons of feed and pond management...to add to the dangers of an-ever pejorative pollution, an unchecked industrial development, plus an expanding population of 300 <u>millions</u> encircling the Gulf of Bohai.[246]

But by 1994 there was a silver lining to the cloud looming over China's aquaculture. That silver lining has been referred to by <u>Seafood Leader</u> as the "<u>other</u> [than shrimp] farmed marvel: scallops."[247] China bay scallops, to be exact. For right on the heels of bad shrimp news came good scallop news. Shellfish market columns in trade periodicals positively glowed with enthusiasm for China bay scallops. <u>Seafood Leader</u> rather breathlessly reported in July/August 1994 that China bays were "flooding into the U.S. market. Imports through March were 7,074 tons -- almost as much as were imported in all of 1993!...[China] <u>bays may be the best seafood value out there</u>...."[248] A letter to the editors of that same summer issue of <u>Seafood Leader</u> waxed rhapsodic:

....."I couldn't ask for anything more, except 'Why have we had this fishery for decades and the Chinese come over here, take a couple dozen scallops back with them and in a matter of a few years, put out tons of a beautiful, quality product for a great price!' I hope the Chinese don't find out too soon how great these scallops are."[249]

The letter writer had it just right -- how the China bay scallop saga unfolded is a whole fish tale in itself. Simply told, in December 1982 Chinese Professor Zhang Fusui hand-carried 120 Nantucket Bay scallops (<u>Argopecten irradians</u>) with him back to his aquaculture lab (Institute of Oceanology, Academica Sinica) at Qingdao on the shores of North China's Yellow Sea. Professor Zhang Fusui had acquired the New England scallops from the U.S. fisheries lab at Woods Hole, Massachusetts.[250] This had been Professor Fusui's third attempt, and during this last effort he had managed to keep 26 out of the 120 scallops alive for the 10,000-mile journey back to China. Eleven years later in 1993, those twenty-six New England scallops had spawned an aquaculture industry producing between 100,000 and 130,000 tons of (now) China bay scallops a year. In 1993, China supplied the U.S. market with more than 6,000 tons of shucked scallop meats. This amount of product made China in 1993 the greatest scallop exporter to the U.S.A. after Canada.[251]

The deep irony was that Nantucket Sound bays had been getting scarce; their recruitment diminished by both industrial pollution and brown algae blooms. Local Yankee catches of bay scallops had sunk to less than 2,000 tons per year; compared to more than 150,000 tons of sea scallops. At the same time, ten thousand miles away in northern China, transplanted American bay scallops survived and thrived in the nutrient-rich warmer waters of the Yellow Sea. (Thus answering the letter writer's question to the Seafood Leader editor. Cf. note 249.) The Chinese had taken their time to learn how to grow their (native) Chinese bay scallop, Chlamys farreri, but the species took two to four years to attain market size. The (new) China bay scallop reached a market size within a mere ten months! This was approximately one half the time required in New England waters to grow the same scallop to harvest size. So from virtually nothing, the Chinese have built a $150 million U.S. (1994 figures) industry employing thousands of workers. The China (American) bay has been a truly phenomenal success story.[252]

Being de jure Communists as well as de facto Confucianists, the Mainland Chinese have developed a sectionalised/specialised aquaculture for China bays which employs extended family coöperatives throughout the seeding, hatchery, and grow-out phases. The brood stock scallops spawn during late March and April and, after "conditioning", are hung in mesh bags -- using a long-line system -- to be harvested late in the year. The scallops are finally processed at local plants which increasingly match (and surpass) in quality those of Japan and the West.[253] Professor Kenneth Chew of the School of Fisheries, at the U. of Washington's Western Regional Aquaculture Center, has estimated that China possesses the capacity to produce 300,000 tons of bay scallops per year.[254] And Dr. Chew should know -- he is indubitably the leading U.S. authority on Chinese aquaculture.

By 1996, Mainland China had gone global as both seafood exporter and importer. Editor Peter Redmayne, of Seattle's Seafood Leader magazine, commented on a trip he had taken to Tokyo during 1994. While in Japan, Redmayne had asked three editors of [as many] seafood dailies what they considered to be the top [fish] story of the day:

"The huge increase in the world salmon supply' said one editor.
"The decline of traditional Japanese wholesale markets like Tsukiji', offered another.

"China', said a third. 'They're going to rewrite the rules of the seafood business. Just wait.'"[255]

In 1995 alone, continued Redmayne, China's production of seafood increased more than 2.5 million tons; a figure almost equal to the entire U.S. commercial seafood harvest. From 1991 to 1996, China's total seafood production doubled to 24 million tons; close to 30% of the world total.[256] Also in 1995, Chinese seafood exports increased almost 50% to $2.5 billion U.S. These were mostly aquacultured species like scallops, sea cucumbers (Japanese namako), abalones, eels, even fugu (blowfish) et al. sold to other fast-growing Asian markets.[257] Another important factor in China's favour is her huge, inexpensive, and skilled labour force. China's workers are making the People's Republic the seafood processing centre of the globe. Fish and shellfish from around the world's seas are being shipped to China, which reprocesses these products and re-exports them as seafood to global markets.[258] China's total seafood trade, which by 1996 probably reached $5 billion U.S. per year, was exceeded only by Japan and the U.S.[259] During the mid-1990s, Chinese annual seafood imports had passed $1 billion U.S. The People's Republic lowered their tariffs on seafood on 1 April 1996, and by A.D. 2001 will have constructed 20,000 supermarkets.[260]

The People's Republic of China held its first ever international seafood exposition, 29 October to 1 November 1996, at Qingdao (Tsingtao), Shandong province. The China Fisheries & Seafood Expo '96 was (quite naturally) organised by China's Ministry of Agriculture. Prior to the exposition, Zhang Yan Xi, the vice minister of Agriculture, announced: "The Chinese government is very committed to making China Fisheries & Seafood Expo a major international exposition'."[261]

Readers can bet their bottom dollar that the ChiComs put their best foot forward to facilitate the development of seafood markets for overseas suppliers -- and, of course, to showcase the People's Republic's own growing (by leaps and bounds) seafood exporting industry. China Fisheries & Seafood Expo '96 housed (ca.) 550 exhibits; (approx.) 300 of them were foreign companies (from Hong Kong and 25 other countries), with the remaining 250 being top domestic seafood exporters.[262]

While about half of Expo '96 featured seafood, there were also exhibits of processing, fishing, and aquaculture equipment. In addition to all the exhibits, a comprehensive conference programme for potential investors had been set up to accentuate business opportunities in all aspects of the Chinese seafood industry.

The Ministry of Agriculture offered foreign visitors to Expo '96 participation in industry tours throughout the Shandong peninsula and the Gulf of Bohai littoral; including day-trips to seafood plants and aquaculture ponds. Expo '96 was the very first time any of the above facilities were open to foreign visitors in the People's Republic of China.[263]

The second annual China Fisheries & Seafood Expo ('97) was held from 4 to 6 November at Beijing, Xin Hua district.[264] (Prior to that much-heralded event, however, Governor Gary Locke of Washington State paid an October [surprise] visit to the People's Republic.) Expo '97 was organised by the China Council for the Promotion of International Trade, an [apparently new] agency of the Ministry of Agriculture.[265] The opening ceremonies came in the form of a grand reception in the Great People's Hall -- right beside Tiananmen Square.[266] (This writer wonders if that supreme irony was lost on the fish market-mad Western exhibitors.) After wining and dining the visitors, the People's government advised the foreigners to acquire Chinese partners; the latter would help the former navigate the treacherous waters of imports officialdom and tariffs bureaucracy. Explicated Lu Zhihua, a representative of the city of Kaifeng's Economic & Technological Development Authority:

"The majority of companies or enterprises are not permitted to engage in foreign trade. Only the Chinese [i.e., the People's] entities who have received official permission can engage in foreign trade.'"[267]

(This was, after all, the Red China of old, but today's Dragon is hungry for fish, markets, and money.)

Although the overall attendance at Expo '97 exceeded that of Expo '96, American participation was noticeably down. Indeed, the U.S. had occupied a majority of exhibitor booths at Expo '96, but held less than half in 1997.[268] China editor of Pacific Fishing, Cooper Moo, attributed flagging American industry interest to: (1) A "been there, done that" attitude after the novelty of a Mainland China exposition wore off; (2) a backing away from initial efforts not immediately paying off; and (3) fully five of the missing U.S. booths resulted from a major shift in California calamari (market squid) production.[269] Shipments of California squid (Loligo opalescens) to China declined sharply in 1997 -- there was less product at higher prices. According to Cooper Moo of Pacific Fishing, the Chinese seafood market has developed on two levels -- the high-niche and the ultra-cheap markets. Seafood produce which can be landed into China for 50 cents a pound, ranks as an ultra-cheap market. So when the price of (usually) plebeian squid rose substantially, the practical

Chinese stopped buying.[270] The Communist Chinese are fast learning the ins and outs of free market ways. The next China Fisheries & Seafood Expo is scheduled for 27-29 October 1998, to be held at Dalian ("The Hong Kong of the North") in Liaoning province, on the Liaodong peninsula, PRC.[271]

As mentioned, Governor Gary Locke of Washington State went on an October 1997 mini-pilgrimage to the People's Republic. There were several reasons for the visit: (1) Gary Locke was on record as America's first governor of Chinese descent; (2) a bilateral trade with the world's largest country in population would be of enormous benefit to the Evergreen State and the U.S. Pacific Northwest; (3) the much-ballyhooed China Fisheries & Seafood Expo '97 was to commence in less than a month. Declared the valiant-for-truth Clintonist governor:

"From open markets, there will come open minds and human rights improvements....[M]uch progress has been made toward greater democracy and human rights in China -- but more progress can be made....Democracy and human rights and freedom are like a genie that has been let out of the bottle.'[272]

The governor was simply toeing the cautious line followed by the current White House (and recent U.S. administrations) vis-à-vis the evil gerontocracy which (still) rules from the Forbidden City at Beijing. Acting as unabashed China trade booster, Governor Locke was also fulfilling a dual rôle: That of proud Chinese-American political celebrity (which included the gubernatorial wife), and dutiful Washington statist liberal Democrat representing -- nay, shilling for -- Boeing and Microsoft corps. Our intrepid governor <u>did</u> raise the thorny topic of human rights violations when meeting with President Jiang Zemin. Governor Locke (to his credit) characterised his confrontational rhetoric, to the Associated Press, (paraphrased) as "gentle prodding, diplomatically couched...[that] didn't generate much of a response from the president."[273]

(Indeed! The reader might envision a chastened Jiang Zemin dragging his feet while laughing all the way to the big Boeing-Microsoft bank!)

<u>The Relentless Chinese Search for Coral Reef
Aquarium Fish, Shark-skins and Shark-fins,
and Sea-life Pharmaceuticals</u>

Yet another marine environmentalist statement appeared on bookstore shelves in early 1998. The latest (and at last look) has been Carl Safina's <u>Song for the Blue Ocean: Encounters Along the World's Coasts and Beneath the Seas</u>.[274] As this writer reads as many books on fishing, sea life, and the Pacific Ocean as he can, Song for the Blue Ocean was quickly purchased and eagerly read. In spite of the eco buzz-words, politically correct newspeak, with occasional baby-boomer references to the Fab Four (the Beatles), Jimi (Hendrix), and Joni (Mitchell) -- Carl Safina's was a serious and solid work. The book examined the bluefin tuna in the North Atlantic, various species of salmon of the North Pacific, and coral reef fishes off Palau and the Philippines in the tropical Western Pacific (and this writer was admittedly pleased to see one of his own works used as a biographical reference.)[275] In his section on the Western Pacific, Carl Safina wrote an entire chapter on Hong Kong and portrayed a local fish market in vivid colours:

"Parrot fishes, calico-patterned crabs, various wrasses, snappers, emperors -- wild creatures from all over the southern Pacific and South Asia crowd the concrete runs and glass tanks. In one pen, several moray eels gasp toward death....
"Here are red fish, yellow fish, pale fish, dark fish, fish with small blue spots on a tan background, black-spotted groupers of some kind, groupers with blue scrawled markings on a beige-saddled background, venom-spined scorpionfish, rabbitfish, blue-spotted red coral trout, various morays and other eels, fish that are black-spotted on red, and red-spotted on pink -- virtually all unknown to me."[276]

Absent, noticed Safina, were the open-ocean species sought and bought by the Japanese, who favour the rich and meaty flesh of tunas and other deep-water fishes. The Chinese have the exact opposite taste in seafood, preferring the flaky white flesh of coral reef fishes.[277] (Safina has reminded us that the whole subject of seafood -- of food in toto -- is vitally important to Chinese culture, tradition, and social status.)[278] The city of Hong Kong's 6 million residents devour 300,000 tons of fish and shellfish a year. That amount averages out to about 100 pounds per person per annum; the average American nibbles a mere 15 pounds a year.[279] And Chinese diners (like the Japanese) are not faint of heart: They like their fish fresh. Safina told a Hong Kong market story of a large Napoleon wrasse too big to be sold whole. Thus the fish mongers sliced the over-sized wrasse open with care, so that freshness-obsessed connoisseurs were reassured by seeing its heart beat -- even after large portions of the wrasse's body had been cut away.[280]   Later in his Hong Kong sojourn, Safina described a walk along Des Voeux Road, a "decidedly organic quarter":

"We eventually come to the shops selling sea creatures: dried scallops, dried abalones, dried sea cucumbers, load upon load of these animals, stacked and stored, bundled and proffered, from the far reaches of Earth. Most of these shops are small, barely too big to be called stalls. Their displays are neatly arranged, well presented, and all brightly lit.
"An adjacent pharmacy shop features parts of many different types of creatures, everything from dried sea horses and pipefishes to seal penises and testicles. Virtually every kind of creature that moves in the sea, it seems, is killed and offered as salubrious, believed -- or at least touted -- to promote health and vigor in various ways."[281]

Towards the end of his stroll through the Hong Kong fish markets (and his chapter on South China), Carl Safina arrived at the shark specialty stalls. Each shop had multi-tiered cases displaying shark-fins sorted by size; from large sacks stuffed to the brim with the tiny fins of small species or young sharks, to the enormous fins of forty-foot basking sharks. The dorsal fin of a basking shark, related Safina, is about three feet high and is corrugated like the bark of a tree. It sells for approximately $200 U.S. a pound.[282] The global shark-fin trade (for soup) targets about one hundred of the nearly four hundred shark species inhabiting the world's seas. Between 1980 and 1990, reported Safina (his style getting shriller), the global shark-fin trade doubled in volume. Sharks compose a mere 1% of the world's fish landings; nonetheless that translates to a draught of fishes -- from 40 million to 70 million sharks, was the "conservative" estimate of Carl Safina.[283] But he does have a point:

"Because shark meat is often worth too little to warrant taking up onboard space reserved for more valuable species, fishers often slice the valuable fins from the living animal, then dump it overboard. Bleeding and unable to control their swimming, they wriggle to the bottom of the sea, where they die."[284]

This writer has lived in both Florida and Hawaii where he has observed sharks at first hand, and shall admit to little sympathy for the blood-thirsty brutes. Helpless shipwrecked sailors or fishermen, encircled by hungry sharks, always come to mind. But, Carl, chill out! This is 1998! The 1990s have been the decade of the Top Predator, the Alpha Male/Female, featured in the hundreds of grisly nature video offerings (as well as at the pinnacles of U.S. political power). In the seas, the great white shark is the unchallenged Top Predator. Its preëminent spokesman has been [Top Author] Peter (Jaws) Benchley, who has made "great white" (and millions of dollars) a North American household phrase. Sharks generally combine the hunting, scavenging, and feeding habits at sea of lions/ hyaenas on land. (Lions and hyaenas are often similar in behaviour, but the hyaena meets humanoid TV-viewer resistance as it is truly terrifying, ugly, repugnant, and has failed all attempts at anthropomorphosis. Even the Jane Goodalls couldn't humanise the hyaena, although they tried their best.) Like the terrestrial lion/hyaena, the corresponding marine shark enjoys élite status as an ultimate survivor, a Top Predator.

But Carl Safina has a point. In almost all cases where sharks have been fished commercially, they have been quickly depleted.[285] Sharks (like rockfish) are slow-growing, late-maturing, long-lived. And sharks are viviparous, bearing living -- but few -- young. So love sharks or hate them, their numbers are very limited in a teeming water-world of milt, roe, and spawning by-the-millions. Sharks are also harvested for their skins that are tanned into valuable leather goods, for their oils, and of late increasingly for their cartilage. The last has been used to manufacture anti-cancer pills, which have enjoyed a current vogue as a health fad.[286] Thus targeted for fins, skins, oils, and cartilage, sharks should appear even to the most hardened sceptic as an endangered species. None of the above bodes well for the shark -- especially in Chinese waters -- but this writer would hesitate to concur with Carl Safina's dire prediction:

"In every restaurant we've been to in Hong Kong, various soups featuring shark fins are an entire section in the menu. Just as buffalo tongues, passenger pigeons, and cranes once graced the menus of fancy restaurants in America."[287]

NOTES

176. Barbara W. Tuchman, <u>Notes from China</u> (New York: Collier Books, 1972), p.72.

177. Jeff Jacoby, 'Night falls on Hong Kong; what will Americans do?', <u>Peninsula Daily News,</u> 3 July 1997, n.p.

178. <u>Ibid.</u>, n.p.

179. Op.cit., Jeff Jacoby. He also commented: "Hideous as apartheid was, it paled next to Chinese communism, which shackles 15 million men and women in ghastly slave labor camps... and tortures priests and monks for teaching the word of God."

180. <u>Ibid.</u>, n.p.

181. Alfred A. Knopf, New York, 1997.

182. 'Our "business" with Beijing,' <u>The New American</u>, Vol.13, No.9, 28 April 1997, p.19.

183. <u>Ibid.</u>, p.19.

184. Alvin and Heidi Toffler quoted in <u>ibid.</u>, p.11. NB: Would the Tofflers repeat their rationalisation at Taipei, R.O.C.?

185. Harry Wu, 'Slave Labor Camps Are Still Integral Part of Chinese Economy', <u>Human Events</u>, Vol.53, No.5, 7 February 1997, p.22.

186. <u>Ibid.</u>, p.22.

187. <u>Ibid.</u>, p.22.

188. <u>Ibid.</u>, p.22.

189. A.M. Rosenthal, 'Persecution of Christians rampant', <u>Peninsula Daily News</u>, 13 February 1997, p.A-8.

190. <u>Ibid.</u>, p.A-8.

191. <u>Ibid.</u>, p.A-8.

192. Nina Shea, <u>In the Lion's Den</u> (Nashville, Tenn.,: Broadman & Holman Publishers, 1997), pp.57-84.

193. Nina Shea cited by <u>The New American</u>, Vol.18, No.11, 26 May 1997, p.32.

194. <u>Ibid.</u>, pp.32-33.

195. <u>Ibid.</u>, p.33.

196. <u>Ibid.</u>, p.34. NB: The literature on evangelical Chinese "house-churches" is extensive. Cf. see Joseph Lam with William Bray, <u>China: The Last Superpower (The Dragon's Hunger for World Conquest)</u>, Leaf Press, Green Forest, Ark., 1997. See also Carl Lawrence with David Wang, <u>The Coming Influence of China</u>, Vision House Publishing, Inc., Gresham, Ore., 1996.

NOTES (cont'd)

197. 'Russia and China... Can a Bear Love a Dragon?', The Economist (London). Cited in the National Times, July/August 1997, p.12.

198. Ibid., pp.12-15.

199. Dhimitrios Gheorghiu, 'Cultural Revolutions' Chronicles, Vol.21, No.9, September 1997, p.7. NB: All brackets and contents this writer's.

200. James Dale Davidson, The Capitalist Manifesto (Baltimore, Md.: Strategic Investment Ltd. Partnership, 1995), p.107.

201. Richard Bernstein and Ross H. Munro (New York: Alfred A. Knopf, 1997).

202. See Appendix B for background information on the Paracel and Spratly Islands.

203. 'Dragon Growling', National Review, 31 July 1995, p.44.

204. Richard Bernstein and Ross H. Munro, p.78.

205. Ibid., p.78.

206. Ibid., p.78.

207. Ibid., p.79. See Appendix B.

208. Tom Clancy, SSN: Strategies of Submarine Warfare (New York: Berkley Books, 1996), p.1.

209. Ibid., frontispiece.

210. Ian Slater, WWIII: South China Sea (New York: Fawcett Gold Medal, 1996), pp.8 ff; 396, passim.

211. The New York Times, November 1994, n.p.; cited by Ian Slater.

212. Richard Bernstein and Ross H. Munro, p.180.

213. Ibid., p.180. Cf. 'Dragon Growling', National Review, 31 July 1995, p.44.

214. 'Weinberger Stresses Urgent Need to Deploy Missile Defense System', Human Events, Vol.52, No.45, 29 November 1996, p.15. NB: All brackets and contents this writer's.

215. Richard Bernstein and Ross H. Munro, p.74.

216. Ibid., p.74.

217. John Naisbitt, Megatrends Asia (New York: Simon & Schuster, 1996), p.17.

218. Ibid., p.18.

219. Ibid., p.234. NB: U.S. pro-China feelings go back to pre-World War II Christian missionary years, which personally touched such prominent Americans as Time-Life's Henry Luce and evangelist Billy Graham.

NOTES (cont'd)

220. Ibid., p.125.

221. Ibid., pp.125-126.

222. Op.cit., Naisbitt, p.126.

223. Ibid., p.126.

224. Op.cit., Naisbitt, p.126.

225. Zbigniew Brzezinski, Out of Control (New York: Charles Scribner's Sons, 1993), p.194.

226. Ibid., p.194.

227. Dennis G. Harter, 'Southern China's Economic Miracle and the Role of U.S. Business', Studium Generale Program Notes, Peninsula College (Port Angeles, Wash.), 23 April 1994.

228. 'Microsoft scales the "great wall" of Chinese television', Peninsula Daily News TV Week, 13-20 January 1996, p.2.

229. 'NBC branches out in Asia from its business channel to a network', Peninsula Daily News TV Week, 21-27 January 1996, p.2.

230. Ibid., p.2.

231. 'China has growing influence on lives of Americans', Peninsula Daily News, 29 October 1997, p.C-1. NB: Brackets and contents this writer's.

232. Ibid., p.C-1.

233. Ibid., p.C-1.

234. Jeff Jacoby, 'Night falls on Hong Kong; what will Americans do?', Peninsula Daily News, 3 July 1997, n.p.

235. Ibid., n.p.

236. Rob Lovitt, 'The Great Fall of China', Seafood Leader, November/December 1994, p.51.

237. Ibid., p.51.

238. Ibid., p.52.

239. Ibid., p.52. NB: Along the way, China produced a record 219,000 tons of farmed shrimp in 1991.

240. Ibid., p.52. NB: A/k/a the Sea of Bohai.

241. Ibid., pp.52-53.

242. Ibid., p.53.

NOTES (cont'd)

243. Ibid., p.54.

244. Ibid., p.54.

245. 'Can China Come Back?', Seafood Leader, November/December 1994, p.56.

246. Ibid., p.56.

247. Op.cit., Rob Lovitt, 'The Great Fall of China' p.58. NB: Brackets this writer's.

248. 'Shellfish Market', Seafood Leader, July/August 1994, p.12. NB: Emphasis in original; brackets this writer's.

249. 'Three Cheers for China Bays'; letter to eds. from Chris DiNunno, Village Fish Market and Restaurants, Punta Gorda, Fla. Ibid., p.6.

250. 'Aquaculture Roundup', Seafood Leader, January/February 1994, p.77.

251. 'Shellfish', Seafood Leader, March/April 1994, p.203.

252. Ibid., above; see notes 250 and 251.

253. 'Aquaculture Roundup', ibid., p.77.

254. Ibid., p.77. For an educated and detailed assessment of the future of China bays, see Kenneth K. Chew, 'Bay Scallop Culture in China Revisited', Aquaculture Magazine, Vol.19, No.3, May/June 1993, pp.69-75. NB: This writer is grateful to John E. Burris, dir. and ceo of the Marine Biological Laboratory, Woods Hole, Mass., for personally forwarding a copy of Dr. Chew's article.

255. Peter Redmayne, ed., 'The Great Call of China', Seafood Leader, Vol.16, No.3, May/June 1996, p.5.

256. Ibid., p.5.

257. Peter Redmayne, 'The Hungry Giant', ibid., p.41.

258. Ibid., pp.41-42

259. Ibid., p.42.

260. Peter Redmayne, 'The Great Call of China' ibid., p.5.

261. 'China Hosts First International Seafood Show', Seafood Leader, Vol.16, No.3, May/June 1996, p.46.

262. Ibid., p.46. Cf. 'China Fisheries & Seafood '96' (show directory), Sea Fare Expositions, Inc., Seattle, Wash., USA.

263. Ibid., p.46.

264. Bob Tkacz, 'Making a mark on the world's biggest market', Alaska Fisherman's Journal, February 1998, p.26.

NOTES (cont'd)

265. Cover (official visitor's guide), 'China Fisheries & Seafood Expo '97', Sea Fare Expositions, Inc., Seattle, Wash. (Sub-heading: "Asia's Largest Seafood and Fisheries Exposition".)

266. Bob Tkacz, 'Making a mark...', Alaska Fisherman's Journal, p.26.

267. Ibid., p.26. NB: Brackets and contents this writer's.

268. Cooper Moo, 'China Hosts Second Annual Fisheries Show' Pacific Fishing, Vol.XIX, No.1, January 1998, p.14.

269. Ibid., p.14. NB: Cooper Moo is (also) export manager at Emerald International Trade in Seattle, Wash. Editor and publisher Peter Redmayne, of Alaska Fisherman's Journal and Seafood Leader, served as "overseas co-organizer" of both China Fisheries & Seafood Expo(s) '96 and '97.

270. Ibid., p.14.

271. Ibid., p.56. Cf. 'China Fisheries & Seafood Expo '97', official visitor's guide.

272. 'Locke in China: "Open markets...open minds",' Peninsula Daily News, 8 October 1997, p.A-8.

273. Ibid., p.A-8. NB: Brackets this writer's.

274. A John Macrae Book, Henry Holt and Company, New York, 1998.

275. 'Book Two', ibid., p.442.

276. Op.cit., Carl Safina, p.385.

277. Ibid., p.385.

278. Ibid., p.394.

279. Ibid., p.386.

280. Ibid., p.386.

281. Op.cit., Carl Safina, p.399.

282. Ibid., p.401.

283. Ibid., p.401.

284. Op.cit., Carl Safina, p.402.

285. Ibid., p.402.

286. Ibid., p.402.

287. Op.cit., Carl Safina, P.403.

Part III: Small Tigers and Little Dragons

"We meet amidst predictions that the Asia Miracle will be succeeded by an Asian Meltdown" -- U.S. Secretary of State Madeleine Albright at Vancouver APEC summit, November 1997

"Indonesia is a kleptocracy....If you're a friend or family member of President Suharto, that's how you make money."
-- Doug Bandow of the Cato Institute (Washington, D.C.), November 1997

"[T]he Korean tiger, it turns out, was a paper tiger -- and it was bad paper, too"
-- Terence Jeffrey of Human Events, on the proposed IMF bailout of South Korea, December 1997

For some years now, the four "newly industrialized countries" (NICs) of Hong Kong (gone as semi-independent entity in 1997), Singapore, South Korea (R.O.K.), and Taiwan (R.O.C.) have been referred to as The Four Tigers of East/South-East Asia. The title bespoke economic empowerment, and, despite the current (1997-1998) Asian monetary crisis, the collective title has been well deserved. The "lesser developed countries" (LDCs) of South-East Asia have been generally identified as Indonesia, Malaysia, Philippines, and Thailand. This leaves out North Korea in East Asia; in South-East Asia the LDC designation omits Burma (Myanmar) and the three lands comprising the former Indo-China -- Vietnam, Laos, and Cambodia (Kampuchea). But for South China Sea fishing, Vietnam (though impoverished) is already an important domestic seafood producer. Presently starving (and wretchedly derelict) North Korea has future potential to enter the Pacific (from the Yellow Sea and Sea of Japan) fisheries as a significant participant -- like her prosperous twin sister to the south. Anyway, this writer will go along with the current-wisdom designation of "LDC" for Indonesia, Malaysia, Philippines, and Thailand... plus Vietnam. But this writer takes the liberty of referring to the above four LDCs as the Four Little Dragons for one good reason: All four countries appear to be largely owned and operated by a brilliant and acquisitive upper/middle class of Overseas Chinese. (Again, comparison of Overseas Chinese in South-East Asia to Jews in Western Europe and North America seems unavoidable.)

In his Megatrends Asia, futurist John Naisbitt has devoted a large section of his book solely on the truly remarkable success of the Overseas Chinese.[288] As Naisbitt has rightly stated, the Overseas Chinese (i.e., outside Mainland China) as an Asian phenomenon, is not fully understood by the West. For "Greater China" consists not only of China, Hong Kong, Singapore, and Taiwan, but the millions of Chinese who live in [and dominate the business and politics of] the LDCs of Indonesia, Malaysia, Philippines, and Thailand. The Overseas Chinese network not only rules the Four Little Dragons, as Naisbitt has posited, but indeed all of South-East Asia.[289] In 1996 there were 57 million Overseas Chinese, 53 million in Asia alone, with substantial commercial nodes headquartered in such Western urban centres as London, England; Los Angeles and San Francisco, Calif.; Vancouver, B.C.[290] Gleaning his numbers from Fujitsu Research in Tokyo, Naisbitt revealed some startling statistics: Overseas Chinese ownership of listed companies in Indonesia was a high 73%; in Malaysia, 61%; in the Philippines, 50%; and in Thailand, a whopping 81%.[291] (Overseas Chinese, in fact, predominate in Far Eastern trade and investment everywhere except Japan and South Korea). And, according to Sinophile Naisbitt, had one counted the overall global activity of the Overseas Chinese as a group, it would have ranked in 1996 third behind the United States and Japan.

As for Overseas Chinese ranking as about the richest aggregate on the world stage, Forbes magazine's annual list of top billionaires in 1996 included fully four of them in the top ten. They were (4th) Lee Shau Kee of Hong Kong, (5th) Tsai Wan-lin of Taiwan, (6th) Li Ka-shing of Hong Kong, and (10th) Tan Yu of the Philippines.[293] The United States in 1996 was still home to one third of the globe's billionaires -- with 149 in all. In East Asia, 41 billionaires hailed from Japan and 20 from Hong Kong. In South-East Asia, Malaysia produced 11 billionaires, Thailand 10, and the Philippines 9. This writer would wager a goodly sum that most -- if not all -- of the South-East Asian billionaires were Overseas (i.e., ethnic) Chinese.[294]

Since 1996, there has been the serious matter of the Asian financial crisis. Still -- the Institute for Management Development (IMD) at Lausanne, Switzerland, published its annual report on international competitiveness for 1998, and the figures showed Singapore and Hong Kong (of Greater China), in second and third place respectively (in 1997, too); right on the heels of number one United States.[295] (A mere five years ago, proud Japan had been the world's second most competetive economy; by 1997 she had plunged to 18th place on the IMD list. And 1998 has turned out to be an even worse year of fiscal disarray for Japan.)[296]

The fifth annual summit of the Asia-Pacific Economic Cooperation forum was kicked off at Canada Place, Vancouver, B.C., on 25 November 1997. Prime Minister Jean Chrétien of Canada would host, with U.S. President Bill Clinton, ChiCom President Jiang Zemin, and Japanese Prime Minister Ryutaro Hashimoto attending along with other Asia/Pacific V.I.P.s. Although there were three additional APEC "economies" (México, Chile, and Papua New Guinea) represented since the 15-member Seattle "Pacific Forum" of 1993, the buoyant "can-do" carnival atmosphere at Seattle had not carried over to relieve the muted dullness present at Vancouver. (The APEC members are officially designated "economies" to avoid the presence of now non-nation Hong Kong and unrecognised Taiwan, "Chinese Taipei"). The avowed purpose of APEC, of course, is to achieve the globalist goal of "'free and open trade and investment'" within the Asia/Pacific region by 2020 A.D.[297]

Things got off to a sombre start with immanent U.S. Secretary of State Madeleine Albright declaring, "'We meet amidst predictions that the Asia Miracle will be succeeded by an Asian Meltdown.'"[298] Madame Secretary's non-diplomatic utterance set the down-tone for the entire conference. But the tactless Ms. Albright had merely expressed a fear shared by all at the summit. In a special report, Chris Wood of the Canadian magazine Maclean's described vividly the extent of the crisis in Thailand, until recently the most flamboyant of the Little Dragons:

"Nowhere has the meltdown been worse than in Thailand. With debt sky-rocketing and the country's accounts drenched in red ink, Thai officials abandoned attempts to protect the value of the baht against the U.S. dollar in July [1997]. Since then, the nation has plunged steadily deeper into political as well as economic crisis. In August, the International Monetary Fund arranged a $21-billion bailout for the floundering Thai economy -- in return for the promise of a balanced budget and other austerity measures. But partisan manoeuvring within the ruling six-party coalition stalled the reforms, while criticism of the government mounted. Finally last week, beleaguered Prime Minister Chavalit Yongchaiyudh handed in his resignation to the King of what was once Siam...."[299]

Thailand was the first of the Asian dominoes to fall, with the Thai contagion spreading like a deadly virus all across South-East and East Asia -- causing stock markets to plummet and local currencies to plunge (versus the American dollar).[300] The easy-going Thais usually like to respond to problems by saying "Mai pen lai!" ("Don't worry!") But the Thais are worried now. After more than ten years of riding a red-hot economy, Thailand today (1998) is in fiscal free-fall. Laid-back Thai finance officials let the overheated go-go economy get out of control, with the resultant summer of 1997 meltdown -- currency crisis and bank closures; closely followed by an International Monetary Fund (IMF) bailout of $21 billion U.S. It could take years for Thailand to regain her financial footing.[301] As mentioned, Prime Minister Chavalit Yongchaiyudh resigned in mid-November 1997.

The next Asian domino to fall was Indonesia, an island sub-continent with an area the size of México, and with a population of circa 200 million (the globe's fourth largest and the Islamic world's highest). Indonesian strongman President Suharto (Soeharto) had -- until late May 1998 -- ruled the huge archipelago since the aftermath of the [September "GESTAPU"] 1965 revolution, the Year of Living Dangerously. During that year, General Suharto ousted nationalist "Bung Karno" Sukarno (Soekarno, 1901-1970) in a complex, apparent coup. Achmed Sukarno was a flamboyant and charismatic figure who dominated post-War Indonesian politics for fully two decades.[302] Since 1965, General Suharto reigned as Indonesia's military dictator until pressed to the wall by the economic crash of 1997, which culminated in the political collapse of 1998. Unlike the Prime Minister of Thailand, however, President Suharto did not go gently into that good night. The President-General did not step down until after a whole week of rioting and violence that killed some 500 persons during May 1998; thereby ending 32 years of autocratic and absolute rule.[303] Vice-President B.J. Habibie was immediately sworn in as new president of Indonesia, and has promised parliamentary elections in 1999[304] and... to loosen the ties of the political/military straitjacket which has for so long bound basic freedoms in Indonesia. In the meantime, Indonesia is a political basket-case as well as an economic disaster.

Back at the APEC summit at Vancouver in November 1997, talky U.S. President Bill Clinton announced that South Korea's request for a $20-billion-plus "rescue package" from the IMF would be the first situation to come under a "crisis response" programme, adopted by the APEC countries shortly before the summit at a meeting at Manila.[305] Indeed, the Associated Press has cited Asia market analysts as expecting the South Korean economy to shrink by more than 1% in 1998 -- this would mean the first fiscal contraction since 1980 for the R.O.K.[306] The Republic of Korea, for gosh sakes! This was the country (along with Japan) to which American Mr. Sam Walton (1918-1992) -- Wal-Mart billionaire extraordinaire -- along with spouse had journeyed to in 1975, where in he had "picked up several ideas."[307]

But South Korea, so recently a paragon of the Asian Miracle, had little to celebrate as she rang in the Year of the Tiger, 1998. Just hours before the end of 1997, American and international banks reluctantly "saved" South Korea from defaulting on loan payments by forking over $15 billion U.S. in short-term debt aid.[308] The previous week, the IMF had made a Christmas Eve pledge to hasten disbursement of $10 billion -- including, for the very first time, the use of U.S. [taxpayer-"contributed"] government funds. <u>This</u>, after the IMF had orchestrated a record $57-billion in early December 1997![309] The catch had been that the R.O.K. government was supposed to only receive IMF largesse on the condition it converted the country's [corrupted] bureaucracy-dominated economy to true free market conditions. Despite some progress in that direction, the South Korean parliament had failed to approve [an agreed to with the U.S.] independence for the central bank; a key and crucial element of financial reform.[310]

President Kim Young Sam left office at the onset of the Year of the Tiger (March, 1998); the incoming president, Kim Dae Jung, has promised to keep interest rates high so as to tempt foreign investors back to cash-starved South Korea. But certain perceptive IMF officials (right for a change) have insisted that only an independent central bank in the R.O.K., protected from domestic political pressure, will restore investor confidence.[311] At the same time, the still-proud-but-now-poor Koreans have to bow and kowtow to both the IMF and to U.S. Treasury Secretary Rubin for financial favours. But then the Clinton administration, which had been keeping the R.O.K. on a short monetary leash, in a reverse decision (as is often its wont) shortly into the new year (1998), decided to "donate" $1.7 billion of U.S. [taxpayer] money to the IMF's total Korean Christmas package. And (oh the humiliation!) the money was to come from the Exchange Stabilization Fund, the source from which the late Great Mexican Bailout had been purloined... and that (still) requires no congressional approval. The avowed Clintonist excuse was that extreme economic disruption in South Korea could prompt a North Korean invasion of the R.O.K. -- thus putting the 37,000 U.S. troops stationed there in harm's way.[312]

This writer, though, believes that conservative commentator Pat Buchanan hit the proverbial nail on the head in January 1998. In a <u>Human Events</u> article, Pat Buchanan listed the winners and losers -- and why -- of the U.S.-IMF Korean Christmas present of $10 billion. The winners are two-fold: 1) American banks ("whose foolish loans will be made good by the IMF bailout")'; and 2) American transnationals ("which can now, like birds of prey, pick off Korea's most priceless assets one by one").[313] The losers, according to Pat Buchanan, are three-fold: South Korea ("much of what its hard-working people built up over two generations will be sold off to scavengers at foreclosures"); Korean workers ("who have seen savings wiped out and pay cut in half in dollars"); and U.S. taxpayers ("who will be put at risk for scores of billions in bailout money, U.S. companies that will be hammered by imports, and U.S. factory workers who will lose high-paying jobs").[314]

Now we know why those protesting Korean workers shown on CNN were wearing "IMFired" T-shirts!

If South Korea is cash-starved, North Korea suffers from real physical starvation -- and from that of the spirit. According to the (Christian) Puebla Institute, a mere half-century ago the North Korean capital city of Pyongyang was known as "Asia's Jerusalem" because of the pervasive Christian presence there. Up to two thousand churches were spread throughout North Korea, and Western missionaries were active everywhere.[315] But then in 1948 Stalinist Kim Il-sung rose to power, and it didn't take long for the Communist dictator to systematically eradicate all vestiges of Christianity. By the early 1960s, Kim's secret police had begun to close all churches and burn all Bibles in North Korea. Priests and clergymen were either executed or imprisoned.[316] When Kim Il-sung died on 8 July 1994, he deified himself ("Great Leader") along with his wimpy, dingy [but dangerous] son, Kim Jong-il ("Beloved Leader"), in a Korean cult of dual personalities. Today, visible Christianity ("superstition") is non-existent as a "hindrance to the socialist revolution." Indeed, hard-scrabble North Korea ranks in 1998 as the hardest-line Communist, most anti-Christian state in the entire world.[317] On the Dedication page of Don Oberdorffer's recent book The Two Koreas, he expressed the hope: "For the people of the two Koreas/May they be one again, and soon."[318]

Another Communist Asian nation which cruelly persecutes Christians (and Buddhists) is Vietnam. The much-heralded renewal of diplomatic and trade relations between the United States and Vietnam in July 1995, has done little to ease the plight of Vietnamese Christians (either Roman Catholic or evangelical Protestant). Nina Shea, author of In the Lion's Den, has told the sad tale (among many) of Christian leader To Dinh Trung, unjustly incarcerated since April 1995 for "'abusing his freedom as a citizen by propagating religion illegally.'"[319] April 1995, the time of To Dinh Trung's arrest and detention, coincided with the 20th anniversary of the Communist "unification" of Vietnam -- and with U.S.-Vietnamese discussions on normalising relations two decades after the Vietnam War.[320]

Nina Shea also reported that "at least thirteen" Vietnamese evangelicals are currently (1997) serving

2-3 year prison sentences for attempting to witness for Christ and preach the Gospel. Churches "authorized" by the Vietnamese government are routinely harassed and fined; their members detained and their property confiscated. Pastors of "unauthorized" house-churches are today persecuted for holding prayer meetings and Bible studies.[321]

Roman Catholics in Vietnam are not faring any better than their Protestant brethren. One example has been Tran Qui Thien, who spent thirteen long years in a Communist "reeducation" camp due to his Christian activities. Appearing before two subcommittees of the U.S. House of Representatives on 8 November 1995, the Reverend Tran Qui Thien declared:

"'No pen will ever be adequate to describe all the acts of terrorism, repression, suppression, murder and imprisonment aimed at the religious leaders and their followers -- purely on religious grounds -- in Vietnam.'"[322]

Boston columnist Don Feder, although a practising Jew of faith, has assailed the celebrated "normalization" of U.S.-Vietnam relations on secular grounds. The whole diplomatic charade, he wrote, was "driven by business interests, period -- not morality, not human rights concerns, not national pride or honor for our fallen heroes."[323]

When President Clinton lifted the trade embargo from Vietnam in 1994, scarcely a year later more than 300 companies had already opened branch offices in the Socialist Republic. U.S. corporations are eager to do business with the globe's thirteenth most populous country. Don Feder predicted that the next step for the mega-firms will be to lobby for most-favoured nation (MFN) status for Vietnam, with subsidies (for themselves) from the Export-Import Bank ("In international relations, money talks, morality walks").[324]

Don Feder reminded his readers that the government of Vietnam which the United States recognised in July 1995, is the same criminal régime responsible for the deaths of more than 55,000 Americans during the war, the nullification of the Paris Peace Accords, the confinement or slaughter of tens of thousand of its fellow compatriots in South Vietnam after 1975, and the mass exodus of a million or more "boat people". That same criminal régime has played a callous mind-game with the United States regarding the sensitive Pow-MIA issue. Twenty years after the fall of Saigon, the Socialist Republic had still not accounted for more than 300 U.S. Armed Forces personnel known to be alive in Vietnam and Vietnamese-controlled areas of Laos.[325]

This is also the very criminal régime that brutally oppresses the 73 million (1995) people of Vietnam, and with which corporate America can hardly wait to do business. Don Feder has concluded that:

"From his student-protest days to this loathsome act [of recognition], Clinton has completed his betrayal of our fighting men. He might as well demolish the Vietnam War Memorial and sell the rubble to Hanoi for building material. Why let sentiment stand in the way of profits?"[326]

Despite the "Asian Meltdown" of Madeleine Albright, the Pacific Age is still very much upon us. It is not the same "Pacific Rising" so extravagantly touted by Simon Winchester during the late 1980s/early 1990s,[327] but the Asia-Pacific Rim economy (and general well-being) is still healthy in the late 1990s. Despite the Asian Meltdown, the South Pacific nations of Australia and New Zealand;[328] the "Greater China" areas of China (incl. Hong Kong), Taiwan, and Singapore; Canada's province of British Columbia; the U.S. states of Alaska, Washington, Oregon, and California; and the South American country of Chile are all alive and thriving. (Only the Pacific state of Hawaii is sick -- 1) Japan's investment speculation bubble burst in the early 1990s, also cutting into tourism; 2) Hawaii's sugar industry disappeared with the closing of five plantations; 3) the Cold War ended, and with it, massive military spending; 4) the present Asian Meltdown has also adversely affected Hawaiian tourism; and 5) a far-left ruling Democratic Party, which has run the whole show in Hawaii for far too long and bled the Aloha State dry.)[329] And Simon Winchester was Pacifically-correct in 1991 when he wrote:

"Los Angeles is set to take over from New York as the busiest port in the world. Hong Kong shot past Rotterdam in 1987 to become the world's busiest container port. Forty per cent of American trade with the Pacific is carried on ships that arrive in the port of San Francisco."[330]

The globe is still undergoing a sociological "Pacific Shift", though different than that envisioned by New Ager William Irwin Thompson in 1985.[331] But the Pacific Aerospace cultural ecology is still intact, and the old Far East is still the new Far West -- although the borders now end with Tibet rather than at Okinawa. The Pacific Age which started as The American Century during the mid-1940s, will continue into the next millennium as The Dragon Century, the century of the Chinese. Australian writer Murray Sayle, who has lived in Japan since 1975, described the New Asia-Pacific Man (and Woman) in an article for the Canadian magazine Maclean's during APEC 1997:

"Ethnic Chinese... now control most of the business of Asia that the Japanese do not, and they have to talk about it somewhere. Increasingly, it is over after-dinner drinks in the bar of one of the Western-style hotels that are now to be found in just about every Asian city, Communist, capitalist, or still deciding. Here you will meet New Asia Pacific Man (and Woman). With their big plans, portable phones, credit cards and the universal language, basic English, they are evolving a business style that Confucius never knew, and Dale Carnegie never influenced...."[332]

Are these yuppy élites a less toxic Asia-Pacific version of William Irwin Thompson's "New Millennial Managers"? Anyway, as repellent as a yuppified army of New Asia-Pacific Men (and Women) might sound, surely the Murray Sayle image is vastly preferable to the grim memory of pith-helmeted North Vietnamese armoured forces rolling into Saigon (now Ho Chi Minh City), April 1975? But Vietnam, China, and even miserable North Korea shall someday be free and democratic members of the Asia-Pacific community. And in the 1990s, Taiwan (the Republic of China) had open and honest elections -- the first political unit of Greater China to ever do so. Economic liberation through the free market is happening today; political liberation can't follow soon enough.

1990s: East Asia Fisheries Update and South-East Asia Seafood Market Overview

"The Pacific's greatest asset is its fish. The shoreline waters of the continents and the more temperate islands yield herring, salmon, sardines, snapper, swordfish, and tuna, as well as shellfish...." The Pacific Rim Almanac, 1991

The two East Asian "Small Tigers" of South Korea and Taiwan received much negative publicity (along with the Tiger, Japan) in global fishing news all during the late 1980s and into the early 1990s. The Big Story back then was driftnet fishing, the notorious "curtains of death" engulfing all sealife in their dreadful path through the Pacific Ocean. There has been much analysis of East Asian driftnet fishing and its ban by the UNO in 1991 (which took effect in 1993) in this (and esp. an earlier) work.[333] Although there exists lasting resentment throughout the Pacific industry for the Korean and Taiwanese driftnet fleets who operated outside international law, both the R.O.K. and the R.O.C. are today generally accepted as bona fide and full-fledged members of the Pacific high-seas fishing community.

Simon Winchester's 1991 work, Pacific Rising, included a map of world fishing in the back of the book; one issued by the UNO's Food and Agricultural Organization (FAO).[334] In the FAO map, it is interesting to note that the approximate gross tonnage of the North Korean fishing fleet (with factory ships) was listed at 858,000. The corresponding tonnage of South Korea's fishing fleet was listed at 886,000.[335]

Either the FAO's numbers were seriously inaccurate, or the contemporary decline of starving North Korea has been more precipitous than previously thought. (The figures for prosperous, more populous South Korea were not surprising.) But the closure of high-seas driftnet fisheries, the refusal by ever more nations to permit distant-water fleets into their exclusive economic zones (EEZs), and with coastal/estaurial pollution affecting many fast-growing Asian countries -- have all been reasons for drastically reduced fish landings by East Asian fleets.[336] With the increased accumulation of national wealth during the early 1990s, both South Korea and Taiwan have lowered their punitive tariffs and become seafood importing countries; for the domestic market and for re-processing and re-sale.[337] A telling 1995 story in Seafood Leader told of seafood-hungry Taiwanese brokers buying salmon entrails from the B.C. provincial government:

"When a [Vancouver] processor gutted some fresh salmon for them [the Taiwanese buyers] to sample, he was amazed when the Taiwanese grabbed the guts and asked for some wine and soy sauce. After whipping up the mixture and eating it, the Taiwanese convinced their host there was a good market for salmon entrails in Taiwan...."[338]

Despite the East Asian trend since the mid-1990s to import seafood, however, the percentage of all-Asia food-fish production in live-weight equivalents was a world high 53.6%.[339] Global food-fish action in 1996 was still in Asia, but focus on seafood production had shifted from East Asian fishing to South-East Asian aquaculture. Indeed, the true 1990s success story in the Asian fisheries has unfolded in the "lesser-developed countries" (LDCs) of South-East Asia. These are the "Little Dragon" nations of Indonesia, Malaysia, Philippines, and Thailand (and, included in this chapter, Vietnam.) The seafood action in South-East Asia has been in shellfish aquaculture -- down south, marine pond shrimp-farming (genus Penaeus spp.) has been every bit so successful as the growing of China bay scallops farther north. Shrimp News International listed the "big six" of world shrimp farming for 1993 and 1994 as (in order): Thailand, Ecuador, Indonesia (which drew even with Ecuador in 1994), India, China, and Vietnam (which surpassed China in 1994).[340]

Aquaculture Magazine, comparing world aquaculture statistics in 1995, listed the "Estimated Global Production of Pond-Raised Shrimp 1991-1994." In the top twelve countries were Thailand (1st place), Indonesia (2nd place), Vietnam (5th place), China (6th place), Philippines (8th place), and Taiwan (9th place).[341] Exemplary figures for Thai-raised shrimp, in head-on metric tons, were : 153,000 for 1991 increasing to 248,000 for 1994 -- a huge percentile growth.[342] And although Indonesia's metric tonnage actually <u>decreased</u> from 140,000 in 1991 down to 100,000 in 1994, the figures were hefty enough to keep the vast archipelago in second place (neck-and-neck with relatively tiny Ecuador).[343] Thailand, in fact, during 1994 set a new world record of farmed shrimp production, with about twice and a half times that of second-place Indonesia and third-place Ecuador combined.[344] And, according to Darryl E. Jory, adjunct prof. at U. Miami Marine School (Coral Gables, Fla.), the giant (black) tiger -- along with <u>P. vannamei</u>, the western white -- has been the aquacultural shrimp of choice in 63% to 76% of global marine pond production.[345] So, as this writer has long suspected, the 1990s has been the decade of the Tiger; the giant black tiger prawn, <u>Penaeus monodon</u>.

<u>Tiger Tales from Thailand</u>

Thailand has served as both paradigm and paragon for farming the giant tiger prawn, during the late 1980s and throughout the 1990s. By the mid-1990s, shrimp ponds had replaced rice paddies in southern Thailand, and farmed tigers had placed Thailand as the globe's leading exporter of seafood.[346] In early 1996, editor Peter Redmayne of <u>Seafood Leader</u> reported that investment capital was "pouring into [neighbouring] Cambodia and Vietnam, both of which are well-suited to follow Thailand's lead."[347] But the tale of tiger-farming in Asia really began in Taiwan in 1968. After biologists in the R.O.C. had learned to induce roe-laden female shrimp to spawn on command, Taiwan farmers were faced with two problems: (1) How to find shrimp-feed that was nutritious as well as appealing; (2) and how to deal with poor water quality resulting from uneaten feed building up on (hence fouling) pond bottoms.[348] By the early 1980s, the Taiwanese had solved both problems, and had developed an integrated and highly specialised aquaculture industry near Kaohsiung, in the southern R.O.C. By the mid-1980s, giant tigers were the seafood rage in Japan and "tiger tails" were beginning to catch on in the U.S.A. By 1988, Taiwanese harvest of farmed giant tigers reached 100,000 tons.[349] But Taiwan, the Republic of China, is a small island nation similar in size to Vancouver Island, B.C. As more and more land in Taiwan was purchased and used for shrimp-farming, more and more water was required to flush out the increasingly crowded (and foetid) ponds. Then came the collapse. In that same year, 1988, prawn harvest decreased from 100,000 to just 30,000 tons -- as disease and death ran rampant through the ponds of Taiwan. In 1995, the industry had yet to recover in the R.O.C. But as an article in <u>Seafood Leader</u> observed at the time:

"...[E]ven before tigers [had] tanked in Taiwan, investors from Taiwan and Japan had started building tiger farms in the Philippines, Indonesia and Thailand, where the weather was warmer, and labor and land were cheaper."[350]

The Seafood Leader should have added that suitable aquacultural land was far more plentiful in South-East Asia. If Taiwan may be likened in size to the Netherlands, the Philippines is as large in area as Arizona, Indonesia comparable to all México, and Thailand bigger than Spain. And the Thai tiger-farmers did learn their lessons from Taiwan. After an initial outbreak of the yellowhead virus in 1994, which threatened to decimate their shrimp ponds, the Thais fought back with medicated feeds; prompting Japanese buyers to reject Thai tigers because of antibiotic residues.[351] Thai tiger-farmers have proven to be clever and resilient, utilising the brains and brawn of both government and business. Faced with poisonous effluents issuing from the four major rivers encompassing Bangkok (including the Chao Phraya), and with pollutants emanating from factories at the country's industrial core, the Thai aquaculturists picked up their operations and equipment and moved all - en masse -- to the southern coast on the Gulf of Siam. Here land was cheap, the water was clean, the drainage was better, there were good roads; there was also access to a nearby container port.[352]

The innovative Thai prawn-farmers readily adapted to new aquaculture technology, used superior feeds, and improved pond-management methods. Thailand's harvest of giant tigers soared to 80,000 tons produced in 1989, and, when the market became saturated with rival shrimp from China and Indonesia, the dollar-smart (rather than penny foolish) Thais virtually gave away black tigers at promotional prices -- thus opening new markets in the West for the giant prawns.[353]

As for being "environmentally-friendly", the Thai government banned the clearing and cutting of mangroves for shrimp-farms so far back as 1989.[354] Not only that, but the government, working along with the private sector, has spent many millions of baht on mangrove reforestation. The Thai government committed itself in 1995 to spending $60 million U.S. on six deepwater pumping stations that would funnel discharge water from shrimp ponds three kilometres out to sea.[355] And as for all those discarded prawn shells, at least one Thai company in 1995 was reported to have planned turning the waste material into... "chitosan, a cellulose-like substance with applications ranging from biodegradable bandages to water filters."[356]

And by early 1996, a leading South-East Asian building manufacturer, Thai Gypsum Co., was marketing a "'Hi-Tech Thai House'", which would presumably pay for itself.[357] This would be accomplished, the high-sales pitch promised, as the Hi-Tech Thai House came with a fish-farm instead of a front lawn; a fish-farm capable of producing 50 tons of fish (such as tilapia) per annum.[358] The eco-friendly Hi-Tech Thai House was powered from a bank of 120 solar cells storing electricity in lead-free vanadium batteries. The Hi-Tech Thai House cum fish-farm was 100% water self-sufficient, due to a closed system generating no waste. Thai Gypsum Co. claimed that aquatic plants, biological filters, and an ozone disinfectant would be employed to treat all the water in a tri-step process. The fish-farm itself consisted of six grow-out ponds in which fish could be raised at densities "'10 times higher than normal.'"[359] The Thai Gypsum Co. was asking a handsome price of $500,000 U.S., and was last seen talking to the Asian Development Bank and World Bank about their eco-friendly Hi-Tech Thai House.[360]

But this writer pens these words on 30 June 1998, and a dreadful post-Asian Meltdown surreality has possessed the "Land of Smiles", the free-living culture of Mai pen lai ("Don't worry!"). As noted the Thai prime minister resigned in mid-November 1997, and Thailand has been in fiscal free-fall ever since. Gone from the Thai urban landscape is the big Mercedes auto, the cellphone, the Rolex watch, the credit-line, the luxury condo overlooking a golf course. Gone are the wretched excesses in Thailand as they are in Japan. The Thais are not just broke but deeply in debt.[361]

Even "Tiger Fever" has cooled in Thailand.[362] American golfer Tiger Woods, who won the Masters Tournament at Augusta (Ga.) in 1997, is half Thai. So when Woods triumphed at the Asian Honda Classic held in Thailand (also) in 1997, all Thais celebrated. Tiger Woods was accorded the full V.I.P. treatment, being ubiquitously filmed and showered with flowers wherever he went. Woods was even awarded a medal by the Prime Minister (now resigned). But not in 1998: Everybody loves a winner -- the Thais are no different -- and Tiger Woods failed to win the Masters Tournament in April 1998. So this time around, Tiger was not lionised in Thailand. The Associated Press wryly commented:

"Back then [in 1997], Thailand was still considered an Asian miracle economy. Now the country is one of the worst-hit in Asia's financial collapse. The government has changed and the national mood is no longer one of celebration."[363]

Will Tiger [prawn] Fever cool down in Thailand? Whether it does or not, chances are that Thailand will continue to be the sole Asian nation able to feed herself.[364] If the Asian Meltdown worsens, self-sufficient Thailand will be in far better shape to cope than supposedly beyond-the-crisis, crowded city-state Singapore.[365] But proving almost as injurious to the shrimp-aquaculture industry in South-East Asia as the economic meltdown, has been the mounting hostility of global environmentalism. The fisheries reader well knows the impact on the industry -- world-wide -- of corporate conservationism; in the forms of dolphin-safe tuna and turtle-safe shrimp, save the rainforests and (now) save the mangroves.[366] Save-the-mangroves groups have directly targeted aquaculture -- especially shrimp-farming in South-East Asia. Eco-critics have asserted (quite correctly) that shrimp ponds pollute the environment with decaying organic feed and waste; along with chemically-based fertilisers, antibiotics, and larvicides.[367] As one Alfredo Quarto of the Seattle-based Mangrove Action Project (MAP) has remarked on shrimp-farming:

"It's basically slash and burn aquaculture....Mangroves are cleared for ponds, the average life of a shrimp pond is three to five years, then they are abandoned and the land is ruined....the long-term future of these places and people is being sacrificed for short-term gains."[368]

According to the UNO's Food and Agricultural Organization (FAO) in 1996, circa half the world's mangroves have disappeared over the last three decades -- and about half of that loss to shrimp-aquaculture.[369] (The other half was from wood-cutting, residential construction, and industrial development.) Environmental reports, published during the mid-1990s, have accused Thailand of wiping out 87% of her mangroves on the way to becoming the world's number farmed shrimp producer (250,000 tons in 1995).[370] But during that same period (1996), the director general of Thailand's Dept. of Fisheries, a Dr. Plodprasop Suraswadi, began seriously addressing the environmental concerns regarding mangroves and aquaculture. The Thai government banned outright the cutting of mangroves, began mangrove re-planting, committed $40 million U.S. to preserve existent mangroves, and (as noted) started a $60 million U.S. construction of six deepwater pumping stations, which would flush shrimp pond discharge water more than a mile out to sea. Tiger prawn exports during the mid-1990s were worth $2 billion U.S. to Thailand, and Dr. Plodprasop Suraswadi expressed his conviction that Thai aquaculture could eventually produce 500,000 tons of farmed shrimp per year <u>without</u> expanding pond acreage.[371]

Dr. Darryl E. Jory, adjunct prof. at the Marine School of the University of Miami, wrote in 1995 that "[w]ell formulated, nutritionally-balanced, environmentally-friendly feeds are critical for the further expansion, economic and ecological viability of the shrimp industry worldwide."[372] Moreover, Professor Jory was optimistic about the future prospects of shrimp aquaculture -- whether in Asia or Latin America. Although Jory emphasised that shrimp ponds are monocultural ecosystems (thus fragile and unstable) and must be carefully managed, he predicted that farmed shrimp could contribute fully half of all global shrimp production by A.D.2000.[373] And the lion's share of that shrimp production might well be giant tiger prawns grown in Thai ponds.

There's a new kid on the block; a new player on the South-East Asia field of aquaculture -- the Socialist Republic of Vietnam. The Vietnamese pursuit of fish-farming has been far more peaceful (and productive) than their politics. Fish tales from Vietnam started trickling into the Western industry press during the early 1990s. <u>Seafood Leader</u> reported in autumn 1993 that Vietnam had been positioning herself as "the Mississippi of Asia", by raising catfish in the vast estuary of the Mekong Delta.[374] But no sophisticated fish ponds or "Hi-Tech" Thai houses were seen there (yet) though; the Vietnamese method of farming was still the fish-raft cage, with the farmer dwelling on a raft above a cage lowered into the river. Despite the U.S.-Vietnam trade embargo being in place during the early 1990s, plenty of Mekong River catfish product was reaching American markets via traders in Hong Kong and Singapore.[375]

By autumn 1995, Vietnam expected to double her exports of seafood to a market value of $1 billion U.S. by 2000 A.D.; increasing production by 31% to 1.6 million tons.[376] At the same time, the newly-founded Vietnamese-Russian Joint Commission for Fisheries agreed to institute official coöperation between their respective fish-processing companies.[377] Vietnam's Vice Minister of Fisheries, Vo Van Trac, said that there were abundant opportunities for foreign investors in Vietnamese aquaculture, processing, preservation, and distribution of fishery products. (Most of the Socialist Republic's processing plants in 1995 were admittedly well-worn, putting out raw or semi-finished product. Prospects looked bright, however, for Vietnam to become a fish re-processing centre.)[378]

By late 1996, the Vietnamese had sent an entire delegation to Seattle, Washington, seeking Pacific

Northwest technological expertise to assist Vietnam's ambitious fishery development plans.[379] Vietnam already ranked in the top twenty fishing nations of the world, with average landings of 700,000 metric tons and exporting seafood with an approximate worth of $550 million U.S. in 1995. Vietnam possesses a coastline stretching more than 2,000 miles, and EEZ (exclusive economic zone) of 400,000 sq.mi., and with a largely undocumented marine [resource] biomass of valuable lobster, grouper, red snapper, sea bass, and tuna populations within her waters.[380] But the Vietnamese fishing fleet of circa 75,000 boats is (1996) small and poorly equipped, and is used to land 90% of the country's coastal catch. Ashore, Vietnam's 170 or so refrigerated processing plants are run-down and low-tech.[381] As a result, Fisheries Minister Dr. Huynh Cong Hoa took his delegation to high-tech Seattle, hoping for possible American assistance in upgrading Vietnam's in-shore flotilla of small boats to a deep-sea fishing fleet of 250 to 300 vessels; suitable for heavy-duty use in territorial and international waters. Dr. Hoa told members of the Seattle Trade Development Alliance that:

"Computer-aided design for the construction of deep-sea trawlers, purse seiners and longline...vessels from 150 to 200 feet is one priority that the United States can play a vital role in....
"American companies would be most useful in the supply of... medicines and technology for our aquaculture industry....Vietnam will concentrate on...the development of offshore and deep-sea fishing fleets combined with establishing a cost-effective aquaculture industry over the next few years....
"We've estimated that the investment needed for fisheries development for the years 1996 through 2000 would be about $6 billion (U.S.).'"[382]

(Dr. Hoa foretold that fishery exports to the United States would increase when the Socialist Republic eventually received Most Favoured Nation trading status.)

Meanwhile, the Vietnamese were off to mixed reviews in 1997. British trade magazine Seafood International reported in September that would-be shrimp farmers in Vietnam had ruined 100 hectares of rice fields: The wanna-be aquaculturists had dug a channel to the sea to provide salinity for the proposed shrimp farm, with the saltwater overflowing into the surrounding area.[383] Furious rice producers in Bac Lieu province, related Seafood International, had lodged a formal protest to the local authorities.[384] But in the very same issue of the magazine, it was also reported that Vietnam had exported "large quantities of crustaceans" to Singapore during the first five months of 1997.[385] The two statist regimes of oppressive (Communist) Vietnam and repressive (capitalist) Singapore, a strongly complementary odd-couple who badly need each other, both eagerly await a large increase in bilateral trade as equal members of ASEAN

(Association of South-East Asian Nations).³⁸⁶

The present Socialist Republic of Vietnam shall some day be a free and democratic member (and important fish producer) in the Asia-Pacific Rim community. Economic liberation through the open market is happening today; political liberation can't follow soon enough.

What is the immediate future of U.S. seafood exports to Asia after the Meltdown? Although the Tiger (Japan), one Small Tiger (South Korea), and two Little Dragons (Thailand and Indonesia) appear the hardest hit, all Asia has been jolted by the financial collapse... which in turn has affected seafood markets. Export statistics for January 1998, reported by Cooper Moo of Pacific Fishing, tell a sad story of reduced exports of U.S. seafood to South-East Asia by more than 50% in value.³⁸⁷ Indonesia, Philippines, Singapore, and Thailand all bought less than half the American fish they had in 1997, and most economists have prognosticated that market recovery could take years rather than months.³⁸⁸ So far (1998), only the "Greater China" political units of Mainland China, Hong Kong, Taiwan, and Singapore are riding out the Asian economic typhoon safely -- so far.³⁸⁹

Farther north, the value of U.S. seafood exports to South Korea decreased 14% in 1997 from 1996; Pacific cod, for example plunged 40% in 1997, about the ratio that the Korean won fell against the American dollar.³⁹⁰ A precipitous drop in the value of the Korean currency, coupled with an economy struggling for its very life, has put the price of Pacific cod (a Korean favourite) out of sight -- and out of reach -- for most Korean consumers. The U.S.-Asia seafood trade editor of Pacific Fishing magazine visited a Seoul wholesale seafood market in early 1998, stopping at a stall that displayed "frozen whole Alaskan cod."³⁹¹ The editor, Cooper Moo, asked the fish-monger how much the price for that particular seafood item had varied during 1998. The cost had doubled came the amused reply³⁹² Pacific Fishing's Cooper Moo, in an Author's Note, post scripted:

"Just published trade statistics show a 12% increase in U.S. seafood exports to Korea in the first two months of 1998."³⁹³

Sometimes after a meltdown the resulting amalgam is stronger than the original ingredients.

"May they [the two Koreas] be one again, and soon."³⁹⁴

NOTES

288. Simon & Schuster, New York, 1996.

289. John Naisbitt, pp.14-15.

290. Ibid., pp.14, 19. NB: The Overseas Chinese community in "Hong-couver", B.C., was 20% of Vancouver's total pop. in 1996.
Ibid., p.48.

291. Ibid., pp.19-20.

292. Ibid., p.20.

293. 'Microsoft chief still richest, but Asian tycoons gaining', Peninsula Daily News, 1 July 1996, p.A-8.

294. Ibid., p.A-8.

295. Bruce Bartlett, 'U.S. Ranks as World's Most Competitive Economy', Human Events, 22 May 1998, p.17.

296. 'U.S. leading in world competitiveness ', Peninsula Daily News, 11 May 1998, p.C-6.

297. Chris Wood, 'Vancouver hosts the APEC extravaganza', Maclean's, Vol. 110, No. 46, 17 November 1997, p.64.

298. 'Pacific leaders to endorse bailout plan', Peninsula Daily News, 25 November 1997, p.A-2.

299. Op.cit., Chris Wood, p.65. NB: Thailand was formerly known as Siam. The baht is the unit of Thai currency. (Brackets this writer's.)

300. Ibid., p.65. Cf. 'Year of Tiger turns into year of recession', Peninsula Daily News, 15 June 1998, p.C-6.

301. Maclean's, Vol. 10, No. 46, 17 November 1997, p.81.

302. Ibid., p.74. NB: The long and complex history of Indonesia, from ancient Srivijaya to modern Irian Jaya, was clearly and carefully explicated by Robert Van Niel, Ph.D., to his students (among them this writer) at the University of Hawai'i at Manoa, 1980-1981. Mahalo nui!

303. 'Suharto steps down', Peninsula Daily News, 21 May 1998, p.C-1.

304. 'Indonesians look to uncertain future', Peninsula Daily News, 31 May 1998, p.C-5.

305. 'Clinton: Committed to salmon talks', Peninsula Daily News, 24 November 1997, p.A-1.

306. 'Year of Tiger turns into year of recession', Peninsula Daily News, 15 June 1998, p.C-6.

307. Sam Walton with John Huey, Made in America, Sam Walton: My Story (New York: Doubleday, 1992), p.158.

308. 'The Year of the IMF', U.S. News & World Report, Vol. 124, No. 1, 12 January 1998, p.42.

NOTES (cont'd)

309. Ibid., p.42.

310. Ibid., pp.42-43.

311. Ibid., p.43.

312. Ibid., p.43.

313. Patrick J. Buchanan, 'Bankers Win, Americans and Koreans Lose in IMF Bailout', Human Events, Vol. 54, No. 2, 16 January 1998, p.5.

314. Op.cit., Pat Buchanan, p.5.

315. Nina Shea, In the Lion's Den (Nashville, Tenn: Broadman & Holman Publishers, 1997), p.68.

316. Ibid., p.68.

317. Ibid., p.68.

318. Don Oberdorfer, The Two Koreas: A Contemporary History (Reading, Mass.: Addison Wesley, 1997).

319. Op.cit., Nina Shea, p.70; citing Vietnamese authorities.

320. Ibid., p.70.

321. Ibid., p.75.

322. Op.cit., Nina Shea, pp.77, 124; quoting Rev. Tran Qui Thien.

323. Don Feder, Who's Afraid of the Religious Right? (Washington, D.C.: Regnery Publishing, Inc., 1996), p.227.

324. Ibid., p.228.

325. Ibid., pp.228-229.

326. Op.cit., Don Feder, pp.229-230. NB: Brackets this writer's.

327. Pacific Rising (New York: Prentice Hall Press, 1991), passim.

328. NB: Author Paul Theroux's "Meganesia". See his The Happy Isles of Oceania (New York: Ballantine Books, 1993 ed.) As Theroux has called Anglophone Australia and New Zealand "Meganesia", so will this writer take demo-geographic license and create "Macronesia -- a cross-cultural Anglo-Hispanic Asian-Pacific combination of Australia and New Zealand, California and Hawaii, Japan and the Philippines.

329. 'Trouble in paradise: Hawaii economy ailing', Peninsula Daily News, 6 April 1998, p.C-6. NB: Political assessment this writer's only.

NOTES (cont'd)

330. Op.cit., Simon Winchester, p.29.

331. Pacific Shift (San Francisco: Sierra Club Books, 1985), passim.

332. Murray Sayle, 'The art of doing business in Asia', Maclean's Vol. 110, No. 46, 17 November 1997, p.71. NB: Surely the obviously Anglocentric Sayle should have included basic Chinese (Cantonese, anyway) along with basic English? How about bilingual pidgin and Pinyin?!

333. Cf. C.D. Bay-Hansen, Fisheries of the Pacific Northwest Coast Vol. 2 (New York: Vantage Press, Inc., 1994), pp.130-145.

334. New York: Prentice Hall Press, pp.464-465.

335. Ibid., p.464. NB: The FAO statistics (typically) put Taiwan's (relatively high) fleet numbers under Mainland China's (relatively low) fleet numbers; listing the total statistical result under [the People's Republic of] China.

336. Daniel Shaw, 'The Far East's Growing Appetite For Seafood Fuels Export Markets', Seafood Leader, July/August 1995, pp.93-94.

337. Ibid., pp.94, 96.

338. 'A Gutsy Move By B.C. Processors', Seafood Leader, September/October 1995, p.75.

339. 'The Action's in Asia', Seafood Leader, Vol. 16, No. 3, May/June 1996, p.22. NB: Source was FAO.

340. Shrimp News International figures cited in Seafood Leader, September/October 1995, p.92.

341. Darryl E. Jory 'Global Situation and Current Megatrends In Marine Shrimp Farming', Aquaculture Magazine, Vol. 21, No. 4, p.76.

342. Ibid., p.76.

343. Ibid., p.76.

344. Ibid., p.77.

345. Ibid., pp.77, 79. NB: Darryl E. Jory cited C.P. Shrimp News (1995) and World Shrimp Farming (1992-1994) as his sources.

346. 'In Asia, Aquaculture Is Food for Thought', Seafood Leader, January/February 1996, p.5.

347. Ibid., p.5. NB: Brackets this writer's.

348. 'Black Tiger Shrimp', Seafood Leader, Vol. 15, No. 2, March/April 1995, p.172.

349. Ibid., p.172.

350. Ibid., p.172. NB: Brackets this writer's.

NOTES (cont'd)

351. Ibid., pp.174, 176.

352. Ibid., p.176.

353. Peter Redmayne, 'Thailand Rides Tigers To the Top', Seafood Leader, Vol. 15, No. 1, January/February 1995, p.62.

354. Ibid., p.126.

355. Ibid., p.126.

356. 'Waste Not, Want Not', Seafood Leader, Vol. 15, No. 5, September/October 1995, p.100.

357. 'High-Tech House Features Fish Farm', Seafood Leader, Vol. 16, No. 1, January/February 1996, p.30.

358. Ibid., p.30. Tilapia (family Cichlidae).

359. Ibid., p.30.

360. Ibid., p.30.

361. Chris Wood, 'Vancouver hosts the APEC extravaganza', Maclean's Vol. 110, No. 46, 17 November 1997, p.64.

362. 'Tiger in Thailand', Peninsula Daily News, n.d., n.p.

363. Ibid., n.p. NB: The writer contacted the local newspaper and searched at the local library, with no luck in finding date or page of this story. Approx. April-May 1998. (Brackets this writer's.)

364. Lowell Ponte, 'Food: America's Secret Weapon, Reader's Digest, May 1982, pp.66-68.

365. "Singapore: Beyond the Asian Crisis", 24 June 1998, KCTS-TV (PBS) Seattle, Wash. NB: Footage featured (mostly) pale, hirsute Seattleite yuppies gaping in the Asian statist paradise.

366. Daniel Shaw, 'Shrimp Under Fire', Seafood Leader, Vol. 16, No. 6, November/December 1996, p.29.

367. Darryl E. Jory, 'Global Situation...', Aquaculture Magazine, pp.80-81.

368. Daniel Shaw, 'Shrimp Under Fire', pp.29-30. See also 'Mangroves at the Roots of the Sea', Mangrove Action Project (Port Angeles, Wash.: Puffin Graphics, 1994).

369. Ibid., p.30. Cf. Carl Safina, Song for the Blue Ocean (New York: Henry Holt and Company, 1998), pp.422-423. See too Robert Elegant (1990), [also] pp.422-423.

370. Ibid., p.32.

371. Ibid., p.32. Cf. note no. 355 above.

372. Darryl E. Jory, 'Global Situation...', Aquaculture Magazine, p.81.

NOTES (cont'd)

373. Ibid., pp.82-83.

374. 'Vietnam Rebuilds With Catfish', Seafood Leader, (Whole Seafood Catalog issue), September/October 1993, p.28.

375. Ibid., p.28.

376. 'World Bites', Pacific Fishing, Vol. XVI, No. 10, October 1995, p.33.

377. Ibid., p.33.

378. Ibid., p.33.

379. Gordon Burridge, 'Vietnam Trolls Seattle For Fishing Technology', Pacific Fishing, Vol. XVIII, No. 1, January 1997, p.18.

380. Ibid., p.18.

381. Ibid., p.18.

382. Ibid., p.18. NB: Dr. Hoa's remarks not listed in sequential order.

383. 'Last Bites', Seafood International, Vol. 12, Iss. 9, September 1997, p.88.

384. Ibid., p.88.

385. 'ASEAN shellfish deal', ibid., p.13.

386. Ibid., p.13.

387. Cooper Moo, 'Asian Woes Spawn Seafood Slump', Pacific Fishing, Vol. XIX, No. 6, June 1988, p.19.

388. Ibid., p.19.

389. Ibid., p.20.

390. Ibid., p.19.

391. Op.cit., Cooper Moo, p.19. NB: Moo is most probably referring to Pacific cod (Gadus macrocephalus), or the related walleye pollock (Theragra chalcogramma), sometimes marketed as "Alaska jumbo cod". Then again, the non-gadoid and unrelated sablefish (Anaplopoma fimbria), is often referred to as "Alaska cod".

392. Ibid., p.19.

393. Op.cit., Cooper Moo, p.20.

394. Don Oberdorfer, The Two Koreas, see note no. 318. (Brackets this writer's.)

Bibliography for Chapter 1

1. Brun, Michel. The Sakhalin Incident. Four Walls Eight Windows, New York, 1996.
2. Brzezinski, Zbigniew. Out of Control. Charles Scribner's Sons, New York, 1993.
3. Clancy, Tom. Debt of Honor. G.B. Putnam's Sons, New York, 1994.
4. Elegant, Robert. Pacific Destiny. Avon Books, New York, 1990.
5. Fallows, James. More Like Us. Houghton Mifflin Company, Boston, 1989.
6. Friedman, George and Meredith LeBard. The Coming War with Japan. St. Martin's Press, New York, 1991.
7. Gibney, Frank. The Pacific Century. Charles Scribner's Sons, New York, 1992.
8. Halberstam, David. The Next Century. Avon Books, New York, 1992 ed.
9. Ishinomori, Shōtarō. Japan Inc. University of California Press, Berkeley and L.A., Calif., 1988. (U.S. ed.) A benkyō manga -- "study comic."
10. Manchester, William. American Caesar. Little, Brown and Company, Boston, Mass., 1978.
11. Naisbitt, John. Global Paradox. Avon Books, New York, 1994.
12. _____ . Megatrends Asia. Simon & Schuster, New York, 1996.
13. Orwell, George. Burmese Days. Harcourt Brace Jovanovich, Publishers, San Diego, Calif., 1974 ed. (Orig. published in 1934.)
14. Reischauer, Edwin O. The Japanese. Harvard University Press, Cambridge, Mass., 1977.
15. Slater, Ian. WWIII: South China Sea. Fawcett Gold Medal, New York, 1996.
16. Toffler, Alvin. Powershift. Bantam Books, New York, 1991.
17. Winchester, Simon. Pacific Rising. Prentice Hall Press, New York, 1991.
18. _____ . Pacific Nightmare. Ivy/Ballantine Books, New York, 1992.

Main Reference

Besher, Alexander, ed. The Pacific Rim Almanac. HarperCollins, New York, 1991.

Acknowledgements

I extend my most sincere thanks -- again after all the years -- to my professors at the University of Hawaii at Manoa (1979-1981): Drs. J.J. Stephan and Paul Varley (visiting), Japan; Drs. Walter Vella and Bob van Niel, South-East Asia; Drs. Tim McNaught and Ian Campbell (visiting), Pacific Islands.

To the memory of Dr. Werner Quast (1931-1997) of Peninsula College, P.A. This entire work owes much to a single Studium Generale lecture I gave at PC in April, 1992...thanks to Dr. Quast. ("Aufwiedersehen, Lausbub!")

Thanks to Ms. Jennifer Dresser and Ms. Sandra Trinter of Waterfront Press, Seattle, for China Seafood Expos information.

My eternal gratitude to Ms. Hiroko Kamevari Sonne, for putting up with me in Boston, 1964-1965.

Flashback: Once in a Blue Moon

"Oh, Heavens, it's [B]lue [M]oon time[!]"
- Associated Press, 30 June 1996

It was the second to last day in June 1996. This writer and his elder son were on their way from Port Angeles past Discovery Bay to the Kitsap Mall in Silverdale, Washington. This writer and son were driving in style -- in the latter's Texas-made 1995 Chevrolet Impala; 5.7 litre, V-8 engine, 4-door Sedan. According to number one son, the weight of the Impala exceeded that of a (Chevrolet) Camaro by more than 1,000 lbs. The automobile was definitely oversized, overweight, and overpowered ... sort of like the United States Army of which this writer's elder son was a proud member. André had been in for nearly seven years to date, but was still only a Sergeant E-5. He had also become a social statistic, having been married and divorced within three years. Father had jokingly warned son not to marry a woman who watched the television show Cops: She was sure to have a mean streak! Short time had proven this father right, which was not so funny.

But the divorce and accompanying imbroglio had brought father and son closer together, and they were now heading towards the Hood Canal Bridge. The big, battleship-grey car shone in the sunshine; it was early summer and André had used much energy and lots of Mother's California Gold Carnauba paste-wax. With the late, great Stevie Ray Vaughn singing and playing "Crossfire" and "Tightrope" over the Impala's tape-deck, father and son discussed life in the 1990s -- race, class, and gender wars; UFOs; the American South; the U.S. Army and the Pax Americana. Was This Man's Army becoming the Clintonised Womyn and Gay People's Liberation Army?

Later that summer during July, TV ads would start appearing for the newly-released "Courage under Fire." This writer knew that he would never shell out hard cash to view yet another pc movie. But "Courage under Fire," after a cursory glance at the reviews, appeared to contain more than its usual share of cw suppositions and pc stereotypes. The movie starred Denzel Washington (Sir Blacque Truth-Seeker), Lou Diamond Phillips (the Hispanic-Composite Soldier-Grunt), and Meg Ryan (the White Woman Warrior-Hero[ine]). Ms. Ryan's previous films had included "The Doors" (1991) and "French Kiss" (1995). Ms. Ryan's unlikely character in "Courage under Fire" was the first hypothetical female nominated for the Medal of Honor; in this case as a medevac pilot[ess] killed under fire during Desert Storm (the Persian Gulf War).

As Owen Gleiberman reported in <u>Entertainment Weekly</u>, "The White House badly wants her to receive the posthumous citation (it will seem 'progressive', and therefore politically savvy)"....[395] And David Ansen of <u>Newsweek</u> editorialised that "[t]he brass [hats] and the White House love its PR potential, and ... they're counting on a speedy coronation."[396] Although the movie's theme entailed Denzel Washington's character, Serling, checking up on conflicting stories of her [Ms. Ryan's] heroic "death", the disturbing assumption was that women in combat is an already given fact.[397] Well, this writer wouldn't have to worry about "Courage under Fire" adding too much to the <u>kultursmog</u>. The movie wasn't a monster hit, and even <u>Newsweek's</u> David Ansen sniffed, "I'd like 'Courage under Fire' more if it weren't quite so sure of its own importance... it seems to want us to pin a medal to <u>its</u> chest.[398] Finally, Owen Gleiberman of <u>Entertainment Weekly</u> summed up:

...."[Director Edward] Zwick, who honed his craft creating the touchy-feely epiphanies of <u>thirtysomething</u>, has never let go of his middlebrow liberal coziness. Meg Ryan has a fine, edgy presence here, but it's hard to escape the feeling that her Karen Walden [the downed medevac pilotess] is being set up as a role model -- in much the same way the film accuses the Army of doing. The real enemy in <u>Courage Under Fire</u> turns out to be Zwick's eternal nemesis: macho insensitivity, and all that it breeds (arrogance, a hard heart, lies).... <u>Courage Under Fire</u> is just a melodrama done with prestige solemnity...."[399]

Once at the Kitsap Mall, father and son visited the sumptious new Barnes & Noble Booksellers store. This writer noticed, with some amusement, that books on UFOlogy and Christianity were both grouped around the general category of Belief Systems.[400] After browsing in adjacent aisles, father and son repaired to the Mall itself for a cup of coffee. But this writer was still mulling over the effeminised "Don't ask, don't tell" Army. The liberal powers-that-be also wanted to sexually integrate the all-male Citadel in South Carolina, and the Virginia Military Institute. Why? War, so bad as it is, was surely a man's "game." An army's function, as Rush Limbaugh bluntly put it, was "for killing people and breaking things." For those enamoured of war, but not able to participate, there was the "Century of Warfare" video -- complete with close-up shots of dead bodies, destroyed cities, and battle scenes -- brought to you at a low price by those family-values folks at Time-Life [Warner].

1996 had so far been a very bad year for Angry White Guys. They were being blamed for all the Black church burnings throughout the South. One wondered if the almost-as-numerous <u>White</u> church fires had been set by the same Angry [Southern] White Guys? But that question, and answer, wouldn't have been grist for the pc network-news mills. The summer of 1996 was a summer of discontent for this writer, but not today, Saturday 29 June. The writer was so elated at seeing his son again after eighteen months, that he actually spent some money at the Mall -- on a millenharian paperback, The End of the Age by an M.J. Agee, and an audio-cassette, Rollin' on the River, by Creedence Clearwater Revival. This was living dangerously!

On the way back to Port Angeles, the writer thought about his chosen home-town. P.A. was an All-American City, and designated the Second National City by President Abraham Lincoln himself in 1862. P.A. had historically been called the Cherbourg (in northern France) of the Pacific, and presently declared the sister city of Mutsu City, Japan. Cars on their way out of town on U.S. 101 would pass Swain's General Store, at which bullets/live ammo were available at the impulse-buying counter. How P.A.! On reaching City Pier off Front Street, the motorist could turn left up Lincoln Street, which bisects Port Angeles the way Broadway does New York City. Half-way up Lincoln to the right was the new Safeway, in so splendid an edifice as the Clallam County Courthouse across the street, or Masonic Temple (erected 1921) farther uptown on the right hand side. On the solid brick walls were written in raised red letters: "SAFEWAY FOOD & DRUG OPEN 24 HOURS." That was all; no enticements or detailed notices of prescriptions, cosmetics, groceries, or beverages. This was P.A., Soviet-style. There's no more room for charm, Buddy. But the new Albertson's had recently relocated out of the P.A. central core, further eroding it, and the already user-unfriendly library had come on-line, was going up-scale, and soon moving out of downtown. Yet letters to the editor of the local newspaper kept wondering why P.A. was failing as central urban magnet.

The next day was Sunday, 30 June 1996, and U.S. presidential candidate Charles Collins was (again) coming to town. Independent Collins had come to Port Angeles previously that year, on 3 March, and this writer had attended the rally at the rickety headquarters of the right-wing Clallam Citizens Coalition. It seemed strange that a political candidate based in Forsyth, Georgia, and Panama City, Florida, would whistle-stop at Port Angeles, Washington, twice in one year. Panama City is about as far from Port Angeles in the contiguous United States as one can get. No matter -- Charles Collins espoused some positions on certain issues of interest to this writer. Collins opposed any call for a Constitutional Convention; wanted to restore "all" of the Bill of Rights; abolish income tax and Internal Revenue Service (IRS), Bureau of Alcohol, Tobacco, and Firearms (BATF)... "and all other abusive, oppressive, and unconstitutional agencies in Federal Government having police powers."[403]

The 30 June meeting, held in the same old downtown building, was lower in attendance than that of 3 March. Folks, even fiery far right-wingers, were less sanguine than during the halcyon, crazy days of the Republican primaries. What chance did Charles Collins -- not half as rich as H. Ross Perot -- have against the well-connected Bill Clinton or Bob Dole? Clinton was the incumbent president, an Oxford University Rhodes scholar, a Skull and Bones Society of Yale member etc.; Dole was a 33rd degree Freemason with ca. thirty-five years spent within the Washington, D.C. "Beltway." This writer got to meet candidate Collins after the meeting, and was bold enough to ask whom he had supported in the last Florida gubernatorial race, surely Jeb (son of George) Bush as opposed to Lawton Chiles? No, Charles Collins had replied in a slow Deep South drawl, regarding this writer with baleful eyes, he had not backed the young GOP [upstart] runner but supported the old Democrat [political hack] sitting governor (1991-). Then Charles Collins grinned, and this writer recognized him for the Old 'Possum he was.

So who would be the more effective in maintaining the Pax Americana? This writer would vote for Dole-Kemp as he had voted for Bush-Quayle, Reagan-Bush, and Ford-Dole before them ... but with reservations. Like many other American political, social, and religious conservatives, this writer couldn't -- and didn't -- expect a U.S. Dream Team. But, most assuredly, the draft-dodging and armed forces-hating Bill Clinton was not the man to keep the American peace in the Asia-Pacific region ... or anywhere else in the years ahead. And how would this writer's own neighbours vote? They talked tough, but invariably voted for Big Government (or none at all). Come hunting season in October, and they would have something bloody hanging on a rack in back of their house. Among the rocks and trucks, the trophy would prove the hunters' skill and self-reliance. Until October, though, there were (now) lion and zebra, hyaena and wildebeeste virtual-reality video games to amuse the bored during off-season.

The writer's daughter was home for the summer from U.Washington ("Grunge Academy"). On one of her packing boxes, in smart red and white letters on a gold and green background, was printed: "Clan MacGregor Blended Scotch Whisky Imported and Bottled by Alexander MacGregor & Company, Edison, N.J.

08837." It looked tempting, but politics or not, it was something this writer could never do ... not even once in a Blue Moon.

NOTES

395. 'Bound by Honor', 26 July 1996, p.32.   J

396. 'Woman under Fire', 15 July 1996, p.59.

397. Entertainment Weekly, 12 July 1996, p.68.

398. 'Woman under Fire', 15 July 1996, p.59.

399. 'Bound by Honor', 26 July 1996, p.33.

400. Not so far-fetched after all! See Appendix L, FutureFish 2000.

401. 'Collins for President '96', Campaign Courier (Forsyth, Ga.), Vol.1, No.2, 12 July 1996, passim.

Bibliography: 51 Years of the Pax Americana

19. Gingrich, Newt and William R. Forstchen. 1945. Baen Publishing Enterprises, Riverdale, N.Y., 1996.

20. Tuchman, Barbara W. Notes from China. Collier Books, New York, 1972. Incl. news-making essay, 'If Mao Had Come to Washington in 1945: An Essay in Alternatives.' (Copyright by Council on Foreign Relations, Inc.)

Chapter 2 :   <u>On the Home (Fishing) Front</u>

Entry: Anatomy of a Treaty, 1993-1998

Part I: Oh! Canada!

Part II: Alaska, the Last (Fishing) Frontier

Part III: The Lower 48, Northwest Coast

<u>Entry: Anatomy of a Treaty, 1993-1998</u>

"When sharing arrangements are in dispute, the incentives for either party to act to rebuild the resource are greatly diminished."
-- Canadian Foreign Minister Lloyd Axworthy in a letter to U.S. Secretary of State Madeleine Albright, 1997

<u>Pacific Salmon Treaty Update: 1983-1993</u>

The last time this writer tackled the tortuous problem of the Pacific Salmon Treaty was in 1983, while attending the University of Victoria, B.C. (The brief study ended up as Chapter V in the writer's first published volume on the North Pacific fisheries.)[1] Since that year, the Pacific Salmon Treaty was re-signed by the United States and Canada in 1985, renegotiated in 1989, and was to be again renegotiated in 1992. Irreconcilable differences between the parties, however, waylaid that latest attempt. Since 1992, efforts to revive the Pacific Salmon Treaty have been yearly, piecemeal, and half-hearted at best. During early June 1993, negotiators from both the U.S. and Canada had check-to-jowl face-offs at Montréal, but (as usual) were not able to resolve any major disagreements. Formal negotiations dissolved to the [low] level of tele-conference calls. The basic sticking points (as they had always been since the very beginning) were: 1) U.S. access to Fraser River sockeye in Puget Sound, and 2) Canadian access to Columbia River coho off Vancouver Island.

But conditions had also become more complicated. For the preceding four years, U.S. fishermen had been restricted to catching 8 million Fraser River sockeye. The figure had been decided on before Canadian fishery enhancement had resulted in huge Fraser River sockeye returns. The Americans, complained the Canadians, now wanted a change in the Treaty guaranteeing them a percentage of the sockeye run rather than a set number. The Canadian position in 1993 was two-fold: 1) <u>Canadian</u> enhancement effort (and time, money) had greatly increased Fraser River sockeye returns, and 2) that the (greedy, unreasonable) Americans should not be entitled to a percentage of the Fraser River sockeye run. Canadian fishing organisations sounded willing to reduce their catch of U.S.-bound coho in return for other considerations, but insisted that they were not going to sacrifice Fraser River sockeye. On that point, the Canadians remained adamant.[2]

But the Canadians were far from being goody two-shoes themselves. One Phil Eby, the executive director of the Fishing Vessel Owners Association of B.C. (FVOA, representing seiners), suggested that a way to bring the U.S. back to the negotiating table was to revert to a plan employed two years before -- to deliberately overfish American coho stocks. And in 1992, a year prior to the re-opened negotiations, the Canadians had attempted to intercept every south-bound Fraser River sockeye salmon (with limited success). Threatened Phil Eby:

"So, unless Canada is prepared to acquiesce to the bully tactics of the U.S., we may be forced to take action where they [the Americans] are most sensitive -- their dwindling stocks of chinook and coho on the West Coast'."[3]

The aborted 1993 non-negotiations left the American's angry and absent; the Canadians confused and concerned.

1994: Alaska versus Canada and the Others -- I

"The sovereign state of Alaska"
-- Jack Nichol, past pres. of the United Fishermen and Allied Workers Union and alternate commissioner for Canada

"Not much the Canadians could do about Alaskan arrogance...."
-- Jacob A. Mikkelborg, Seattle attorney

Unlike the demersal Pacific halibut, Pacific salmon ignore political boundaries when migrating from their open ocean home to their natal streams. These natal streams and rivers extend all the way up the North American west coast from Northern California through Southeast Alaska -- the "Rain Coast" of Ecotopia. This huge geographic arc encompasses four western U.S. states and a single Canadian province, British Columbia (and, to some extent, Yukon Territory). Canada, then, is at a distinct political disadvantage, and as a result has always preferred to negotiate on a national rather than a regional basis. So, instead of the federal Canadian Department of Fisheries and Oceans (DFO) dealing directly with its counterpart in Washington, D.C., Ottawa has had to wrangle with separate local departments of fish and game in the State of Alaska or

Washington State. As Pat Chamut, director of the DFO Pacific Region, and a commissioner to the Pacific Salmon Commission, has put it: "Canada sees the treaty as a treaty between Canada and the United States, not between Canada and Alaska and Canada and Washington state.'"[4]

The Pacific Salmon Treaty is vitally important to both the U.S. and Canada as it provides a structure for dealing with the three paramount issues of 1) resource conservation, 2) equity in harvest equivalent to each party's "production of salmon originating in its waters'", and 3) the "desirability in most cases of reducing interceptions '".[5] The rub comes when there is disagreement as to who is intercepting whose fish. Therefore, when the Treaty was reauthorised for the years 1989-1992, a provision called for the Americans to fish no more than a total of 7 million Fraser River sockeye, and no more than 7.2 million Fraser River pink salmon during that four-year period. But, according to Canadian commissioners, American fishermen exceeded the allotted catch by over 360,000 salmon. This happened, contended Canada, because the U.S. refused to include interception of fish off Noyes Island in Southeast Alaska (directly south of Iphigenia Bay and northwest of the town of Craig and Lulu Island). Those salmon return to the Skeena and Nass River system in northern B.C., and all the way down to the Fraser River in southern B.C. For his side, Canadian Commissioner Pat Chamut has admonished:

"Alaskans are taking a very large benefit that should be gained by our fishermen.... We have a situation where the Alaskans are taking increasingly large slices of Canadian production, and elsewhere [Washington state] they want an increasing share of Fraser River sockeye at the same time as they want a reduction [in Canadian catch of diminished U.S. chinook and coho] on the south coast.'"[6]

Answering back for Southeast Alaska has been Mel Seibel, special assistant to the Alaska Dept. of Fish and Game (ADF&G) assigned to the Pacific Salmon Commission:

"...[T]here is some concern that the numbers [of interceptions] have gone up. We have had very strong pink runs. We have increased those fisheries. Sockeye from the Skeena and Nass have increased. Also, the Fraser contributes to that. As we harvest pinks we do catch more sockeye but the proportion has not changed.'"[7]

As noted earlier, Pacific salmon (genus <u>Oncorhynchus</u> spp.) differ from Pacific halibut (<u>Hippoglossus stenolepis</u>) in that they are specifically claimed by their habitat of national origin, in this case the rivers of the United States and Canada. As a result, the International Pacific Halibut Commission, which has been a resounding success in conservation practise since its inception in 1923 (as the International Fisheries

Commission), hasn't had to labour sorting out national rights to the fishery resource -- or to spend millions of dollars maintaining halibut stocks.[8] (Conveniently for the IPHC, the Pacific halibut is considered by both parties as a single-stock fish that happens to spawn in waters on either side of the Can-Am border.)

Both U.S. Commissioner Mel Seibel and Canadian Commissioner Pat Chamut have expressed support for the Pacific Salmon Treaty. Mel Seibel (of the ADF&G) has said, "I can't accept the allegation that the treaty system is broken'"; and Pat Chamut (of the DFO Pacific Region) has concurred: "It [the Treaty] has lots of warts...but we are far better off with it than without it.'"[9]

In 1994 Jacob A. Mikkelborg was an attorney with the Seattle law firm of Mikkelborg, Broz, Wells & Fryer, and a frequent contributor to Western Viking periodical. (He still may be that in 1998; 1994 was this writer's last subscription year to WV.) As a member of what the writer calls the "Greater Ballard [Norwegian-American] Community", Jacob Mikkelborg had some interesting observations on the Northwest "Fish War". During mid-1994, Canada laid a punitive tax on U.S. vessels exercising their "right of innocent passage" through the B.C.-controlled Inside Passage, that links the Lower 48 to the State of Alaska. Feelings among American fishermen south of the 49th parallel were predictably high and hostile. But Jacob Mikkelborg took a totally unexpected tact in analysing the problem. For rather than blame the plight of Lower 48 fishermen (i.e., mostly Seattle-based from Washington State) in-transit on the Canadians, Mikkelborg shifted the burden squarely onto the broad shoulders of the Last Frontier:

"Aided and abetted by geography and the doctrine of territorial jurisdiction headland to headland encompassing all the waters in between, Alaska became a marine super-state and 'The Great Land'. Alaska acquired territorial jurisdiction over a vast expanse of sea, waters once international. Fisheries on those high seas, almost by default, came under Alaskan territorial jurisdiction. As political awareness grew with state development, regulation increased and traditional fisheries by non Alaskans met Alaska based competition chafing under Federal Constitutional restraints.
"Alaska's jurisdiction offshore, out at sea, on the high seas all the way to the Russian border, now is taken for granted. If you doubt this, just try to sell a cargo of King Crab [sic] harvested in the far reaches of the Bering Sea at a port other than an Alaskan port. You will be met by a swat team of Alaska State Troopers, a delegation of Alaskan Assistant Attorneys General, warrants for your arrest, extradition, fines in many thousands of dollars. Your next fisheries permit will be slow coming from guess who? AF&G -- short for Alaska Fish and Game."[10]

Jacob Mikkelborg might have felt quite differently had he hailed from Petersburg, Alaska ("Lille Norge"/"Little Norway"), rather than Seattle, Washington, but his point was well taken. Mikkelborg referred to a growing attitude in Alaska of "Alaska for Alaskans"; a belief that state officials should exercise complete control over Alaskan fisheries. This sentiment in turn has led to a..."summary rejection of Canadian concerns with over-interception of sockeye salmon migrating past places like Noyes Island in Alaskan waters, bound for Canadian river systems.[11] But the full impact of official Canadian indignation and Alaska jurisdictional arrogance, would fall on those Lower 48 (i.e., Washington State) fishermen transversing the B.C.-regulated Inside Passage, to fish Alaskan waters. The same impact would adversely affect those sailing back from Alaskan waters, to market fishery products. Thus, concluded Mikkelborg:

"....[N]onAlaskan fishermen were left holding the bag. Canadian desperation to this point has no effect on the Alaska based fishery but every effect on [a] Washington based interstate fishery accustomed to innocent passage between states. Anyone want to buy a salmon boat, suitable for interstate commerce via the Inside Passage?"[12]

(In an article three weeks later, 22 July 1994, Jacob A. Mikkelborg reported that north-bound U.S. boats would <u>not</u> be required to stop at Nanaimo, B.C., paying $1500 Cn. [$1100 U.S.] to the Canadian authorities. Despite the behind-the-scenes tax moritorium, however, the mutual Alaska-Lower 48 misgivings remained.)[13]

1994-1996: The Era of Brian Tobin

Peripatetic Brian

"....[M]y anxiety was heightened when I learned that the Canadian Fisheries Minister, Mr. Brian Tobin, has been visiting southwest England to thank Cornish fishermen for their support of Canada's fishermen in their dispute with the European Union. His visit was in turn disrupted by demonstrators against Canadian seal hunting."[14]
-- John O'Sullivan, ed., <u>National Review</u>, September 1995

If Newt Gingrich was <u>Time</u>'s Man of the Year in the United States for 1995, in Canada 1995 was the Especial Era of Brian Tobin. The soon-to-be peripatetic Canadian Fisheries Minister, Brian "the Tobinator" Tobin, erupted onto the world stage during Canada's "turbot tempest" stand-off with the European Union (i.e., Spain) in 1995. And Brian was everywhere, playing fish conservationist to the max, from the Grand Banks of (his native) Newfoundland to the not-so-grand canyons of New York City; wherein are situated the not-so-hallowed halls of the UNO at Turtle Bay. But it was Brian Tobin's outrageous eco-posturing and unmitigated <u>chutzpah</u> that won the day -- and (most of) Canada's heart. "Tobin's War" was so effectively waged, that the proud Spanish were forced to back down. The "turbot tussle" turned politically ugly and highly partisan. Indeed, many E.U. members sympathised (notably the United Kingdom and Éire, but also various Northern

European political leaders) with Canada. The notorious Spanish temper was only restrained by the iron grip of the Eurocrats at Brussels. (At the time, this writer predicted that "Brian Tobin will find the free-wheeling, ebullient Alaskans on Canada's other side a decidedly different kettle of fish.")[15]

All throughout 1995, while trying to grapple with the colossal ego and intricate vagaries of the inner Brian Tobin, the writer received a <u>National Geographic</u> article by the post; sent by the wife of an old Norwegian-American friend (still) living in North Jersey. (The writer had long since eschewed reading the pompous, pc texts of the <u>National Geographic</u> years ago, but will still look at the splendid photographs on occasion.) And the article by Michael Parfit was like many a doomsday, Chicken Little tale of polluted seas and vanishing marine resources...but wait! For there, sandwiched within the turgid prose, was a picture-perfect mini-portrait of Brian Tobin; to wit:

"Later I meet the prime mover in Canada's war of nerves with the outside world. He is Brian Tobin, the minister of fisheries and oceans. We talk in a hotel room above the stone-circled harbor at St. John's, Newfoundland. Tobin, who will be stuck with fish metaphors as long as he has this job, has been described as a tough guppy: 'Small, colorful, and furtive'. He doesn't look furtive to me. He looks more like the kind of fish you'd find gnawing cattle down to [the] bones in the Amazon."[16]

1995 had been so totally the Year of Brian in Canada, that on 8 January 1996 Tobin -- still federal fisheries minister -- announced his intention to run for the premiership of his home province, Newfoundland. By 19 January 1996, Liberal Premier Clyde Wells had passed the political baton on to his successor, and Brian Tobin would officially become premier of Newfoundland a week later.[17] But Fisheries Minister Tobin, known as "Captain Canada" to his (mostly) admiring compatriots, could boast of some real accomplishments (according to your national p.o.v.) during his two-year tenure: 1) The imposition of a strict conservatory programme for Atlantic groundfish; 2) a long-term aid package for laid-off Maritimer and Newfie fishermen; 3) a brand-new Canada Oceans Act; 4) a revised and rewritten Fisheries Act (the original dating almost back to Confederation in 1867); 5) coming out on top of the turbot imbroglio over the E.U. (i.e., Spain) in 1995; and (6) the collecting of $438,000 Cdn. in transit license fees from American fishermen travelling through B.C.'s Inside Passage.[18]

Eventual reimbursement of the fees or not -- "Captain Canada", with his melodramatic style and panache, was perceived by his countrymen as tweaking the tail-feathers of the imperial U.S. bald eagle. With

Brian Tobin, Canada's last three fisheries ministers have all been from Newfoundland (John Crosbie, Ross Reid), but "the Tobinator" left over-sized shoes (or hat?) to fill.[19] (Those shoes would soon be worn by one David Anderson, previously Canada's minister of revenue.) Probably the very best eulogy (among many) to Brian Tobin's fisheries ministry was provided by a letter to the West Coast industry paper, <u>The Fisherman</u>:

....."I am a [Canadian] fisherman so my awards should have an aquatic name...Hmmm, let's see...something scaly and cold-blooded. I've got it -- here is my leading candidate for the 1st Annual Fishy Characters award.
"The Turbot Award. The winner is, by unanimous decision, Brian Tobin. Anyone who's actually seen a turbot knows that they show a lot of teeth but little else of significance.'"[20]

Reviews are still mixed regarding his overall act, but love him or hate him, Brian Tobin forced the world to sit up and take notice of Fisheries Canada. CBC News commentator Rex Murphy aptly described Brian Tobin as "a charged particle of emotional and political energy'."[21] For many Canadians that particle was positively charged; to most Americans (and all Spaniards) that particle was negatively charged. Meanwhile, the perennial renegotiating haggle over the Pacific Salmon Treaty with the United States, would be left to Canada's former chief taxman David Anderson, and to New Zealand mediator Christopher Beeby.

1996: Alaska versus Canada and the Others -- II

"Alaska is a little bit tired of being the scapegoat for poor Canadian management. We've constrained our fisheries and Canada hasn't been doing the same thing."
-- Alaska Treaty Commissioner David Benton

By the spring of 1996, with Brian Tobin gone, Canada's Dept. of Fisheries and Oceans had ceased to be the sole focus of American industry ire. The hassling now entailed the sovereign state of Alaska "wrasslin'" the Lower 48 states of Washington and Oregon, along with the Treaty Northwest tribes. The "wrestling" was to avoid lengthy and costly court litigation, one of America's favourite (and most wasteful) past-times of the 1990s. Washington and Oregon withdrew a conservation proposal which would have shut down the Southeast (Alaska) chinook fishery, but as of May 1996 there was still no official plan.[22] Alaska's Department of Fish and Game (ADF&G) offered the Northwest Treaty tribes a proposal to negotiate a lasting and equitable agreement for sharing salmon stocks. The offer, of course, was an attempt by the ADF&G to avoid extension of the hated Judge Boldt decision to include Alaska; the dreaded decision arbitrarily handed down during the mid-1970s, that had awarded Native Tribes ("First Nations") in Washington and Oregon fully one half of all harvestable salmon.[23]

Meanwhile, south of the border, British Columbia Premier Glen Clark was loudly calling for a ban on U.S. Navy submarine passage through Canadian waters...<u>and</u> for a revival of transit fees on U.S. fishing vessels in transit through the Inside Passage. Alaska's David Benton countered by accusing Canada of grossly mismanaging her own fisheries, of decimating B.C. coho stocks, and the (Canadian) federal government of having "misplaced'" some $65 million Cdn. of buy-back money. These funds had been paid by Canadian fishermen to reduce the number of permits issued in the commercial industry sector.[24] (Not surprisingly, little progress was expected by either party at the international level for 1996.)

To add to Treaty woes in 1996, scientists as well as politicians (from both sides of the Can-Am border) were now engaged in numbers-crunching. The numbers were related (as they always have been) to the three critical areas of 1) salmon (i.e., chinook) management science, 2) environmental events and factors, and 3) equity, equity, equity.[25] Alaska has boasted that the "Great Land" employs a fundamentally different approach to fishery management science than Canada -- an "abundance-based'" fishery management system. Whereas Canada, the Lower 48 states, and the Northwest Treaty tribes base their proposed catch limits on pre-season forecasts, Alaska fishery managers (i.e., the ADF&G) adjust harvest ceilings <u>in season</u>, based on data collected while fishing on the grounds.[26] Going on the record as sceptical of the Alaska in-season abundance-based management model, has been Dr. Brian Riddell, a Canadian federal fisheries scientist and co-chairman of the Treaty's Chinook Technical Committee:

"In principle, the idea is sound....What we found to be flawed was the index of harvest rate and how the estimate of abundance in the fishing region is calculated in-season...."It's a fisherman's job to find fish.... So as abundance declines, the index based on fishery performance is very likely to be biased high because of the ability of the fisherman to respond to the gradual reduction in the abundance of fish. Their catch rates will stay up higher than a proportional change or reduction in abundance.'"[27]

In 1992 and 1993, the waters of British Columbia were once again warmed by the effects of "El Niño", as they were in 1983, with the same, resultant invasion of chub mackerel that devastated juvenile chinook stocks.[28] Those massive mackerel predations of the early 1990s, have greatly reduced the ocean survival rate of the Robertson Creek hatchery fish. And those salmon, Dr. Riddell has maintained, have come from one of Canada's most productive hatcheries for the <u>Alaskan</u> fisheries. Furthermore, Dr. Riddell has claimed,

Alaska fishermen continue to catch 25% to 28% of West Coast Vancouver Island wild stocks.[29] To ensure more wild stock escapement, B.C. Fisheries Minister David Zirnhelt journeyed to Alaska in April 1996 to: 1) Admonish the Alaskans that U.S. interception of Canadian salmon had <u>increased</u> by 50% since 1985 (the year of the Treaty re-signing); 2) that Canadian interception of American salmon had <u>decreased</u> by 25%, costing B.C. fishermen $500,000; 3) and to tirelessly preach the gospel of equity, equity, equity.[30] (The reader might well imagine how much of a rhetorical success the B.C. Minister's finger-pointing speeches were up in Alaska!) But whether or not interception equity could be achieved -- or even agreed to -- remained to be seen.

Dr. Brian Riddell has reminded both parties that Canada took steps to protect her West Coast Vancouver Island chinook stocks in 1995 by lowering B.C. fishery harvest ceiling by a full 50%; that Alaska was requested to do the same, but only cut the state's catch of chinook from 260,000 to 230,000 fish. Unhappy with Alaska's apparent lack of effort, in 1995 B.C. slapped the infamous $1,500 Cdn. transit tax on American boats from the Lower 48 traversing the Inside Passage. Again, New Democratic Party Premier Glen Clark loudly called for a renewed imposition of the feared transit fee, with added sanctions on U.S. vessels, and yet more onerous customs, immigration, and safety inspections. NDP ideologue or not, Glen Clark was facing an up-and-coming provincial election.[31] And in a Canadian election (except in Québec), Americans are always fair game.

By 26 June, at any rate, the U.S. sector of the Pacific Salmon Commission (PSC) had agreed on chinook catch numbers for the 1996 season. The chinook harvest levels for Southeast Alaska (the American area of hottest dispute) were set at 140,000 to 155,000 fish.[32] The agreement between Alaska, the Lower 48 states, and the Northwest Treaty tribes not only set chinook harvest numbers for 1996, but caused a paradigm shift in management style to an Alaska-type in-season, on-the-grounds, abundance-based management model -- effective in 1997.[33] Canada's (almost aghast) response, as voiced by Minister of Fisheries and Oceans Fred Mifflin in a 27 June press release, was immediate:

"Despite Canada's strict curtailment of chinook harvests in B.C. waters, it seems that the U.S. is more interested in maintaining its fisheries in Alaska than it is in maintaining the health of chinook stocks'."[34]

Canada had closed all targeted chinook fisheries in B.C. for 1996, and requested that the Alaskans reduce their Southeast chinook fishery in 1996 by a full 50% -- as Canada had in 1995. The Canadians were referring to the same PSC (Pacific Salmon Commission) numbers as used by the Americans, and had thereby reckoned on an estimated 1996 Southeast chinook harvest of 120,000 salmon. Cutting the chinook in half to 60,000 fish, the Canadians had reasoned, would ensure present conservation and continued future recruitment for all parties.[35] Our old Canadian acquaintance, Dr. Brian Riddell of the PSC's Chinook Technical Committee, was pessimistic about Alaskan compliance:

"Our management plan was premised on not getting cooperation in 1996....What we are expecting to see in our terminal areas already takes into account no change in the harvest rate in Alaska.'[36]

Canada has moved by stating that regulations will be implemented requiring U.S. vessels, travelling through Canadian waters, to notify the Department of Fisheries and Oceans (DFO) to obtain clearance for passage. As U.S. vessels will be further required to stow their gear -- although American fishermen generally stow their gear anyway -- the Mickey Mouse regulation was simply another small way to tweak the tail-feathers of the U.S. bald eagle.[37] As Washingtonian Jacob Mikkelborg wrote in 1994, "non-Alaskan fishermen [have been] left holding the bag."[38]

1997: Canada versus Alaska and the Others

...."[C]onsolidation of the U.S. stance has trumped Canada's divide-and-conquer card."
-- Bob Tkacz, Alaska Fisherman's Journal Treaty analyst

In the Alaska Fisherman's Journal of January 1997, editor Bob Tkacz commented that the 1996 consensus among Alaska, the Lower 48 states, and the Northwest Treaty tribes on U.S. chinook management was a "breakthrough".[39] It was a breakthrough, according to editor Tkacz, because the Lower 48 states and the Northwest Treaty tribes had finally bought in to the Alaska-style abundance-based system. And, opined editor Tkacz, "This consolidation of the U.S. stance has trumped Canada's divide-and-conquer card."[40] Bob Tkacz's scorn for Canada's call for equity notwithstanding, Canadian officials have stuck to their guns -- unequivocally maintaining that there would be further talks on chinook management unless there were an equity agreement.[41]

And the Canadian's have insisted that any agreement on equity would depend on a research document -- produced by and in possession of New Zealand mediator Christopher Beeby -- to be accepted as a fallback position. Finally, Canada was peeved that the U.S. delegation had not steered Alaska's chinook harvest proposal through the Treaty's proper grievance channel, known as Article 12. Bob Tkacz concluded his 'Salmon Treaty '97' analysis by claiming that a Canadian source had told him "to expect his country to use the U.S. refusal to release the Beeby document -- and the Article 12 issue -- as twin public relations clubs to batter Alaska this [1997] year."[42]

But 1997 did not turn out to be another year of wretched rhetorical excess or miserable diplomatic blunder by representatives of either side. With governmental scientists and official negotiators bogged down in a slough of Treaty fine point, American and Canadian <u>fishermen</u> have finally had the opportunity to sit down and talk with each other. These "stakeholders" in the salmon resource held meetings during the week of 10 February 1997 at Portland, Oregon.[43] Divided into northern (B.C. and Alaska) and southern (B.C., the Lower 48 states, and the Northwest Treaty tribes) fishery panels, the stakeholders endeavoured to reach regional salmon management accords which would coincide with the basic Treaty values of resource conservation and harvest equity. The fishing industry panels had a month to think, talk, and work things out before reporting their progress (or lack of) back to the U.S. State Department and the Canadian foreign ministry.[44] (Although semi-unofficial, the week-long stakeholder conference was both urgent and significant. For if official Treaty talks failed to show progress, fiery B.C. Premier Glen Clark had threatened to provoke "an international incident'." Quick to respond, Senator Frank Murkowski and state Senator Jerry Mackie proposed that Alaska drop Prince Rupert, B.C., from the Washington-Alaska ferry run.)[45]

A firm foundation, however, had been laid down for the idea that fishermen could solve fishery problems. At the Sitka Salmon Summit in August 1996, Alaska Governor Tony Knowles recommended that the issue of harvest equity should be tackled as a regional fisheries issue (hence northern panel/southern panel). The governors of Washington and Oregon ("the Lower 48 states") both liked the regional fisheries concept; as did the U.S. federal government.[46] And Alaska Governor Knowles, at a stakeholders' meeting at Juneau in May 1997, expressed his personal confidence in the stakeholders' approach to shared salmon fishery problems:

"People most affected by issues are often the most likely to come up with creative and lasting solutions. The ability of stakeholders to resolve problems among themselves, where bureaucrats and elected officials have failed, is sometimes amazing.'"[47]

This writer concurs with a hearty "Amen" to that. Indeed, while their federal U.S. and Canadian counterparts were getting ever more deeply mired in Treaty disputation, stakeholders from both sides of the border were slowly but surely reaching common (fishing) grounds. Since stakeholder talks started in February 1997, informal agreement had been arrived at regarding the Noyes Island, Tree Point, and Canadian Area[s] 1 and 4 fisheries. By late April, stakeholders from both countries expected to have a formal accord ready for official acceptance by their respective government.[48] In late April, Joy Thorkelson, B.C. northern stakeholders chair and business manager for the UFAWU, announced:

"I think we have a general basis for agreement....I am hopeful that we will be able to plug in the right numbers and if we aren't...maybe we can have at least a range of numbers that both sides consider agreeable.
"I don't know why there would be any concern that the Canadian government isn't going to implement what the stakeholders say. The Canadian government will, unless we come up with something completely off the wall.'"[49]

But "coming...completely off the wall" was exactly what Canada's chief federal negotiator, Ives Fortier, had done, according to <u>Alaska Fisherman's Journal</u> analyst Bob Tkacz. For Ives Fortier had gone "off the wall" by stopping, during early May, all further stakeholder meetings and temporarily halting official Can-Am Treaty talks. Fortier accused the United States of having not acted in good faith at the negotiations, after his State Department counterpart had taken a southern stakeholders' proposal to Washington, D.C., for U.S. federal approval. British Columbia stakeholders, outnumbered and overwhelmed by their opposite U.S. numbers, had accepted Alaska's abundance-based management system -- from which northern sector salmon recruitment numbers were to be taken. Now that the Treaty talks had broken down yet again, Canada returned to her earlier [immutable] position of equity, equity, equity.[50]

The stalled negotiations brought the fire-eating ideologues of both sides roaring from out of the woodwork. Combustible B.C. Premier Glen Clark proclaimed the cancellation of a contract with the U.S. Navy, which had permitted the Americans to utilise facilities off Vancouver Island for testing submarine torpedoes. Within just days, there followed the seizure of four U.S. fishing vessels which had not radio-broadcast their entry into B.C. waters as required by law. With possible fines of up to $350,000 for every offending boat, rational Port Hardy, B.C., Judge Brian Saunderson (eventually) fined the U.S. detainees less than $300 each.[51]

Alaska Senator Frank Murkowski reacted to the threat of further seizures by suggesting that the U.S. Navy might provide an armed escort for American fishing boats traversing Canadian waters. And (the other) Alaska Senator Ted Stevens, chairman of the Senate Appropriations Committee, went on record saying that he would not support continued voluntary U.S. financial aid, in cleaning up old Cold War radar warning sites in Canada's far north.[52] Ted Stevens' statement prompted Glen Clark to submit a letter of protest to the Alaska senator, complaining that Stevens' (political) position was "very similar to the continuing destruction of Pacific salmon pursued by your [American] nation'"; that he, Premier Clark, would urge Prime Minister Jean Chrétien to freeze "all expenditures on cross border programs providing environmental benefits to the United States'."[53] (According to Alaska Fisherman's Journal, July 1997, Alaska Treaty Commissioner David Benton expected stakeholder meetings to continue during the autumn of 1997, and Canadian federal negotiator Ives Fortier said that he would employ B.C. stakeholders in an advisory capacity.)

1997: The Summer of Impasse

"There's a major and fundamental flaw in the treaty. And that is that the salmon runs of the Northwest coastal rivers are declining, so Canadians are able to catch fewer U.S.-bound fish, while, on the other hand, the Fraser has remained a great engine of salmon production, which has allowed Americans to catch more Canadian-bound fish. And that's fundamental disparity, both in the number of fish caught and in the income produced caused the two countries to grow further and further apart [sic]."[54]
-- Joel Connelly, Seattle Post-Intelligencer, 1 August 1997, reporting on southern sector fisheries

It is not very often that fish tales are deemed news-worthy enough to be aired over the big "mainstream" American television networks; certainly not by the élite powers that be at so-called "public" television (PBS). Indeed, the last time there was a major fish news-story broadcast on PBS (to this writer's knowledge), had been on 3 November 1994. The feature had dealt with the collapse of the Atlantic Northeast Coast fisheries (with that topic examined at length in this work).[55] There had been, at the time in 1994, the predictable and typical PBS dark references to "The Tragedy of the Commons" (Garret Hardin, 1968). But politics rather than ecologism was the theme of the 1997 fish news-story, and PBS mainly concerned itself with the Pacific Salmon Treaty per se, along with Can-Am dynamics. 'Focus -- Salmon Feud' was reported by Rod Minott of KCTS-Seattle on 1 August 1997, from the Port of Prince Rupert, B.C. Rod Minott was seen interviewing one Des Nobels, Canadian salmon fisherman, a fortnight after about two hundred B.C. fishing vessels had blockaded the Alaska ferry Malaspina. Some of the ensuing interview from Prince Rupert went like this:

"Rod Minott: Fishermen blame the poor catch on several things, including Canadian officials who curtailed the amount of fishing that would be allowed this year, a policy meant to conserve shrinking stocks of salmon. But, most of all, the fishermen blame Americans. Canadians allege that thousands of sockeye salmon bound for Canadian rivers this year have ended up in the nets of U.S. fishermen just up the coast from here in Southeast Alaska. Des Nobels calls that alleged over fishing piracy on the high seas.

"Des Nobels: Yes. This is a fish war. This is a salmon war. I don't know how you could portray it otherwise. I'm besieged. My neighbors are besieged, and we're dealing with a siege mentality, which puts you in a war scenario, if that's the term you want to use. This is no longer a dispute. We've been disputing for four years. The dispute is over. The war is on.

"Rod Minott: Two weeks ago, Nobels and other Canadian fishermen retaliated against the U.S. Several hundred [ca. 200] fishing boats trapped the Alaskan ferry Malaspina in the Port of Prince Rupert. The radio call went out for them to join the blockade.

"[Canadian] Spokesman [in flashback]: Come on out, guys. We need your support. Get yourself in the lines here. Show these big [American] boys that we're not putting up with their crap.

"Rod Minott: At one point, some of the protestors burned an American flag. Tempers flared on both sides.

"[Canadian] Fisherman [in flashback]: They're pirating our fish. We're not trying to pirate anything. They can have their damned ferry back. Give us our fish back. It's very important that that boat stays right there.

"Rod Minott: Ferry passengers voiced outrage.

"[American] Ferry Passenger [in flashback]: It would seem to me that it is a borderline terrorist act.

"Rod Minott: For three days the fishermen defied a Canadian court order to release the U.S. ship. A breakthrough came after a personal plea by Canada's fisheries' minister to end the standoff. Two hours later, the Malaspina was released. But the dispute over Pacific salmon was far from over...."[56]

And so it went. Oddly enough, the individual most responsible in setting the stage for the Malaspina hijacking, B.C. Premier Glen Clark, came up with possibly the best (if not a novel) final solution to the Pacific salmon dispute:

"Premier Glen Clark: I mean, really, this is a common property resource that we share. It's in both our countries' interest to conserve the species where we can both benefit from that, not to mention just the fact that we want this species to survive. And so a rational thing to do would be to allow somebody independent, jointly approved by both countries, to come in and make the determination."[57]

(Let's hope that possible mediation by a third party would preclude New Zealand's Christopher Beeby or, worse, the United Nations.)

Some Canadians, especially in British Columbia, hail Premier Glen "Hornblower" Clark as the spiritual successor to former federal fisheries minister Brian "Captain Canada" Tobin, as flag-waving Maple Leaf super-patriot. <u>Maclean's</u> magazine cover-story for the week of 4 August 1997 was 'Darn Yankees! Our fish. Their nets. Guess who's winning?' The sardonic query is answered on the <u>Maclean's</u> cover itself. A wiry, evil-eyed, corn-cob-pipe-smoking Uncle Sam is hauling in a gillnet full of Pacific salmon...along with a captured, cowering Premier Clark; who is vainly trying to protect the caught salmon, while holding aloft a mini-Maple Leaf flag. The political cartoon was apt -- it expressed Canada's sense of utter helplessness in the face of raw American power. Canada's very impotence prompted <u>Maclean's</u> editor Robert Lewis to rather proudly comment:

"Enter Glen Clark, the bull-in-a-china shop of a B.C. premier. He not only raised the heat on the issue, but must have secretly delighted the very federal officials who spent the week apologizing for his behavior. Without him, the wimpy feds would have had no card to play in Washington. Last week, at least, they could point to their own loose cannon on the deck -- just as Bill Clinton has to deal with fiery and unpredictable legislators, including Alaskan Republicans in Congress, who want to bash Canadians. Clark was nothing if not bold, welcoming the blockade of a U.S. ferry boat in Canadian waters and belittling the federal efforts. Glen Clark is no Lester Pearson. But it was refreshing to see a Canadian standing up to be counted. Raw domestic politics proved useful in international affairs."[58]

In reality, both Americans and Canadians would have to tone down their heated words, spoken in haste during the Summer of Impasse 1997. Prime Minster Jean Chrétien, returning from summer holidays, immediately sent Foreign Affairs Minister Lloyd Axworthy to Washington, D.C., to mend neighbour fences. After only a single day of wrangling with American officials, Minister Axworthy ruefully admitted to Canadian news sources that he had gotten virtually nowhere with his U.S. counterparts. Thus Minister Axworthy returned to Ottawa with nothing more than an agreement to re-start the already stalled re-negotiations of the failed 1985 Pacific Salmon Treaty.[59] So a weary Minister Axworthy, hoping to "lower the temperature '" in the dispute, announced that Canada had given up trying to convince the United States of direct, step-by-step, Treaty negotiations which would get down to binding agreements.[60]

Instead, two "eminent persons" would now meet with the various parties in the salmon dispute and seek new methods of how, once more again, to re-open meaningful talks. Those two "eminent persons" turned out to be (American) Seattle businessman William ("Bill") Ruckelshaus, and (Canadian) retiring U. British

Columbia president, David Strangway. Messrs. Ruckelshaus and Strangway were to report by the end of 1997 directly to their respective president and prime minister with their findings. Meanwhile, during the waning months of 1997, the United States and Canada would stay in close touch daily, thereby more able to stifle any incipient incident like that at Prince Rupert. Minister Axworthy asserted that "This is now taken on by both the President and Prime Minister as a matter of high priority '."[61] (Fish war hawks, on both sides of the border, immediately denounced the concept of salmon diplomacy by "eminent persons".)

Pacific Salmon Treaty 1998: Back to Square One, All Over Again

"If the [T]ragedy of the Commons is to be averted, rules must be established for the preservation of the fish and time is not on their side."
-- From report by William Ruckelshaus and David Strangway

Ah, the Tragedy of the Commons -- the writer just knew the dreaded T-phrase would come up again before FutureFish was completed! But this time around the infamous phrase has also been used against the spectre of eco-alarmism rearing its ugly head in the Pacific Salmon Treaty talks. John van Amerongen, editor-in-chief of the Alaska Fisherman's Journal, expressed the industry's concern in a near-perfect February 1998 piece:

"....Perhaps Mr. Ruckelshaus and Mr. Strangway were too caught up in the mythical '[T]ragedy of the Commons', which they cite so dramatically in their report....
"Why they feel the need to play the heartstrings of the uninformed escapes us. So, commercial fishermen on both sides of the border don't have rules and they can head out willy-nilly any time they want to plunder the resource into extinction? This column is way too short to begin to list the rules or to ponder the sense of that statement.
"What we can say is that the sticking point is not the conservation of the Commons, but in the horse-trading -- and that's best left up to the real horse traders, not the politicians."[62]

Editor van Amerongen also brought to readers' attention the disturbing fact that Ruckelshaus and Strangway had recommended (to their federal superiors) disassembling the bilateral Can-Am stakeholder meetings. (Unbelievable! To dismiss those most concerned with salmon survival and recruitment is like shooting the Treaty in its one good foot!) Originally, John van Amerogen has reminded us, the two "eminent persons" had acknowledged the importance of stakeholders' (i.e. fishermen) input, both northern (Alaska-B.C.) and southern (B.C.-the Lower 48 states and the Northwest Treaty tribes) sector fishery panels. So now, in early 1998, the stakeholder meetings were to be dismantled and to "turn salmon treaty negotiations back to the same people who have been unable to reach agreement for the past four years".⁶³ And, according to van Amerongen, the stakeholder talks were making far more progress than the official Treaty talks -- particularly in the northern sector. The <u>Alaska Fisherman's Journal</u> editor has written that one Alaska fishery panel delegate assured him that Canadian gillnetters would have caught hundreds of thousands more sockeye, and Alaskans hundreds of thousands less, had an offer from Alaska been accepted by Canada last spring."⁶⁴ The last might have been merely another Alaskan tall tale, but Editor John van Amerongen has asked (almost plaintively): "So why scuttle the prospect for similar compromise and go back to [S]quare [O]ne?"⁶⁵

During late March/early April 1998 they were at it again, meeting for four days at Washington, D.C., they -- the United States and Canada -- each have a new champion, a chief negotiator, to resolve the old problem of reviewing provisions of the Pacific Salmon Treaty. The Americans have appointed a Mr. Robert Owen as U.S. top-gun, who recently acted as an arbitrator in Bosnia-Herzegovina. He would be flanked by officials from the U.S. federal government and various state big-shots from Alaska, Washington, and Oregon. Canadian Federal Fisheries Minister David Anderson has anointed the provincial advisor to the British Columbia premier on salmon, a Mr. Dennis Brown, along with law professor Donald McRae, of the University of Ottawa, as Canada's chief negotiator.⁶⁶ Despite this convocation of combined North American brains and talent, a late March Associated Press article evinced little confidence for future progress of the Treaty talks:

"The inability of Canada and the United States to renew provisions of the treaty for several years -- particularly concerning conservation and dividing the number of fish caught -- has resulted in increasingly bad feelings among the principals."⁶⁷

The combined Can-Am federal, state, and provincial top-guns and big-shots tried again, convening for three days in April at Vancouver, B.C., with (again) no agreement being reached. And, at last reading, they all got together yet one more time for three days in May at Portland, Oregon. Both the United States and Canada in mid-May were hoping to set catch limits for the summer season of '98.⁶⁸ Observed Ben Spiess of <u>Pacific Fishing</u> magazine, "While limits may ease tensions on the fishing grounds, the real problem -- finding a fair split of fish between Canada and the U.S. -- remains unresolved."⁶⁹ Remarked Bob Thorstenson of the Alaska delegation, "In terms of the differences over what's an equitable split between the countries, things haven't changed much'."⁷⁰ Concluded Ken Duffey, also of the Alaska delegation, "It's going to take Canada getting past their definition of equity for this to work'."[71] Rebutted Athana Mentzelopoulos, spokesperson for the Canadian minister of fisheries in Ottawa, "The bottom line for the negotiations is flexibility'."[72]

This writer is firmly convinced, after having followed the tortuous path of the Pacific Salmon Treaty for fully fifteen years, that American and Canadian fisherfolk (i.e., the northern and southern sector fishery panels, the stakeholders) ought to determine the ultimate fate of the salmon resource. These commercial user groups, including the First Nations and Northwest Treaty tribes, are the true stakeholders in the salmon's future conservation. It is their turn to speak on an official level, then to act with binding authority, as the North Americans most concerned with the Pacific salmon's abundance and recruitment. Only then shall there be a fair and equitable sharing of the available salmon resource between the various North American user groups. The career politicians, top-gun officials, and big-shot bureaucrats from Washington, D.C., and Ottawa have failed their fisherfolk and the Pacific salmon long enough. The time has come for the true-blue salmon stakeholders to decide. And it will be the stakeholders who save the salmon.

NOTES

1. Ch. V, 'Canadian-American Relations and the Pacific Salmon, 1937-1983', Fisheries of the Pacific Northwest Coast, Vol. 1 (New York: Vantage Press, Inc., 1991), pp.109-123. NB: For those readers with little knowledge of the Pacific Salmon Treaty, this is a good place to start!

2. T.J. Doherty, 'U.S.-Canada Salmon Treaty Stalled', Pacific Fishing, August 1993, p.26.

3. Ibid., p.27. Quoted is Phil Eby of FVOA. (Brackets this writer's.)

4. T.J. Doherty, 'Pacific Salmon Treaty: An End to Coastwide Skirmishes?' Pacific Fishing, April 1994, p.32. Quoted is Pat Chamut, director of the DFO Pacific Region.

5. Ibid., p.32. NB: Treaty language is cited.

6. Ibid., p.32. Quoted is Pat Chamut, Canadian Commissioner.

7. Ibid., p.34. Quoted is Mel Seibel of the ADF&G.

8. Ibid., p.35. NB: See C.D. Bay-Hansen, Ch. IV, 'The Pacific Halibut', ibid., Vol.1, pp.53-78.

9. Ibid., p.36 (or last page). Quoted are Mel Seibel and Pat Chamut.

10. Jacob A. Mikkelborg, 'The Northwest "Fish War" of '94', Western Viking, Vol.105, No.26, 1 July 1994, pp.1-2.

11. Ibid., p.2. NB: Noyes Island in Southeast; near Craig, Alaska.

12. Ibid., p.2.

13. Jacob A. Mikkelborg, 'NW Fish War', Western Viking, Vol.105, No.29, 22 July 1994, p.5.

14. John O'Sullivan, [then] ed., 'Peripatetic Brian', National Review, Vol. XLVIII, No.17, 25 September 1995, p.6.

15. NB: Brian Tobin, Canada's former fisheries minister, got a good write-up earlier in FutureFish 2000 Chapter 2, Part III, Section A, 'The Collapse of the Atlantic Northwest Fisheries'.

16. Michael Parfit, 'Diminishing Returns: Exploiting the Ocean's Bounty', National Geographic, Vol.188, No.5, November 1995, p.19.

17. Marjorie Simmins, 'Canada's Brian Tobin steps down....leaving many of the country's fisheries problems unresolved', The Fishermen's News, Vol.52, No.2, February 1996, p.8.

18. Ibid., p.8.

19. Jerry Vovcsko, 'Filling Tobin's shoes', ibid., p.9.

20. The Fisherman, 24 January 1996. Cited in 'Bye, Bye, Brian', Alaska Fisherman's Journal, Vol.19, No.3, March 1996, p.2.

NOTES (cont'd)

21. Jerry Vovcsko, The Fisherman's News, 'Filling Tobin's shoes' p.9.

22. Bob Tkacz, 'Little Progress Reported with Canada', Alaska Fisherman's Journal, Vol.19, No.6, June 1996, p.18.

23. Ibid., p.18.

24. Ibid., p.18.

25. Pete Figueroa, '"Science" Hinders Salmon Treaty Talks', ibid., p.40.

26. Ibid., p.40.

27. Ibid., p.40. Quoted is Dr. Brian Riddell, Canadian co-chairman of the Treaty's Chinook Technical Committee.

28. NB: For invasions of chub mackerel (Scomber japonicus) into B.C. waters, see C.D. Bay-Hansen, ibid., Vol.1, Ch.VII, 'Pelagic Fishes'; pp.120, 144-146.

29. Pete Figueroa, '"Science" Hinders Salmon Treaty Talks', p.40.

30. Ibid., p.40.

31. Ibid., p.40.

32. Sam Smith, 'Pacific Salmon Treaty update', The Fishermen's News, August 1996, p.7.

33. Ibid., p.7. Author's note: This was to commence the next year, 1997, and the reader might already foresee Canadian displeasure with the Alaskan abundance-style plan.

34. Ibid., p.7. Quoted is Fred Mifflin of the DFO.

35. Ibid., p.7. NB: PSC numbers were gleaned from the bi-national Chinook Technical Committee.

36. Ibid., p.7. Quoted is Dr. Brian Riddell.

37. Ibid., p.11.

38. See note no. 12.

39. Bob Tkacz, 'Salmon Treaty '97', Alaskan Fisherman's Journal, Vol.20, No.1, January 1997, p.24.

40. Ibid., p.24. Author's note: Was editor Tkacz just grand-standing to his fellow Americans, or being intentionally uncharitable to Canada? Or both?

41. Ibid., p.88.

42. Ibid., p.88. Brackets this writer's.

43. Bob Tkacz, 'Stakeholders Get their Chance', Alaska Fisherman's Journal, Vol.20, No.3, March 1997, p.5.

NOTES (cont'd)

44. Ibid., p.5.

45. Ibid., p.5.

46. Ibid., p.5.

47. Charlie Ess and Brad Warren, 'Fishermen Tackle U.S.-Canada Salmon Accord', Pacific Fishing, Vol. XVIII, No.6, June 1997, p.39. Quoted is Alaska Governor Tony Knowles.

48. Bob Tkacz, 'Stakeholder Meetings Squashed by Canada', Alaska Fisherman's Journal, July 1997, p.8.

49. Ibid., p.8. Quoted is Joy Thorkelson of the UFAWU.

50. Ibid., p.8.

51. Ibid., p.9.

52. Ibid., p.9.

53. Ibid., p.9. Quoted is B.C. Premier Glen Clark. (Brackets this writer's.)

54. Rod Minott, KCTS-Seattle, 'Focus -- Salmon Feud', The Newshour with Jim Lehrer, WNET New York, N.Y., 1 August 1997, p.11.

55. See Chapter 2, Part III, Section A in FutureFish 2000.

56. Rod Minott, 'Focus -- Salmon Feud', p.10. (Brackets this writer's.)

57. Ibid., p.12. Quoted is B.C. Premier Glen Clark.

58. Robert Lewis, 'From the Editor: Politics and foreign policy', Maclean's Vol.110, No.31, 4 August 1997, p.2. NB: Lester B. Pearson, Nobel peace prize winner of 1957, served as prime minister of Canada from 1963-1968.

59. Chris Wood, 'Darn Yankees', ibid., p.12.

60. Ibid., p.14. Quoted is Foreign Affairs Minister Lloyd Axworthy.

61. Ibid., p.14. Quoted is Foreign Affairs Minister Lloyd Axworthy.

62. John van Amerongen, 'There's a tragedy here, but it's not the Commons', Alaska Fisherman's Journal, Vol.21, No.2, February 1998, p.16. NB: For an in-depth Analysis of "The Tragedy of the Commons", see Chapter 2, Part III, Section B in FutureFish 2000.

63. Op.cit., John van Amerongen, p.16.

64. Ibid., p.16.

65. Ibid., p.16.

NOTES (cont'd)

66. 'Salmon talks resume', Peninsula Daily News, 29 March 1998, p.1-7.

67. Ibid., p.A-7.

68. Ben Spiess, 'Salmon Treaty Wrangling Focuses on "Equity"', Pacific Fishing, Vol.XIX, No.6, June 1998, p.14.

69. Op.cit., Ben Spiess, p.14.

70. Ibid., p.14. Quoted is Bob Thorstenson of the Alaska delegation.

71. Ibid., p.14. Quoted is Kevin Duffey of the Alaska delegation.

72. Ibid., p.14. Quoted is Athana Mentzelopoulos, spokesperson for the Canadian minister of fisheries.

Part I: Oh, Canada!

....."But above the American Left and Right is Canada, a place free of the American dream and the European nightmare. No longer a colony, not yet an independent national power, Canada, like Switzerland, is The Peaceable Kingdom to which those weary of conflict go to escape the burden of a national destiny. As the Canadian historian John Conway has remarked: 'America is Faustian and Dionysian: [Apollonian] Canada is not."[73]

-- William Irwin Thompson, At the Edge of History, 1972

The thoughts and views of New Ager William Irwin Thompson are always interesting and entertaining, although mostly off-base and usually off-the-wall. In 1972, Thompson selected an inaccurate quote concerning Canada, too. For far from being Apollonian during the early 1970s, Canada appeared chthonian (dark, damp, forbidding) to this writer and his spouse and child living in Montréal, Québec, from December 1969 to March 1971. Canada seemed truly the land God gave Cain; a land fraught with dangers. The year 1970 was highlighted for the Bay-Hansen family by the birth of second son, Petter Moritz; but disturbing images of kidnapping and murder, Canadian minister Pierre La Porte and British envoy James Cross, Paul Rose and the FLQ also jolt the memory. A young man of twenty-six then and fresh out of the U.S. Army, the writer had ardently supported Québec sovereignty -- although sincerely deploring the terrorist tactics of the FLQ (Front de Libération du Québec).

On reflection these twenty-eight years later, the writer had backed the idea of an independent French Canada from a sense of historical and cultural justice. After all, New France had once encompassed a major portion of North America. Today, the only visible political remnants of the French presence on the vast continent are Québec and Louisiana -- arguably Canada's most interesting province and America's most interesting state. Early on (1789-1799), the frightful and appalling excesses of the French Revolution gave the Northeastern élites of the young American republic all the excuse they needed to return to their natural Anglophilia. Americans have forgotten that when the thirteen colonies confirmed their independence in 1781 at Yorktown (in southern Virginia), there were more French than American soldiers present -- and with a vigilant French navy awaiting right offshore.[74] In 1917, U.S. General "Black Jack" Pershing landed in Europe with the American expeditionary force and grandly declared, "Lafayette, we are here'."[75] That delayed statement of U.S. gratitude to France was, as Clyde Wilson of Chronicles has perceptively written, "...a nice touch. But everybody knew we had come to save the Brits and not the Frogs."[76]

The Frogs -- the French -- are an easy target in an increasingly Anglophone -- and Anglocentric -- globalist world of popular culture. For individuals like this writer, who was educated in England (U.K.) from 1952-1955, and later in New England (U.S.A.) from 1959-1965, Francophobia was an easy cultural path to take. After all, the French always lost to the English on the battlefield and at sea -- in Europe, Africa, South Asia, and North America. After the smoke of the nineteenth century finally cleared, the Union Jack flew proudly over all Canada; from Cairo to Cape Town, South Africa; over all India; Australia and New Zealand. The <u>tricouleur</u> hung limply over tiny Cayenne; over dusty places in North and West Africa; Indo-China, French Polynesia, and a few small islands in the Western Pacific. The French were defeated by the British at Québec in 1763 and at Waterloo in 1815. (To the adolescent male mind, winning is everything; losing nothing.) At the same time, the Roman Law of France was adjudged inferior to the English Common Law of Great Britain. (How lucky we Americans were that the Founding Fathers were transplanted Englishmen, who brought British law with them, along with the philosophies of John Locke, Thomas Hobbes et al.) In the Anglophone schools, academies, and colleges attended by the writer during his youth, the superiority of English institutions to those of France was drilled over and over, year after year, into this (then) young skull filled with mush.

But the young British or (Northeast) American student -- to become truly cultivated -- must learn the French language. As Thomas Fleming of <u>Chronicles</u> magazine has verily written, "French is an international language of European civilization, the language of one of the two greatest literatures since the fall of Rome."[77] (The other great literature, of course, being English.) Having had French language and culture forced down his throat since age eight years, this writer later regurgitated the whole mess during high school years. He also rejected, outright, the French existentialism of Camus, Sartre, and Genet; subsequently despised the deconstuctionist "French Thought" of Foucault, Lacan, and Derrida. All throughout his twenties, the writer was willing to join the Anglophile world in blaming the Frogs for everything from ineffectual politics and paranoid linguistics in Europe to botched terrorism and paranoid linguistics in North America.

During his thirties, however, the writer's Anglophilia started taking some severe head-shots and body-blows. The post-Swinging London and Mini-England of the 1970s was an ugly place of screaming tabloids and soccer hooliganism; the officially-celebrated Fab Four (the Beatles) were stale, house-warm, and John Lennon

(Member of the British Empire) took himself far too seriously. (The writer had never much liked the Early Beatles either; far preferring Eric Burdon and the Animals.) The United Kingdom and the United States, since the mid-1960s, had evolved into the evil (twin towers of London and Los Angeles) headquarters of both rock music and the global counterculture. During the 1970s and 1980s, the rapidly-expanding New Age movement was also strongest within the Anglo-American cultural rubric; spawning such New Paradigm sophists as David Spangler and Benjamin Creme, Marilyn Ferguson and William Irwin Thompson.

By the 1980s and 1990s, there was the all-too-visible British royal family -- the Windsors, "the royals" -- as an extra English irritant in Anglocentric popular culture. When the divorced Princess Diana perished in a "mysterious" Paris auto crash on 30 August 1997, the Anglophone/Anglophile world literally went bonkers and starkers. Elton John, yet another fading British rock star living in the U.S.A. (in his case, Atlanta, Ga.,) composed a special song for the deceased Diana, and flew from "Hot-lanta" to London to perform her eulogy (and make millions of U.S. dollars on the CD version.) On 8 February 1998, Queen Elizabeth II (partially as royal atonement) knighted the former Reggie Dwight as Sir Elton John; a bizarre case of the Queen knighting a queen under a shadow of the already-growing cult of Lady Diana, "the people's princess." Everybody who was Somebody in Gollywood attended the Westminster Abbey memorial service.

Far more serious has been the spread of the English language (mostly via the U.S.A.) as the linguistic vehicle for the cancerous spread of politically-correct newspeak. Recently, French author Jean Raspail (The Camp of the Saints, Scribners: 1975) has tactfully philosophised on the spread of [dumbed down] English as the sole global idiom:

"[If] English would have stayed in New England...the world would not be at the mercy of a single hegemonic language which is losing its original beauty in the global melting pot....[78]
...."Today..the recent phenomenon of globalization spreads like a tidal wave, fed by irresistible American expansion....I am convinced that the United States did not intend it to be so, but this inevitable globalization, which first overtook the old divided world of Europe, is naturally modeled upon the most powerful and dynamic nation in the West, the United States. The game is one-sided."[79]

At an October 1993 Studium Generale at Peninsula College (Port Angeles, Wash.), the topic was the 'Effect of NAFTA on Canadian/U.S. Relations'. The programme was presented by Dr. Mary Ann Hendryson of Western Wash. U. at Bellingham, and was mainly concerned with NAFTA (North American Free Trade

Association), Can-Am bilateral economics, and the recently-selected new prime minister of Canada, Kim Campbell. Economics outside the periphery of the fishing industry usually bore this writer, but Dr. Hendryson's characterisation of the Anglo-American-Canadian intra-family dynamics was both interesting and accurate. For Dr. Hendryson portrayed America and Canada as the two opposing daughters of a common mother, Britannia. Whereas the elder sister, America, was described as large, independent, and rebellious; the younger sister, Canada, was depicted as small, dependent, and obedient.[80] Just over two years after Dr. Hendryson's lecture, on 30 October 1995, the sovereignty referendum was held in Québec. (Canada would prove, as in 1970, that she had become a domineering matriarch herself, who could chastise her would-be-runaway child, Québec.) The "<u>Non</u>" votes (against secession from Canada) barely squeaked by the "<u>Oui</u>" votes (for Québec sovereignty) with a razor-thin plurality of 50.6%.[81] But this time, instead of sending federal troops into the wayward province, Canada used guile to discipline Québec. Late in the referendum campaign, the Canadian government purposefully rushed through citizenship status for recent arrivals to Québec province. <u>Monday Magazine</u> of Victoria, B.C. (not exactly an organ of the <u>Parti Québécois!</u>), quoted one Thérèse André, artist and sovereignist, as saying after the referendum went down to defeat:

"It's a scandal. It's a theft....
"What this means is that a [Canadian] passport is worth as much as a box of Cracker Jack's....
"[T]he [federal] machine managed to steal the referendum result since the number of new citizens in the last three months was greater than the final 50,000 or so votes that separated the Oui from the winning Non.
"We don't believe in the result. These people ["New Canadians"] received their papers with a letter saying it is their duty to conserve Canadian unity.
"And because many arrived from totalitarian countries, they interpreted their citizenship as being dependent on voting No.'."[82]

In some ways the Canadian government's victory was a Pyrrhic victory. The referendum results showed conclusively that fully half of all Quebecers, with a majority of Francophones, were (in 1995) disloyal to Ottawa. And Québec is easily Canada's largest province (bigger than Alaska), and second in size of population (after Ontario). But even as Prime Minister Jean Chrétien and the ruling Liberal Party would have to scramble to mend fences, Québec (i.e., the Frogs) lost the Canadian culture war too. The word was out from "First Nations" Crees, that were Québec to separate from Canada, they and other Native Tribes (in turn) would secede from <u>Québec Libre</u>. The shameless temptation of Ottawa largesse and honeypot federal entitlements were far too strong to resist; with the pc excuse being that Québec habitants were merely (other) White exploiters, and that French was just another Euro-colonial language.[83]

What had occurred in Canada two years previously, the election of 1993, had been every bit so divisive of national political and social fabric as the Québec referendum. For the election of 1993 had triggered the complete collapse of the "Tories", the Progressive Conservative Party of Canada. With the political ghost of former Conservative Prime Minister Brian Mulroney looking on, the Tories chose former Justice Minister Kim Campbell[84] as their standard-bearer, over frizzy-haired and charismatic Québec Conservative, Jean Charest. It was a stupid political blunder (the P.C. striving to be pc), but 1992 and 1993 were also the glory years of Fleetwood Mac's Bill Clinton and Al Gore south of the border. Kim Campbell, a blonde two-time divorcée from countercultural Vancouver, B.C., achieved infamy for doing the Twist (in a black cocktail dress) at an official Canadian cultural event during the summer of 1993...as interim prime minister and Conservative Party leader.[86] Kim Campbell's opponent was perennial Liberal Party hack, Quebecer Jean Chrétien. Although used-up and worn-out, Chrétien appeared solid as a rock compared to the dancing double-divorcée from flaky Vancouver, who looked very weak on "family values".[87]

In his fine book on Canadian politics, The Anxious Years (1996), Jeffrey Simpson referred to Jean Chrétien as "'yesterday's man'"; not once, but three times.[88] "Yesterday's man" had become a favourite epithet of the Tories for old "Grit" Party hound-dog, Jean Chrétien. Despite Campbell's gender and New Woman image, the Left-leaning Canadian press stuck to their accustomed Liberal Party/Big Government guns. Media bias, coupled with Chrétien's political savvy, contributed in large part to the shattering Tory defeat of 1993. Kim Campbell even lost her own seat in her home riding of Vancouver Centre.[89] The Progressive Conservative Party was totally devastated, with normally Tory votes going to the Bloc Québécois in the East, and to the Reform Party in the West.[90] It would be up to Québec's Jean Charest to put the PC Party humpty-dumpty pieces back together again. Author Jeffrey Simpson, columnist for the venerable Toronto newspaper The Globe And Mail, has perhaps best portrayed the political Jean Chrétien as Prime Minister in December 1995:

"Presiding...was the ultimate pragmatist, Jean Chrétien, who turned his former critics' taunts about being 'yesterday's man' into political virtue. Canada is Number One, he cried. Sure, we have a deficit problem, but let's not get our knickers in a knot over it. National unity. What, me worry? The United Nations says Canada is the 'best country in the world' in which to live. Me and the premiers, we go to China together as Team Canada, and we sign billions of dollars of contracts. I fly here and there and everywhere, and I get along with everyone. I crack up Bill Clinton and make [French President] Jacques Chirac smile. And I get along with the premiers. I keep all my Cabinet ministers in place. No scandals! My name isn't [Brian] Mulroney. And the people love me. If you doubt it, check the latest Gallup poll. What, me worry?"[91]

But in May 1997, before the soon-to-be-held Election '97, Canadian national media were still asking the question: "Is Jean Chrétien up to the job?"[92] The Canadian current wisdom responded "Surprise -- the answer [didn't] turn out to be either 'No' or 'Non'."[93] For despite Chrétien's shortcomings as leader in the Canadian public eye, Rightist sources were already calling for conservatives (Tories in the East and Reform in the West) to coöperate in the election <u>following</u> the up-and-coming Election '97. Western conservatives had already, apparently, conceded Election '97 to the Liberals, and could do nought else save vote to prevent the Bloc Québécois from forming the official opposition...again.[94] As the editor of Victoria, B.C.'s <u>Monday Magazine</u> has wryly observed:

....."[E]veryone else has been simply uninterested. Canadians won't bet on win, place and show until they see an actual horse race....
"The federal Liberals have taken this [zig and zag, left or right] approach into [E]lection '97 following the success of other Dodge City politicians like Bill Clinton and British P.M. Tony Blair. In power, such politicians are blandly inaccessible. In mid-campaign, they have no offense, but endless defense."[95]

If the Canadian federal government seems reluctant to let Québec go, Ottawa appears more than eager to divest Canada of the humongous Northwest Territories, north of the adjoining ten provinces. The N.W.T. are composed of the District of Mackenzie in the west and the District of Keewatin in the east, stretching across the vast upper continent from the Yukon border to the Davis Strait -- which separates Canada from Danish Greenland. Combined, if the huge islands of the far north District of Franklin are counted, the N.W.T. comprise a total land mass of well over a million square miles; more than twice the size of Québec, Canada's largest province. But Ottawa looks happy to amicably separate from a relatively untapped area the size of Argentina.

A mere 65,000 human beings live in the N.W.T.; mostly Inuits (i.e. Eskimos) in the east, and 50% Aboriginals (i.e., Indians) and 50% Whites in the west.[96] The 85% Inuit eastern half of N.W.T. -- the District of Keewatin -- with a population of 26,000, will be known as Nunavut; the western half of N.W.T. has yet (mid-1998) to find an ethnically-correct, non-European name.[97] The District of Mackenzie is therefore still called "Western Arctic", and both Districts will be officially cut loose from Mother Canada on All Fools Day, 1999.[98] (Sure as Christmas, to this writer, is that the 50-50 Aboriginal/White racial split in Western Arctic does not augur well for future allocation of political power. Perhaps that's a partial reason for Canada's voluntary no-fault divorce of the Northwest Territories, with their ice, rocks, and tundra? Or is it to compensate for Ottawa's death-grip on Québec, with her sparkling cities, green hills and verdant valleys, and St. Lawrence Seaway? Go figure!)

Beautiful, Supranational British Columbia

"We [Canadians] have a lot of transatlantic institutions but we don't have any transpacific ones. They need to be encouraged.'"
-- Earl Drake, former Canadian ambassador to China, on APEC summit 1997[99]

"Canadians should recognize the Pacific Region not as the Far East but as the New West."
-- Pierre Trudeau, former Canadian Prime Minister, late 1960s

At the dawn of the 1990s decade, the Asia-Pacific region accounted for approximately 43% of Canada's non-U.S. trade; almost 10% higher than total trade with Europe.[100] But for all her new-found Pacific Rim enthusiasm, Canada's Trade to Asia remained quite modest, with exports amounting to a meagre 10% of total (with Japan alone taking 5%). As the editors wrote in The Pacific Rim Almanac of 1991, Canada's Pacific Rim "trade fever' [was] still more prospect than...bonanza."[101] Japan's foreign investment in Canada, however, had been significant, with total investment in 1986 at $2.3. billion. In 1991, educated guessers estimated direct Japanese investment in Canada to be at $4 billion; with total bilateral Canada-Japan trade at $18 billion. This was twice as large as Canada's overall trade with Britain, and three times as much as her trade with Germany. Trade figures between Canada-Japan was expected to reach the $40 billion mark by A.D. 2000.[102]

In 1991, The Pacific Rim Almanac (accurately) predicted an intensified interest in Canada (i.e., British Columbia) by Hong Kong, resulting in a flight of capital and emigrants (i.e., rich refugees) in anticipation of the Hong Kong Handover of 1 July 1997.[103] Indeed, so numerous have the Chinese become in Vancouver,

B.C., that now (1998) some hometown wags call their city "Hong-couver". During the 1990s, Canada has seriously tried to become a Pacific Rim nation; by expanding her export-development programmes, opening new offices throughout the Asia-Pacific region, and strengthening her investment-promotion activities.[104] And in the autumn of 1989, on the very eve of the 1990s, the Canadian Cabinet approved a set of (quite) expensive initiatives (rather) grandly designated "Pacific 2000". The entire programme involved a federal commitment of a pending $65 million through 1994. Some of the "Pacific 2000" package contained components with self-explicatory labels: 1). "Pacific Business Strategy" (at $14 million); 2). "Japan Science and Technology Fund" (at $25.1 million); 3). "Asian Languages and Awareness Fund" ($14.7 million); and 4). "Pacific 2000 Projects Fund" ($11.4 million).[105]

This writer has stated that Canada has seriously tried to join the Pacific Rim community, that culminated in the APEC summit held during November 1997 at Vancouver, B.C. But Chinatowns -- no matter how exponentially big -- do not a Pacific Rim nation make. For Canada, unlike the United States, Australia, and New Zealand, is not really a Pacific Rim nation. British Columbia does not have the preponderance vis-à-vis Canada in the way California does to the United States. In Canada, the city of Vancouver and B.C. province may be number three respectively, but the U.S. states of Oregon, Washington, Alaska, and Hawaii further extend American claims as Pacific Rim nation. The U.S. Territory of Guam, and the Commonwealth of the Northern Mariana Islands (including Saipan, Tinian, and Rota), remain as U.S. Micronesian spoils from the Spanish-American War (1898) and the Great Pacific War (1941-1945). Guam and the N.M.I. also extend the political borders of the United States to the far Western Pacific.

What extends the westward (and natural) geographical borders of Canada, are British Columbia's 200 mile EEZ and the millions of magnificent Pacific salmon swimming out to sea from the province's mighty rain and snow-fed rivers. Indeed, with the warming Japan Current washing her shores on the west, and cut off to the east by the Rocky Mountains from the rest of Canada, the Pacific province often seems -- and feels -- very distant from the federal government at Ottawa. Separation from Canada (of a different kind than Québec separatism) has been a recurrent theme in B.C. political history. (When this writer and family moved to Victoria, B.C., in August 1982, a "Western Canada Concept" political pamphlet awaited in our new postbox.)

If Québec ever does secede from Canada, British Columbia might secede too; or, along with some western provinces, seek admission into the United States. Another scenario, as suggested by futurists Alvin and Heidi Toffler, could be the formation of a completely new political entity (recall Ecotopia?!) uniting Canada's western provinces with a number of U.S. northwestern states and possibly Alaska.[106] The Tofflers, as some other futurists, have foreseen a geopolitical and economic formula of sure success:

"Such a federation or confederation could start life with vast resources, including Alaskan oil [and fish!]; Albertan natural gas and wheat; Washington State's nuclear, aerospace, and software industries; Oregon's timber and high-tech industries; giant port and transport facilities serving the Asia Pacific trade; plus a highly educated work force. It could, at least in theory, become an instant economic giant with a massive trade surplus -- a key player in the world economy."[107]

No fear to be a French-speaking Canadian sometime-resident (or occasional tourist) in the backwoods of Beautiful British Columbia! You are first an Easterner, and therefore despised with typical Western truculence. But you are also a "Franco-Colombien" (if male); you are the reason that Canadian cereal boxes are bilingually labeled! In some parts of Supernatural British Columbia, all outsiders are suspiciously regarded. During early 1984, this writer took a fish-business trip to Port Hardy, B.C. (on the northern tip of Vancouver Island), ending up at the Thunderbird Inn. This was definitely the Great White North, proof of which came in the form of a human tooth embedded in this writer's T-Bird room floor-carpet. The writer was still drinking in 1984, and swiftly repaired to the T-Bird bar off the motel lobby. The angry-eyed local tipplers parked there didn't like the writer's drink order -- sweet Vermouth straight up (a Hvaler, Norway, mainstay) -- and even less his lingering New York accent. The writer quit Port Hardy after a couple of days convinced that Up-Island British Columbians were among the most unfriendly and impolite people he'd ever encountered.

Things didn't get much better (either) during a late 1984 fish-business trip to Campbell River (midway up the east side of Vancouver Island), "The Salmon Fishing Capital of the World". The would-be-entrepreneur, whom the writer met while there, was (also) originally from Norway, as was his grim-faced spouse. Not once, in two days of visiting, did the writer see that couple smile or hear them laugh. The whole journey turned out to be a Strindbergian experience in a deeply depressing town, where (mostly Native) local topers drank openly and abandonedly, in a land God truly gave to Cain.

The writer had sojourned extensively in Vancouver, B.C., during 1971, and had revisited in 1982. At first look, Vancouver was a stunning, world-class city -- on a par with London, Paris, or Montréal. At last look, Vancouver's kitschy-<u>chic</u> West End was swarming with prostitutes, and American-type street people had even taken over classy Robson Street ("<u>Strasse</u>") after dark. Sixteen years later (1998), Vancouver has devolved into Amsterdam-on-the-Fraser. A combination of Left Coast counterculturalism, and Euro-Canadian-style permissiveness, has created the northern Babylon-by-the-Bay (the southern being San Franciso). Today Vancouver, B.C., reigns as the legalised pot capital of North America. In 1998, columnist Don Feder has reported, Vancouver boasts the largest needle give-away programme in North America.[108] No less than 2 million syringes have been distributed each year since the programme started in 1988, with HIV prevalence among intravenous drug users skyrocketing as a result; rising from between 1% and 2% all the way up to 23%. Deaths from drug overdoses have increased 500%, and now (in 1998) Vancouver, B.C., holds the dubious distinction of having the highest heroin death rate in North America.[109]

In 1998, pc B.C. also became the very first province in Canada to <u>voluntarily</u> grant pension benefits to homosexual "couples" who are public employees.[110] In response, Focus on the Family Canada spokesman John Sclater said that he was more worried by the B.C. government's definition of a wife and family, rather than extended pension benefits to gays. Stated Sclater:

"We obviously have a government here that really desires to be seen to be out in front of everybody in terms of redefining spouse, marriage and family....
"We think we're into very unchartered waters here. We think there's something rooted in the biology and dynamic of a heterosexual couple. We have to uphold the ideal that most people say works best.'"[111]

This writer was not surprised when the B.C. provincial government started paving the way in 1997 for same-sex partners to be defined as "spouses". An example of how far British Columbia has gone officially pc, was amply proven on a CBC newscast aired 16 March 1998. The B.C. Teachers Union was shown holding an "anti-homophobia, anti-heterosexualism" rally (sit-in). The Baby-Boomer and Thirtysomething teachers were unanimously in accord and overwhelmingly Anglo-European in ethnicity; the handful of ["How <u>dare</u> they?"] dissenters were mostly New Canadians, with a sprinkling of East Asians. A few more enlightening years in the Great White North would change their [narrow] minds, was the pervasive sentiment.

The writer and his family lived from 1982-1987 in Victoria, B.C., on Vancouver Island -- Canada's "Pacific Island". Despite the great displays of <u>faux</u>-English High Tea at the Empress Hotel, and the pretentious "Tweed Curtain" separating Oak Bay from Downtown, Victoria really wasn't very English at all; indeed, British Columbia wasn't very British save in name. But both the capital and the province worked very hard at developing a cultural state of hybrid Anglo-Canadianism. (The big government broadcast medium, CBC, aided in this process; as did PBS, its élitist Anglo-American <u>doppelgänger</u> south of the border.) When the few home-grown institutions failed in Anglophone Canada (as they often do), Dear Olde England -- along with her old and tired accoutrements -- was trotted forth as cultural rôle model. But since the mid-1960s, Great Britain had been mini-England, and has reigned from that era (along with California) as world headquarters of the global counterculture.

So when the writer read during the summer of 1998 of a fast-rising suicide rate on Vancouver Island, he was not surprised then either. The rash of suicides have been attributed to job losses in the wood products and fishing industries.[112] One of the hardest hit of the Up-Island communities has been Port Hardy, B.C. Throughout the winter of 1997 and spring of 1998, there have been four suicides and fully 29 suicide attempts in Port Hardy; a town of 6,000 which usually averages no more than a single suicide per year.[113] According to the coroner covering far-northern Vancouver Island, the suicides were all men ranging in age from the early 20s to the mid-40s[114] (the ages for MTV and Monty Python). This writer is absolutely convinced that, although wide-spread unemployment in forestry and fishing triggered the incidents of [attempted and completed] suicide at Port Hardy, a meaningless and valueless Godless cultural ecology first set the stage for this latest tragedy.

But take heart! The euphoric state of Ecotopia is nigh! Coming soon, to radio-listeners throughout the depressed northeastern Up-Island area, will be the first all-cetacean station![115] Starting Wednesday, 22 July 1998, the eco-sophists at the Vancouver Aquarium were slated to broadcast the live sounds of killer whales communicating within their pods. The whistles, squeals, chirps, and clicks have been monitored via a subaquatic microphone at Telegraph Cove near Robson Bight since 1984, and are now set to go big-time LIVE over channel CJKW-FM 88.5.[116] ORCA-FM is only the first step in a grand plan, WhaleLink, which aspires to mount more hydrophones along the coastal sea-roads travelled by "whale-killers", from the Queen Charlottes in the north to the San Juans in the south.[117] As whale watching, now with whale listening, might soon be the sole marine activity permitted by law in the Can-Am Ecotopia, let us pray that ORCA-FM shall convey its soothing eco-message that less is more; to convince those legions of unpaid and out-of-work, formerly proud members of the Northwest Coast's forestry and fishing industries. Those unemployed workers shall soon be replaced by Spandex-clad, cellphone-carrying, hiking and biking, whale-watching (and listening) eco-tourists.[118]

Meanwhile, who will fish for a living, fish for the tables of consumers, providing food for their families and for us?

NOTES

73. William Irwin Thompson, <u>At the Edge of History</u> (New York: Harper Colophon Books, 1972), p.104. Thompson cited William Kilbourn, <u>Canada: A Guide to the Peaceable Kingdom</u> (Toronto: 1970). NB: Brackets this writer's.

74. Clyde Wilson, editorial, <u>Chronicles</u>, Vol. 22, No. 4, April 1998, p.6.

75. <u>Ibid</u>., p.6. Quoted is U.S. General John J. Pershing (1860-1948), commander of the American Expeditionary Force in World War I.

76. Op.cit., Clyde Wilson, p.6.

77. Thomas Fleming, 'The Heart's Geography' <u>ibid</u>., p.12.

78. Jean Raspail, 'Defending Civilization', <u>ibid</u>., p.15.

79. <u>Ibid</u>., p.16.

80. Dr. Mary Ann Hendryson, 'The Effect of NAFTA on Canadian/U.S. Relations', <u>Studium Generale Program Notes</u> (Peninsula College, Port Angeles, Wash.), 7 October 1993.

81. Lyle Stewart, 'Fear and loathing on the referendum trail', <u>Monday Magazine</u> (Victoria, B.C.), 2-8 November 1995, p.4.

82. <u>Ibid</u>., p.4. Quoted is sovereignist Thérèse André. NB: Brackets this writer's.

83. C.D. Bay-Hansen, Letter to the Editor, 'Canada dysfunctional', <u>Peninsula Daily News</u>, 26 November 1995, p.A-10.

84. Jeffrey Simpson, <u>The Anxious Years: Politics in the Age of Mulroney and Chrétien</u> (Toronto: Lester Publishing Limited, 1996), p.184.

85. Kim Campbell, <u>Time and Chance: The Political Memoirs of Canada's First Woman Prime Minister</u> (Toronto: Seal Books, 1997), p.378.

86. <u>Ibid</u>., p.338.

87. Jeffrey Simpson, p.186.

88. <u>Ibid</u>., p.79.

89. Kim Campbell, p.397.

90. Jeffrey Simpson, p.125.

91. <u>Ibid</u>., p.85. NB: Simpson depicted Chrétien in a 20 December piece written several weeks after the Quebec referendum of 30 October 1995. (Brackets this writer's.)

92. James MacKinnon, 'Swing party', <u>Monday Magazine</u> (Victoria, B.C.), Vol.23, Issue 20, 8-14 May 1997, p.6.

NOTES (cont'd)

93. Ibid., p.6. Brackets this writer's.

94. Ibid., p.6.

95. Op.cit., James MacKinnon, p.6. Brackets this writer's.

96. 'Dividing northernmost Canada no small task', Peninsula Daily News, 14 June 1998, p.C-4.

97. Ibid., p.C-4.

98. Ibid., p.C-4.

99. Earl Drake, now (1997) a business consultant and academician in Vancouver, B.C., cited by Chris Wood, 'Vancouver hosts the APEC extravaganza', Maclean's, Vol.110, no.46, 17 November 1997, p.66

100. Alexander Besher, ed., The Pacific Rim Almanac (New York: HarperCollins Publishers, 1991), p.191.

101. Ibid., pp.191-192.

102. Ibid., p.192.

103. Ibid., p.192.

104. Ibid., p.192.

105. Ibid., pp.192-193. Source for The Pacific Rim Almanac: External Affairs and International Trade Canada.

106. Alvin and Heidi Toffler, War and Anti-War: Survival at the Dawn of the 21st Century (Boston: Little, Brown and Company, 1993), p.207.

107. Ibid., p.207. Brackets this writer's.

108. Don Feder, 'Needle-Exchange Programs Encourage Drug Addiction', Human Events, 29 May 1998, p.16.

109. Ibid., p.16.

110. 'B.C. grants same-sex benefits', Peninsula Daily News, 24 June 1998, p.A-8.

111. Ibid., p.A-8.

112. 'Suicide rate jumps on Vancouver Island', Peninsula Daily News, 17 June 1998, p.A-3.

113. Ibid., p.A-3.

114. Ibid., p.A-3.

115. 'It'll be killer radio when orca whales go live', Peninsula Daily News, 16 July 1998, p.A-6.

NOTES (concl'd)

116. Ibid., p.A-6.

117. Ibid., p.A-6.

118. See Chapter 1, Part III, in FutureFish 2000.

Part II: Alaska, the Last (Fishing) Frontier

"Alaska is now ours [U.S. 200-mi.EEZ], and we've taken good care of her, too. While fisheries collapse all around us, Alaska's resources are stronger than ever: more salmon, more cod, more halibut, more flatfish... It hasn't all been perfect, but Alaska is one fishery of which we can be thankful and proud."
-- Roger Fitzgerald, Seafood Leader, November/December 1995[119]

Thus spake 'Fitzgerald on [Alaska] Fish' in late 1995. And whatever one might think of Alaskan tall [fish]tales, Seafood Leader's Fitzgerald was right: ...."It hasn't all been perfect, but Alaska is one fishery of which we [Americans] can be thankful and proud." Even though Roger Fitzgerald entitled his editorial 'Still the Last Best Place', the inference was that Alaska was Still the Last Best Place...to fish.

During 1995 and 1996, there was much tough talk (and writing) about America's Last Best Place, but it usually meant the Cowboy West; in particular the Treasure State of Montana. Indeed, one William Kittredge (along with an Annick Smith) in 1995 edited The Last Best Place: A Montana Anthology.[120] The grandiose claim in the title of the Montana Anthology, however, was amply disproven during the following year, 1996 -- the year of the Unabom[b]er,[121] the Freemen, and the Fondas (Ted and Jane). Such Montana places as Lincoln, Roundup, Jordan, and Brusett became (temporarily) infamous in the American popular culture of 1996.

All of which has left Alaska as the Last Best Place. Since 1995, anyway (and decades before), Alaska has surely been the Last Best Place to Fish. But before extolling the fisheries of the Great Land, this writer must cite a negative note. Not all visitors to the state of Alaska have been so overwhelmed by the Last Frontier as the locals assume. British author Simon Winchester was decidedly underwhelmed by the "dreary little town"[122] of Ketchikan, Ak., in which he sojourned during the 1980s:

"'Spread wings of a prostrate eagle', the town's [Ketchikan] name supposedly meant, according to a brochure which obligingly published a translation of Tlingit Indian vernacular: but it looked too dreary a place to have much association with eagles, what with the endless drizzling from a slate-grey sky and the all-pervading smell of sour lumber steam, glue and fish. There were perhaps a dozen streets, all of which either ended in the sea or petered out into the pine forests with which Revillagigedo Island was wholly covered. There was a pulp mill at one end of town, belching steam; there was a sawmill at the other with its pepperpot-shaped sawdust furnace oozing smoke; a clutch of orange-painted fishing boats hugged the small dockside, clustered beside a rusty old coaster settled deep in the water from which fresh supplies -- boxes of oranges and bananas, cheese and chocolate -- were being unloaded. It looked like a frontier town, connected only by air and by sea to a country of which, for most of the time, it can hardly have seemed an integral part."[123]

But snobby Englishman Simon Winchester shouldn't have let outward appearances fool him. Their dreary little towns aside, Alaskans are savvy and plenty smart. And like other Americans, residents of the 49th state are sophisticated in the ways of law and finance. (See Chapter 3, Part I). They are accustomed to affluent and educated outsiders who come to Alaska thinking they can hoodwink the local "rubes". A perfect example of rôle-reversal occurred in the aftermath of the great Exxon Valdez oil spill of Good Friday, 1989 (twenty-five years to the day after the Alaska earthquake of 1964). As most readers will recall, the Exxon Valdez oil tanker was run aground on Bligh Reef in Prince William Sound, South-Central Alaska. The negligent and intoxicated skipper, Joseph Hazelwood, thereby caused the most destructive oil spill in history: 11 million gallons of crude spilled into Prince William Sound, polluting about 1,500 miles of Alaska shoreline.[124] There followed, naturally, a great hue and cry, with the inevitable fingers of blame pointing at the evil Exxon Corp. Throughout the early 1990s, ecological disaster wing-nuts and environmental Cassandras had their say and (dooms)day in the sun, all at Big Oil's expense. Finally, on Friday 16 September 1994 at Anchorage, an eleven-member jury returned a verdict after more than twelve days of deliberations: The federal jury verdict ordered Exxon Corp. to pay $5 billion U.S. in punitive damages for the Exxon Valdez oil spill -- the second largest award ever in a civil case.[125]

The watching, waiting, and listening local-"yokels" (savvy Alaska Natives, smart commercial fishermen, and sophisticated property owners), estimated at between 12,000 to 14,000 in number, had originally demanded three times the amount of $5 billion that Exxon was to be penalised. The Alaskan plaintiffs, reported the Associated Press, had based their figures on Exxon Corp.'s own statements of annual profits.[126] If the verdict/award handed down on 16 September 1994 was beginning to look like a legal lottery, a prior Alaska courtroom shenanigan was assuming the appearance of outright judicial chicanery. An article, appearing 2 August 1994 in the Kodiak Daily Mirror, stage-whispered that:

"Jurors in the Exxon Valdez case won't be disturbed with information on a surprisingly robust pink salmon run in Prince William Sound, a federal judge ruled Monday.
"U.S. Di[s]trict Judge H. Russel Holland said in his order that he would deal with the issue after the jury decided how much commercial fishermen should be awarded for actual damage done to their fisheries by the 1989 oil spill....
"Preliminary numbers show that 150 boats at the Solomon Gulch Hatchery near Valdez have netted 14 million pink salmon. Only 5.5 million fish were expected after last year's [1993] harvest of 5.8 million, the area's lowest since 1978.
"Shhhhh!'"[127]

Indeed, the Associated Press had reported the comeback of pinks in Prince William Sound so early as July 1994.[128] And new environmental information would be forthcoming to further dampen Alaska plaintiff high spirits. A New York Times article of December 1995 reported that a new federal investigation had found that a high percentage of the tar balls, still sticking to the coastline and islands in and around Prince William Sound, did not originate from the Exxon Valdez oil spill.[129] Federal investigators discovered that the few but widely scattered tar balls were, in fact, from California-imported oil.[130] Dr. Keith A. Kvenvolden, of the U.S. Geological Survey, said that the California tar balls most probably had resulted from the Alaska earthquake of 1964, which had destroyed much of the port city of Valdez, along with storage tanks containing paving asphalt and marine fuel oil.[131] Dr. Kvenvolden explained that, although the California tar balls posed no real ecological danger, his research had proven the longevity of oil spill "residence time[s]".[132] But, coupled with his (highly) likely earthquake hypothesis, the forthright Dr. Kvenvolden concluded: "A lot of Alaskans didn't want to hear what we had to say, because they were legal claimants against Exxon.'"[133]

Question: Had the Exxon Valdez oil spill devolved into a shameless legal lottery for Alaskan plaintiffs? Even acerbic neo-Conservative humourist, P.J. O'Rourke, commented in 1994 that "Exxon had to spend $2.2 billion cleaning up after the Valdez. It paid an additional $800 million to Alaska and the federal government and, as of this writing, still faces $1.5 billion in civil lawsuits. That's $4.5 billion Exxon could have spent reducing the price of home heating for the poor...."[134]

By mid-1995, Exxon Corp. funds were already being put to good use for the people and state of Alaska. Environmentalist periodical, Common Ground, reported in its May/June 1995 issue that the permanent protection of Kodiak Island ecosystems was now owed, "ironically", to the 1989 Exxon Valdez oil spill[135] The Exxon Valdez Oil Spill Trustee Council, which controlled the use of fines paid by Exxon Corp., approved release of $89 million (get the rough justice?) to purchase 270,000 acres from Native corporations (no doubt represented by countercultural non-Native attorneys) on Kodiak Island. The Conservation Fund (certainly infested with ec Big Green-agenda lawyers) "helped negotiate" the acquisitions, announced Common Ground, which were then to be added to Kodiak National Wildlife Refuge.[136] Surely, in reasonable retrospect, the Exxon Valdez oil spill of 1989 might be seen as one of the best things to ever happen to the people and state of Alaska? If little can be learned from the aftermath of the 1989 oil spill -- with its tawdry legal high-jinks -- the neighbouring Canadians and Americans of the Lower 48 are now convinced of the high level of "Sourdough" sophistication. If nothing else, Alaskans came of legal age during the first half of the 1990s.

In 1998, this writer has lived 14 years in Washington State (1975-1978, 1987-1998); 4½ years in British Columbia (1982-1987); and exactly zero sum years in Alaska. Except for a single trip to Anchorage in 1980, the writer knows only of 1990s Alaska from books and magazines on the fishing industry; and from fishermen friends transplanted to Port Angeles, Washington. There were also the Norwegian-American "Alaska-menn" residing in Ballard (north Seattle) during the 1970s, with whom the writer was personally acquainted (see `FutureFish 2000`). There are, too, the writers ex-"partners" in the fish business -- one from Petersburg, the other from Fairbanks -- who have remained virtually invisible since 1990, after being soundly trounced in a fraud arbitration court case at Port Angeles. Alaskan tellers of tall tales notwithstanding, the Great Land has always loomed as an extra special place to this writer. Alaska the Last Frontier retains a revered niche in the continuing annals of Norwegian Americana, imbued with a near-mythic Norse-folk quality.[137] And, as columnist Roger Fitzgerald has observed, "Alaska is one fishery of which we can be thankful and proud."[138] Alaska means far more, then, than just a "fishery", but Alaska most assuredly reigns as Still the Last Best Place to fish off North America. (Maybe off both the Americas! Regarding South America, this writer can only think of the pure, cold waters off southern Chíle as a suitable contender for the intercontinental Last Best [Fishing] Place.)

Much has been written on the Alaska fisheries. There is nothing this writer could possibly add to the history, lore, or wealth of stories about the Pacific salmon, Pacific halibut, or king crab fisheries of Alaska. (Alaska has dominated the Northwest Coast fisheries for so long, that Pacific salmon is often known as Alaska salmon, Pacific [walleye] pollock as Alaska pollock, sablefish as Alaska [black] cod etc.etc.) Only those who have themselves experienced the Alaska fisheries (there are many and some write well) are able to add their own personal dimension to arguably the toughest and most dangerous work at sea anywhere (hence the sky-high pay). As merely a student of the Alaska fisheries, this writer owes a debt of gratitude to his teachers -- authors/fishermen Robert J. Browning, Francis E. Caldwell, Russ Hofvendahl, William B. McCloskey, Jr., Joe

Upton, Spike Walker and others. In 1978 the writer had the opportunity of migrating to Petersburg ("Little Norway"), Alaska, and enter the seafood business in some capacity; he chose instead to earn an M.A. in Pacific history at the University of Hawaii at Manoa. So twenty years later, it is as historian rather than fisherman or processor, that this writer adds his 2¢ worth of words on the Alaska fisheries. (And has taken the liberty of including his own two previous works on the Pacific Northwest Coast fisheries in the Bibliography. However modest, together the two slim volumes took seven years to research and write; done on the writer's own time, effort, and money.)

In the North Atlantic, New England has a fished-out George's Bank; the Maritimes and Newfoundland have the not-so Grand Banks, and the Danes still fish the icy waters around Greenland. In the North Pacific, the Russians fish the Sea of Okhotsk and Bering Sea off Kamchatka; the Japanese (until 1945) used to fish, unimpeded, the teeming pelagic waters around Nan'yō gunto, the Micronesian islands in the western and mid-Pacific. With "Americanization" of the northeastern Pacific since 1976-1977, and a 34,000-mi. coastline (nearly 2/3 that of the Lower 48) with hundreds of fjords, inlets, islands and skerries directly on the warm-water Japan Current (kuroshio)... Alaska is the U.S. commercial fisherman's dream-come-true. The only stumbling block to Alaska fishermen is control; international, federal, and local -- the renowned Alaska Department of Fish and Game. The ADF&G manages all fisheries in state waters, with management in the EEZ (3-200 mi. offshore) handled by the NMFS (National Marine Fisheries Service) and the NPFMC (North Pacific Fisheries Management Council).[139]

The primary fish of Alaska -- both commercial and sport -- is, and always has been, the king (chinook, spring, tyee) salmon. All salmon management with Canada is, as has been noted, negotiated through the PSC (Pacific Salmon Commission).[140] Beyond the 200 mi.-EEZ, salmon harvests are now authorised by NPAFC (the North Pacific Anodromous Fish Commission), formed in 1993 to replace the old INPFC (International North Pacific Fisheries Commission), which regulated salmon from 1957 to 1992.[141] The new Commission consists of Canada, Japan, the Russian Federation, and the United States (viz. the Pacific salmon-producing countries). The NPAFC Convention prohibits high-seas salmon fishing and trafficking in illegally-caught salmon. The NPAFC was further empowered in 1993 (activated 1 January) by United Nations General

Assembly (UNGA) Resolution 46/215, that bans -- outright -- any large scale pelagic driftnet fishing in the Pacific (or in any of the world's oceans and seas).[142]

Alaska has appeared flooded with salmon throughout the 1990s, despite the sad statistics (emanating from the U.S. Department of Commerce) of Alaska salmon becoming "overutilized".[143] While southern British Columbia and the Pacific Northwest states bemoan the loss of salmon habitat and recruitment, Alaska harvests so much salmon that U.S. seafood corporations have a hard time selling the overflow. The glut of Alaska salmon has come at a time when America's no. 1 seafood customer, Japan, is strapped for cash. Japanese consumers are looking for lower-priced protein-rich food -- like beef.[144] For Alaska, this is a potentially catastrophic market development. The Japanese eat circa 80 lbs. of fish per person a year (Americans consume approx. 15 lbs. per capita), and, after squid (ika), their next favourite is salmon (sake, zake). In 1994, Japan imported 300,000 tons of salmon, and they consumed close to a record 500,000 tons of salmon the following year.[145]

The Japanese consumer, however, couldn't have been buying Alaska salmon during 1995, for the simple reason that the U.S. seafood corporations in Alaska -- along with the U.S. subsidiaries of Japanese seafood corporations in Alaska -- weren't able to sell the record haul of Bristol Bay [red] sockeye (beni-zake) in Japan.[146] And prices (a perennial problem) were more than 40% lower than they had been the previous year. The sudden downturn crash caught Alaska processors completely unawares. They had already paid many fishermen the large amounts of money anticipated, so now faced huge losses if forced to sell to Japan at then-current bottom-dollar prices.[147] The Associated Press vividly described the situation in the summer of 1995:

1) "Usually, by this time in the season, some 30 to 50 percent of the Alaska red [sockeye] salmon harvest already is sold in Japan, providing a surge of cash that Alaska processors use to see them through the seasons.
 2) "This season, processors say about 10 percent [had] been sold. Trampers full of Alaska fish...[arrived] in Japan, and [U.S.] industry officials...[scurried] over in largely futile efforts to find some place to sell the fish. Lacking buyers they [were]...forced to stockpile the salmon in expensive cold storage and hope the market turn[ed] up.
 3) ...."Last year [1994] the Bristol Bay run surged late, and Japanese buyers -- concerned the season would be a bust -- pushed prices sharply higher. Fishermen benefited with prices that started at 50 cents a pound and ended up at just shy of $1 a pound.
 "This year [1995], processors paid 70 cents a pound."[148]

Alaska salmon was even the topic of a lead article on the front page of a 4 September 1996 edition of The Wall Street Journal.[149] To his credit, the WSJ staff reporter got right down to the three main obstacles to marketing Alaska salmon:

    1. Eco-frightened consumers. Although commercial fishermen caught a record 217 million salmon during the four-month fishing season of 1995, many eco-scared U.S. consumers weren't buying salmon after hearing reports of salmon demise in Lower 48 rivers like the Snake and the Columbia. Meanwhile, "Alaska is awash in salmon"[150] (which remained unsold).

    2. Competing salmon farms. The rapid growth of salmon-farming in Norway, Canada, Chile, and the U.S. states of Maine and Washington. Fish-farming has translated to fewer Alaska fishermen, plus a smaller market share for those remaining in the industry. A dozen years ago, for example, Japan opened her own chum salmon-hatchery programme. By 1995, the Japanese had produced 70 million lbs. of chum -- more than thrice the Alaska catch of that year.[151] (So home-grown chum was what the Japanese were buying in 1995!)

    3. Plummeting prices. In 1996, approximately half the world's sales of salmon came from farmed fish, but Alaska continued to have huge runs of pink salmon. The very volume of pinks drove down prices while worsening the salmon glut: "So many fish were landed that four million pounds of salmon ended up parked in refrigerator vans on the Seattle waterfront after cold-storage operators ran out of space."[152]

On a subject related to cause no. 1 (above) many enviro-conscious American consumers mistakenly assume that Alaska salmon are endangered too. During March 1996, the Alaska Seafood Marketing Institute (ASMI) took a poll among shoppers around the United States, finding that so many as 2.8 million households believed that salmon -- everywhere -- tottered on the very edge of eco-extinction. One Grant Trask, a Petersburg fisherman who took part in a winter 1995-1996 Alaska salmon promotion trip, told The Wall Street Journal that he was "dismayed" to read an article by a Midwestern food writer urging her ec readers to purchase farmed salmon. Retorted an angry Grant Trask, "'We're awash in salmon and she [the food writer] is telling readers the world is running out of them.'"[153] Gunnar Knapp, director of U. Alaska's Salmon Market Information Service, has confirmed the salmon irony: "Someone does a report on the Columbia River salmon disappearing, and people end up thinking that the salmon they're looking at on the supermarket shelf is the last one in America.'"[154]

During the 1990s, Alaska landed the all-time record number of salmon in 1991 with 189 million fish; the value of the state-wide harvest (314,200t) in 1992 estimated at U.S. $575 million.[155] Indeed, the total value of all fishery products exported from the state of Alaska in 1992 was U.S. $1.884 billion (not including the $652 million worth of seafood exported from Washington State -- the [sea]lion's share of which was caught in Alaskan waters).[156] Fishing in Alaska has always been a traditional part of the Last Frontier's heritage; in the 1990s, fishing has contributed greatly to the outdoor recreation, [sea]food supply, and economic well-being of the 49th state. The fishing industry has also been, according to the NOAA (1992), the largest non-governmental employer in the state of Alaska.[157] The Alaska fisheries were also given an enormous boost after the safe passage through Congress of the Magnuson Fisheries Conservation and Management Act (MFCMA) of 1976. Until "Americanization" (the now U.S. 200-mi. EEZ), the Alaska groundfish fisheries -- except for that of the Pacific halibut -- were dominated by foreign vessels; hook, line, and sinker.[158] Implementation in 1977 of MFCMA completely skewed the dynamics of the North Pacific fisheries. By 1992, the combined [domestic] ex-vessel revenues of the Alaska groundfish fisheries had reached an amount in excess of $U.S. 655 million.[159]

When this writer lived in Ballard during the mid-late 1970s, Alaska "bottomfish" (excepting the mighty Pacific halibut) were discussed by Seattle fishermen in a deprecating or offhand manner. (Only the halibut skippers retired on Sunset Hill seemed to know anything about "bottomfish", or consider them significant.) Pacific (grey) cod, Pacific (walleye) pollock, sablefish (Alaska/black cod), rockfish, flatfish -- none of the above figured very prominently in tall Alaska fish tales. The Magnuson Act changed all that. As noted, of the $1.884 billion earned by Alaska fishery products exported in 1992, fully <u>one third</u> was from the lowly groundfish.[160] Not only that, but the very success of the all-American "factory-trawler" fleet (many of the vessels being the huge at-sea processor "mother-ships") had added to the growing problem of resource waste in the Alaska industry. The Associated Press reported in 1994 that a study, conducted for the ADF&G by a Juneau consulting firm, concluded that over 740 million lbs. of edible fish (mostly Pacific cod, Pacific pollock, and other groundfish) had been summarily dumped overboard (largely by a "voracious" trawler fleet) during 1993.[161] The discarded groundfish, the study asserted, could have provided 46 million fish dinners -- a full meal-a-day for every single resident of the state of Alaska for nine whole months![162] Obviously, for better or for worse, the Alaska groundfish fisheries were booming in the 1990s. (Indeed, so much fishing pressure had been put on Pacific pollock [by the Russians, Poles, Chinese, and Koreans], that coöperative multilateral research and negotiations closed the pollock fishery in the international zone of the central Bering Sea [i.e., the "Donut Hole"], for 1993 and 1994, to allow stocks to rebuild.)[163]

The reader (if not a fisherman) might have noticed from NOAA numbers (see note no. 160), that the

1990s groundfish bonanza has been found in the Bering Sea/Aleutian Islands rather than in the Gulf of Alaska. This is because nearly half the Bering Sea overlays the Pacific continental shelf of North America, an area officially known as the Eastern Bering Sea (EBS).[164] Almost half a million square miles in magnitude, the EBS continental shelf is one of the globe's most productive marine resource regions; recruiting 15 million tons and more of groundfish, plus millions of tons of salmon, crab, and other species.[165] According to the North Pacific Fishery Management Council (NPFMC) in 1995, the Eastern Bering Sea and Aleutian Islands (BSAI) had a biomass of 6.8 million tons of Pacific pollock; 2.38 million tons of yellowfin sole (Pleuronectes asper); 2.23 million tons of rock sole (Pleuronectes bilneatus); 967,000 tons of Pacific cod; 725,000 tons of flathead sole (Hippoglossoides elassodon); and 623,000 tons of Alaska plaice (Pleuronectes quadrituberculatus).[166] Flatfishes are the species of Bering Sea abundance, despite the problems of by-catch associated with some species (viz. rock sole), poor quality of others (viz. arrowtooth flounder/turbot), lack of market acceptance, and meaningless blanket groundfish quotas (i.e., the "2 million-ton cap").[167] In 1995 John Iani, vp of UniSea Inc. of Redmond, Wash., and Dutch Harbor, Ak., simply put it all together:

"We've done a real good job of harvesting and marketing pollock. Now we need to work on flatfish and harvest what's out there...."[168]
"The fish [yellowfin sole] is out there; let's use it.'"[169]

The most famous flatfish of them all, the great Pacific halibut (Hippolossus stenolepis), has long supported a proud and important traditional fishery in the waters off the Pacific Northwest Coast. This writer has extensively researched and written on H. stenolepis, both the resource and the fishery, in a former work.[170] The NOAA reported in 1993 that the Pacific halibut, managed by the International Pacific Halibut Commission (IPHC), was fully utilised and "generally in good condition"; with recent U.S. harvests that averaged 31,000t worth $70 million ex-vessel.[171] Other catches (than commercial) of Pacific halibut in 1992 were 4,000t taken in the sport fishery; unreported catches totaling 650t taken for personal use; a "wasted mortality" of 1,450t due to discard and fishing by lost gear (i.e., "ghost nets" etc.); and an "incidental catch mortality" of 9,260t taken by fishermen targeting other species.[172] In 1992, approximately 6,273 commercial U.S. and 435 Canadian vessels were licensed to fish Pacific halibut, with exploitable stocks having apparently peaked at 200,000t during 1986-1988 (NOAA figures).[173] The Pacific halibut during the 1990s has been fully utilised and vigilantly watched by the IPHC.

And so we come to the king crab fisheries of Alaska, a topic which has been very well documented over the last twenty years. This writer has specified the king crab fisheries of Alaska. The almost tiny-by-comparison Dungeness crab (Cancer magister) of more southerly fisheries, is a genus Cancer crab. The two Tanner crabs, Chionoecetes opilio and C. bairdi, serve as small-fry substitutes for their fellow lithode [Lithodidae] crab, the much grander king crab. There must be few adult Pacific Northwesterners living who are not familiar with the Alaska king crab fishery, from the boom-and-bust 1970s to the present. An article appearing in the Victoria Times-Colonist (reprinted from the Los Angeles Times) in 1985, described vividly the go-for-broke atmosphere surrounding the rise and fall of the 1970s Alaska king crab fishery:

"In the 1970s, the crab harvest was unprecedented. Deckhands poured champagne and snorted cocaine. Fishermen chartered Lear jets. Retailers did a brisk business in earrings dangling with king crab teeth.... The cause of the rich harvests...began in 1972, an abnormally cold year in which the crab larvae flourished. By the late 1970s, the offspring had grown to adults and jostled each other on the crowded ocean bottom. Word of the bounty spread and fortune hunters headed...to cash in on the ocean's offering....

"Cage after cage hoisted from the ocean floor brimmed with the enormous creatures, beloved as the filet mignon of crabs. Bells regularly sounded in harbor taverns, signaling free drinks for the house. Crab vessels rose fourfold in value and deckhands boasted earnings of more than $100,000 a season....However, even as the harvests grew bigger and bigger, a shift in nature... already had begun to alter the ocean in a way that would bring this economy crashing down. The Alaskan king crab's population plummeted. The statewide catch...collapsed....Crab fishing was prohibited in Kodiak in 1982 and halted throughout most of the Gulf of Alaska and the Bering Sea in 1983'"[174]

The Alaska king crab fisheries still retain their on-the-edge aura of boom and bust; king "Cowboy" crabbers still relish their bad-ass image as well as their big pay. Despite all the bragging, however, there is still no denying the roughness, toughness, and dangers of the Alaska king crab fisheries. In 1992 king crab recruitment levels were still low compared to those of the 1970s, and landings comprised a mere 26% of the total Alaska crab catch.[175] Nonetheless the king and -- and now far more significant -- Tanner [snow] crab fisheries, have been the top shellfish fisheries of Alaska in the early 1990s, with the combined crab harvest of 1992 worth U.S. $305 million.[176] As with groundfish, the Eastern Bering Sea (EBS) overlying the Pacific continental shelf, has served as a vast cornucopia producing huge amounts of Tanner crab.[177] And, again as with groundfish, the Bering Sea/Aleutian Islands (BSAI), region has been a giant marine nursery for the extant king crab stocks; recruitment has been slow but steady since 1985.[178] This writer strongly suspects, however, that Alaska landings of Bering Sea king crab will never again reach the peak of 74,000t attained in 1980.[179] The future crab fisheries of Alaska will mainly target the two Tanner "snow" crabs, C. opilio and C. bairdi, even though both these species were (also) in 1992 "fully utilized".[180]

The importance of the Bering Sea crab fisheries has had a profound effect on Alaska and Pacific Northwest politics. By the mid-1990s, the controversy regarding ITQs (individual transferable quotas) versus CDQs (community development quotas) raged throughout Alaska and the Lower 48. On the pro-ITQ side were the [mostly] Seattle-based fishermen and the factory-trawler companies; on the pro-CDQ side were (1) the State of Alaska and the (2) "Alaska political establishment"; (3) Alaska shore-based processors and (4) Alaska Native corporations. Arni Thomson of the Alaska Crab Coalition commented in April 1996 that, backing the Native corporations, were..."their support industry of consultants, politicians, and bureaucrats."[181] CDQs, conceded Thomson, could serve legitimate social and economic goals, but abuses could also prove "very costly to those who [were] not the recipients of the special allocations of federal fishery resources. Unlimited CDQs, not ITQs, pose[d] a major threat to the long-term viability of small, independent fishing businesses."[182] Arni Thomson, Executive Director of the Alaska Crab Coalition, put his case convincingly for ITQs:

...."What should be of significance to the American public, and what has been lost in the rhetoric of Alaska-based anti-ITQ/pro-CDQ arguments, is the national importance of these resources. Since about two-thirds of this product is taken in federal waters off Alaska, a more equitable system of allocation is justified than Alaskan interests have in mind. The system of allocation that Alaska prefers for the Bering Sea shellfish and groundfish fisheries could lead to wholesale displacement of highly efficient, independent fishermen in favor of joint ventures between vertically-integrated seafood corporations and the Alaska native CDQ recipients."[183]

One member of the "Alaska political establishment" who has worked hard for the ensured control of Alaska resources by Alaska residents -- including permits and quotas to fish in federal waters -- is Senator Frank Murkowski (R--Ak.). (Readers will recall the Alaska senator's heroic efforts several years ago to ban high-seas driftnets and combat high-seas salmon piracy. Cf. FF2000). In an appearance before the Alaska legislature on 17 January 1996, Senator Murkowski assured state legislators that the federal Magnuson Act (MFCMA) -- due to be reauthorised that year -- would "see major changes beneficial to Alaskans, including community development quotas [CDQs], by-catch and discard reductions, and the establishment of new ground rules for fishery management policy with respect to individual transferable quotas [ITQs]'"[184]

(This writer has always been pro-ITQs except in the case of Alaska. ITQs have worked extremely well, for instance, in the abalone and lobster fisheries of New Zealand. But if these same NZ dive fisheries were opened to better-capitalised and more numerous Australians, for example, the marine resources of New Zealand would soon be taken over and stripped bare by the crowding outsiders. ITQs could benefit Alaska fisheries and fishermen, too -- but fishermen resident in Alaska.)[185]

By early 1997, the 20-year-old Magnuson Act had become officially reauthorised. The new Magnuson Act (MFCMA) imposed reductions in by-catch and discards, set specific fishing limits, and made available federal boat buy-back programmes.[186] The new Magnuson Act also mandated a moratorium on new ITQ plans through the year 2000, and ordered a study of existent ITQ plans by the National Academy of Sciences.[187] Furthermore, the renewed MFCMA "set aside" 7.5% of Bering Sea crab, cod, and flatfish for exclusive use by Alaska Native villages via the CDQ system; plus made permanent the Bering Sea pollock CDQ system (with a hitherto expiration date of 1995).[188] The unanimous (100 to 0) Senate vote for passage hardly reflected the long and bitter political battle between Senator Ted Stevens (R--Ak.) and Senator Slade Gorton (R--Wash.); the latter standing loyally by his [mostly] Seattle-based factory-trawler fleet -- crewed by the men and women who had traditionally caught and processed the Pacific pollock, and who unanimously supported ITQs.[189]

Senator Ted Stevens, like Senator Frank Murkowski, is another member of the "Alaska political establishment" who knows and cares about the fisheries of Alaska. Since coming to Capitol Hill in 1968, Ted Stevens has involved himself with every major fisheries-issue legislation in the Senate for the last three decades.[190] In the words of Pacific Fishing's Joel Gay, Ted Stevens was there for: "Creation of the 200-mile exclusive economic zone [EEZ] and subsequent efforts to 'Americanize' it; marketing legislation designed to get Americans to eat more seafood; inspection legislation aimed at ensuring the quality of the seafood that Americans eat; banning high-seas driftnets; and closing loopholes that allowed foreign companies to reflag their factory trawlers as American."[191]

By early 1997, Senator Stevens had helped pilot the reauthorised Magnuson Act through Congress, and had assumed his post as chairman of the Appropriations Committee.[192] Certainly Senator Ted Stevens, in his new position, will act as a powerful advocate for the U.S. fishing industry in general, as he has for the Alaska industry in particular.

Before relinquishing the subject of "Alaska's political establishment", honourable mention must be made of Representative Don Young (R--Ak.). The feisty congressman first attracted international attention by publickly chastising Canadian fishermen for their "'outlandish behavior'" at Prince Rupert harbour, in July 1997.[193] "'If the Canadians think they've gained anything through these goon tactics, they're sadly mistaken'", Young had added bluntly.[194] Congressman Young and Senators Murkowski and Stevens, have been dubbed "The Three Musketeers" by the Canadian mass-media. During the bilateral salmon treaty talks of 1997-1998, the terrible Alaska triumvirate were cordially loathed by Canadian negotiators for their unyielding Alaska-first stance. Young, Murkowski, and Stevens were also feared by the Canadians for their raw political power in Washington, D.C. As the Canadian magazine Maclean's explained to their readers:

....."[T]hey are all long-serving Republicans who chair key congressional committees. Young heads the resources committee of the House of Representatives; Murkowski chairs energy and natural resources in the Senate; and Stevens presides over Senate appropriations. That makes all three heavy hitters, and they were quick to send a message to British Columbia's fishermen: you picked the wrong state to mess with."[195]

Surely, with congressional watchdogs such as Young, Murkowski, and Stevens, the Alaska fisheries are safely guarded from Canadian, Lower 48, or U.S. federal encroachments? The answer is: Almost but not quite. The lead-in to a June 1997 Pacific Fishing article ominously stated: "For nearly 40 years, Alaska had a tight grip on the management of its fisheries. But a monster in the state's bylaws has come of age, and if the state can't decide which users should get first crack at the resources, the feds have vowed to take over."[196]

"Users" (above) means fish-user groups, and there are three main user groups in Alaska -- commercial, sport, and Native (subsistence). But the word "subsistence" in 1990s Alaska is fraught with racial -- hence political -- overtones. On the one side, commercial fishermen in Alaska catch fully 58% of all U. S. seafood production, with the industry employing 1/6 of the state's entire private sector.[197] Salmon alone in 1995 earned more than U.S. $461 million at the docks, with charter and sport fishing generating unquantifiable but big bucks.[198] (By now, we all know how much more a sport-caught fish is worth than a commercially caught fish!) On the other side, Native denizens of the 200 plus rural villages of coastal and interior Alaska claim that their very survival is largely dependent on more than 500,000 salmon a year (besides millions of pounds of crab, halibut, waterfowl, and caribou). And in many of the villages lining the Yukon River, for example, unemployment surpasses 60% (and the Yukon drains more than 1/3 of all Alaska).[199] The "monster" phrase buried deep in federal laws and Alaskan by-laws reads "[to conduct] customary trade", which permits "subsistence" fish to be sold for hard cash. Therein lies the rub -- the razor-thin line between subsistence and commercial use. And unless Alaska sorts this all out, the U.S. federal government will take control of the state's subsistence fisheries in October 1997.[200] Thus might an entire sector of the proud state of Alaska's fisheries revert to the disastrous management policies under a Washington, D.C.-imposed Territorial-type (pre-1959) status.

Count on it!

NOTES

119. Roger Fitzgerald, 'Fitzgerald on Fish', <u>Seafood Leader</u>, Vol.15, No.6, November/December 1995, p.34. (Brackets this writer's.)

120. Seattle and London: University of Washington Press, 1995 ed. NB: William Kittredge came out in 1996 with <u>Who Owns the West?</u> (San Francisco: Mercury House), another book exploiting the theme of "the new heartland nation".

121. NB: For a portrait of Ted Kaczynski, see Appendix J, in `FutureFish 2000`.

122. Simon Winchester, <u>Pacific Rising: The Emergence of a New World Culture</u> (New York: Prentice Hall Press, New York, 1991), p.206.

123. <u>Ibid.</u>, pp.202-203. Brackets this writer's.

124. 'Exxon vows to appeal $5 billion damage verdict', <u>Peninsula Daily News</u>, 18 September 1994, p.A-5.

125. <u>Ibid.</u>, p.A-5.

126. <u>Ibid.</u>, p.A-5.

127. Cited in 'For Your Eyes Only', <u>Alaska Fisherman's Journal</u>, Vol.17, No.9, September 1994, p.2. NB: Emphasis and brackets this writer's. (PS: Jury tampering? Withholding of evidence?)

128. 'Pink salmon rebounds in Prince William Sound', <u>Peninsula Daily News</u>, 21 July 1994, p.A-2.

129. Cited in 'For Whom the Tar Balls Toll', <u>Alaska Fisherman's Journal</u>, Vol.18, No.12, December 1995, p.2.

130. <u>Ibid.</u>, p.2.

131. <u>Ibid.</u>, p.2.

132. <u>Ibid.</u>, p.2.

133. <u>Ibid.</u>, p.2.

134. P.J. O'Rourke, <u>All the Trouble in the World</u> (New York: The Atlantic Monthly Press, 1994), pp.160-161.

135. 'Kodiak: from oil spill to Eden', <u>Common Ground</u>, Vol.6, No.4, May/June 1995, p.3.

136. <u>Ibid.</u>, p.3.

137. See the works of 1920s Danish-American sailor, writer and "Wobbly" philosopher Hjalmar Rutzebeck. See Bibliography, Ch.3, Pt.I, in `FutureFish 2000`.

138. Roger Fitzgerald, "Still the Last Best Place", cf. note no. 119.

139. <u>Our Living Oceans</u>, NOAA Tech. Memo., NMFS-F/SPO-15 (Silver Spring, Md.: U.S. Dept. of Comm., 1993), p.79.

NOTES (cont'd)

140. Cf. Part I in this chapter.

141. Our Living Oceans, p. 79.

142. Ibid., p.79. See also Ch.1, Pt. II; Ch.2, Pt. III, Sec. C, in FutureFish 2000.

143. Ibid., p.80. NB: These statistics applied to Bristol Bay chinook (Oncorhynchus tshawytscha) in 1990-1991, and for Yukon River chinook. In 1993, even-year pinks, (O. gorbuscha) were below 1970-1990 harvests, and wild sockeye (O. nerka), pinks, and chum (O. keta) had declined in Prince William Sound.

144. Peter Redmayne, ed., 'Japan: In the 1990s, Selling Seafood Is Not So Simple', Seafood Leader, July/August 1994, pp.49-50.

145. Buck and Shelly Laukitis, 'Fishing For Opportunity in Japan', Pacific Fishing, August 1995, p.32.

146. 'Japan salmon market "worst in 20 years"', Peninsula Daily News 6 August 1995, p.D-1.

147. Ibid., p.D-1.

148. Ibid., p.D-1.

149. Chicopee, Massachusetts Eastern Edition.

150. Bill Richards 'Fishermen in Alaska, Awash in Salmon, Strive to Stay Afloat', Wall Street Journal, 4 September 1996, p.A-1.

151. Ibid., pp.A-1, A-8.

152. Ibid., p.A-1.

153. Ibid., p.A-8. Quoted is Grant Trask of Petersburg, Ak.

154. Ibid., p.A-8. Quoted is Gunnar Knapp of U. Alaska.

155. Our Living Oceans, p.79. (NOAA figures)

156. Ibid., p.14.

157. Ibid., p.14.

158. Cf. Ch.1, Pt. II; Ch.2, Pt. III, Sec. C, in FutureFish 2000.

159. Our Living Oceans, p.15.

160. Ibid., p.15. NB: NOAA figures for the Bering Sea/Aleutian Islands region were $522 million; for the Gulf of Alaska, $133 million).

161. 'Report: 740 million pounds of fish dumped', Peninsula Daily News, 30 August 1994, p.A-1.

NOTES (cont'd)

162. Ibid., p.A-1. NB: The study was authored by Larry Cotter of Pacific Associates of Juneau, Ak. Cotter estimated that 507 million lbs. of perfectly good fish had been discarded during 1992.

163. Our Living Oceans, pp.23, 105-106. Pacific [Alaska, walleye] pollock (Theragra chalcogramma); Pacific [grey, true] cod (Gadus macrocephalus); sablefish [black/Alaska cod] (Anoplopoma fimbria).

164. Rob Lovitt, 'Big Sea, Big Changes', Seafood Leader, Vol.15, No.6, November/December 1995, p.98.

165. Ibid., p.98. Cf. Our Living Oceans, p. 104.

166. Ibid., p.98. Cf. Our Living Oceans, pp.105-107.

167. Ibid., p.100. NB: Arrowtooth flounder or turbot (Atheresthes stomias).

168. Ibid., p.102.

169. Ibid., p.104.

170. Ch.IV, 'The Pacific Halibut', Fisheries of the Pacific Northwest Coast, Vol.1 (New York: Vantage Press, Inc., 1991), pp.53-78.

171. Our Living Oceans, p.15.

172. Ibid., p.104.

173. Ibid., p.104.

174. Maura Dolan in the Los Angeles Times, cited in the Victoria Times-Colonist, 1 November 1985, p.D-14.

175. Our Living Oceans, pp.15., 110. NB: There are two other species of king crab besides the renowned red king crab (Paralithodes camtschatica): Blue king crab (P. platypus), and golden/brown king crab (Lithodes aequispina).

176. Ibid., p.15. NB: "Tanner" crab is a misnomer. Chionoecetes tanneri has never been found in Alaskan waters; the "Tanner" crabs in the Alaska fisheries are C. opilio and C. bairdi. Cf. Robert J. Browning, Fisheries of the North Pacific: History, Species, Gear & Processes (Anchorage: Alaska Northwest Publishing Co., 1974 ed.), p.22.

177. Ibid., pp.110-111. NB: In fact 99% of all Tanner crab, with C. opilio making up 54% of value in 1992.

178. Ibid., pp.110-111.

179. Ibid., p.110.

180. Ibid., p.110.

181. Arni Thomson, 'Alaska's Vision Of The Future Is Blurred', Pacific Fishing, April 1996, p.23. NB: Arni Thomson is (1996) Executive Director of The Alaska Crab Coalition.

NOTES (cont'd)

182. Ibid., p.23. Brackets this writer's.

183. Ibid., p.24. Emphases this writer's.

184. Ibid., p.24. Quoted is Senator Frank Murkowski. (Brackets this writer's.)

185. For comparison of NZ ITQs to case of California sea urchin fishery, see Mike Radon, 'Under ITQs We All Win, Letters, ibid., p.25.

186. 'Magnuson Act Finally Reborn', Seafood Leader, Vol.17, No.1, January/February 1997, p.20.

187. Ibid., p.20.

188. Ibid., p.20.

189. Ibid., p.20.

190. Joel Gay, 'Sen Ted Stevens: Pity Those Who Ignore Him', Pacific Fishing, Vol. XVIII, No.3, March 1997, p.33.

191. Ibid., p.33.

192. Ibid., p.33.

193. Andrew Phillips, 'Hard-baked Alaska: "Get a grip, Canada"', Maclean's, Vol.110, No.31, 4 August 1997, p.15. NB: The blockading of the Alaska ferry Malaspina by B.C. fishing boats. See Part I.

194. Ibid., p.15.

195. Ibid., p.15.

196. Charlie Ess, 'Feds Poised To Take Over Alaska Fisheries', Pacific Fishing, Vol.XVIII, No.6, June 1997, p.36.

197. Ibid., p.36.

198. Ibid., p.36.

199. Ibid., p.36.

200. Ibid., p.36. NB: That deadline has subsequently been changed to 1 December 1998. Cf. 'Lawmakers' inaction sets stage for takeover' Peninsula Daily News, 27 July 1998, p.A-6. See also Allen Baker, 'Alaska Fleets Brace for Federal Subsistence Management', Pacific Fishing, Vol. XIX, No.9, September 1998, p.14.

Part III: The Lower 48, Northwest Coast

"[The reauthorised Magnuson Act] means Washington fishermen have to pay non-working Alaskans for the right to fish in Alaska."
--Sen. Slade Gorton (R--Wash.), 1996

"[The American Fisheries Act] would eliminate a good portion of the Seattle-based factory trawler fleet....It does absolutely nothing to reduce the amount of fish caught in the North Pacific. It only changes who catches and processes the fish."
--Sen. Slade Gorton (R--Wash.), 1998

"When Greenpeace and I agree, there is trepidation on both sides."
--Sen. Ted Stevens (R--Ak.), 1998

NB: According to fisheries expert Carl Safina, there would be no commercial salmon fishing in Washington waters during 1998. [201]

On 16 July 1994 a dubious milestone was reached in the evolutionary history of Big Blue oceanic regulation. For on that day, fully 135 miles of Washington State outer coastline was effectively turned over to the U.S. federal government.[202] The new Olympic Coast National Marine Sanctuary stretches from the Evergreen State's (far northern) Cape Flattery halfway down the western shore to Copalis Beach (near Ocean City), and 40 miles out to sea; an area encompassing 3,300 square miles -- twice the size of Yosemite National Park in California.[203] But Washington is no California. The Evergreen State is one of the smallest and most densely populated states west of the Mississippi river. Washington state and U.S. federal officials proudly announced then that the Olympic Coast National Marine Sanctuary was [just] the first in the Pacific Northwest, and the fourteenth in the United States.[204] The Olympic Coast National Marine Sanctuary was formally dedicated on 16-17 July 1994, by big-wig attendees Secretary of Commerce Ron Brown, U.S. Representative Norm Dicks, and Washington Governor Mike Lowry.[205] Also present was a certain Tom Hyde of the Center for Marine Conservation (from the other Washington), who triumphantly crowed:

"By adding protection to both offshore and nearshore resources, the [S]anctuary complements other protected and recognized areas including national wildlife refuges, wilderness, the Olympic National Park, Biosphere Reserve and World Heritage Site."[206]

"Protected" and "wilderness" have become Big Green (and Big Blue), buzzwords for shoving Joe Camel and locking Joe Six-pack out of federal land and sea [p]reserves. These very same restricted-activity areas, however, are open to eco-sensitive élites and are aggressively patrolled by armed New Woman park/sanctuary rangers. These New Woman rangers can be easily identified by their upper-body strength, "granny" glasses, condescending attitude (especially towards yuppy male ectomorphic eco-tourists), and are often accompanied by an intimidating German Shepherd dog (to further scare the lumpen proletariat). Even the enthusiastic editor of The Daily World, of nearby Grays Harbor (Wash.), admitted to the closed nature of the Olympic Coast National Marine Sanctuary:

....."It's a mixing zone for migrating whales, dolphins and porpoises, and home to some of the largest colonies of birds in North America....
"Most people who visit it [the Sanctuary], will never see the rocky offshore islands and migrating whales except in pictures. It's not the sort of place one goes boating, unless, like present-day commerc[i]al fishermen or Indian tribes that have lived off the ocean's bounty for thousands of years, it's the way they make their living."[207]

But this writer was already beginning to smell an aquatic rat. Still sitting on the board of The Bay Foundation in 1994, the writer received copies of a letter trumpeting the dedication of the Olympic Coast National Marine Sanctuary dated 3 August 1994; sent from the Center for Marine Conservation (Washington, D.C.) to The Bay Foundation (New York, N.Y.). The self-congratulating epistle was signed by one Roger E. McManus, President, Center for Marine Conservation, and secreted that treacly, pompous essence exuded only by self-perceived Very Important Persons:

"Dear [Programme Administratrix]:
"I am [sic] recently returned from a trip that included two days on Washington state's magnificent Olympic Coast. I always feel lucky to have an opportunity to visit this rugged, almost primitive area, but this time I was especially pleased to be there for the designation of the Olympic Coast National Marine Sanctuary...."[208]

Roger E. McManus went on to express how his organisation (the CMC) had been..."the principal non-governmental voice for protection in 10 other of the 14 marine sanctuary designations thus far"; how the CMC had been the leading private organisation supporting the NOAA's National Marine Sanctuary Program.[209] In his letter, President McManus grandly handed out kudos to two CMC disciples present at the Dedication Celebration, who had accepted two governmental awards on behalf of the Center for Marine Conservation: [CMC] Pacific Coast Habitat Conservation directress, Rachel Saunders, and [CMC] Washington Habitat Conservation specialist, Tom Hyde.[210] (If the Center for Marine Conservation rings a bell in the reader's mind, it is because the powerful and influential CMC has figured prominently in a past chapter in this work.)[211]

It was the CMC's afore-mentioned Rachel Saunders who told The Daily World of Grays Harbor, Washington, that sanctuary designation would not prohibit commercial or recreational fishing; indeed, Ms.

Saunders was optimistic that a "pristine" Olympic Coast would be good for fishermen -- as well as being conducive to [eco-] tourism.[212] But this writer flat-out disbelieves the hollow assurances of Ms. Saunders and the Center for Marine Conservation. It is the writer's firm belief that the CMC, the NOAA, et al. verbally approve of fishing -- of any kind -- for <u>now</u> in the Olympic Coast National Marine Sanctuary so as not to unduly alarm the private citizens of Washington State. Step by step during the years ahead, the U.S. federal government shall at first curtail and then finally ban commercial fishing altogether. (Count on it!) The same fate eventually awaits sport fishing, but at a much later date. What the powers-that-be ruling the Olympic Coast National Marine Sanctuary really wish for the general population of Washington State is <u>eco-tourism</u>: viz; "ecological awareness....educational programs....promoting activities such as whale watching and tide pooling as low impact adventures...."[213] The obedient masses must be herded onto shuttle buses and guided through the wonders of the great outdoors by official (and officious) tour-group leaders. This is the grand eco-plan for Century 21. In the not-so-distant future, no fishing -- of any kind (except by eco-friendly Native "fishers", of course) -- will be permitted to sully or disturb the pristine nature of Olympic Coast National Marine Sanctuary. Since 1994, the U.S. eco-feds have been working on a scheme to take over (i.e., nationalise) the inner Straits of Juan de Fuca.[214]

On 6 November 1998 this writer visited Olympic Coast National Marine Sanctuary offices in the Federal Building at Port Angeles, Washington. Mr. George A. Galasso (now a civilian, see note 202) was polite and pleasant on a personal level, but would not get into particulars concerning the fate of the Northwest Straits Marine Conservation Initiative. Mr. Galasso was understandably more concerned with the marine resource than user-group fishing rights. Before leaving, and having received NOAA literature, the writer asked about the local activities of the Center for Marine Conservation. "They're no longer around here", answered Mr. Galasso shortly. The writer shut the door knowing that, although the people of the Olympic Peninsula were quit the attentions of the CMC, we had not seen or heard the last of the Northwest Straits Marine Conservation Initiative.

But there is a happy mid-story dénouement to this story in progress. On 9 November 1998, the writer telephoned Mr. Michael Murray, NOAA, at the Northwest Straits Seattle office. Mr. Murray turned out to be both agreeable and helpful. Neither he nor his superiors (said he) at NOAA had played an activist rôle vis-à-vis the Northwest Straits Conservation Initiative for the "last year and a half".[215] Rather, sorry-faced Democrat Senator Patty Murray, along with rock-ribbed Republican Representative Jack Metcalf, had recommended a state citizens' advisory commission to work in tandem with NOAA and local officials. According to Mr. Murray, the present (1998) Northwest Straits Advisory Commission is purely an advisory body composed of user-group representatives and other (environmentally) interested parties. And no action may be taken regarding the Northwest Straits Conservation Initiative -- ensured the earnest Michael Murray -- unless Congress passes a specific act. Despite the beneficial wisdom of the NOAA working with local citizenry to protect the marine resource, there are power-hungry men (and women) in high eco-federal places who are not so informed or as benign (for <u>now</u>) as the National Oceanic and Atmospheric Administration.

High-handed government and independent commercial fishermen sometimes butt heads dramatically. On 19 May 1996, Morro Bay (Calif.) fisherman Jim Blaes refused to allow U.S. Coast Guard officers to board his 38-foot vessel, <u>Helja</u>.[216] The armed and warrantless officers of the U.S. Coastguard cutter, <u>Pt.Chico</u>, had notified Blaes of their determination to board the <u>Helja</u> for an at-sea safety inspection.[217] Blaes refused to comply with what he felt to be simply a random boarding. While the Coast Guard was attempting to board the <u>Helja</u>, Blaes got on his telephone and called the White House, radio talk shows, and other media.[218] Surmised Blaes later, "The media kind of saved my butt....I had had enough....I said they couldn't board me armed, that only one unarmed man could come aboard. They said it was against Coast Guard policy. One officer asked me if I intended to be a martyr.'"[219]

But within an hour of the attempted boarding by the Coast Guard, the <u>Pt Chico</u> requested permission from the Long Beach District Command Center to fire warning shots across the bow of the <u>Helja</u>.[220] The Coast Guard request was denied, and there ensued a three day stand-off, from 19-22 May 1996.[221] Jim Blaes was subsequently charged and acquitted of five felonies, although he was convicted of two misdemeanours -- resisting a federal officer under U.S. Coast Guard regulations and under the Magnuson Fisheries Act (a single charge under two counts).[222] One fortuitous outcome was that Blaes, who had been sentenced by a San Jose federal court to parole until completing 100 hours of community service, was permitted to donate fresh fish to a San Luis Obispo area food bank (thence soup kitchens), with every pound of "red snapper and rock cod" counting as one hour.[223] Another felicitous result was that Morro Bay area fishermen reported fewer random "safety" boardings by the Coast Guard. "'I want to end this armed Coast Guard harassment once and for all,'" Jim Blaes announced, while appealing his misdemeanour convictions.[224]

Might Northwest Coast fishermen look forward to a kinder, gentler, less arrogant U.S. Coast Guard? John Warner, fisherman and owner of Basin Tackle Shop in Charleston, Oregon, has opined, "They have changed their policies quite a bit toward fishermen...but they've got a long way to go to repair the damage that they caused with the Blaes situation'."[225] And Alan Dujenski, a Coast Guard veteran now a marine safety specialist for Acordia Northwest in Seattle, has added: "Things are getting better, but it's because the symptoms are being treated -- the problem's not being solved. If you think boardings at sea are going to improve safety, you're wrong....They're parading these boardings as being safety, and they're not. They're enforcement boardings.'"[226] Two years after the stand-off, Jim Blaes has concluded: "There have been lots of changes, but if the laws don't change, then it's all going to be for naught. As long as warrantless searches are permitted, nothing's been accomplished.'"[227]

If the swollen, heavily-armed, U.S. federal government agencies are not reined in and cut down to size, they will continue to act in an arbitrary and abusive manner. The Jim Blaes stand-off might be just the first of many such accelerating incidents in the Northwest Coast fisheries.

This writer has already covered (from an Alaskan perspective) the U.S. Senate fight for fish in the far North Pacific; between Alaska senators Frank Murkowski and Ted Stevens, and Slade Gorton and Patty Murray of Washington State. The (fish) bone of contention between the parties has been the reauthorised Magnuson Act (MFCMA) of 1997, which mandated a moratorium on new ITQs, plus fish and shellfish set asides for Alaska Natives under the CDQ system. It has been estimated that during 1995, two years before the Magnuson Act was renewed, 55 to 60 factory trawlers based in Seattle harvested more than 1/3 of all seafood taken from American waters; 12 of the fishing vessels were owned by the Tyson seafood conglomerate.[228] "Tyson Foods" and "factory trawlers" are eco-buzzwords, and, for their own eco-reasons, Greenpeace has supported Murkowski, Stevens, and CDQs; whereas the Environmental Defense Fund and the Center for Marine Conservation have lined up behind Gorton, Murray, and ITQs.[229] (The alliances have surely created some strange bedfellows! Simply put, Greenpeace dislikes Tyson Foods and the Seattle-based trawler fleet. The EDF and CMC dislikes the preponderance of Japanese-owned Alaska-based seafood companies.) But everybody knows that the basic issue (with Magnuson as red herring) has boiled down to whom Alaska's seafood bounty truly "belongs" -- Alaska Natives, Alaskan fishermen, and Alaska-based shore processors, or to Lower 48 (seafood) corporate giants and the Seattle-based fishing fleets? During the summer of 1996, all Alaska and greater Seattle were bracing for the pending reauthorisation of MCFMA.

Both Washington senators have decried Ted Stevens' bill to renew the Magnuson Act, accusing the reauthorised MFCMA of heavily favouring Alaska (which it does). As Senator Gorton has expressed in exasperation, '"It means Washington fishermen have to pay non-working Alaskans for the right to fish in Alaska.'"[230] Senator Murray, on her part, has attacked some of the (pro-Alaska) language in the CDQ provisions, saying it was "a blatant violation'" of National Standard #4 of the (original) Magnuson Act, which states that "management measures shall not discriminate between residents of different states.'"[231] (Senator Murray occupies the late Warren G. Magnuson's senatorial seat.) The conservative but pragmatic Senator Stevens, however, has also co-opted the environmental issue from the ultra eco-liberal Patty Murray. Ted Stevens has emphasised often (and effectively) that the main opposition to his bill was from the Seattle-based fishing fleet (viz. <u>Tyson, trawlers</u>) which have resisted the by-catch provisos contained in the bill; provisos that curb wasteful fishing practices.[232] The politically-astute Alaska senator in 1995 enunciated the fact that 60 (some counted 55) factory trawlers, fishing in Alaska's Bering Sea, discarded almost so much fish and shellfish as was landed in the entire New England lobster fishery, Atlantic mackerel fishery, Gulf of Mexico shrimp fishery, Pacific sablefish (black cod) fishery, and North Pacific halibut fishery <u>combined</u>![233] Said Stevens of his bill:

> "We hope that this bill will bring an end to the inexcusable amount of waste that is occurring in the fisheries off Alaska and in other parts of the country....If this bill doesn't work...[Stevens would press for legislation that would] eliminate these vessels [factory trawlers] that are destroying the reproductive capability in the North Pacific.'"[234]

To add insult (by Alaska) to injury (to Lower 48 fishermen), the Supreme Court decided on Tuesday, 20 January 1997, to allow Alaska to continue collecting fishing license and permit fees from out-of-state vessels (i.e., Seattle factory trawlers).[235] Commercial fishing licenses in the Great Land cost thrice as much for residents of the Lower 48 than they do for Alaskans (eg. crab permits @ $750 for outsiders; $250 for insiders).[236] The higher Alaska fees were challenged by a group of out-of-state fishing crew members in a 1984 class-action lawsuit, which contended that the higher fee unlawfully discriminated against interstate commerce, and denied non-residents the same privileges granted Alaskans.[237] The end result -- before the lawsuit finally failed at the Supreme Court -- was that an Alaska Superior Court ruled that higher fees depended on residency rather than the movement of goods across (Alaska) state lines; hence there had been no discrimination against interstate commerce.[238]

But things would get even worse in 1998 for non-resident Alaska fishermen. On Thursday, 26 March 1998, a Senate subcommittee was scheduled to hear changes proposed in the 1987 Anti-Reflagging Act by Senator Ted Sevens (who else?!).[289] As Pacific Fishing's Joel Gay once wrote of the Alaska Republican, "Pity those who ignore him."[240] The Anti-Reflagging Act of 1987 (twice discussed at length throughout these pages) mandated that all U.S. fishing vessels be "Americanized" by requiring them to be solely owned, operated, and controlled by U.S. citizens.[241] But backers of Stevens' bill maintained that a Coast Guard [mis]interpretation of the Anti-Reflagging Act let slip in, through a loop-hole, a small flotilla of foreign-built and foreign-owned boats to fish U.S. (i.e., Alaska) waters.[242] Stevens' bill would prohibit any new factory trawlers over 165 feet in length, and apply strict new ownership requirements that would essentially ban any foreign-owned vessels from fishing American waters.[243] This decision will hit home immediately -- Seattle-based American Seafoods Co. is not only the largest fishing-processing presence in the Bering Sea, but is owned by the Norwegian holding company, Aker-RGI.[244] But, at the same time, many of the smaller American-owned boats sell their catch to shore-based Alaska processing plants, the largest of which are Japanese.[245] At stake is about 2 million tons of groundfish (mostly Pacific pollock) harvested each year from the Bering Sea.[246] It is truly an American fishing industry dilemma. Ted Stevens' bill, a/k/a the American Fisheries Act, though controversial and hated by many, has become a dire necessity.

Terry Leitzell, vice-president for legal and government affairs for Victor Seafoods, has clinched the matter and put it to rest: "Our members will tell lawmakers how their plans to build vessels at American shipyards and employ Americans in the fishery were shattered when foreign companies took advantage of the Coast Guard-created loophole '."[247]

Even though there has been a precipitous decline in commercial fishing in Washington waters, fish (viz. Pacific salmon) have always played a principal part in Evergreen State politics and still do. So when former Representative Jolene Unsoeld (D.--Wash.) was refused confirmation in February 1998 (for the first time in three decades), for the governor's nominee to continue serving on the Fish and Wildlife Commission, Pacific Fishing editor Brad Warren sprang gallantly to her defense. Besides Washington fishermen, the Pacific salmon had always had a friend in the person of Congresswoman Unsoeld on the Fish and Wildlife Commission, wrote editor Warren:

"Commercial fishermen and conservationists united behind Unsoeld's efforts to make sure fish-grabbing [didn't] masquerade as conservation, and to provide adequate water and habitat for wild salmon. While she served in Congress, Unsoeld dug in to reduce the harm to fish done by federal dams; she initiated legislation that now puts decision-making for fish-passage facilities in the hands of resource management agencies instead of the Federal Energy Regulatory Commission, an outfit that had a long history of approving projects that wrecked salmon runs."[248]

Former Rep. Jolene Unsoeld has appeared in a former book of this writer's and in this work.[249] Along with then-U.S. Secretary of Commerce, Robert Mosbacher, Senator Frank Murkowski (R--Ak.), Representative W.J. Tauzin (D--La.), and fellow Evergreen Congressman John Miller (R--Wash.), Jolene Unsoeld was in the forefront -- throughout the late 1980s and early 1990s -- of the world-wide battle to ban driftnet fishing. The final closing of "the curtains of death" in 1992 owed much to the Congresswoman's serious efforts.[250] This writer remains grateful to Jolene Unsoeld for her rôle in exposing the bogus "conservatory" nature of Initiative Measure 640 in 1995; in reality an agendum to "shut down commercial fishing while allowing sport anglers to catch and release wild salmon to death."[251] Initiative Measure 640 went down to defeat, but the fight ended in a somewhat Pyrrhic victory as Referendum 45 passed; a measure which lengthened the already-sticky reach of the state's Fish and Wildlife Commission.[252] The Washington sport fishing lobby's influence was greatly expanded over the Commission, with their coup in denying (however vicariously) Jolene Unsoeld's reconfirmation to the Commission.[253]

There had been more than enough fishy politics in Washington State throughout 1995-1996 to keep

Evergreen voters preoccupied. Very different from Brad Warren's forthright praise in 1998 of Jolene Unsoeld, had been The Fishermen's News editor Sam Smith's (crocodile) tear-jerking tribute of 1996 to his former columnist, Don Stuart, of 'Salmon for Washington'.[254] In early autumn 1996, Stuart was embarking on his sure-fire political rise; slated by editor Smith of The Fishermen's News as obvious shoo-in as certain Democrat candidate to face that reprehensible Republican, conservative Rick White.[255] But something (not so) funny happened on the way to the forum -- Don Stuart was ignominiously trounced by one Jeffrey Coopersmith in the September 1996 Democrat primary; ignominious in that Coopersmith was a Washington, D.C., transplant who had resided in Washington State only since 1991.[256] Not just that, but Don Stuart was a bona fide former commercial fisherman, a local Seattleite one-of-us guys; self-described in his own words as "someone who is genuinely middle class."[257] But Don 'Salmon for Washington' Stuart tried his level best to appeal to 1) Big Labour big-wigs, 2) post-Feminism gender feminists, 3) New Knowledge sector-purveyors, 4) and far Left eco-élites who dominate the electorate of the 1990s Emerald City. Quoth the wanna-be candidate:

"There are several key endorsements that I really had to fight to win; the state Labor Council and a whole host of labor unions have given me their sole endorsements. That's a big achievement. In addition to that, the National Organization for Women, Washington Federation and American Federation of Teachers, the Washington Education Association and National Education Association. Those are very significant endorsements. Plus, I think I'm going to have some very good support from the environmental community."[258]

Oh, wow! But in spite of all the liberal heavyweight backing, Seattle old-timer Don Stuart was outfoxed by Seattle newcomer Jeff Coopersmith (who was Third Wave to Stuart's Second Wave). Coopersmith, in turn, would go on to lose to Congressman Rick White in the general election of 1996. (Evangelical Christian Rick White would be eventually unseated by becoming literally trashed -- in a political garbage-compactor operated by state Democrats -- during the disastrous mid-term elections of 1998). But endeavouring to be all things to all politically-correct Washingtonians, accompanied by his bragging rights as an erstwhile commercial fisherman, didn't work for Don Stuart. Perhaps he might have garnered some cross-over votes if the wanna-be candidate had remained true to his ("middle-class") core constituency. Don Stuart appeared to have been rejected by the very same top-drawer New Seattleites he had so assiduously courted. Surely there was some political/poetic justice to that... and a moral lesson too.

Washington State is in truth not a very fish-oriented place any more. The Associated Press reported, from a study done in 1998, that the Evergreen State has more residents working in high-tech jobs than in any other American state, with 35% of total economic employment in technology.[259] High-tech jobs in the digital world (Microsoft), aerospace (Boeing), and biotechnology employs more than 266,000 state residents, earning $60.6 billion U.S. in annual sales revenue and $13.9 billion U.S. in annual labour income.[260] And as every single high-tech position spawns 2.36 others, the cumulative effect on Washington's economy translates into more than 895,000 jobs, $105.7 billion in labour income, and $2.9 billion in state tax money.[261] In greater Seattle and environs, the Second Wave sport-fishing lunch-bucket worker has been replaced by the Third Wave whale-watching compu-nerd technocrat. Non-candidate Don 'Salmon for Washington' Stuart learned that lesson the hard way.

This writer has also learned his lessons the hard way, but about the fishing industry; both in (some say the People's Province of) British Columbia and (some say the Soviet State of) Washington. In 1984, after a brief trip to Westport, the writer soon found out from a spokesman (with a Norwegian name) for the local crabbers association that the Dungeness crab (Cancer magister) resource was "fully utilized". Not only that, implied the spokesman (with the Norwegian name) not unkindly, there was no point in coming down to Westport to purchase a boat and gear, invest in a crabbing license, or search for a partner... or even expect to be hired on as crew.

Two years later, while shucking shrimp and picking crab at Port Angeles, the writer seriously discussed harvesting, processing, and marketing underutilised "goose" (leaf) barnacles (Pollicipes polymerus) with his senior partners in the seafood business. There was even a potential percebes-buyer in Spain standing by. But all the high hopes of 1986 came to nought in 1987 after the arrival of a single-paged, tersely-phrased letter from the Washington Department of Fisheries (WDF); with an attached list of seemingly endless environmental rules, regulations, and reasons why we -- Pisces Seafoods -- couldn't possibly harvest goose barnacles in the State of Washington.

Another non-traditional seafood species which interested this former processor was the giant sea cucumber (Parastichopus californicus), found in abundance during 1987 in Washington waters.[262] But it was in 1987 that this writer became a former seafood processor, and interest in bêche-de-mer remained purely academic. A decade later, the writer read an article in the British Seafood International magazine about North American commercial fisheries for sea cucumbers. Washington State has a fishery, but it is tiny, tenuous, and rigourously managed by the WDF:

"Washington divided state waters into four harvest areas in 1990, each of which goes unfished for 3.5 years following a six-month harvest. The state also limited entry into the fishery at that time. This management change occurred as a result of a spike in landings when fishermen recognised the creature's marketability. Landings began in 1970 and remained below 10 tonnes a year through 1977.
"After that, landings rose to just under 200 tonnes until 1987. Then landings jumped for three years, culminating in 1989 with 1,000 tonnes, prompting conservation measures."[263]

Was there really a 1990 dearth of "sea-rollops" in Washington waters? Only state marine biologists knew for sure. But one fact has become crystal clear: Even marginal, peripheral, non-traditional fisheries of underutilised species in Washington State are relentlessly supervised by the WDF authorities at Olympia.

Despite the onus of government, many "new" and underutilised (non-traditional) species are, nevertheless, gaining in fishing attention as (traditional) salmon and groundfish fisheries continue to shrink under federal and state restrictions. The aforementioned sea cucumbers, Pacific hagfish ("slime eels", Order Hyperotreta), and the opalescent "market" squid (Loligo opalescens) are sought in Northwest waters by a "handful of boats... often at heavy expense and limited success".[264] An April 1998 Associated Press story reported from Astoria, Oregon, on a successful spot prawn (Pandalus platyceros) fishery operating up and down the Northwest Coast since 1993.[265] Big spot prawns, also called "sweet prawns", are trap-fished on rocky bottoms from Alaska to México, and sold live to buyers for Asian markets, groceries, and restaurants in large North American West Coast cities.[266] ("Dancing shrimp" [Japanese odori -- large prawns consumed live -- are a popular specialty item in Chinese and South-East Asian [viz., Overseas Chinese] cuisine.) In Oregon, a mere 16 permits have been issued in 1998 for spot prawns; ostensibly to protect both the marine resource and "pioneering fishermen from sudden pressure on [the] fishing grounds."[267] The particular Oregon fish tale told by the Associated Press has been an upbeat one, for five years anyway. The shrimp fishing vessel's owner said that the key to his continuing success -- in the frontier fishery of spot prawns -- was a "'lot of electronics, a lot of money and a captain and crew that is [sic] very good'."[268]

Either have all the above or migrate to Alaska. And that is exactly what a certain spot prawn fisherman, who had supplied Pisces Seafoods with the tasty crustaceans, did in 1987. Frustrated and infuriated by WDF rules, regulations, and micro-management, arguably one of the best shrimp fishermen on (the south side of) the Strait of Juan de Fuca pulled up stakes in Sequim and headed for Alaska, summer 1987.

Whale-watching anyone?

Slouching Toward Cascadia?

..."[T]he 'lotus-land' stereotype associated with Cascadia is, in fact, quite accurate... [F]or Cascadians, environmentalism has become a sort of secular religion...
"In the new global economy, national borders are less and less important....Regions are on their own....[T]he message is to think globally, but act locally (or regionally). In short, the modern world -- especially technology -- has transformed notions of territory, space and nation".
-- Professor Alan F.J. Artebise of U. British Columbia, on "Cascadian" culture

In 1994, a thought-provoking book was published -- Breakup: The Coming End of Canada and the Stakes for America. The author was one Lansing Lamont, Time magazine's chief Canadian correspondent during the 1970s and managing director for Canadian Affairs at the Americas Society.[269] Lamont's scenario unfolded in a fairly predictable manner: Québec sovereigntists declare independence, after which follows the quick collapse of Canada. British Columbia, stranded out on the Left Coast, decides that her future lay south with Seattle, Portland, and San Francisco; not east with Edmonton, Saskatoon, or Winnipeg.[270] (The Pacific province has never been very enthusiastic about Ottawa, either.) But dissimilar to some prognostications of a politically-independent Ecotopia (composed of northern California, coastal Oregon and Washington, western B.C. and southeast Alaska), B.C., Yukon, and then the Prairie provinces leave a shattered Canada to join the United States.[271] Lansing Lamont foresaw a grand North American Pacific alliance whose members, along with the former Canadian political units, would include Alaska, the entire Pacific Northwest, and Montana. This proposed grand Left Coast alliance ("Cascadia") has been described by Justin Raimondo in Chronicles magazine as "a postindustrial economic powerhouse rich in raw materials with a gross annual product of some $280 billion. An independent Cascadia would be the ninth wealthiest nation on earth...."[272]

An independent Cascadia, however, was not in Lansing Lamont's cards. Rather, Lamont envisioned "Cascadia or Pacifica, as the new region [is] variously tagged, constitut[es] a perfect vehicle for the gradual assimilation of Canada's drifting provinces into the American matrix".[273] So it's the American Century to some, after all, with Canadian protests notwithstanding.

Downtown Seattle is the headquarters of a quasi-self-imaged libertarian and free-market-oriented think tank, the Discovery Institute. According to Justin Raimondo of Chronicles, the Discovery Institute's "Cascadia

Task Force" was set up as an activist arm to simply erase the existing Can-Am border, thus (so much as possible) implementing an expansionist/globalist statism.[274] To prove his point, Raimondo has (as exemplar) pointed to a major project of the Cascadia Task Force, which has been lobbying for an augmentation of Amtrak service from Vancouver, B.C. to Eugene, Ore. In the words of the Discovery Institute, "[I]mproving intermodal connections through public-private partnerships.'"[275] In the pro-private sector opinion of Justin Raimondo, the Cascadia Project -- funded by the cities of Seattle, Surrey (B.C.), the port of Tacoma, and the Henry M. Jackson Foundation -- is closed public-sector mercantilism instead of open free-market economics.[276] Raimondo has also indicated a projected "map of Cascadia", posted on the Discovery Institute's website, as further damning evidence:

"...[I]t defines 'Main Street Cascadia' as Portland, Seattle, Victoria, and Vancouver [B.C.]. This is surrounded by a shaded area deemed 'Cascadia' proper, which includes two states, Washington and Oregon, and all of British Columbia. An even larger swath of territory, stretching from Alaska to Idaho, and including Montana, is designated the 'Pacific Northwest Economics Region', clearly meant to delineate the outer reaches of the Cascadian empire."[277]

And will this hypothetical Cascadian empire be presided over by a pot-smoking New Government of Youth, as visualised by New Ager William Irwin Thompson in 1972?[278] Probably not exactly, but there would be a pronounced neo-pagan subcultural tinge to any Seattle-ruled region ("Cascadia"), whether or not the future (evil) "empire" contained the wackier areas of Northern California ("Pacifica"). The difference in William Irwin Thompson's addled but accurate prophesy has been the digital technology of cyberspace; an admirably suited although very diverse countercultural vehicle from psychedelic T.V. Thompson was prescient in his 1985 book, Pacific Shift, about an entirely new global Anglo-Californicated counterculture.[279] In 1985, however, Thompson hadn't reckoned on the meteoric late 1980s-early 1990s rise of Asia and Asians in North America.

During the 1990s, the dynamic Asian component has been transposed into political power -- a Chinese Canadian lieutenant-governor of British Columbia, and a Chinese American governor of Washington State. In no place on earth has digital technology been more enthusiastically embraced than in today's Asia; South, East, and South-East. The reader has only to travel to Northern California's "Silicon Valley" or to Redmond, Washington, to notice the large numbers of new Asian-American technicians working in cybertechnology. The downside (to this evangelical Christian, anyway) of the Asian trend is a re-importation of Eastern religions into an already post-Christian, "alternative" eco-lifestyle. (On a scale; many Asians are Spirit-filled Christians.) Mixed in with a Luciferian cyberspace ecology, the result is a potent New Age sorcerers' blend.[280]

But isn't the syncretic New Age cum-Eastern religion merely a manifestation of the Pacific Way, of which Cascadia -- either independent or under the U.S. aegis -- is merely an integral, synergic part of the greater global whole? And isn't the latter state-of-being exactly what the hip and privileged denizens of Ecotopia-Cascadia-Pacifica have striven for ever since the 1967 Summer of Love?

NOTES

201. Carl Safina, <u>Song For the Blue Ocean: Encounters Along the World's Coasts and Beneath the Seas</u> (New York: Henry Holt and Company, 1998), p.125.

202. Lt. George A. Galasso, NOAA, 'Protecting Marine Areas of National Significance', <u>Studium Generale Program Notes</u>, 29 September ,199[4], Peninsula College (Port Angeles, Wash.), p.1.

203. <u>Ibid.</u>, p.1. Cf. Doug Barker, 'Beauty on a grand scale: Marine sanctuary dedication Saturday,' <u>The Daily World</u> (Grays Harbor, Wash.), 14 July 1994, p.A-1.

204. Lt. George A. Galasso, NOAA, <u>Studium Generale Program Notes</u>, p.1.

205. Todd Jacobs, 'Sanctuary offers public a voice', <u>The Olympic Coast National Marine Sanctuary Dedication Celebration</u>, Center for Marine Conservation (Washington, D.C.), 16-17 July 1994, p.3.

206. Tom Hyde, 'The crafting of a national treasure -- our nation's newest marine sanctuary, <u>ibid.</u>, p.5.

207. Doug Barker, 'Beauty on a grand scale', <u>The Daily World</u>, p.A-1. NB: This writer thought dolphins and porpoises were the same species of marine mammal? (Brackets and emphasis this writer's.)

208. Roger E. McManus, Center for Marine Conservation fund-raising letter to The Bay Foundation, 3 August 1994, first page.

209. <u>Ibid.</u>, first page.

210. <u>Ibid.</u>, first page. NB: The very same Tom Hyde as mentioned in note 206.

211. See Chapter II, <u>FF 2000</u>, 'Century 21 -- Time for Big Green Oceanic Regulation?' NB: Perhaps the title should be partially changed to '...Big <u>Blue</u> Oceanic Regulation?'! (PS: Outfits like the powerful and prestigious CMC owe their very existence to generous institutions like The Bay Foundation.)

212. Doug Barker, 'Beauty on a grand scale', <u>The Daily World</u>, p.A-8.

213. Rick Hert, 'Sanctuary offers new opportunities ' <u>The Olympic Coast National Marine Sanctuary</u>, p.7. NB: If the reader suspects that the CMC might be commercial fishermen-friendly, write for their literature at: Center for Marine Conservation
        1725 De Sales Street, N.W.
        Washington D.C. 20036
        Attn. Roger E. McManus, pres.
        Tel. (202) 429-5609

214. This was the strenuously-resisted Northwest Straits Marine Conservation Initiative. On 16 September 1998, Senator Patty Murray (D.--Wash.) introduced a bill to establish the Northwest Straits Advisory Commission. In governmental parlance: "[T]he term 'Northwest Straits' means the marine waters of the Strait of Juan de Fuca and of Puget Sound from the Canadian border to the south end of Whidbey Island." (Sec. 11. Definitions, Title IV, <u>Northwest Straits Marine Conservation Initiative</u>, p.9.) The Advisory Commission works with environmental monitoring programmes and submits annual advisory reports.

215. Mike Murray, NOAA, via telephone call, 9 November 1998.

NOTES (cont'd)

216. Jo McIntyre, 'Fisherman Tangles With Coast Guard', <u>Pacific Fishing</u>, July 1996, p.20.

217. <u>Ibid.</u>, p.20.

218. <u>Ibid.</u>, p.20.

219. <u>Ibid.</u>, p.20.

220. Christine Hansen, 'Jim Blaes: Case Closed?', <u>Pacific Fishing</u>, Vol. XIX, No.5, May 1998, p.56.

221. <u>Ibid.</u>, p.56.

222. <u>Ibid.</u>, p.56.

223. 'Blaes Pays Up' <u>Pacific Fishing</u>, Vol. XVIII, No.6, June 1997, p.14.

224. <u>Ibid.</u>, p.14.

225. Christine Hansen, 'Jim Blaes: Case Closed?', <u>Pacific Fishing</u>, p.57.

226. <u>Ibid.</u>, p.57.

227. <u>Ibid.</u>, p.57.

228. 'Fish fight hooks NW lawmakers', <u>Peninsula Daily News</u>, 7 May 1996, p.A-5. NB: The Magnuson Act was named for the late Senator Warren G. Magnuson (D--Wash.).

229. <u>Ibid.</u>, p.A-5.

230. 'Fishing quota nets debate', <u>Peninsula Daily News</u>, 24 May 1996, p.A-5.

231. John van Amerongen, 'Industry Awaits "New" Magnuson Act', <u>Alaska Fisherman's Journal</u>, Vol.19, No.6, June 1996, p.22.

232. 'Fishing quota nets debate', <u>Peninsula Daily News</u>, p.A-5.

233. 'Fishing quota bill nets mixed support', <u>Peninsula Daily News</u>, 20 September 1996, p.A-6.

234. <u>Ibid.</u>, p.A-6. Quoted is Sen. Ted Stevens of Alaska. (Brackets this writer's.)

235. 'High court OKs high Alaska fishing fee', <u>Peninsula Daily News</u>, 22 January 1997, p.A-1.

236. <u>Ibid.</u>, p.A-1.

237. <u>Ibid.</u>, p.A-1.

238. <u>Ibid.</u>, p.A-1.

239. 'Doubt cast on fish bill', <u>Peninsula Daily News</u>, 26 March 1998, p.A-7.

NOTES (cont'd)

240. Joel Gay, 'Rainmakers: Sen. Ted Stevens', Pacific Fishing, Vol. XVIII, No.3, March 1997, p.33.

241. 'Doubt cast on fish bill', Peninsula Daily News, p.A-7.

242. Ibid., p.A-7.

243. 'Northwesterners joust over Alaskan attack on trawlers', Peninsula Daily News, 27 March 1998, p.A-2.

244. Ibid., p.A-2. NB: U.S. ownership requirement would now be raised to 75% rather than at a 50% minimum.

245. Ibid., p.A-2. Cf. Ch.1, Pt.II in FutureFish 2000; and Ch.1, Pt.II, in this work.

246. 'Doubt cast on fish bill', Peninsula Daily News, p.A-7.

247. Ibid., p.A-7.

248. Brad Warren 'Fishy Politics', Pacific Fishing, Vol. XIX, No.3, March 1998, p.5. (Brackets this writer's.)

249. Fisheries of the Pacific Northwest Coast, Vol. 2 (New York: Vantage Press, Inc., 1994). Cf. 'Post script: The Drift-Net Fleets and High Seas Pacific Salmon Interception '. pp.130-146.

250. NB: Rep. "Billy"Tauzin of Louisiana chaired the House Merchant Marine and Fisheries Committee, of which Reps. Unsoeld and Miller were members during the late 1980s. That House subcommittee had (1989) jurisdiction over the U.S. Coast Guard.

251. Brad Warren, 'Fishy Politics', Pacific Fishing, p.5.

252. Ibid., p.5. NB: For more on Initiative Measure 640, cf. Ch.2, Pt.III, Sec. C in FutureFish 2000.

253. Ibid., p.5.

254. Sam Smith, 'The Candidate ', The Fishermen's News, August 1996, p.20.

255. Ibid., p.20.

256. Ibid., p.20.

257. Ibid., p.20. Quoted from Don Stuart.

258. Ibid., p.20. Quoted is Don Stuart.

259. 'Study: Washington ranks No.1 in high-tech jobs.', Peninsula Daily News, 7 June 1998, p.E-1. NB: The study was taken by the Technology Alliance, a non-profit group founded by lawyer William Gates, Jr., father of embattled cyberdweeb Bill Gates III.

260. Ibid., p.E-1.

NOTES (cont'd)

261. Ibid., p.E-1.

262. NB: For biology, history, and fishery in Washington State waters of goose barnacles and sea cucumbers, see C.D. Bay-Hansen, Vol.2, pp.77-89.

263. Nancy Griffin, 'Sea cucumbers in the frame', Seafood International (Farnham, Surrey), Vol.12, Iss.9, September 1997, p.75. NB: For Asian markets for trepang, see source note 262.

264. 'Fishermen seek alternatives', Peninsula Daily News, 26 April 1998, p.A-7.

265. Ibid., p.A-7. (Large prawns are known as kuruma-ebi in Japanese.)

266. Cf. C.D. Bay-Hansen, Vol.1, pp.159-166.

267. 'Fishermen seek alternatives', Peninsula Daily News, p.A-7.

268. Ibid., p.A-7. NB: Interviewed by the AP was Paul Daniels of Garibaldi, Ore., owner of the fishing vessel Lady Rosemary.

269. Justin Raimondo, 'The Road to Cascadia', Chronicles, Vol.21, No.7, July 1997, p.24.

270. Ibid., p.24. Ernest Callenbach, Ecotopia: The Notebooks and Reports of William Weston (New York: Bantam Books, 1990), passim.

271. Ibid., pp.24-25. Cf. Joel Garreau, The Nine Nations of North America (New York: Avon Books, 1981), pp.245-287.

272. Ibid., p.25. See also Ch.1, Pt.III; Ch.2, Pt.I, in FutureFish 2000.

273. Ibid., p.25. Quoted is author Lansing Lamont. (Brackets this writer's.)

274. Ibid., p.25.

275. Ibid., p.25.

276. Ibid., p.25.

277. Op.cit., Justin Raimondo, p.25.

278. William Irwin Thompson, At the Edge of History (New York: Harper Collophon Books, 1972), p.169, ff.

279. William Irwin Thompson, Pacific Shift (San Francisco: Sierra Club Books, 1985), passim. NB: But in a Pacific cyberspace cultural ecology rather than in a Pacific aerospace cultural ecology.

280. William Irwin Thompson, Darkness and Scattered Light (Garden City, N.Y.: Anchor Press/Doubleday, 1978), p.117, ff.

Bibliography for Chapter 2

1. Bay-Hansen, C.D. Fisheries of the Pacific Northwest Coast, Vol.1. Vantage Press, Inc., New York, N.Y., 1991.

2. _____. Fisheries of the Pacific Northwest Coast, Vol.2. Vantage Press, Inc., New York, N.Y., 1994.

3. Caldwell, Donna and Francis. The Ebb and the Flood: A History of the Halibut Producers Cooperative. Waterfront Press, Seattle, 1980.

4. Eppenbach, Sarah, Alaska's Southeast. The Global Pequot Press, Old Saybrook, Conn., 1994 (5th) ed.

5. Hadman, Ballard. As the Sailor Loves the Sea. Harper & Brothers Publishing, New York, 1951.

6. Kahrs, Jeffrey. The Deep Sea Fishermen's Union of the Pacific: A Short History. DSFU, Seattle, 1983. (87pp)

7. Rustad, Dorothy. I Married a Fisherman. Alaska Northwest Publishing Co., Edmonds, Wash., 1986.

8. Stevens, Homer and Rolf Knight. Homer Stevens: A Life in Fishing. Harbour Publishing, Madeira Park, B.C., 1992.

9. Upton, Joe. Inside Passage: Seafaring Adventures Along the Coast of British Columbia. Alaska Northwest Books, Bothell, Wash., 1992.

10. Walker, Spike. Nights of Ice. St. Martin's Press, New York, 1997.

Main References

A. Campbell, Kim. Time and Chance: The Political Memoirs of Canada's First Woman Prime Minister. Seal Books, McClelland-Bantam, Inc., Toronto, 1997.

B. Simpson, Jeffrey. The Anxious Years: Politics in the Age of Mulroney and Chrétien. The Globe and Mail, Lester Publishing Ltd., Toronto, 1996.

Acknowledgements

My thanks and no-thanks go to those at the University of Victoria, British Columbia, who helped and hindered me during 1983-1984: Thanks to helpful prof. Dr. Jim Hendrickson and fellow (older) student, Stonewall (sic) Jackson; no-thanks are due to academic harpy, Dr. Patricia Roy, and administrative gorgon, Ms. Charlotte Girard. "O.K. Thanks then, eh, and goodday!"

FISH, BIRDS, AND MARINE ANIMALS OF THE OLYMPIC PENINSULA
Chuck Rondeau

"Coho salmon"

"Steelhead"

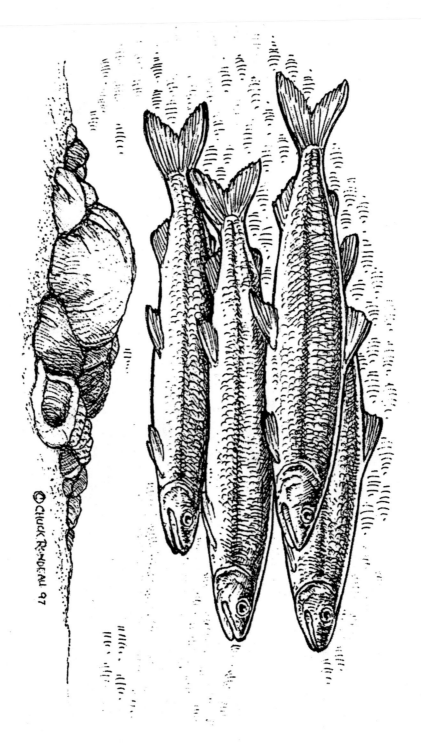
"Smelt"

"Albacore"

"Calico rockfish"

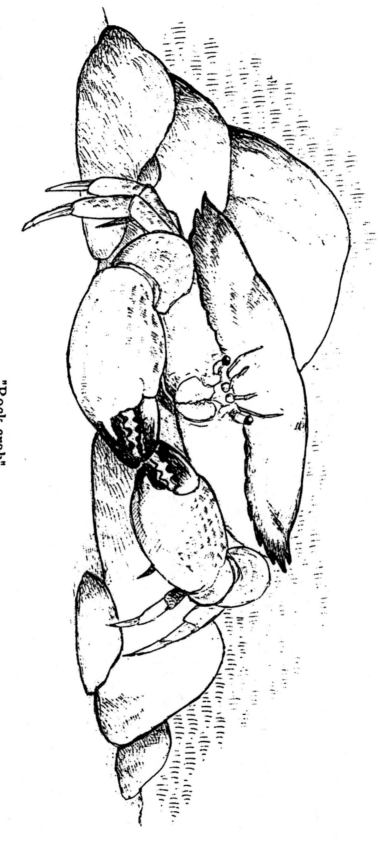

"Rock crab"

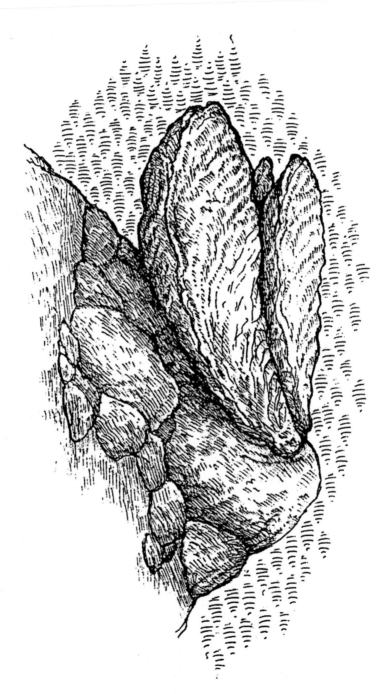

"Pacific oysters"

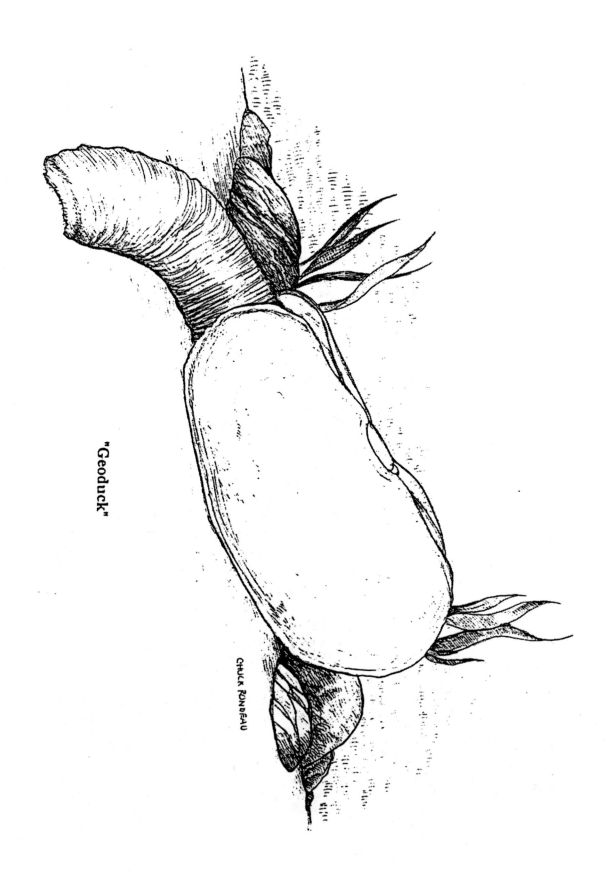
"Geoduck"
CHUCK RONDEAU

"Gooseneck barnacle"

"Blue heron"

"Ravens"

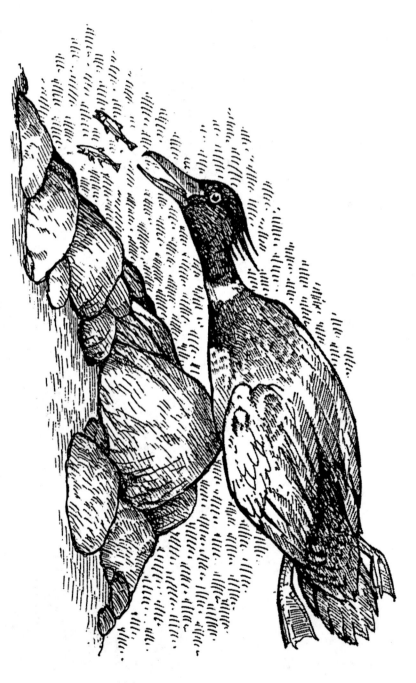

"Merganser"

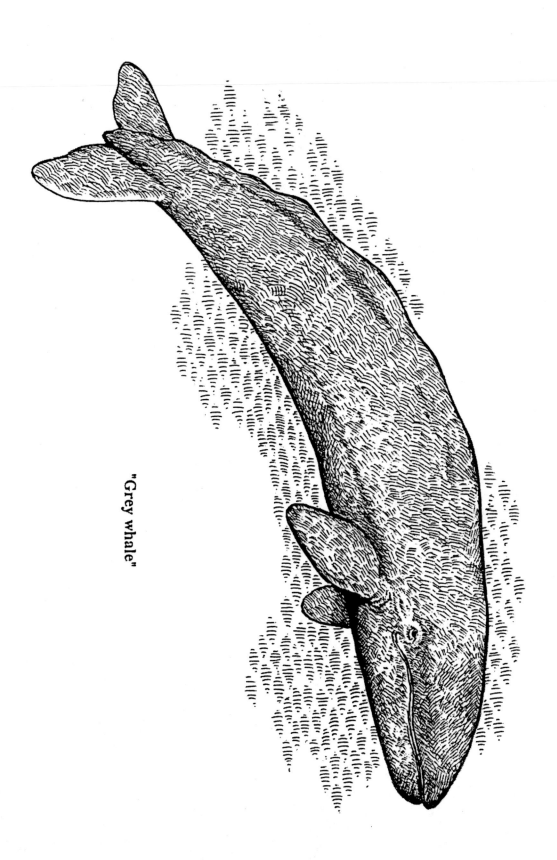
"Grey whale"

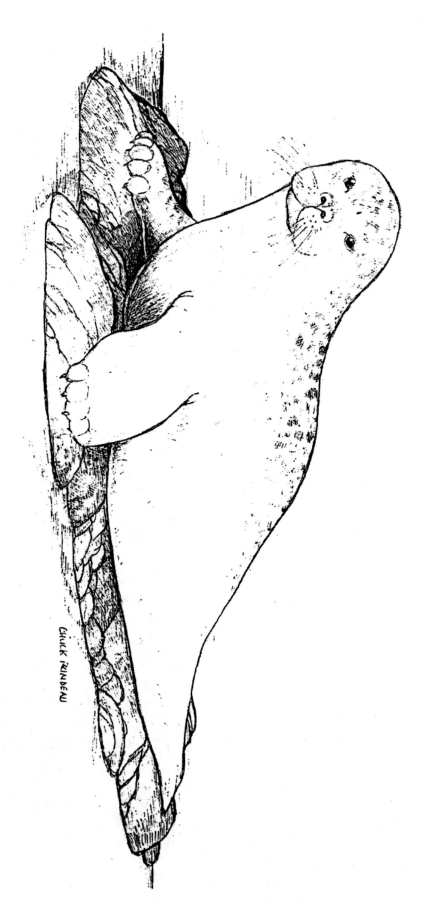

"Harbour seal"

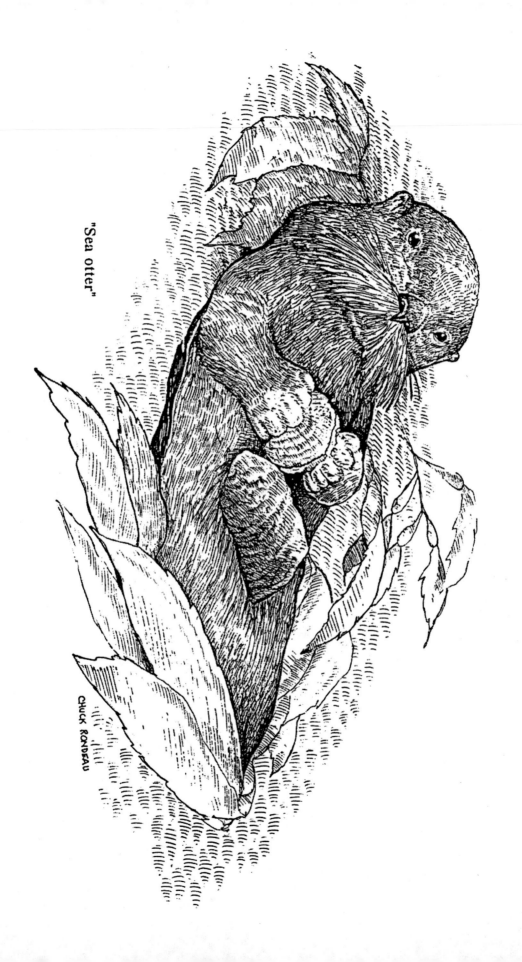

"Sea otter"

CHUCK RONDEAU

Chapter 3 : The Summing Up

Part   I: U. S. Elections 1996 -- A Retrospection

Part II: Future [Fish] Imperfect

Part III: "Ha det, og takk for all fisken!"
              ("So long, and thanks for all the fish!")

Part I: U. S. Elections 1996 -- A Retrospection

"Every country has the government it deserves" -- Joseph de Maistre

"In a city [Washington, D. C.] without watchdogs, the fox is the overseer" -- Ancient Sumerian proverb

"A wise fellow who is worthless always charms the rabble" -- Euripides

It is late January 1999 and all over the United States, from Buffalo (N. Y.) to Beverly Hills (Calif.), Clintonmania is running riot. Although an entire year has elapsed since the Monicagate story broke cover, Bill Clinton has never been more popular with the American people. At this writing President Clinton has already been impeached in the House for perjury and obstruction of justice, and is being "tried" for the two Articles of Impeachment in the U. S. Senate. Nevertheless, approval of President Clinton's job performance has never been greater than in current CNN/USA Today polls. Indeed, O. J. (Onan Jefferson) Clinton has never walked taller in the eyes of the U. S. electorate. A commensurable outcome has been that the presidential paramour, Monica Lewinsky, has erupted from the Oval Office (some say Oral Orifice) as the Modern Material Girl, temporarily replacing the shop-worn Madonna. As cynical critic H.L. Mencken (1880-1956) once observed, "No one ever went broke underestimating the intelligence of the American people."

And who would have "thunk" it? In the 1990s, though, anything (and everything) has seemed possible. This writer was born at the tail-end of the Silent Generation (1922-1945), coming of age along with many Baby Boomers (1946-1964) during the countercultural 1960s.[1] The Clintonian 1990s have changed the very language of the (counter) culture. All tobacco products, from chew to coronas, have been reduced to a "drug delivery system." The grievous sin of adultery has devolved into "alternative reproductive strategy". Bestiality -- human sex with animals -- is now defined as mere "interspecies communication." And a sex-change operation may today be described as "gender reassignment surgery." The debased, future-shock disposition of contemporary

American English suits perfectly the nightmarish visions of fear, chaos, and predation dispersed by the neo-Darwinian scientists of the New Physics. The talky boys and girly men at the Clinton White House are the chief practitioners of parsed, weasel-word doublespeak; effectively employing lingual légerdemain to defend the nefarious Clintonist agenda, while utilising verbal sleight of tongue to personally destroy their Boss Hogg's political enemies.

Has there ever been such a lethal tag-team trio in American politics as Clinton/Clinton-Gore? At the top of the U. S. power food-chain is POTUS (the official acronym for President of The United States), a compulsive liar, coupled to FLOTUS (the official acronym for First Lady of the United States), a pathological fraud. (The First Couple could conceivably be called the Duchess and the Whitewater Fox.)[2] Next comes VPOTUS (the official acronym for Vice President of the United States), the Vice Perpetrator in dialing-for [campaign]-dollars. (Let's call him "V[I]PER".) Bill Clinton is <u>Huckleberry Finn's</u> King, Al Gore is the Duke, and they have become as all-American as chop suey and pizza pie. With their political flapdoodle and environmental flim-flam, POTUS and VPER are updated sleazy caricatures straight out of Sam Clemens. Slick "trickerations"[3] of all kinds have been concocted by the Clinton White House, mostly attributable to the Great Arkansas Razorback himself at the epicentre. The transcendalist and essayist, Ralph Waldo Emerson (1803-1882), warned his countrymen of political demagoguery: "The louder he talked of his honor, the faster we counted our spoons."

But the American people have ignored Emerson's sage advice -- they turned out in droves (those that did) and reëlected Billy Liar (again) in 1996. Thus modern lawyer-author Vince Bugliosi's paraphrase, "You can fool most of the [American] people most of the time," was (again) proven right. POTUS, FLOTUS, and VPER, however, did their electoral homework during the <u>first</u> Clinton/Clinton-Gore administration, carefully heeding H.L. Mencken's dictum: "The whole aim of practical [i.e., pragmatic] politics is to keep the populace alarmed, and hence clamorous to be led to safety, by menacing it with an endless series of hobgoblins, all of them imaginary."

POTUS accomplished this by cleverly spreading fears of Social Security going bankrupt; FLOTUS (the single blot) tried but failed to establish/enforce Medicare/Clintoncare on America; but VPER succeeded in disseminating eco-panic with his sky-is-falling environmentalism.[4] Not only that, but the terrible troika of Clinton/Clinton-Gore managed to convince clueless American voters that the Republicans 1) actually <u>preferred</u> polluted air, contaminated soil, and dirty water; 2)<u>wanted</u> lonely and destitute senior citizens to subsist on cat-food; 3)<u>were threatening to take away</u> the hot school lunch programme from needy minority children. The tactics, of course, worked every time. All that the hapless House Majority Whip, Rep. Tom De Lay (R--Tex), could do was to honestly but lamely respond, "A lie unanswered becomes the truth." But the GOP --The Stupid Party -- failed to get its message out into the American mainstream, thereby unable to counter effectively. The Democrat prevarications of 1996 remained mostly unchallenged, and Clinton/Clinton-Gore retained power almost by default.

It was palaeo-Conservative columnist, Samuel Francis, who first called the GOP "The Stupid Party." It was also from the acerbic pen of Sam Francis, in <u>Chronicles</u> magazine, that we first read of Senator Bob Dole (R--Kansas) referred to as "an incompetent cripple", and Representative Tom De Lay described as "a harmless drudge." But the utter contempt of Sam Francis for the Republican non-campaign in 1996 was entirely justified. For the very same GOP campaign managers ("spin controllers") who lost the election for George Bush and Dan Quayle in 1992, funked and flubbed the race for Bob Dole and Jack Kemp in 1996. So early as May 1996, the feckless Republicans were sneeringly featured in a "Doonesbury" comic strip. Ultra-cool cartoonist Garry Trudeau (always on the Left side of pc) drew an anonymous voice-bubble, issuing from an upper White House window, containing the current wisdom: "The G.O.P. seems to have spent the last year redefining itself as the party of assault weapons, tobacco, pollution, medicare cuts, and government shutdowns."[5]

For instance, The Stupid Party couldn't even cash in on the song "Soul Man." "Soul Man" was a monster smash from 1967, sung by Sam & Dave; it also rhymed with "Dole Man." All the GOP "spin-meisters" had to do during the summer of 1996, was to call Sam Hayes and Dave Porter out of their retirement, offer them some decent campaign war-chest money (and free publicity for a possible show-biz comeback), and hit the campaign trail running -- and singing "I'm a <u>Dole</u> Man." The Republican candidate would be accompanied by Sam & Dave and their old sidemen.[6] This writer still believes that Sam & Dave and Bob might have turned the demographic tide that summer -- appealing to more Boomer, Buster, minority, and female voters -- showing the venerable Kansas senator to be the humourous, fun-loving, and likeable man that he is. But the people-unfriendly "spin-doctors" of The Stupid Party, naturally, never thought of anything like that. So they just went right on portraying Bob Dole as the angry, aggrieved, wounded World War II AmVet who would clean up the White House. But presenting permanently pissed off, <u>old</u> White Guys for political office doesn't win elections in 1990s America -- Janet Jackson's "Rhythm Nation" -- even if the candidate is a bona fide hero. Thus the crafted public image of Bob Dole lost the U. S. electorate too.

U. S. mass-media venom was especially reserved for a singular candidate in The Stupid Party primaries, one Pat Buchanan; a man so socially conservative, that he infuriated the pin-stripe liberal élites of Beltway, D.C., while embarrassing country club moderates in the big-top Republican tent. (Another socially conservative firebrand who found no place in the GOP circus tent was Alan Keyes. But Ambassador Keyes is an African-American, so the pc mass-media ignored him -- with malign neglect -- to political death.) In 1998, a Robert G. Sepic from Norman, Oklahoma, wrote a letter to <u>Chronicles</u> which exactly described U. S. political party differences: "The Republicans are but the Fabian wing of the Socialist Party, the Democrats being the Jacobins of the same."[7] But there was a particular animus against Pat Buchanan that went beyond the usual Jacobin/Fabian-collaborationist press. Buchanan fared no better in the plastic medium of television. In a February 1996 rendition of ABC-TV's <u>Nightline</u>, Buchanan was chillingly depicted as a Right-wing extremist and anti-Semitic bully.[8] An article appearing in a conservative periodical a year after Election '96, related further evidence of the trashing of Pat Buchanan:

..."Already any politician who dares to stray outside the Demopublican guidelines is viciously smeared with the sort of demented rhetoric laid upon Pat Buchanan, who in the Establishment press, was transformed from a popular, sensible TV commentator to 'America's ultimate Neanderthal hunter-gatherer who tiptoes not so lightly past the paranoid New World Order fantasies of the militia loonies'...'a jolly thug, saying mean things with a smile'...'has a lock on the clenched fist and camouflage fringe'...'Heil, Buchanan, etc.etc.'"[9]

But it was the politically-incorrect CNN <u>Crossfire</u> host who had electrified the somnolent 1992 Republican Convention at Houston, Texas. From the podium, overlooking a restive sea of cowering Stupid Party delegates, Pat Buchanan thundered:

"This election is about much more than who gets what. It is about who we are. It is about what we believe, it is about what we stand for as Americans. There is a religious war going on in our country for the soul of America. It is a cultural war as critical to the kind of nation we will one day be as was the Cold War itself."[10]

Pat Buchanan could have made the same speech during Election '96 with the identical result: The American People voted for the morally reprehensible Slick Willy anyway. What is deeply disturbing, however, is that The American People were quite aware in 1992 that their President-to-be was a scalawag and mountebank; by 1996, The American People knew full well that their Coward-in-Chief was also a charlatan and scoundrel. The great-big-fat-stupid, self-indulgent, dumbed-down American People of the 1990s reëlected Clinocchio anyway. (Whatever happened to all those Americans who voted <u>twice</u> for Ronald Reagan during the 1980s?)

Despite his present "trial" (January/February 1999) by the U. S. Senate, O.J. Clinton -- the hippest and the slickest of the sly cats -- will "walk", as the original O.J. (Simpson) walked. And the teflon Chief Executive has never ridden on a higher horse (if the liberal media polls are to be believed) in the unseeing eyes of The American People. An accurate reason for the Boy Clinton political victories of 1992 and 1996 have been given by Thomas Roeser, a thoughtful editor at <u>Chronicles</u>:

"Bill Clinton may be the most dishonored President in American history, but who is to blame for his ascension to the White House? George Bush, who waged an incompetent campaign for reelection? Bob Dole, tongue-tied and incomprehensible, unburdened by principle? Yes -- but the fullest answer is more simple.
"At the heart of it, Bill Clinton was elected because he's us. Sadly, as his sky-high job approval numbers show, he is the cultural distillation of what we have become, the microcosm of our nation."[11]

As a result, the ticket of Clinton/Clinton-Gore has successfully practised the politics of division, marginalising, and polarisation. The Great Empath (who feels <u>your</u> pain) and his cohorts have played the "race card" masterfully. "Divide and Conquer" has assumed a whole new meaning under the minorities/ womyn/ seniors electoral spoils system for officially-sanctioned entitlement groups; as instigated by POTUS, FLOTUS, and VPER. African-American Representative J.C. Watts (R--Okla.), a staunch Conservative, decried the Clinton/Clinton-Gore campaign flacks, who were sent out to garner Black votes and to ensure Black loyalty, as "race hustlers" and "poverty pimps."

The Clintonian "politics of meaning" has translated into "public values" in lieu of personal morals; hence Chinagate, (FBI) Filegate, Monicagate, Travelgate, Whitewatergate and Jane Doe numbers one through five. Shameless to the end throughout both his scandal-ridden administrations, President Clinton declared at a White House "Millennium Evening", 25 January 1999. '"You all know I am a walking apostle of hope'".[12] Before mentally relegating William Jefferson [Blythe] Clinton to the trash-heap of history, this writer must ask -- and answer -- one question. Q: How does Paula Corbin Jones <u>know</u> that the intrusive presidential phallus slopes (to the Left!) from Peyronie's Disease? A: The Artful Dodger/Flasher got up front and personal in that Little Rock hotel room, and Mrs. Jones has been telling the truth all along. At the time of Election '96, the writer scribbled an angry letter to his local newspaper, the <u>Peninsula Daily News</u>:

Port Angeles, Wn.
5 November 1996

Dear Mr./Ms. Editor!

After Bill Clinton is reëlected in 1996 -- be prepared:

1. For Hillary Care redux, with huge, "managed" HMO centres administered by newly-appointed bureaucrats and staffed by strange "health care providers";

2. For nicotine to be declared a dangerously-addictive drug by the FDA, and its use in any form --- i.e., as a "drug-delivery system" -- proscribed by law even in private cars and homes;

3. For yet more draconian environmental laws passed in the name of "biodiversity", accompanied by yet more vast federal land grabs (as recently in anti-Clinton Utah);

4. For the further empowerment of litigation lawyers, in place of much-needed tort reform;

5. For the complete triumph of the NEA, with yet more politicisation and dumbing-down of U.S. no-choice public education;

6. For America's young people to be increasingly schooled in New Age values and "alternative life-styles", as religious institutions and military academies steadily lose their tax-free status due to stricter standards of political correctness;

7. For the further advancement of the radical gay agenda in the name of gender awareness, along with mandated HIV/AIDS education -- i.e., condom distribution and needle exchange;

8. For the further implementation of Affirmative Action, "Mend it, don't end it," with its biased double standards in the name of "multiculturalism";

9. For the continued disarming of the civilian population, along with the concurrent expansion of trigger-happy federal agencies such as the FBI and [B]ATF;

10. For the entire Olympic Peninsula, with adjoining waterways, to be arbitrarily declared an eco-Disneyland by Al Gore, Bruce Babbit, and the EPA, where no-one will hunt, fish, log, or grow trees ever again.

Granted, the letter to the editor looks a bit strident, sounds a little shrill, more than two years later. But four more years of Clintonism was bad food for dismal thought. And this writer clearly recalls 5 November 1996. It was a crisp and sunny autumnal day, and the writer walked dejectedly through downtown Port Angeles, down Railroad Avenue to City Pier; GOP cap (still) clapped defiantly on balding head. The fall leaves lay brown and sere, and the writer distinctly remember three things on his mind: 1) Mr. Newt's (Gingrich) "new paradigm" had been sheer bunkum; 2) that The Stupid Party no longer deserved this voter's allegiance; and 3) Creedence Clearwater Revival's "Keep on Chooglin'" contained sound advice as well as being a good song.

A Good Norwegian Bites the Prairie Dust in Election '96

"A Scandinavian economist once proudly said to free-market advocate Milton Friedman, 'In Scandinavia we have no poverty.' And Milton Friedman replied, 'That's interesting, because in America among Scandinavians, we have no poverty either'".[13]

Barely mentioned -- this writer caught the election results in passing on PBS (of all channels!) -- was the defeat of Republican Senator Larry Pressler of South Dakota. The loss to American politics of Larry Pressler, the unbribable Congressman and incorruptible Senator, was a severe blow indeed, and upset this voter so much as the undignified demise of the Dole-Kemp ticket. There were two reasons for the writer's dolour. The first reason was political: Larry Pressler was a true-blue Conservative with an unblemished congressional record; unerringly voting on the Right side of issues. The second reason was personal: Larry Pressler was a real Scandihoovian boy who hailed from South Dakota, representing the heart and soul of Norwegian America. And Larry Pressler was just plain likeable, in the way that the late Hubert H. Humphrey (also of Norwegian descent) had been likeable, in a sense that transcended partisan or party affiliation. And with American politicians of Norwegian descent the voter didn't have much choice. There had been the late, blustery-boring Henry M. [Poop] "Scoop" Jackson (D--Wash.), or the cold, sour-faced Walter F. ("Fritz") Mondale (D--Minn.). But Larry Pressler was different.

Larry Lee Pressler was born 29 March 1942 in Sioux Falls, S.D. This is Norwegian "køntri" ("country"), the Northern Great Plains bread-basket dubbed "Frokostlandet" ("the Breakfast land") by Norwegian writer-folksinger Erik Bye (himself Brooklyn-born). Whenever the late Olav V of Norway visited

America, he would speak of stopping at "norske-distriktene" ("the Norwegian districts"), which meant the North Central states and the Pacific Northwest. The states of the Upper Midwest are famous in Norwegian-American song and story; in the lives and works of authors such as Thorstein Veblen, Knut Hamsun, and O.E. Rølvaag. The latter, Ole Edvart Rølvaag (1876-1931), was the great chronicler of Norwegian Americana, and especially of South Dakota. This writer grew to maturity reading the great Rølvaag trilogy of Giants in the Earth (I de Dage, 1924), Peder Victorious (1929), and Their Father's God (1931). But O.E. Rølvaag was more than the poet of the Dakota prairie; he helped found the Norwegian American Historical Society in 1925. Rølvaag recognised the clash between transplanted and native cultures in the United States, and advocated the then-radical notion that American culture and society would be richer if immigrants kept their old customs alive in the new land.

Sioux Falls, South Dakota, still serves as the epicentre of ethnic Norwegian "Ole and Lena" jokes, published and distributed by local humourist, "Red" Stangeland. But next-door Minnesota is crushingly pc. The international headquarters of the Sons of Norway fraternal organization is located at Minneapolis, "Moscow on the Mississippi." An elderly S/N Lodge member once proudly informed this writer, that a Sioux Falls high school was the first in the nation to allow gay couples to attend Prom Night. Whether the apocryphal story was true or not, is there something rotten in the Heartland?

There is, according to the 1996 movie, Fargo (starring William H. Macy and Frances McDormand). The writer watched Fargo on television, aired between Thanksgiving and Christmas 1998 over the Turner Broadcasting System. Was Fargo, a grim offering directed/produced by the Coen brothers, intended as a requiem for Norwegian America? It surely looked that way! Every dark Scandinavian stereotype was there -- from the lowering Hardangerfele (fiddle) music introducing the film's lugubrious milieu; to the flat, frozen, featureless landscape drive-bys (including Brainerd, Minn., with statue of Paul Bunyan and Babe, the blue ox); to the greedy, weak, and cruel Bergmannesque character of Fargo's main male persona, "Jerry Lundegaard" (William H. Macy). There occurred no tragic incidents; only pathetic happenings. And the viewer found no heroes in Fargo except heroic State Trooperette "Marge Gunderson" (Frances McDormand), who brings in the remaining crazed killer-at-large to justice. To rub in some heavy-handed Feminism, the Coen brothers have "Officer Gunderson" very pregnant at the time of capturing the (type-cast, blond-haired) armed and dangerous second goon. But "Marge Gunderson", in turn, is proud of her loutish husband, "Norm", whose mallard etching has won a contest; and has been accepted by the U. S. Postal Service as the model for a future 3¢ stamp. Although well acted and strangely compelling, Fargo left this viewer with a nagging, disquieting, almost existential angst that all is not well in the Heartland, in Scandihoovian køntri.

But good-guy Larry Pressler came out of the icy Great Plains. This writer would like to report that Larry Pressler is a product of one of the many Lutheran colleges that still dot the Upper Midwest, but that would be untrue. Like Bill Clinton, Larry Pressler studied as a Rhodes Scholar at Oxford University; unlike Bill Clinton, Larry Pressler is a Vietnam veteran who served in the U. S. Navy from 1966 to 1968, before attending Harvard Law School in 1971. Larry Pressler continued to serve his country both in the House and in the Senate, throughout the '70s, '80s, and '90s. Relating a small segment of Larry Pressler's story reminds the writer of another, earlier movie -- Frank Capra's pre-World War II "Mr. Smith Goes to Washington," starring the late, great James Stewart. In that cinematic classic, Jimmy Stewart played an innocent abroad from the Western prairies, an honest congressman in Washington, D. C., whose political naïveté even frustrates his more worldy gal-Friday, "Saunders" (Jean Arthur). But "Mr. Jefferson Smith's" determined idealism prevails; his long, brave filibuster on the congressional floor eventually exposes the crooked politicoes in the back "smoke-filled rooms." At the time of the film's 1939 release, Joseph Kennedy, the U. S. ambassador to Great Britain, is quoted as having said, "I feel that to show this film in foreign countries will do inestimable harm to American prestige all over the world.'"[14] (Joe Kennedy Sr. did his own irreparable harm which has been well documented.)[15]

For more than twenty years, Larry Pressler was this writer's favoured "Mr. Smith"-type candidate to eventually run for president, win through character and achievement, and take office as the first Norwegian-descended First Executive.   Larry Lee Pressler was not reëlected Senator from South Dakota. (But neighbouring North Dakota sent two of the most relentlessly Clintonist senators back to Washington, D.C. Pressler himself was replaced by a Clinton apologist. Is there "values clarification" in the Heartland?) William Jefferson Clinton, however, was reëlected President of the United States. Had the egregious Joe Kennedy been right after all? Are there in 1999, sixty years later, any "Mr. Smiths" left standing (tall) in the U. S. House or Senate?

The American Century, 1898-1998

America is..."the last, best hope of earth"
--Abraham Lincoln, 16th President of the United States
"The late, great United States"
--Michael Reagan, conservative radio talk-show host

Alexis de Tocqueville (1805-1859), the famous French author and statesman, extensively toured the United States during the early nineteenth century. After his travels, Tocqueville remarked about the young republic, "America is great because America is good." At the time, the perceptive French visitor had noticed -- and also commented on -- the strong Christian faith of the American people. More than a century and a half later, Americans still feel they are a special people, but not because of their religious piety. Throughout this work there have been numerous references to "The American Century"; a phrase coined by <u>Time-Life, Inc.</u> titan Henry Luce, more than a half century ago. There are (too) many confused Americans in 1999 who equate the amount/degree of "smart bombs", digitisation status, bull markets, and bad movies produced in Hollywood, with American greatness. Instead of educated foreign visitors praising America, it is American citizens patting themselves on their collective, T-shirt logo'ed "We're Number One", back. A fitting example of this is <u>The American Century</u>, a ponderous (700 plus pages), expensive ($50.00 and up), self-congratulatory coffee-table tome by Liberal-leftist Harold Evans.[16] The deep irony -- to this writer -- is that 1898 and 1998 have been watershed years in U. S. history. In 1898, after her (not-so) "splendid little war" (coined by future Secretary of State John Hay) with Spain, America embarked on her rocky path to world leadership and global supremacy. By 1998, America had become an international laughing stock because of her unstable, vacillating foreign policy, and the scandalous antics of her adulterous, perjurious president. And the mid-term elections of 1998 proved (yet again!) that U. S. voters generally identified with the public values, while ignoring the personal immorality, of their president. The shifting paradigm of 1990s American thought and lifestyles assured victory for Clintonian candidates. The Great Empath himself, mainstream America now firmly believed, did it all for minorities (both racial and sexual), women, seniors, and especially "the <u>children</u>." Thus,

The American Century in truth has lasted a mere <u>half</u> century -- with its zenith in 1945 and its nadir in 1998.

How has this collapse in social mores, the root cause of American decline since World War II, come about? This writer is convinced that the answer is simple, and may be given in a sort of "Reverse Decalogue" (a backasswards Ten Commandments, if you [the reader] will):

1) Absolute truth has been replaced by situational ethics,
2) equality has become egalitarianism,
3) freedom has become license,
4) intellect has been replaced by will or opinion,
5) money has replaced manners and celebrity achievement,
6) personal morals have been replaced by public values,
7) reality itself has become social construct,
8) reason has become emotion,
9) responsibilities have become rights,
10) transcendent standards have become relativist values.

And U. S. pluralism has fragmented into particularism (viz. "multiculturalism"), or as VPER Gore once said, "<u>E Pluribus Unum</u>' means 'Out of one, many'" (or words to that Freudian Slip effect).

NOTES

1. NB: This writer has characterised recent decades as the Fighting '40s, the Fabulous '50s, the Sick '60s, the Silly '70s, the AIDS '80s, and the Gay '90s. PS: Baby Busters (1965-1983)

2. Inspired by the 1976 film comedy, "The Duchess and the Dirtwater Fox", starring George Segal and Goldie Hawn.

3. An effective word used by former heavyweight boxing champion Larry Holmes (world title holder 1978-1985), in reference to promoter Don King's financial wheeling, dealing, and cheating. Cf. <u>Larry Holmes: Against the Odds</u> (New York: St. Martin's Press, 1998), with Phil Berger, <u>passim.</u>

4. Albert (Arnold) Gore, Jr., <u>Earth in the Balance: Ecology and the Human Spirit</u> (New York: Plume, 1993).

5. <u>Peninsula Daily News</u>, 'Comics', 12 May 1996, p.2.

6. NB: "Soul Man" and "Hold on! I'm a-comin'" by Sam Hayes and Dave Porter. Flashback Records, Los Angeles, Calif., 1997. (Orig. 1967, Atlantic Recording Corp.)

7. Robert G. Sepic, letter to editor, <u>Chronicles</u>, October, 1998.

8. Ted Koppel, 'Hometown Boy', <u>ABC NEWS Nightline</u>, February 1996.

9. 'A New Kind of Government: The elected Dictatorship', <u>McAlvany Intelligence Advisor</u> (Phoenix, Ariz.), Fall 1997, p.10. NB: Hence the irrational fear of, and inexplicable antipathy for, Pat Buchanan by otherwise charitable Jacobin Ted Koppel, and as-a-rule reasonable Fabian, George Will. (PS: After sitting aghast through the Walter Winchell-type "hatchet job" on Buchanan by ABC News, this writer stopped watching <u>Nightline</u> altogether. This, after having been a faithful viewer since the programme's inception since about 1980.

10. Exerpt from Pat Buchanan's speech at GOP Convention in Houston Texas, August, 1992.

11. Thomas Roeser, ed., "Clinton Portrait", <u>Chronicles</u>, Vol. 22, No. 11, November 1998, 'Cultural Revolutions' page.

12. Maureen Dowd, 'Holy Father and Very Prodigal Son', <u>Peninsula Daily News</u>, 28 January 1999, p.A-4. NB: Both POTUS and FLOTUS are lawyers, and have been "walking apostle[s] of hope" for litigation lawyers both within the Clinton White House and without. Tort reform, anyone?

13. P.J. O'Rourke, <u>Eat the Rich</u> (New York: Atlantic Monthly Press, 1998), p..69.

14. Leslie Halliwell, <u>Halliwell's Film Guide</u> (New York: Harper & Row, Publishers, 1989, ed.), pp. 686-687.

15. Cf. Ronald Kessler, <u>Sins of the Father</u> (New York: Warner Books, Inc., 1996), <u>passim</u>.

16. Harold Evans, <u>The American Century</u> (New York: Alfred A. Knopf, 1998).

Bibliography for Part I

1. Aldrich, Gary. Unlimited Access. Regnery Publishing, Inc., Washington, D.C., 1996.

2. Atkinson, G.L. The New Totalitarians. Atkinson Associates Press, Clinton, Md., 1996.

3. Bork, Robert H. Slouching Toward Gomorrah. Harper Collins Publishers, New York, N.Y. 1996.

4. Feder, Don. Who's Afraid of the Religious Right? Regnery Publishing, Inc., Washington, D.C., 1996.

5. Gingrich, Newt. Renewing America. Harper Collins Publishers, New York, N.Y. 1996.

6. Lewin, Leonard C. Report from Iron Mountain. Simon & Schuster, Inc., New York, 1996. (Orig. published in 1967)

7. Olasky, Marvin. The Tragedy of American Compassion. Regnery Publishing, Inc., Washington, D.C., 1995 ed.

8. Walter, Jess. Every Knee Shall Bow. Harper Collins Publishers, New York, N.Y., 1996.

Part II: Future (Fish) Imperfect

"Fishermen will stand by it [the Dead Sea] from En Gedi to En Egla'im; there will be places for spreading their nets. Their fish will be of the same kinds as the fish of the Great Sea [the Mediterranean], exceeding many."
--Ezekiel 47:10, Holy Bible (NKJV)

...[T]he key to survival is enlightened local control of natural resource use."
--Carl Safina, Song for the Blue Ocean (1998)

Was the Old Testament prophet Ezekiel describing world fishing in the New Millennium? Or will nothing less than a God-given miracle save the globe's oceans, seas, and the fishes therein? Ever since the United Nations declared 1998 Year of the Ocean, a plethora of articles and stories in various print media -- plus a variety of televised programme "specials" on the oceans -- have drummed into the reading/viewing public an oft-repeated eco-dirge: Global waterways are dying and the fish are perishing with them. Besides industrial pollution, international fishing has been the culprit to blame. As a consequence, much world attention has been focused (with much nonsense written and reported) on the international fisheries, with their perceived [purely] deleterious effects on the planet's marine life. An irony, seemingly not noticed by academia, has been that the vast majority of those loudly decrying the demise of the marine resource due to international overfishing, would proudly identify themselves as Darwinian Evolutionists.

Yet the knowing minds at the British Museum or the Smithsonian Institution would be boggled to learn that it was no less a personage than British scientific philosopher Thomas H. Huxley (1825-1895), that arch-paragon of Evolutionism, who announced at the 1883 International Fisheries Exhibition that: "Any tendency to over-fishing will meet with its natural check in the diminution of the supply...this check will always come into operation long before anything like permanent exhaustion has occurred.'"[17]

Thus it was that "Darwin's bulldog", a darling still of the intellectual élites, had pooh-poohed the very idea of overfishing as a reason for depletion of the marine resource. T.H. Huxley, who had been appointed to three British fishing commissions, made the above remarks at the International Fisheries Exhibition held at London, England; attended by the world's major fishing nations at the time.[18] T.H. Huxley had served as primary interpreter and exponent of the Darwinian lesson that nature acted as a great, indomitable, force that held the key to solving all life's problems. Darwinist eco-theory, along with high hopes for scientific progress, were fashionable credoes in the nineteenth century salons of the Western intelligentsia. Hence T.H. Huxley's address, a self-confident explication of why prevailing fears of overfishing were scientifically irrational, went virtually unchallenged by academia in 1883.[19]

But times have changed -- and so has the technology of fishing. More than a hundred years after the International Fisheries Exhibition, Darwinism has remained an atrophied orthodoxy -- but fisheries technology has made a quantum leap. Mark Kurlansky, author of Cod: A Biography of the Fish That Changed the World, has described the harvesting of the Atlantic cod during the 1990s:

"[T]echnology continued to focus on the goal of catching more fish. Factory ships grew to 450 feet or larger, with 4,000-ton-capacity or more, powered by twin diesel engines of more than 6,000 horsepower, pulling trawls with openings large enough to swallow jumbo jets. The trawler hauled its huge net every four hours, twenty-four hours a day. Pier fishing, a technique often practiced by the Spanish fleets out of Vigo [Galicia][20], suspended a huge trawl between two factory ships. One operated the trawl, and the other processed the fish. After the net was hauled up, the vessels switched roles and continued, so that the fishing never stopped."[21]

Kurlansky also wrote of "rockhoppers" and "ticklerchains" as extra aids (as if needed!) in trawling for codfish. "Rockhoppers" are large discs, which have replaced regular rollers along the bottom of nets, that hop up when hitting a rock; thus making it possible to drag close to a rough bottom without damaging the net. "Ticklerchains" stir up the sea bottom, causing noise and scattering dust. The ticklerchains then act like bush-beaters to drive game-birds from out of cover -- in this case the frightened cod, cowering in protective nooks and crannies along the bottom -- into the awaiting nets. The results of these methods have contributed to making "[t]he ocean floor left behind...a desert."[22]

The computerised, high-tech/high-sea fishing fleets of today catch more fish, more efficiently. And international governments have encouraged -- via subsidies -- their fishing industries to expand faster than the marine resource can reproduce itself. In June 1997, the World Wide Fund for Nature and the U.N. Environment Program blamed "the global fisheries crises" on the $50 billion-plus paid in subsidies by the international governments to their fishing industries.[23] The WWF/U.N.O. logic (sound in this instance) perceived subsidies as falsely inflating the profitability of fisheries, thereby engendering new investment in vessels and men...adding to the vessels and men already fishing for a much-diminished marine resource. Asserted WWF International Director Claude Martin: "They send the wrong economic signal to participants in depleted fisheries by creating incentives for...high levels of fishing.'"[24] International governments, asseverated the WWF/U.N.O., continue to pour money into their overcapitalised and overcompetitive fishing fleets, which continue to lose money while depleting fish stocks. According to the WWF/U.N.O. report, the capacity of the world's fishing fleets has grown five-fold, while the productivity of the world's major fishing areas has declined.[25]

By October 1997, the National Marine Fisheries Service (NMFS) submitted to Congress a list of 96 species of fin-fish and shellfish considered to be over-harvested by American commercial and sport fishermen.[26] The species named by the NMFS included: "Pacific salmon, American lobster, bluefin tuna, several types of flounder, Atlantic sea scallops, several Atlantic sharks, swordfish and pink shrimp."[27] (The NMFS report was required by the reauthorised Magnuson [Stevens] Fishery Conservation and Management Act [MFCMA]; the latter so renamed due to the Herculean labours of Alaska Senator Ted Stevens.) The NMFS director of sustainable fisheries, Gary Matlock, has flatly stated, "It's time to eliminate the problem of overfishing'."[28] The NMFS used data from the eight regional fishery management councils around the United States, which, in 1997, had 39 different management plans at work for species such as groundfish in New England, shrimp in states along the Gulf of Mexico, and salmon in the Pacific Northwest.[29]

Five additional fishery management council plans were under development in 1997, with ten more in the offing; each containing a mandatory ten-year time-limit to rebuild failing fish stocks. Although the total worth of the U.S. commercial industry was reckoned in 1997 at $3.5 billion, not everyone in the industry was happy with the NMFS report submitted to Congress -- or with the NMFS-mandated outcomes.[30] Commented Richard Gutting, Jr., a vice-president of the National Fisheries Institute (which represents well over one thousand seafood companies): "It's fair to say that a likely outcome will be further cutbacks in harvest quotas, further restrictions on the size of the catch, the seasons and the places where fishermen can operate, the type of gear they can use.'"[31]

The Pacific Fishery Management Council had been busy doing its ground[fish]work for some time before the 1997 NMFS report, having anticipated drastic federal fishing reduction mandates. When the Council was to meet in September 1997 at Portland, Oregon, it had already considered cuts of up to 65% for sablefish (black cod), 35% for long-spined thornyheads, 20% for short-spined thornyheads, and 19% for Dover sole during the coming 1998 season.[32] (Both thornyheads are marketed primarily in Japan.) According to the Associated Press, deepwater species earned $48.3 million for American fishermen in 1995;"about half the value of fish landed on the West Coast outside the huge Pacific whiting fishery."[33] The Council in 1997 also mulled over a decision to reduce ling cod landings by 60%, plus a 34% cutback in widow rockfish (sold as "snapper").[34]

For the Pacific Fishery Management Council, those were truly draconian steps. Ling cod, black cod, Dover sole, and rockfish ("snapper") are all popular seafood items on West Coast dinner tables. Not only that, but placing bag-limits on ling cod and rockfish would also seriously annoy the numerous West Coast recreational fishermen and the rich, powerful sport-angler lobby.[35] By the end of 1997, the U.S. federal government had declared its intention to activate restrictions on commercial landings for 83 species, among them West Coast favourites ling cod, black cod, Dover sole, and rockfish ("snapper").[36] The U.S. Department of Commerce then announced that groundfish hauls would be lowered by a whopping 65%; commencing effectively 1 January 1998.[37]

Richard Methot, director of the Seattle-based fish-monitoring division of the NMFS, has rationalised the drastically-reduced groundfish catch (35%) remaining to American fishermen by saying that the strict new limits "may be a case of erring on the side of safety, but they are being put in place to assure that we have fish for the future.'"[38] Rod Moore, executive-director of the Portland-based West Coast Seafood Processors Association, has countered with, "It's going to be a rough year.... We're looking at an overall harvest reduction of about one-third. Not everyone's going to survive that.'"[39]

"Erring on the side of safety" has never been the concern of biologist/activist and first class eco-irritant, one Elliott Norse; a party often mentioned and cited throughout this work.[40] Formerly chief scientist at Washington (D.C.)-based Center for Marine Conservation, Elliott Norse founded the Marine Conservation Biology Institute (MCBI) at Redmond, Washington, in 1996.[41] Dr. Norse is the proud possessor of a Ph.D. in marine ecology; has worked for the Environmental Protection Agency (the dreaded EPA), the President's Council on Environmental Quality, the Ecological Society of America, and (of course) the busy-body Center for Marine Conservation.[42] Besides his 1993-published Global Marine Biological Diversity,[43] Dr. Norse has written or edited two other works calling for "biological diversity" -- a catchy eco-buzz phrase long credited to that worthy. According to Pacific Fishing magazine, Dr. Norse started up the MCBI with "seed money" donated by the National Oceanic and Atmospheric Administration (NOAA) and from "several [other] sources"; no doubt the same fat foundations which liberally finance such eco-nuisance organisations as the Center for Marine Conservation.[44]

Dr. Norse's most recent eco-bomb was dropped at a 15 December 1998 press conference, held at Boston, Mass. The good doctor's presentation coincided with the American Oceans Campaign (1998 being the U.N.O.-designated Year of the Ocean), and an article co-authored by him appeared in the Journal of Conservation Biology. The article in question had paralleled trawling at sea to clear-cutting on land, with "clear-cutting the ocean" posing a major threat to both economic sustainability and biological diversity.[45] The reaction of the U.S. industry was strong and immediate. Brad Warren, editor of Pacific Fishing, wrote a measured reply the following February (1999):

"Historically, such [trawling] restrictions have been enacted mainly for three reasons: (1) to protect established non-trawl fishing interests from trawler competition; (2) to reduce conflict between fishing fleets (which has sometimes flared into episodes of ramming and even arson); and (3) to ease pressure on fish stocks.
"Norse and some of his like-minded colleagues are pursuing a fourth and more recent objective: to preserve biodiversity. They argue that bottom trawling -- really more like farming than logging -- plows the seafloor, making it unsuitable habitat for many species. Those species need protection from the ravages of human food production, Norse and his associates contend."[46]

Not all scientists agree with Dr. Norse and his cohorts. After the December 1998 press conference at Boston, a prominent fisheries-gear researcher at the Massachusetts Institute of Technology Sea Grant programme, one Cliff Goudey, was quoted in FishNet USA as observing, "Dr. Norse appeared to have left his scientific credibility at the door.'"[47] And according to The Washington Post, NMFS deputy director Andrew Rosenberg has commented on the scientific findings of Elliott Norse and co. by remarking that, "They haven't shown that productivity of the ecosystem has been demonstrably changed.'"[48]

The response from the West Coast industry has been simultaneously predictable and heated. John Gauvin, director of the Groundfish Forum, has indignantly stated, "In the North Pacific, there are more areas closed to trawling than all of Georges Bank [off New England] is open.... The environmental groups have incredible P.R. machines'."[49] And the previously-cited Rod Moore, executive-director of the West Coast Seafood Processors Association, has ventured that, "My impression is that there are those who will try to keep the issue alive, but the group in the environmental community doing most of the trawl bashing at the moment is more of a fringe group. The real issue they're [Norse and co.] trying to get at is marine [p]reserves.'"[50] Having read Dr. Norse's ideas on Big Blue regulation in <u>Global Marine Biological Diversity</u> (1992), this writer agrees with Rod Moore et al. and says "Amen"! But the last word cited in behalf of the local industry must be that of the thoughtful editor of <u>Pacific Fishing</u>, Brad Warren, who has concluded:

"Catchy metaphors about clear-cutting may draw public attention to the ocean's health, and that's not bad. But can this new group of trawl critics really shed light on the hard questions about how to feed a hungry world without wearing out its resources? Is it wise to call for more fishing closures when you don't even know what closures already exist? Norse and his associates might do better by boning up on existing protected areas, and studying their effects. Additional marine preserves are worth a try if they hold real promise as a remedy to some fisheries conservation problems. Like any other fashionable but costly cure, though, they shouldn't be prescribed just because they sound good.'"[51]

Indeed, Elliott Norse and (in the words of the Associated Press) "1,600 scientists from around the world" had kicked off the Year of the Ocean by issuing dire warnings (in early January 1998) about rapid, "unprecedented damage to oceans" they ascribed to the usual suspects -- overfishing, pollution, and coastal development.[52] Dr. Norse and his fellow ec scaremongers trumpeted abroad their consensual wish that U.S. lawmakers establish new coastal [p]reserves, further empower the Endangered Species Act and Clean Water Act; that President Clinton hold (viz., <u>stage</u>) a White House conference on oceans.[53] But for every action there is a reaction. An interesting alternative to Big Green oceanic regulation (disturbing images of heavily-armed, blue-helmeted eco-police come to mind) has been proposed by the free-market think-tank Cato Institute; a solution that is certain to cause a wailing and gnashing of teeth among Dr. Norse and his MCBI minions at Redmond. For the Cato Institute has suggested that, as global fish populations become depleted, the answer lies in <u>"privatizing"</u> the fisheries![54] Jerry Taylor, the Cato Institute's director of natural resource studies and senior editor of its <u>Regulation Magazine</u>, has made plain his organisation's position:

"[T]he environmentalists actually seem to have put their finger on a problem.... But for the most part, we have an economic problem, not a resource problem.... The best remedy is not more habitat protection and marine sanctuaries and government management. It is to privatize the [marine] resources.'"[55]

In a reverse interpretation to the usual neo-Marxist construct of the "tragedy of the commons", Jerry

Taylor as Cato Institute spokesman, has freely admitted that as no one [single world government or global corporation] actually owns the vast ocean resource, international fishing fleets benefit from overfishing in the short term -- while losing [the marine resource] in the long run.[56] Taylor has concluded that "The incentives are wrong and in many cases the result is damaged ecosystems'."[57]

It is close to three decades since Wesley Marx came out with his alarmist (albeit well researched, well written) The Frail Ocean.[58] The dutifully eco-concerned Marx, perhaps seeing himself as a natural[ist] successor to the late-lamented Rachel Carson (of The Silent Spring), was the first of three eco-authors/ocean environmentalists envisioned by this writer to issue a hypothetical State of the Oceans, 2000 A.D. -- as hope for the new millennium.[59] At this writing, early March 1999, no sign has yet appeared that Wesley Marx will be forthcoming with a revisionist follow-up to The Frail Ocean of 1970. Indeed, the eco-worried Marx might well have perished during the thirty-year interim from extreme eco-high anxiety.

Another eco-author/ocean environmentalist whom this writer was hoping could make a positive eco-statement on Oceans 2000, had been Jacques-Yves Cousteau. But as the reader knows, the good Captain Jacques expired 25 June 1997 at Paris, France. Last and maybe least, this writer imagined the possibility of old Norwegian Kon-Tiki hero, Thor Heyerdahl, as giving Oceans 2000 a passing (if not clean) bill of health. But even when this writer first "proposed" the make-believe State of the Oceans, 2000 A.D. back in 1997, Thor Heyerdahl -- for all his good qualities and great achievements -- was the third choice. The thought was then that Thor Heyerdahl had been living far too comfortably, for too long, in his senior celebrity status to attempt a reversal of a lifetime of eco-pessimism by issuing a positive, declarative statement on Oceans 2000. Thor Heyerdahl remains a real hero...but still a poor choice.

It is now March 1999, and obviously the envisioned eco-revisionist State of the Oceans, 2000 A.D. has neither been written nor will be. However, this ever-searching writer has found something close -- different, certainly, from the needed high optimism for our young people; but a ray of hope for Oceans 2000 nonetheless. That eco-statement has been Song for the Blue Ocean by Carl Safina (1998), and it is this writer's considered opinion that this book ought to receive far more attention than it has so far.[60] Although not agreeing with all of Carl Safina's concepts ("'the precautionary principle'" taken too far), or sharing his enthusiasm for the U.N.O. (which says..."we must err on the side of caution"), this writer found Song for the Blue Ocean sober but up-beat. Safina's basic eco-statement was thus:

"In the years since I began these travels in the early 1990s, the two possible fates for the world's oceans have...been duking it out, and the forces of good have landed a few solid punches."[61]

Safina went on to emphasise that the reauthorised Magnuson Act (MFCMA) of 1996-1997 resulted from the voluntary coöperation of over 125 American marine resource user-groups.[62] These ranged from "sportfishing clubs to scuba groups to conservation organizations to commercial fishing associations to scientific societies (perhaps the largest coalition ever assembled to work on an environmental issue in America)...."[63] To his great credit in this era of eco-pessimism, Safina has chosen three inspirational exemplars of environmental self-policing, corporate eco-leadership, and transnational volunteerism to sum up his book:

1) U.S. long-line vessels fishing off Alaska have actually requested regulations requiring gear which won't snag albatrosses or other marine birds;
2) giant Euro-conglomerate, Unilever, the continent's largest purchaser of seafood, has willingly forfeited millions of dollars by ending its practise of mass-buying small fish to spare the greater sea food-web;
3) Japan (the globe's favourite eco-villain) has freely offered to pay for the building of a new Coral Reef Centre for the tiny Republic of Palau (Belau, Micronesia), so that the local people and foreign tourists can see and learn about the unique coral splendour found in still-living Palauan reefs.[64]

Finally, Carl Safina has referred to our own Pacific Northwest as the ideal model in solving the eco-problems of world fisheries: ..."[T]he key to survival is enlightened local control of natural resource use."[65] And enlightened local control of natural resource use has been an underlying theme throughout this work.

The cover story of the March 1999 edition of Discover magazine (Vol. 20, No.3) was titled, 'Wonder Drugs from the Deep', and subtitled, "Potent new cures from the ocean that will revolutionize medicine." The ocean has provided sustenance and healing for humankind since history began. In latter years, the ocean and seas have served as a primary source of nutritional (sea)food and powerful medicines on a massive scale; of which early Man -- whether in Norway or on the Northwest Coast, in Japan or on the Pacific Islands -- could merely dream. Today, technology is literally plunging into planet Earth's final frontier -- the depths of the ocean and the seas -- in modern Man's quest for disease-fighting weapons, and to stave off death. In the Darwinian struggle for life between the waves, millions (and millions!) of marine organisms produce specialised chemicals which have multiple uses -- from discouraging predators to restraining growth in competitors. Many of these naturally-produced chemical compounds are potential pharmaceuticals beneficial to Man. During the 1990s, marine scientists have increased exponentially their search of the underwater drugstore/chemists' shop that is The Deep.[66]

The whole realm of marine pharmacology holds especial interest for this writer, who has over the years kept a separate notebook on the subject. In carefully reviewing these copious notes, the writer can but simply reiterate that maintaining a healthy ocean ensures the continuing improved health and extended longevity of humankind. Listed below are just a few of those stirring instances:

In The Frail Ocean, Wesley Marx wrote (so far back as 1970) that the adhesive substance secreted by barnacles to anchor themselves to rocks, was being investigated as a potential bacteria-fighting dentifrice.[67] In the same work, Marx also mentioned that medical scientist C.P. Li, of the National Institute of Health (NIH), had found that "paolin" -- a constituent isolated from abalone "liquour" -- would inhibit polio in laboratory rats.[68]

In early 1994, Oregon marine chemist William H. Gerwick, a professor of pharmacy at Oregon State University, discovered an anti-cancer ingredient in a tropical alga, Lyngbya majuscula.[69] Gerwick and an OSU team first found the alga off Curaçao, near Venezuela in 1991.[70] The OSU scientists returned in late 1993 to collect further specimens of L. majuscula, when tests showed that a component they called "curacin 'A'" inhibited cancer cells from dividing.[71] Said William Gerwick -- who received pending patent approval for curacin 'A' --"It shows some selectively for colon cancer and perhaps for breast cancer.'"[72]

Another promising anti-cancer compound from the sea which received much attention in 1994, from both science and medicine, was "bryostatin '1'."[73] This chemical component is extracted from bryozoans, the minuscule and spongelike marine growths attaching themselves in colonies to docks, boat-bottoms, and even seaweed. George R. Pettit, an organic chemist and director of the Cancer Research Institute at Arizona State University, has stated that bryostatin '1' appears to be effective in treating melanoma, lymphomas, leukemias, and ovarian cancer.[74] The marine environment is "incredibly diverse biologically and is enormously complex'", has pronounced William Fenical, director of the Marine Research Division at the Scripps Institution of Oceanography, LaJolla, Calif.:

"But I think the ocean is 25 to 50 years behind the study of [the] terrestrial environment. We need to turn to the marine environment for the discovery of new drugs.'"[75]

As if to reassure his California colleague (quoted above), Matthew Suffness, programme director of the NCI's National Cooperative Natural Product Drug Discovery Group at Bethesda, Md.,....'[T]he marine natural products area is becoming more and more prominent in science.'"[76]

Also reported in early 1994, was the development by an anonymous Victoria (B.C.) biochemist of a technology to concentrate and manufacture fish-oil, cost effectively.[77] This fish-oil would be the Omega-3 type oil, containing the kind of (unsaturated) fatty acid that is good for the body; the exact opposite of Omega-6 oil, containing the kind of (saturated) fatty acid that is bad for the body. Being Canadian, however, has meant that the biologist/budding fish-oil salesman has been buffeted about between the B.C. provincial government (which spend millions of $ Cdn. per year via Pharmacare on cholesterol-reducing pharmaceuticals), and Health and Welfare Canada (which regulates drug availability) in his efforts to market the salutary product.[78] The director of the federal Health Protection Branch at Victoria said in 1994 that Omega-3 was officially treated as a drug, rather than as a food, as a result of the scientific claims made by the fish-oil's proponents.[79] (This writer sincerely hopes that Canadian bureaucracy has since permitted the Victoria biochemist/Omega-3 manufacturer to pursue his goal.)

Scallop-processing waste might aid in mankind's war against cancer. In 1995, Japanese scientists discovered that stock remaining from scallop-steaming contained glycogen poly-sugar, a recognised anti-cancer agent.[80] The Japanese scientists then experimented with the efficacy of scallop-waste ingredients by injecting lab guinea-pigs with 200 mg of glycogen poly-sugar. The lab animals, according to Seafood Leader, were "100% cured."[81]

Starting in 1996, the cancer-fighting compound "ecteinascidin" -- found in the lowly sea squirt (genus Ascidium) -- would be tried in human tests. So declared scientists at the University of Illinois, after ecteinascidin proved successful in battling cancer in lab animals.[82] Enthused Kenneth Rinehart, a professor of chemistry who worked on the study, "The results on mice with cancerous tumors have been dramatic.'"[83]

In 1996, the Catherine Atzen Day Spa in downtown New York City celebrated its tenth anniversary of business. The Day Spa's formula for success has included a base ingredient pharmaceutical, "Integral DNA b.e.",inherent in Catherine Atzen cosmetic products.[84] Despite its fancy monicker, there is nothing secret about Catherine Atzen's base ingredient: For Integral DNA b.e. is nothing more (or less) than the DNA molecules present in salmon roe. Catherine Atzen Laboratories have been utilising salmon DNA for years to treat clients suffering from burns, post-surgery scars, premature aging, and damage from exposure to the sun.[85] Now (in 1996), Catherine Atzen has incorporated Integral DNA b.e. into her line of skin-care products. Because the salmon DNA acts as a polymer -- meaning it retains water -- Integral DNA b.e. keeps the skin moist and protects it from the sun by absorbing ultra-violet radiation. Or so Catherine Atzen Laboratories/Day Spa has claimed, adding that Integral DNA b.e. could actually "restructure'" facial skin cells.[86] But that might be just another fish story!

And in 1998, Radio Australia announced that scientists in New Zealand were in process of synthesising a protein that would close wounds without surgical stitching.[87] That protein was obtained from "down under" mussels (see Appendices); from the strong, natural glue which fastens mussels to rocks and docks, piers and pilings. A team of chemists from the University of Auckland has been studying the mussel's adhesive protein, which is secreted by the bivalve's foot, for possible use in dentistry for fixing teeth.[88] NZ researchers also hope that the adhesive properties of the mussel protein will eventually enable them to bind together human tissues and cells.[89]

All throughout the 1990s, marine scientists have been combing the oceans and seas in search of naturally-produced chemicals with potential benefit to Man. Since 1995, planet Earth's final frontier, The Deep, has been expanded. For during that year a new oceanographic map, thirty times more detailed than any previous map, was co-developed by Scripps Institution researcher David T. Sandwell and the NOAA's Walter H.F. Smith.[90] Sandwell and Smith used a combination of recently declassified U.S. Navy information and formerly secret European Space Agency data to co-develop their map.[91] The complicated data and details of the new oceanographic cartography were provided to scientists over the Internet, and installed portions of the Scripps/NOAA map showed a newly-discovered sea floor rise in the South Pacific.[92] As marine life tends to gather or cluster in shallow areas, compu-savvy New Zealand fishermen were already -- by late 1995 -- catching fish and lobsters off the new South Pacific sea floor rise![93]

In 1996, scientists from 13 institutions in the United Kingdom, United States, and México were involved in a strange oceanographic experiment: 1600 miles west of the Galápagos Islands, the international team dumped a half metric ton of iron into the tropical Pacific; changing the usually blue water to a murky green that procreated a 30-times propagation in phytoplankton.[94] Presto! The half metric ton of iron (the missing ingredient, now found) had turned a previously lifeless stretch of Pacific Ocean into a biological marine oasis...literally overnight. According to an Associated Press report, "[P]lankton bloomed by the ton", and squid, sensing a food source, were quick to arrive at the scene; closely followed by white-tipped sharks and sea turtles.[95] Also, as the algae grew, the vegetation used carbon dioxide as building blocks -- carbon dioxide taken directly from the earth's atmosphere into the ocean.

Present at the site was Bill Cochlan, a U.British Columbia-trained oceanographer, who estimated that the tiny algae plants pulled about 2,500 metric tons of carbon dioxide out of the atmosphere.[96] (Some scientists believe that carbon dioxide, much of it produced by burning forests and fossil fuels, has been largely responsible for "global warming.") Bill Cochlan and his fellow scientists have not advocated the widespread iron fertilisation of the ocean -- the so-called "Geritol solution" to global warming -- and have voiced concern that cash-strapped governments and hungry fishing fleets will use iron-dumping to artificially raise fish stock levels.[97] Bill Cochlan has, however, figured that a <u>single</u> supertanker, disseminating iron into the oceans of the world, could negate the ill effects of the carbon dioxide pumped into the atmosphere by 5,000 supertankers burning oil.[98] Sometime soon, in Century 21, the "Geritol solution" might be one clear, clean economic response to both global warming and a depleted marine resource.

While entire new areas of the global ocean are being discovered, whole new ways are being found to employ the amazing varieties of sea-life. Throughout the 1980s, this writer researched the potential of underutilised marine species as both seafood for Man and as agrifeed for his livestock. Krill -- euphausiids -- those tiny crustaceans composing animal plankton came often to mind, but the writer never did formally expound on the topic of krill-as-fish-feed in either of his published works (1991, 1994). Then an October 1997 article in <u>Forbes</u> magazine reintroduced the subject of krill-as-feed. In fact, Biozyme Systems founder David Saxby was introduced to krill as human food at a Ukrainian restaurant in mid-1996.[99] Saxby, of West

Vancouver, B.C., didn't much like the taste or texture of the pinkish 2-inch crustaceans, but right off considered krill's commercial potential as an agrifeed. On returning home to Canada from Ukraine, the (then) 57-year-old engineer found that salmon, trout, goldfish, cats, and even piglets loved the shrimp-like krill.[100] Thus a brand new aspect of B.C. mariculture was introduced, and, by late 1997, David Saxby's Biozyme Systems was mass-producing krill hydrolysates.[101]

According to Forbes, krill hydrolysates are essentially partly digested proteins. Biozyme Systems has (1997) a patented process of "...stewing the krill, shell and all, in an enzyme cocktail in order to chop long protein molecules into peptides and amino acids."[102] Out of the Biozyme Systems process is yielded the end product in either a liquid form or powder. It is this liquid/powder that David Saxby markets to manufacturers of fish-feed, which they then add as a gustatory supplement to bulk fish meal (usually anchovy).[103] And the krill (Euphausia superba) additive works: In a recent UBC study, rainbow trout feeding on pellets flavoured with krill hydrolysates, were 67% heavier after 100 days than those fed on plain pellets for the same time-period.[104] Forbes projected that Biozyme's 1997 sales would reach $3 million; Saxby foresaw $45 million in sales of his product for 1998.

Sometimes Necessity really is the mother of Invention. In 1990, when the widely-feared Environmental Protection Agency (EPA) ordered seafood processors in the western Alaska communities of Kodiak, Dutch Harbor, and Akutan to quit dumping fish waste, the seafood processors were compelled to install fish meal plants to reduce tons of discard.[105] But, from what must have seemed at the time a near-impossible task, by 1997 the Alaska processors had turned an onerous mandate into a thriving new industry. Nancy Brown of Pacific Fishing magazine reported in June 1997 that: "Millions of investment dollars for equipment and research are producing a high-protein fishmeal and fish oil from groundfish and salmon carcasses; new plants are making better product and opening new markets."[106]

During the initial break-up of Yugoslavia in the early and mid-1990s, 1700 Croatian refugees -- a day --received sustenance from a nutritious fish powder. That high-protein fish powder was produced and supplied by International Seafoods of Alaska (ISA) at Kodiak.[107] Another local outfit, the Kodiak Fishmeal Company, installed a new facility in 1995; the outcome of seven Kodiak processors getting together to fund the $15-

million plant as partners in the new operation. The plant manager there told Pacific Fishing that state-of-the art equipment had almost trebled fish meal production to 800 tons per diem.[108] And Trident Seafoods at Akutan (in the Aleutian Islands) also opened a new facility in 1995, increasing production capacity two-fold to 1500 tons per day. (During the following year, the old Akutan plant was moved, in toto, to Sand Point [Alaska], with fully half the production capacity of the new plant.)[109]

Summing up, Pacific Fishing reported in mid-1997 that over 2 million metric tons of whitefish would "cross the docks" in Bering Sea/Aleutian Islands (BSAI) and Gulf of Alaska harbours during that year. As today, the vast bulk of discard would be reduced to fishmeal and by-products such as fish oil, bone meal, and fertiliser. Further regulations were set by various federal, state, and local bodies (for 1998) applying to the Bering Sea/Aleutian Islands and Gulf of Alaska areas, which included Pacific cod, Pacific (Alaska/walleye) pollock; retention of rock sole and yellowfin sole for these area fishermen would most likely be enforced (according to Pacific Fisherman) by A.D. 2003.[110] Retention of so much whitefish, of course, means that the vast bulk will eventually be reduced to fishmeal in Alaska plants. While decrying throughout this work the terrible waste of usable marine resources, the writer worries how the American seafood dinner of Year 2003 might appear. Perhaps the great American fish-feast of Century 21 will resemble the one so hilariously described in 1998 by satirical funnyman Joe Queenan:

"The Red Lobster menu consisted almost entirely of batter cunningly fused with marginally aquatic foodstuffs and configured into clever geometric structures. I immediately began to suspect that the kitchen at Red Lobster consisted of one gigantic vat of grease in which plastic cookie molds resembling various types of food were inserted to create a structural resemblance to the specific item ordered. This was the only way to determine whether you were eating Buffalo wings or crabcakes. Technically, my dinner -- the Admiral's Feast -- was a dazzling assortment of butterfly shrimp, fish filet, scallops, and some mysterious crablike entity. But in reality, everything tasted like Kentucky Fried Chicken. Even the French fries.
"Red Lobster was a consummate bad experience.... The food tasted like baked, microwaved, reheated, overcooked, deep-fried loin of grease.
Admiral's Feast, my ass."[111]

NOTES

17. Mark Kurlansky, <u>Cod: A Biography of the Fish That Changed the World</u> (New York: Walker and Company, 1997), pp.121-122.

18. <u>Ibid.</u>, pp.121-122.

19. <u>Ibid.</u>, pp.121-122.

20. See <u>FF 2000</u> for more on the Spanish factory-trawler fleets in the North Atlantic.

21. Op.cit., Mark Kurlansky, p.140. NB: Brackets this writer's.

22. <u>Ibid.</u>, p.140.

23. 'Subsidies get blame for fish crisis', <u>Peninsula Daily News</u>, 3 June 1997, p.B-4.

24. <u>Ibid.</u>, p.B-4. Quoted is Claude Martin of the WWF. (Author's note: "High levels of fishing" are <u>why</u> fishermen fish, Claude! But Mr. Martin has a valid point -- government subsidies to the fishing industry should be slashed.)

25. <u>Ibid.</u>, p.B-4.

26. 'Plenty of fish in sea?', <u>Peninsula Daily News</u>, 7 October 1997, p.C-1. NB: As the U.S. federal government has counted 279 fish species total in American waters, the 96 listed as overfished comprise an entire 1/3 of all species!

27. <u>Ibid.</u>, p.C-1. (Author's note: Surely the NMFS has not included the vast shoals of "Pacific salmon" off Alaska?!)

28. <u>Ibid.</u>, p.C-1. Quoted is Gary Matlock of the NMFS.

29. <u>Ibid.</u>, p.C-1.

30. <u>Ibid.</u>, p.C-1.

31. <u>Ibid.</u>, p.C-1. Quoted is Dick Gutting of the NFI.

32. 'Cuts loom for limits of Pacific groundfish?', <u>Peninsula Daily News</u>, 21 August 1997, p.A-6.

33. <u>Ibid.</u>, p.A-6. This writer assumes that the Associated Press meant <u>groundfish</u> when reporting "fish", and meant the Lower 48 states of Washington, Oregon, and California when reporting "West Coast".

34. <u>Ibid.</u>, p.A-6.

35. <u>Ibid.</u>, p.A-6.

36. 'U.S. to restrict commercial fishing', <u>Peninsula Daily News</u>, 29 December 1997, p.A-2.

37. <u>Ibid.</u>, p.A-2.

38. <u>Ibid.</u>, p.A-2. Quoted is Richard Methot of the NMFS.

NOTES (cont'd.)

39. Ibid., p.A-2. Quoted is Rod Moore of the WCSPA.

40. Cf. FutureFish 2000, Ch.2, 'Century 21 -- Time for Big Green Oceanic Regulation?' NB: Source for Elliott Norse's eco-agenda was his (edited) Global Marine Biological Diversity: A Strategy for Building Conservation into Decision Making (Redmond, Wash.: Center for Marine Conservation, 1992).

41. Author's note: Redmond, Wash., is also home to neighbouring Microsoft Corp.; surely a compatible coincidence!

42. Susan Chambers, 'Report Blasting Bottom Trawling Draws Yawns, Criticism', Pacific Fishing, Vol.XX, No.2, February 1999, p.57.

43. Elliott A. Norse, ed., Global Marine Biological Diversity (Washington, D.C. and Covelo, Calif., 1993).

44. Susan Chambers, 'Report Blasting Bottom Trawling Draws Yawns, Criticism', ibid., p.57. (Author's note: The Center for Marine Conservation, with Elliott Norse as chief eco-spokesperson, was a big favourite of a couple of The Bay Foundation board members during this writer's three years of service there, 1992-1995.)

45. Ibid., p.35.

46. Brad Warren, 'Bottom Trawling = Clear Cutting?', Pacific Fishing, Vol.XX, No.2, February 1999, p.5.

47. Ibid., p.5. Quoted is Cliff Goudey of MIT Sea Grant.

48. Susan Chambers, ibid., p.57.

49. Ibid., p.57. Brackets this writer's.

50. Ibid., p.35. Brackets this writer's.

51. Brad Warren, 'Bottom Trawling = Clear Cutting?', ibid., p.5.

52. 'Scientists sound alarm on ocean health', Peninsula Daily News, 25 January 1998, p.A-5.

53. Ibid., p.A-5.

54. Ibid., p.A-5.

55. Ibid., p.A-5. Quoted is Larry Taylor of the Cato Institute. (Brackets this writer's.)

56. Ibid., p.A-5. Brackets and contents this writer's.

57. Ibid., p.A-5.

58. Ballantine Books, Inc., New York, N.Y., 1970.

59. See Ch.3, Pt.III, Sec.C in FutureFish 2000.

NOTES (cont'd.)

60. :<u>Encounters Along the World's Coasts and Beneath the Seas</u> (New York: Henry Holt and Company). NB: Safina's wide coverage includes eastern North Pacific and southern Western Pacific waters.

61. <u>Ibid.</u>, p.437.

62. <u>Ibid.</u>, p.437.

63. Op.cit., Carl Safina, p.437. NB: The world high-seas driftnet crisis of the late 1980s helped trigger U.S. fisheries self-reform during the 1990s. (Parenthesis Safina's)

64. <u>Ibid.</u>, p.438. NB: Palau is also known as Belau, Bilau.

65. <u>Ibid.</u>, p.424. Emphasis this writer's.

66. 'Medicine: Scientists find health-giving chemicals in marine compounds ', <u>Peninsula Daily News</u>, 20 February 1994, p.C-1.

67. Wesley Marx, <u>The Frail Ocean</u> (New York: Ballantine Books, 1970), p.109.

68. <u>Ibid.</u>, p.109.

69. 'Ocean yields bounty of care', <u>Peninsula Daily News</u>, 20 February 1994, p.C-1.

70. <u>Ibid.</u>, p.C-1.

71. <u>Ibid.</u>, p.C-1.

72. <u>Ibid.</u>, p.C-1. Quoted is William H. Gerwick of OSU.

73. 'Medicine: Scientists find health-giving chemicals in marine compounds ', <u>ibid.</u>, p.C-1.

74. <u>Ibid.</u>, p.C-1.

75. <u>Ibid.</u>, p.C-1. Quoted is William Fenical of the Scripps Institution of Oceanography.

76. <u>Ibid.</u>, p.C-1. Quoted is Matthew Suffness of the National Cancer Institute.

77. Terry Moran, 'Finding the difference between fish oil and snake oil', <u>Victoria Regional News</u>, (B.C.), 9 March 1994, p.R-12.

78. <u>Ibid.</u>, p.R-12.

79. <u>Ibid.</u>, p.R-12.

80. 'Miracle Mollusc?', <u>Seafood Leader</u>, September/October 1995, p.86.

81. <u>Ibid.</u>, p.86.

82. 'Of Mice and Sea Squirts', <u>Seafood Leader</u>, September/October 1996, p.90.

NOTES (cont'd.)

83. Ibid., p.90. Quoted is Kenneth Rinehart of U. Illinois.

84. Jenifer Miller, 'From Caviar to Cosmetics: Here's Roe on Your Face', Alaska Fisherman's Journal, Vol.19, No.7, July 1996, p.25.

85. Ibid., p.25.

86. Ibid., p.25.

87. 'Mussels to Mend Muscles? Possibility Is Pondered', Pacific Magazine, Vol.23, No.2, Iss.128, March/April 1998, p.45.

88. Ibid., p.45.

89. Ibid., p.45.

90. 'New map throws light on ocean floor', Peninsula Daily News, 5 November 1995, p.C-9.

91. Ibid., p.C-9.

92. Ibid., p.C-9.

93. Ibid., p.C-9.

94. 'Sea blooms with "Geritol solution"', Peninsula Daily News, 3 October 1996, p.A-2.

95. Ibid., p.A-1.

96. Ibid., p.A-2.

97. Ibid., p.A-2.

98. Ibid., p.A-2. NB: Emphases this writer's. (Author's note: Results of the milestone experiment were detailed in Nature, October 1996.)

99. Bruce Upbin, 'Don't tell the whale lovers', Forbes, Vol.160, No.9, 20 October 1997, p.153.

100. Ibid., p.153.

101. Ibid., p.156.

102. Ibid., p.156.

103. Ibid., p.156.

104. Ibid., p.156.

105. Nancy Brown, 'From Waste To Wonder Food', Pacific Fishing, Vol.XVIII, No.6, June 1997, p.16.

106. Ibid., p.16.

NOTES (cont'd.)

107. Ibid., p.16.

108. Ibid., p.16.

109. Ibid., p.16.

110. Ibid., pp.16, 17.

111. Joe Queenan, Red Lobster, White Trash, and the Blue Lagoon (New York: Hyperion, 1998), p.11.

Bibliography for Part II

9. Benchley, Peter et al.; Judith Gradwohl, ed., Ocean Planet. Harry N. Abrams, Inc., New York, 1995; with Times Mirror Magazine, Inc., in association with the Smithsonian Institution.

10. Matsen, Brad and Ray Troll. Planet Ocean. Ten Speed Press, Berkeley, Calif., 1994. (Warning! The latter book is a palaeontological rock-video print out! All it has in common with the former work, besides its obversed title, is that they both perpetuate the musty, dusty neo-Darwinism of the British Museum [Natural History], London, U.K.; and the Smithsonian Institution, Washington, D.C.)

Part III: "Ha det, og takk for all fisken!"
           ("So long, and thanks for all the fish!")

"Seattle was growing too fast to take time out for niceties. Seattle bragged and boasted and showed its muscles to the world, in the manner of a strong and growing child...."
                                        --Gordon Newell, <u>Totem Tales of Old Seattle</u> (1956), describing the 1890s Seattle

<u>Norwegian America: Ballard and Beyond</u>

"It's very unpopular these days to study northern European and white cultures'."
                                        --Frankie Shackelford, associate professor at Augsburg College, Minneapolis, Minn.[112]

On the March 1999 issue of glossy <u>Seattle Bride</u> magazine, front cover, was pictured the projectedly perfect Blonde Babe of every male Seattleite's connubial longing. Very different from the Blonde Bride of the 1950s, the millennium-end Seattle Babe had none of the by-gone era's standards of femininity or wholesomeness. No indeed; the Seattle Bride of 1999 had a pinched, petulant mouth a-top an anaemic, anorexic body, and was all sharp angles rather than shapely curves. She was coldly sexy in an unsmiling way. Even though Seattle Bride's body type and facial expression have changed over the decades, Blonde Babe's ethnicity has remained unchanged. For she could be a hypothetical WASP, "Bunny Bixler", from Bellevue; or then again she could be one imaginary Scandinavian, "Musa Sarasen", from Ballard. (The more race/ethnicity-conscious America grows, the paler the Blonde Babe of sexual phantasy seems to get.) As Scandinavians and their culture have long been absorbed into the white-bread American mainstream, the Seattle Bride/Blonde Babe could just so easily have been "Musa Sarasen" as "Bunny Bixler". They are, for our purposes, completely interchangeable.

Except in the greater urban areas of Minneapolis/St. Paul or Seattle/Tacoma where they culturally predominate, Scandinavians are indistinguishable from other Americans of Northern European extraction. In Norwegian America, the locus and focus of our story, leading community lights from Ballard to Bay Ridge have distinguished themselves as [counter] cultural (i.e.,Establishment) enablers,[113] politically-correct fixers,[114] and <u>really good</u> fishermen. Or there were, in the Ballard (north Seattle) neighbourhood in which I partially came of age during the mid-late 1970s. Having not lived in Ballard since 1978 (or visited since the early 1980s), I read with great interest Roger Fitzgerald's sad eulogy of Ballard, in his 'In Search of the Simple

Life' column of April 1997, <u>Alaska Fisherman's Journal</u>:

"I'm moving to Montana sometime in the very near future. The decline of culture, notably in Ballard, is more than I can handle. Art shops are appearing along Ballard Avenue. The bars are being replaced by tidy little restaurants serving veggie burgers. Or bought up by outsiders who destroy decades of tradition by cleaning them up. The average age on Ballard Avenue is somewhere in the mid '20s, and I haven't seen a drunken fisherm[a]n staggering down Ballard Avenue in months. Clearly, the neighborhood is going to hell."[115]

All the places mentioned by Roger Fitzgerald -- among them Hattie's Hat, the Smoke Shop, and the Sunset Tavern -- rang a bell in my mind. I had either heard of them or frequented them myself. I had personally found a home at such drinking establishments as Pete's Place, the Vasa Grill (both on Ballard Avenue), and the Valhalla Tavern (on Market Street). Roger Fitzgerald's poignant article brought back some swarming memories of my own life in Ballard, mostly bad, but I had learned a lot about the North Pacific fishing industry of the 1970s.[116] And those years in Ballard had forcefully dispelled -- once and for all -- any romantic notions about fishing and fishermen; and for me, the almost mystical aura surrounding Norwegian fishermen. For all my life I had learned that whether one fished in the Barents or Bering Sea -- becoming a Norwegian fisherman was the highest station to which you could aspire. A perfect version lasted in my imaginings throughout boyhood; a picture image retained far into manhood.

Generally, fishermen were just like folks from any other working sector of society -- despite all their boasting about boats owned and bucks earned. They struggled with marriages and mortgages; with drink and drugs. Many fishermen were defiant atheists; others were devout Christians. Some fishermen had married several times, and boasted of having a girl in every port. But there were those who were fiercely monogamous, and fished with their spouse (or significant other) as him-and-her team. Not only couldn't fisherfolk ever be socio-culturally categorised with any degree of accuracy, the North Pacific industry itself had undergone a sea-change during the last quarter-century. In his column Roger Fitzgerald rather plaintively asks, "But what does Seattle know about bars? They drink Starbucks; in Ballard we drink whiskey."[117]

But the now-grizzled editor, who is himself a product of the 1960s, must sense, deep-down, the imminent inundation by the Third Wave which is already flooding over the North Pacific industry. The New [Third] Wave industry will be headed by techno-weasels, New Womyn, and compu-wimps. Their <u>tsunami</u> shall wash away and displace Roger Fitzgerald's "old lunch crowd" of fishermen, boatyard workers, and "old salts".[118]

There will be plenty of braggadocio and telling of tall fish tales in the millennial Ballard of the coming century. But it will be by techno-weasels, New Womyn, and compu-wimps over Starbucks coffee in a New Ager-owned retro Hattie's Hat. (Roger Fitzgerald and his Second Wave fishermen friends shall have to console themselves by sharing a bottle -- or a "joint" -- outside, under the stars, at Bergen Place.) And a goodly number of future industry big-shots celebrating at retro Hattie's Hat will be New Norse Northwest men and women. I wonder what my old drinking buddies Gus (wherever he is) and Pinky (may he rest in peace) would think of the new millennial Ballard?

And so we come to the strange, New Norse Northwest saga of Kennewick Man. In July 1996, two college students unearthed a near-complete human skeleton along the banks of the Columbia River at Kennewick.[118] Kennewick is situated in the Tri-Cities (the others are Richland and Pasco) area of Eastern Washington State. On official examination, the skeleton was pronounced to be 9,300 years old (some said 9,200) -- the oldest human remains ever discovered in the Pacific Northwest.[120] Later analysis of the skeleton, namely by local forensic anthropologist James Chatters of Richland, found that the skull exhibited "Caucasoid features, including a long, narrow face, a slight overbite and a prominent Kirk Douglas-type chin."[121]

James Chatters, quite naturally, assumed that the skeleton was that of an early White (Euro-American) settler...until Chatters came upon a really ancient spearhead-point lodged in the hip. Subsequent carbon-dating had determined that the skeleton was 92-93 centuries old. This discovery was fraught with politically-incorrect danger: Here was the distinct possibility that (1) Caucasoids had lived in the Pacific Northwest for more than nine millennia, and that (2) the earliest humans dwelling in our region might not have been the ancestors of present-day Native Americans.[122] Yet another perspective supported a possible theory that the skull was of a type "more closely resembl[ing] people who live in [non-Mongoloid] central and [non-Mongoloid] southern Asia."[123]

Kennewick Man's bones had by mid-1998 become controversial in four ways -- legally, scientifically, culturally, and religiously. According to U.S. federal law, if the skeleton proved to be that of an ancient Indian, it must be immediately reinterred.[124] The Umatilla people in eastern Oregon, of the mid-Columbia tribes, have taken the lead in assuming responsibility for the re-burying of Kennewick Man; citing a 1990 U.S.

federal graves protection law.[125] But the scientific "community" has filed a lawsuit (as of April 1998) at Portland, Ore., to win the right to further study both Kennewick Man and the area along the Columbia River where he was unearthed in July 1996. The exact ancestry of Kennewick Man, contend the scientists, must be ascertained.[126] To complicate matters, the U.S. Army Corps of Engineers had initially confiscated the skeleton of Kennewick Man because the bones were discovered on federal land. But in March 1998, the Engineers halted work on an erosion protection project at the area of the burial site along the Columbia River. Congress (i.e., Senator Slade Gorton and Representative Richard "Doc" Hastings, both Republicans) had successfully blocked the project so as to save the burial site for further scientific study.[127]

Meanwhile, by mid-February 1998, anthropologist James Chatters of Richland, together with sculptor Tom McClelland, produced a clay model of how Kennewick Man might have appeared when he roamed the banks of the Columbia River basin more than nine millennia ago.[128] The Associated Press has described the imagined facial features of Kennewick Man's bust: "He has a narrow chin, prominent cheekbones, a long face, a prominent nose, a size 15 neck and forehead that slopes to its apex far to the back of his head."[129] James Chatters likened their recreation to actor Patrick Stewart (Capt. Jean-Luc Picard of Star Trek: The Next Generation) without a wig; Tom McClelland regarded their recreation as "really kind of an Everyman'."[130]

The Associated Press article, 'Separated at birth?', was subtitled "Experts say Kennewick Man's reconstructed face looks like famous 'Star Trek' character", included a photograph of the clay model. To my jaundiced eye, reconstructed Kennewick Man looks about so much like Capt. Picard as the late Golda Meir (1898-1978) did Petunia Pig. No, indeed; reconstructed Kennewick Man most closely resembles, in my humble opinion, a Scandinavian Same, a Lapplander from the far north of Europe. The Lapps are First Peoples, non-Indo-European certainly, probably Ural-Altaic, of possible Tungusic origins from Central Asia. But one ethnic human being reconstructed Kennewick Man does not resemble is a Pacific Northwest Native American!

There have been other North Americans of Scandinavian origin (or descent) who have seen Kennewick Man in the same (Northern) light but taken the whole affair far more seriously. Anyway, so early as August 1997, when the remains of Kennewick Man were still locked in a U.S. government laboratory vault at Richland, enter the Asatru Folk Assembly; the only religious group besides several Northwest Native tribes granted access to the controversial bones.[131] The Asatru Folk Assembly are North American neo-Pagan followers of the Norse god Thor ("the Thunderer"), and practise a pre-Christian tribal religion which reveres the Earth and Sky deities of ancient North Germanic Europe.[132] (Left Coast New Agers come in many multicultural forms.) But while the U.S. federal government in all its wisdom decided whether to (again) bury or (further) study the bones, members of the Asatru Folk Assembly were permitted, on 27 August 1997, access to ("proximity, not contact") Kennewick Man's closed coffin by the Army Corps of Engineers.[133] The neo-Pagan sect's leadership has claimed that Kennewick Man's remains are those of a North European ancestor, and want the bones re-studied.[134] (The controversial matter is at present in U.S. federal court and is most likely many months away from being resolved.) The Asatru Folk Assembly planned a ceremony to erect a commemorative rune-stone, on 20 March 1998 (vernal equinox), at the site of Kennewick Man's Columbia River resting place.[135]

There is more to this neo-Pagan sect than merely the compulsion to certify a common ancestry with Kennewick Man. For the Asatru Folk Assembly really and truly worship Thor as a god, as I would guess it does the entire pantheon of Norse deities (the Æsir). In a final Associated Press article from a series on Kennewick Man, an accompanying newsphoto shows a bearded man from the Assembly conducting a 29 October 1998 convocation honouring the arrival of the bones at U.W.'s Burke Museum in Seattle.[136] The bearded man's lips are open in reverent oration, his eyes are raised skyward, and he bears aloft a short, stout hammer -- a modern version of mjølner, Thor's (Åsa-Tor) hammer. The bearded man is wearing a cape, fastened at the throat by a faux-Viking brooch, and on his sweatered chest is pinned a miniature mjølner in pewter.[137]

Kennewick Man's 90 century-old remains will rest comfortably at the Burke Museum at Seattle, under the vigilant care of the U.S. Department of the Interior; until the federal courts decide whether "them bones" shall undergo further study, or be returned to a coalition of Northwest Native tribes for reburial.[138] Until then, we have the witty remarks of a J. Nelson of Ocean Shores, Wash., to put the Kennewick Man into a more earthly perspective:

"There has been much ado about Kennewick Man. the 9,200-year-old remains of a Caucasian-featured man found recently along the Columbia River near Kennewick that scientists have stated could be of

Norwegian ancestry.

"As I have some Norwegian blood, I am greatly interested in the outcome of this debate. If it is proven that the Norwegians were here in America before the Indians, then I would be entitled to be self-governed, as are the Indians.

"Reservation land would have to be allocated to the Norwegians. I would be able to dig clams on anyone's property at any time. The Boldt decision, governing fishing rights, would have to be reversed and upheld to the benefit of all Norwegians.

"I could legally own and operate a gambling casino, sell fireworks and cigarettes without having to pay taxes. The BIA (Bureau of Indian Affairs) would no longer be needed. Perhaps a new entity, the NWF (Norwegians Were First), would be established to oversee a timely transition of Indian entitlements.

"It is no wonder the Indian nations want to claim these remains as their own and do not want further studies done that could prove they were not here first."[139]

(I have cited J. Nelson's amusingly sardonic letter in full, but surely the whole question is moot: Our Viking ancestors were around 900 years ago -- not 9,000!)

Another strange New Norse Northwest saga of the Gay '90s has been the ascent and descent (in the current wisdom) of U.S. Army Colonel Margarethe Cammermeyer. Colonel Cammermeyer, a 56 year-old (1998) retired Army nurse who was awarded the silver star in Vietnam, achieved instant pc celebrity status in 1992. For that was the year during which Col. Cammermeyer was discharged from the Washington National Guard; fired for announcing her bisexual orientation to her military superiors in a 1989 security clearance interview.[140] (This occurred prior to the "Don't' ask, don't tell" Clintonist Womyn and Gay People's Liberation Army.) The canny Cammermeyer, sniffing the shifting winds of social change, legally resisted her discharge from the Washington National Guard on the grounds of military bias regarding her sexual orientation; thereby causing a gay rights ruckus that attracted nationwide attention.[141] As noted, Cammermeyer rose to the level of New American Heroine -- especially amongst gays, feminists, and liberal Democrats of all persuasions and proclivities. A book, a gratuitous auto-hagiography of Cammermeyer's Brave Battle versus Heterosexual Homophobia, quickly followed.[142] Hollywood, always primed to take up the baton of a preachy pc cause, soon came out with "Serving in Silence" the T.V. movie, starring Glenn Close, a mid-90s archdeaconess of radical chic.[143]

But fame and fortune were not enough for Cammermeyer: By 1997 she harboured political ambitions and in 1998 actually ran (with much media hoop-la) for Congress, in an unsuccessful attempt to unseat rock-solid incumbent, Rep. Jack Metcalf (R). Cammermeyer lost, despite the enthusiastic support of Hollywood and the powerful gay lobby in a traditionally "tolerant" state. During her 1998 campaign, Cammermeyer insisted that her reasons for running for Congress were (1) issues of health, (2) interest in education, and acknowledged that:

"I understand there is a preoccupation with sexuality in this country. Being gay is just one aspect of that. The issue is how we get beyond that....
"Why is it that my sexual orientation should overshadow my profession and my work as a citizen?....
"Now, why am I running for Congress?'"[144]

Why, indeed, except to flaunt political solidarity with other lesbian candidates of 1998, like Tammy Baldwin of Wisconsin, Christine Kehoe of California, and Susan Tracy of Massachusetts -- all fellow Democrats?[145] (After all, Cammermeyer's sole claim to public[in]fame is as a self-outed gay guardsperson.) The six-foot tall Cammermeyer is a divorced mother of four, grandmother, Washington State's proudest bisexual woman...and a native of Norway.[146] Would all these qualifying criteria be sufficient for the Sons of Norway to anoint Cammermeyer 1998 "Norwegian of the Year"? I didn't check, but wouldn't have been shocked if S/N International had. If so, what did living veterans of the famous Norwegian-American 99th Infantry Battalion (the only U.S. ski battalion of World War II) think of Cammermeyer's selection? But Cammermeyer did lose to Metcalf, Washington State voters rejected "anti-[gay] discrimination" Initiative 688, and "the Republican legislature" (according to the Associated Press) passed a ban on "same sex marriage" (my quotes).[147] But for this Norwegian-American, it was all too close for [cultural] comfort.

In A.D. 1999, are Norwegian (and Norwegian-American) institutions shaky and questionable? For me, all confidence in our institutions started to unravel with my loss of respect for <u>Western Viking</u> in 1994, the year of my 50th birthday and <u>Jubilaeum</u> trip back to Norway. My lasting impression has been that if the United States has devolved into Janet Jackson's "Rhythm Nation", then Norway (and each of her Scandinavian neighbours) surely qualifies as "Babeland, the Emasculation Nation." In 1997, I left the Scandinavian-based Evangelical Lutheran Church in America for ecclesiastical reasons, because the Church in which I had been confirmed, had become little more than a touchy feel-good, sham New Age <u>ashram</u> (starring your spiritual mentors, Pastor Charlie and J.C. Superstar). Two years later, I am poised to drop out of Sons of Norway at the end of 1999 (unthinkable!). Lodge meetings, for this member, have become once-per-month, painfully-boring, evenings of 1) moving heavy furniture around -- twice (setting up and breaking down), and 2) listening to crabby, brain-dead Seniors indignantly defend the policies and (lack of) character of (their) President Clinton. (See Appendix E.)

It's not that I haven't tried. I have taught no less than three S/N Norwegian-language classes (one at our local college); served in various official capacities, including five years as financial secretary; in twelve years missed but one business meeting (that of Monday, 14 May 1988, my 44th birthday). But Norwegian cultural institutions are failing me, and I am failing as a cultural Norwegian (see Appendix E). There are most assuredly plenty of solid Norwegian-American citizens out there in the Upper Midwest and in the Pacific Northwest, but they are mostly elderly, trusting souls who are easily snookered by the disingenuous candy-floss they are fed by such organs as the Norway Times, Western Viking, and S/N Viking. These are honest, decent folks like my wife's aunt (tante) and uncle (onkel), both Norwegian-born and now deceased, who had lived carefully and worked hard in Wisconsin and Illinois for most of the century. Tante Aagot believed, as my wife has uncharitably put it, that she could "bake lefser up to heaven" -- thereby gaining assured entry to the Pearly Gates. As it was, Tante and Onkel quit this mortal coil giving a substantial amount of their hard-earned -- and diligently saved -- funds of a lifetime, to an ELCA Church which had long stopped caring about their spiritual welfare.

The unkindest cultural cut of all came from a "Euro-quotient" test that I voluntarily took from some egghead magazine some years ago. The test asked me my opinion on everything from religious faith, AIDS and homosexuality; on smoking, eating and drinking; my attitude about family structure, social dynamics etc. My highest score, ca. 87.5%, concurred most often with the [culturally conservative] Greeks; my lowest score ca. 12.5%, agreed least often with the [super permissive] Danes, who were the only Scandinavians represented. I was surprised to say the least. Maybe, after all these years, I had finally become an unhyphenated (i.e., just plain) American. Then it shouldn't bother me too much when (and if) I visit Petersburg, Alaska, during the next millennium (say year 2009), that the welcoming signs on the dock express in three languages: "WELCOME TO LITTLE NORWAY!" -- "VELKOMMEN TIL LILLE NORGE!" -- "¡BIENVENIDOS A LA NORUEGA CHICA! (And are those lefser or tortillas that S/N member Sigrid Gomez is making?)

Scandinavia: Babeland and Beyond

"An Italian can seduce a woman by talking. A Finn shows what he wants with his body."
--Tango dance-teacher Åke Blomquist[148]
"Norway itself is like an Askeladden ['Cinderfella'] fairy tale"
--Liv Dahl, Sons of Norway Heritage Programs Manager[149]

When President Bill Clinton journeyed in 1997 to Finland to attend the Helsinki Summit with President Boris Yeltsin, one televised image remained stuck in my mind. It was of a smirking "Boy" Clinton surrounded by a giggling gaggle of ageing, fading, blonde harridans in power suits. Where oh where, I kept asking myself, were the heroes of Viipuri -- or their male offspring? (Little Finland had bravely fought the futile Winter War against the mighty U.S.S.R. during those dark months of 1939-1940.) For there didn't appear to be one single, solitary, Finnish man in sight at the Helsinki summit. Was the whole scene an example of Scandinavian feminism, or was the female Finnish head of state, and her all-woman cortège, simply ga-ga over America's puerile and profligate president?

Another televised image which has remained -- and rankled -- in my mind, was from a 31 March 1998 edition of <u>Public Eye with Bryant Gumbel</u> (CBS-TV). One of the segments presented by the supremely obnoxious host was titled 'Beautiful Women of Iceland'. In a men's journal-type Babeland story, CBS cameras lingered leeringly on (mostly) fresh-faced teenage girls in Reykjavik, Iceland. Planeloads of package-tour U.S. East Coast bachelors were shown roaming the streets of Reykjavik, hoping to get (quickly and personally) acquainted with a local Blonde Babe. (The implication was that wild, willing Icelandic women were readily available for almost instantaneous sexual intercourse.) Among those looking to impress was the ubiquitous Jerry ("The Little Creep")[150] Seinfeld, being chauffeured about in a stretch limo. The only Icelandic male interviewed in the <u>Public Eye</u> piece was an obviously gay "beauty expert". By the programme's end, I was beginning to feel as if I were the sole Scandinavian (heterosexual) man remaining on planet Earth.

Scandinavians have themselves to thank, of course, for the Babeland stereotype of the Nordic countries.[151] Scandinavian men outwardly (but not always inwardly) don't appear to care. Scandinavian <u>machismo</u> consists -- at best -- of a grave mien, a stiff upper lip; at worst, it manifests a lumpen indifference coupled with an emotional sterility. Supposed Scandinavian stolidity might easily be mistaken for a sheer, stone stupidity. Except in the films of Ingmar Bergman, the only recognisably Scandinavian men since the crew of the Kon-Tiki (1947), seem to be those eternally bowing-and-scraping pale-male sycophants at the United Nations Organization.[152]

Writing of the UNO, that institution has served as a geopolitical Shangri-La for Scandinavians, both male and female. The first two secretaries-general were Scandinavians -- Norwegian Trygve Lie (1946-1952) and Swede Dag Hammarskjöld (1953-1961). I had been brought up since my adoption to venerate both men and revere them as Scandinavian rôle models. And my own pervasive sentiment had always been that the UN had never been in better, more honest (i.e., Scandinavian), hands than during those first fifteen years of existence. It has only been recently, after a second, longer look, that I discovered that neither Trygve Lie nor Dag Hammarskjöld had been so high-minded after all. (Indeed, I have praised both men in FutureFish 2000, Ch. 2.) Lie was a dedicated Socialist his entire life, a labour lawyer/mouthpiece, and political boss-man of the Norwegian Social Democrat Party (Arbeider Partiet, an ideological offshoot of the earlier Communist International) -- before assuming his top post at the UN in 1946. If anything, Hammarskjöld (in my mind) has fared just as poorly in historical hindsight. Not only did he push hard for Red China's early admission to the United Nations, Hammarskjöld had been in charge of UN affairs for his government in 1951 when Sweden refused to support even a mild resolution, censuring the ChiCom incursion into a peaceful Tibet.[153]

But Lie and Hammarskjöld are now archaic figures in the past history of the United Nations Organization. Scandinavian men are not so preëminent in UN affairs anymore; they act instead as passive attendants to the more powerful (mostly) Third World secretaries-general since 1961, the year of Dag Hammarskjöld's ["mysterious"] plane-crash death on the way to the breakaway (1960-1963) province of Katanga. (But that's another story.) Scandinavian feminist New Norse Womyn, however, have not been servile bystanders at the United Nations since then -- far from it. Two New Norse Womyn who have been extremely active in behalf of the would/wanna-be world government are both Norwegian. They are shrill feminist Liv Ullmann and strident environmentalist Gro Harlem Brundtland. (These New Norse Womyn make one miss those Old Norse men.)

Although a Norwegian, Liv Ullmann was born in Japan, raised in Canada and New York, and studied drama in England. Indeed, the Ullmann family did not return to Norway until after World War II.[154] Actress Liv Ullmann gained world-wide fame, during the 1960s and 1970s, by starring in Ingmar Bergman films. Ms. Ullmann has appeared in more than 50 movies made in Europe, Israel, North America, Argentina and Australia; has won two Golden Globes, and been nominated for two Tony Awards and three Academy Awards. Directrix Liv Ullmann made her début behind the camera with her 1993 film "Sophie"; later directing "Kristin Lavransdatter". Authoress Liv Ullmann wrote her autobiography, Changing, which was published in 1976; her second book, Choice, was released in Europe and North America during 1984.[155]

Globalist busybody Liv Ullmann presently serves as vice-president of the International Rescue Committee, and is co-founder of Women for Refugee Children World Wide. Since 1980, Ms. Ullmann has been a "Goodwill Ambassador" for UNICEF (United Nations International Children's Emergency Fund). Her alleged interest in children's issues notwithstanding, Ms. Ullmann was actively involved in developing the [some believe extremist and anti-family] U.N. Convention on the Rights of the Child.[156] Besides seeing her co-starring in "The Emigrants" (1970) on Norwegian and Swedish TV (see note 152), I can remember hearing Liv Ullmann on The Tonight Show Starring Johnny Carson; screaming, screaming and screaming, throughout the 1970s and 1980s, about her favourite gender feminism cum child liberation issues. And because of her great public-persona [nuisance] value, the know-nothing Norwegian government awarded Liv Ullmann the prestigious Order of St. Olav in 1994.[157]

Unlike Liv Ullmann, who has successfully combined hypocritical glamour ("40 Carats", 1973) with feminist activism, Gro Harlem Brundtland (trained as a medical doctor) remains a strictly political New Norse Woman. After wearing out her cowed and governmentally-witless countrypersons as Norway's long term prime minister, Dr. Brundtland retired from domestic politics in 1996; by 1997 she awaited appointment as the next UN assistant secretary general. (There were some Gro-groupies, early in 1997, who expected Dr. Brundtland to actually succeed Egyptian Boutros Boutros-Ghali as UN secretary general).[158] But that was not be (either), although Dr. Brundtland is a close personal comrade of Ghanaian Secretary General Kofi Annan (who is married to his own Scandinavian trophy-bride, culled from Sweden's prominent Wallenberg family).[159]

By late August 1997, however, the Norwegian government had officially nominated Dr. Brundtland as candidate for the post of director-general of the World Health Organization, and submitted a formal letter to the current WHO director-general, Dr. Hirashi Sakajima of Japan.[160] (Surely, with a background in both health and politics, the post of director-general of WHO should be powerful enough to partially assuage even Dr. Brundtland's enormous ambitions?) Anyway, in early 1998, erstwhile doctor and former prime minister Gro Harlem Brundtland was elected (after four rounds) by the UNO as new director-general of WHO; she was to succeed Dr. Nakajima in July 1998.[161]

Back in Norway, all was not so felicitous on the home front as at Geneva and New York. Mr. Arne Olav Brundtland had published his memoirs, Gift Med Gro ("Married to Gro"), during the spring of 1997, and his book related in detail the mental problems -- and subsequent suicide in September 1992 -- of the Brundtlands' 25 year-old son, Jørgen.[162] (The young man's daughter, Julie, was born shortly after her father's tragic death.)[163] What has bothered me as a parent, is that a mere three months prior to Jørgen Brundtland's self-inflicted death, his mother's voice was among the loudest of those heard at the Rio Earth Summit of June 1992. Was the young Brundtland's despondency -- and mental illness -- (at least) partially caused by an-always absent mother; a New Woman with plenty of time to spend on her own career? Arne Olav Brundtland, a security expert at Norway's Institute of Foreign Policy, will never equal his illustrious wife's achievements. But at least he has written a true-to-life book about Norwegian society which has shaken his normally smug, settled ("satt"), and self-assured compatriots -- and that's a major achievement in itself.[164] Meanwhile, Gro Harlem Brundtland has already gone on to bigger and better things in her rise to geopolitical power. New Womyn appear to have no-fault motherhood insurance.

I have long accepted that the vast majority of Norwegians (and Norwegian North Americans) accept and approve of Statsminister Gro's global gallivantings, but nothing could have prepared me for the fawning tone of a 13 May 1997 ABC-TV puff-piece interview on Good Morning America, co-hosted by Joan Lunden:

....."Well, for centuries, I mean [like], Scandinavia has been really known, all these countries, for their innovative and their progressive social systems. But when it comes to protecting [feminist] women's rights and [fatherless] children's rights, Norway could really teach most other countries a thing or two. They [the "rights", i.e, privileges] are the top priorities here. Largely responsible for this, former Prime Minister Gro Harlem Brundtland, and she is the first woman [in Norway] to hold that post...

"She's been very instrumental in pioneering some of these sweeping changes that have really greatly improved the quality of life for [sexually permissive] women and for [spoilt rotten] children in Norway. Nice to have you [Dr. Brundtland] here. I think most [American] women, when they hear that, they just want to to pack up and come right over here [to Norway]. But those have been sweeping changes that really have improved life here for [New] women and [father-free] children.... And they [the Norwegians] also have the lowest crime rate in the world. This is a very, very interesting country that we [Americans] could learn a little bit [i.e., a lot] from. Hopefully, we can get some of those [Socialist feminist] programs instituted in America. Thank you for having us [backward Americans] here [in your enlightened country].[165]

Oh, wow! Such abject cultural grovelling! But Joan Lunden won't have long to wait for the U.S.A. to follow Norway's Socialist/feminist example. Those in the Hillary/Billary White House, inside the D.C. Beltway, up in Manhattan, and down in Hollywood are all working overtime, and using their considerable money and power, to ensure American compliance with United Nations norms and Norwegian (Scandinavian) standards. These same nonsensical standards prompted three idiotic Norwegian legislators in late January 1998 to recommend that President Clinton receive the Alfred B. Nobel Peace Prize.[166] Their recommendation coincided exactly with the megaseptic stench starting to issue from the newly-uncovered Monica Lewinsky scandal.

Another matter preoccupying the attention of Norway's ruling political class since 1998 has been something called the Meena Communications Initiative. "Meena" is a pc cartoon character, a ten-year-old South Asian girl who stands up to [male] bullies, tells her [male] father that she must attend school (too), and "raises the consciousness" of her gender-biased [male] brother.[167] Meena is featured on Asian television, broadcast throughout villages on portable projectors, heard on the radio, and one million Meena comic books have already been disseminated throughout Bangla Desh.[168] The executive producer of this cultural nutrasweetness is Canadian Christian Clark, who (unsurprisingly) once wrote for Sesame Street, and has (modestly) assessed his Meena propaganda package as "a historic breakthrough for using media and popular culture for children's rights.'"[169] The Meena Communications Initiative, a child of UNICEF, is mainly funded by the Norwegian government; with a budget of $6 million, the MCI looks to extend its sticky fingers farther afield by producing similar series for Africa and South America.[170]

What's wrong with this picture? Who is the UNO to teach its public values to the unenlightened masses of India, Pakistan, and Nepal? And who is the government of Norway to fund (from tax-payer money) and distribute its approved version of Scandinavian-type feminism? Isn't the First World lecturing the Third

World a case of cultural condescension? And who is Norway to lecture any other country? Are not the Scandinavian nations arguably the three most religiously pagan, culturally profane, and sexually promiscuous societies anywhere? The Norwegian welfare state apparatus is like a "rich, educated nymphette who's bored."[171] Rather than tackling the tough moral problems in her own society, Norway amuses herself with culturally-destructive playthings like the Meena Communications Initiative. The MCI assuages Norway's Socialist conscience (for living so well), and makes her look so virtuous to the other feminist First World Nations (who also live well but not quite so well).

Thus once again, a Scandinavian Babeland has been in the forefront of UNO global social-engineering. But how do New Norse women personally behave these days at home and abroad? Well, sort of like Old Norse women as depicted in American author Michael Crichton's <u>Eaters of the Dead</u> -- half Babe/half feminist. In Crichton's thriller, a visiting Arab Muslim courtier described Old Norse women in Viking times:

"[T]he women show no deference, or any demure behavior; they are never veiled [i.e., modest], and they relieve themselves in public places, as suits their urge. Similarly they will make bold advances to any man who catches their fancy; as if they were men themselves; and the warriors [Norse men] never chide them for this...."[172]

Is there any difference now? Debatably, the above description quite accurately reflects the general behaviour of many Scandinavian women more than a millennium later. I first encountered New Norse women as a young adult (during my salad days) in 1966, when I had first set foot on Norway in nine years. (1966 was also the year of my wedding, to a <u>good</u> Norwegian girl; a rarity then and a scarcity today.) By the mid-1960s New Norse women had earned a notorious international reputation for themselves -- which they are still busy living down to. Bored with their own men, New Norse women have ranged far and wide for more novel, more exciting, sexual partners. (They are the blonde, bikinied Beach Bunnies you see on every strand from the Greek Isles to the Hawaiian Islands.)

And the New norse men are still standing around, unchiding. Outhustled in their own land for (their own) money, women, and work, the soft, Socialist-raised New Norse men are like tame, white laboratory rats who are far too timourous to stem the incursive tide of wild brown gutter and black sewer rats. The lean and tough, hungry and horny, East Europeans and North Africans easily shoulder aside the pampered, battened and fattened, New Norse men aside to get at what they want -- and get: Money women, and work.

Not all New Norse men are just standing around while their country suffers full-scale invasion (and occupation), by what French author Jean Raspail (The Camp of the Saints, 1975 ed.) has called "the eater[s] of turds". Resistance to the hordes of foreigners, state-enforced egalitarianism, and a hopelessly [ef]feminised culture, has taken a bizarre and dangerous turn in Norway. I first got wind of the Norwegian "Black Metal" scene during the spring of 1992. My wife, daughter, and younger son were heading "home" to Norway that summer, and I was to "hold the fort" back in Port Angeles. (My elder son was still in the U.S. Army.) But even before my family left, we read of stave-church (stavkirke) burnings starting in May at Stortveit, and in early June at Fantoft near Bergen.[173] Stave churches are not only the priceless architectural treasures from Norway's rich past, they remain as mute witnesses to the immutable fact that Norway was once a Christian country.

My family and I dismissed the church burnings as the vile acts of demented, drugged-up young punks in Norway copy-catting the church burnings of demented, drugged-up young punks in America. Ironically, I was reading Sigrid Unset's Den Brennende Busk (".The Burning Bush") all throughout the summer of 1992, the moving story of a young Norwegian man's Christian journey during the World War I era. (This particular book was by the more mature, devoutedly Christian Sigrid Undset -- not the flapperish, sophomoric feminist Sigrid Undset worshipped by actress/U.N. official Liv Ullmann.) Later that summer, I learned with growing concern of further stave-church burnings in the Old Country: On 1 August at Revheim, on 21 August at Holmenkollen, on 1 September at Ormøya, and on 13 September at Skjold.[174] The torchings of Norway's stave-churches continued on, after my family had returned, into the autumn and on into the 1992 Christmas season: In early October at Hauketo, and during late December at Åsane and Sarpsborg.[175] (I was horrified and stunned, as I cared about Norway and still do. See Appendix D.)

The co-authors of Lords of Chaos: The Bloody Rise of the Satanic Metal Underground (Venice, Calif.: Feral House, 1998), Michael Moynihan and Didrik Søderlind, have attributed fully one-third of the Norwegian stave-church burnings to the Black Metal movement. And a single, satanic Norwegian youth, armed only with a tin of petroleum and a box of matches, can (in one fell swoop) literally burn down the national symbol of Norway's cultural, historical, and religious heritage. Lords of Chaos closely examines the life and thought of one Varg Vikernes, a/k/a "Count Grishnackh", a leading satanic (dark) light and composer/musician of the Black Metal movement. Count Grishnackh's band, Burzum, took its name from British proto-New Ager J.R.R. Tolkien's (The Hobbit, 1937, etc.) coined phantasy word "burzum", meaning "darkness".[176]

Count Grishnackh expressed the view, to authors Moynihan and Søderlind, that Norway must shuck the "alien shackles of Christianity"; that the Christian [Lutheran] Church was responsible for destroying everything beautiful in true [heathen] Norwegian culture. As Count Grishnackh, Varg Vikernes treated Black Metal fans to a musical "Burzum Tour '92", the very same year of the stave-church burnings. Pro-Count Grishnackh fliers advertised the tour as "Coming Soon to a Church near You'".[177]

Moynihan and Søderlind formally interviewed Black Metal rocker "Ihsahn", lead vocalist with the Norwegian band, Emperor. Among the questions asked was:

Q.: "[Is] it [the extreme Black Metal scene] a reaction against strict Christianity in Norway?"
A.: "The Norwegian State Church is not strict at all. I think it's quite funny, we have female Bishops and priests, and we have homosexual priests and homosexual marriages, which is very much against what's said in the Bible. The State Church in Norway is very liberal.'"[178]

Ihshan had also been interviewed by Pål Mathiesen, contributing editor of the Norwegian cultural newspaper, Morgenbladet. Moynihan and Søderlind cited some of Mathiesen's questions to Ihsahn in Lords of Chaos. For instance, Mathiesen had asked Ihsahn during their interview about the place of the Norwegian State Church in national life. Ihsahn's answer had been direct and unequivocal:

"...[D]estroy it [the Lutheran State Church] because it's weak, because it doesn't have a right to live anymore, because it's just good to everybody. It doesn't have the power to judge anymore. It's just some sort of [S]ocial [D]emocratic Christianity, and these people [satanists] despise that kind of weakness.'"[179]

Perhaps the best statement in Moynihan and Søderlind's Lords of Chaos about the Norwegian government/State Church vis-à-vis the satanist torchings, was made by the co-authors themselves:

"A wave of violence erupts across an otherwise tranquil landscape, a ...humanistic society which has always ensured that its citizens are well-fed and finely educated. Whatever their future station in life, they [young Norwegians] can count on comfort, security, and all the other benefits of a country with one of the highest standards of living in the world ...yet the young dream of murder, blood sacrifice, revenge.
"The Christian religion plays no great part in their lives, though its secular counterpart, the system of [S]ocial [D]emocracy, offers them great opportunity. They [the church burners] reward such benefaction with a curse of fire, basking in the glory of destruction. Nothing excites more than the thought of chapel and vicarage ablaze, illuminating the ...night sky with darting flames ...thrusting upward. The beauty [stavkirken] is in ruins; a temple laid to waste becomes an aesthetic victory."[180]

It's not as if the aspiring young satanists are rebelling against a repressive State Church -- quite the opposite, as contemptuously noted by Ihsahn of Emperor. I strongly believe that the entire phenomenon of the Black Metal scene, to the easy-as-pie sexual availability of Blonde Babes, has occurred in a culture which (never having been particularly Christian anyway) is totally "unchurched". And that is also the case (in varying degree) for most European countries, especially those of traditionally Protestant Northern Europe, where there exists an established State Church. Culture editor Gene Edward Veith of the American Christian magazine, World, has commented wryly on that moribund institution:

"England, of course, with Germany and most of the other European nations, has a tradition of the [S]tate [C]hurch. It is true that the established churches are not doing their jobs. Few people attend them; the tax-supported clergy have little incentive to build congregations, and their theology has become as worldly and vacuously liberal as the societies upon which they prey."[181]

Scandinavians of North America have historically been more "churched" than folks back in the Old Country. The past proliferation of Lutheran schools and colleges in central Canada and the U.S. upper midwest -- all without governmental sponsorship -- attest to that. But the Scandinavian Lutheran Church in North America (the ELCA / ELCC) has changed too, as I have expressed throughout this work. My local ELCA pastor's message, to paraphrase Gene Edward Veith of World magazine, had ..."become as worldly and vacuously liberal as the [communicants] upon [whom he preyed]".[182] So I left the ELCA, forever, during July 1997 -- but the vast majority of my fellow Sons of Norway members have remained; albeit with some serving in silence. Unlike Scandinavian State Church pastors "who have little incentive to build congregations",[183] North American Lutheran clergy (like other mainstream "clergypersons") must build and expand their congregations to keep their particular congregation viable -- and their position too. What follows, then, becomes a cult of [their] personality.

For example, my former ELCA pastor at Port Angeles, has kept his own congregation largely intact by (1) staying strictly on the cutting edge of political/religious-correctness; (2) spoon-feeding his parishioners a steady diet of spiritual (New Age) Angel-food for the soul; and (3) preaching a Garrison Keillorised Gospel of Heaven and Nice. (Only the most openly tolerant and obviously compassionate need apply.) The overall results have been identical: Serious evangelical Lutherans in Scandinavia and North America have been leaving the State Church and ELCA/ELCC in steady trickles, to seek heartier Christian sustenance elsewhere.

In the present Norwegian social culture of America, scorn for evangelical Christianity and church attendance is a virtual given. At the memorial service of an old S/N friend, a past Lodge president loudly boasted to me (in Norwegian), "I only go to church three times a year -- Christmas, Easter, and if there's a family confirmation, wedding or funeral!" My dual-edged reply (in the same language) was, "That's <u>still</u> too many times for a <u>real</u> Norwegian, Mr. President!" In Ballard's <u>Western Viking</u> periodical, there were always listed psalms and prayers, in Norwegian, for those dying old folks at the Norse Home on Phinney Ridge. Turn to the food page ('<u>Retter med Petter</u>') at Christmas-time, and the editor would show you how to bake all those great Norwegian goodies for the Christmas season (<u>Julesesongen</u>). Simultaneously, the food editor would nudgingly insinuate that a truly typical Norwegian <u>Jul</u> didn't include celebration of the Christ-mass.

Several years ago, when I travelled to New York City every four months to attend board meetings of The Bay Foundation, we held a meeting at the Norwegian Seaman's Church; now moved to mid-town Manhattan (at Turtle Bay ...closer to the U.N.!). I remember asking the impassive pastor (rather impudently) if his home ministries (<u>indremisjon</u>) kept him really busy, as Norway has such a very low percentage of Christian believers? The priest just stared, nonplussed: He was, after all, nothing more than a civil servant hired and paid by the Norwegian Welfare State. But his coldly impersonal stare, though brief, told me that he resented my inference. (At least my rudeness had elicited an emotional reaction, however slight.)

I know of at least one Swedish pastor, anyway, who returned to his native land from America, to spread the Word of God among his heathen compatriots. Björn Dahlin had quit Stockholm, Sweden, as a young man, full of anger and violence. After soldiering on Cyprus and then serving in Vietnam with the U.S. Marine Corps, Björn Dahlin had settled afterward in western Idaho. Then Björn Dahlin underwent a dramatic "born-again" experience during the early 1970s, irrevocably converting him to Christianity and changing his life forever. With time, now-Pastor Björn Dahlin felt an irresistible call to preach the Gospel in his native land -- which he eventually did. Why preach the Gospel in settled Sweden rather than in the American "wild west"? Dr. Jan Dahlin (no relation), in the Foreword to Pastor Björn Dahlin's dramatic life story, has explained why:

"You may ask, is Sweden, one of the most civilized societies in the world, really a mission field? Sweden, like the United States, has sent missionaries over the years to the undeveloped countries of the world and accomplished many good things for the Lord. However, through Sweden enjoys a material prosperity among the highest in the world, quite broadly distributed among its people, and over a century and a half of peace, in a spiritual sense, 'There is a famine' (Amos 8:11)."[184]

There might be a spiritual famine in Scandinavia and Northern Europe, but there's no dearth of ravenous intellectual hunger for the secular works (and persona) of nihilistic Norwegian writer, Knut Hamsun (1859-1952).[185] According to the Norway Digest of November 1997, "Knut Hamsun is being read and discussed as never before in Europe."[186] There seems to be especial enthusiasm for Hamsun in Germany and the Czech Republic, with various Hamsun seminars conducted at Praha and Brno, and a jumbo Hamsun exhibit shown at Berlin.[187] Besides filming several German-language versions of Hamsun's grim stories, the inspired Germans had by 1997 published no less than three new translations of Hamsun's grey works[188] -- no small honour for the leading Nordic apostle of suicidal tendencies. (Hamsun was influenced by both Nietzsche and Strindberg. See Appendix C.) Although Hamsun's best known novelle, Hunger ("Sult"), appeared in 1890, it contains the same [frighteningly familiar] post-modern sense of hopelessness so prevalent during the 1990s. Indeed, Hamsun's Hunger painted a portrait of downtown Oslo of that time which is eerily similar to the nocturnal Oslo sentrum of 1999:

"It was about eleven. The street was rather dark, people were wandering all over, silent couples and noisy groups mingled. The great hour had begun, the mating time when the secret exchanges took place and the joyful adventures began....[O]ne or two quick sensual laughs, swelling breasts, rapid heavy breathing; down by the Grand Hotel....The whole street was a warm swamp, with mists rising from it.
...."The sexual energy visibly in all the gestures of those going by, even in the dim flame of the gas lamps, and the motionless steamy night had all begun to affect me -- this air filled with whispers, embraces, hesitant confessions, half-pronounced words, tiny squeals. Even the cats were making love with high-pitched shrieks in the door of Blomquist's Café....
...."I shrugged my shoulders in disgust and looked contemptuously after them as they went by, couple after couple. These babyish, aimless...students who think they are being rakish and Continental every time they manage to pat a girl on the breast! These bachelors, bank clerks, butchers, philanderers...or those fat sows from the cattle market who flop down in the nearest doorway for a glass of beer!
...."I gave a long spit over the sidewalk without bothering about whom it might hit, became furious, full of contempt for these people rubbing against each other and paring off right before my eyes...."[189]

Such Hamsunian contempt! (Such Hamsunian covetousness?) And such an unflattering picture of my hometown! Oslo bei Nacht is not very different in the 1990s than it was a century ago, with Carl Johan Street being Oslo's "main drag" now as it was then. But Knut Hamsun, his artistic anger aside, couldn't carelessly spit on Karl Johansgaten today. For hunting the heavily made-up modern Oslo girls (the Blonde

Babes), and hustling drugs, are the ear-studded, satanic home-grown punks and, increasingly since the 1970s, the dread-locked Third World "eater[s] of turds". Both varieties of human vermin, domestic and imported, are invariably vicious, violent, and high on dangerous narkotika ("stoff"). Both predators and prey (still) "hang out" on Carl Johan Street, and start appearing -- like Knut Hamsun's nightcrawlers a century ago -- at just about 11 p.m. every evening (during the warmer months). Some things never change!

Human life in the modern Scandinavian welfare state has little philosophical or religious significance, and the chronic alcoholism, manic depression, and suicidal tendencies inherent in Nordic culture (and me), have become more pronounced in the spiritually airless hothouse of Social[ist] Democracy. One rarely sees the Norwegian word bør ("ought") anymore relating to personal [i.e., moral] behaviour; if at all with public [i.e., "values"] conduct. (Billy Liar's Clintonist America is catching up fast. If Norway ever gets rid of her useless gaggle of Royals, the Lyin'King could emerge from political retirement in the U.S. to become the first President of Norway! The great Empath perfectly embodies the triumph of public values over personal morals. And if Hillary Rodham Clinton is still with Bill, she could become the First Lady of Norway! But HRC is not Blonde enough, and her stumpy legs are far too lumpy for her to qualify as First Norwegian Beach Bunny in bikini. But HRC's multiple years with the Children's Defense Fund should be sufficient for her to become the Lyin'King's lyin'queen).

In Esquire magazine's 'Dubious Achievement Awards of 1997', there was a Scandinavian winner in the meaning-of-[human] life category: "Sweden began using the energy produced by cremating dead people to heat homes."[190] (Perhaps Pastor Björn Dahlin of McCall, Idaho, ought to launch another Christian mission to his obviously still-heathen homeland. But this one must be heavy-combat duty. Pastor Björn might like to take along Hollywood Swedish tough-guys, Bo Svenson and Dolph Lundren... if he needs them. Former U.S. Marine Vietnam vet Björn Dahlin is pretty rugged himself. Norway, incidentally, is the only country in the world -- that I know of -- in which inoffensive evangelist Billy Graham has actually encountered open hostility. Dr. Graham was received with boos, catcalls, and worse. Twisted twin-sister Sweden might prove as inimical to the Gospel in the next century.)

My own nightmare of Norwegian-American hell would be to have to listen to the pompous voice, and have to look at the pontifical face, of deceased CBS-TV news editor Eric Sevareid... forever! The only Norwegian-American antidote to this eternal cultural punishment, for me, would be to contemplate the great deeds of U.S. agronomist Norman Borlaug (progenitor of the "Green Revolution" and 1970 winner of the Alfred B. Nobel Peace Prize, when it still meant something). Some of the best, and some of the worst, Americans have been Norwegian.

Port Angeles, Wash. -- U.S.A.: "Adjø" from the End of the Road

"The New Rating Guide to Life in America's Small Cities ranked Port Angeles the seventh-best small city for the seventh year in a row."
--Peninsula Daily News, 30 October 1997[191]

Yessiree, when the All-American City of Port Angeles placed seventh in The New Rating Guide, 1997 edition (Prometheus Books, Amherst, N.Y.), our little city went all a-twitter. Author Kevin Heubusch judged 184 small U.S. cities upon several criteria:

"...[C]limate, environment, diversions, economics, education, community assets, health care, housing, public safety, transportation and urban proximity. The [G]uide noted that few cities, large or small, can match Port Angeles' breathtaking location, natural beauty, income level, and crime and school graduation rates."[192]

As if to prove the point to sceptical locals, Sports Afield, in its February 1998 edition, rated Port Angeles the very best outdoor sports-town in all Washington State. The article in the nationally-distributed magazine went on to praise the natural beauty of the Olympic Peninsula, and lauded the Sol Duc, Bogachiel, and Hoh rivers as "a [g]ateway to the best year-round steelhead fishing in North America'."[193] The Sports Afield piece listed other P.A.-area outdoor activities ("diversions") such as hiking, camping, rafting, sea-kayaking, and ...sea-lion watching.[194]

By 1999, The New Rating Guide to Life in America's Small Cities had promoted Port Angeles to sixth place in its U.S. mini-urban Top Ten. Disgruntled P.A. cynics -- mostly transplanted Californians already sick of our wet-winter weather lasting from October to May -- still couldn't (or wouldn't) believe the figures. They reminded their more sanguine fellow Port Angelans that all guides observe people, places, and things in their own way. Nevertheless, according to The New Rating Guide, the No. 1 small city in all America is Mount Vernon, Washington.[195] But what makes Mount Vernon the No. 1 small city in the entire U.S.A. or, indeed, in Washington State? Most best-place guides have traditionally restricted themselves to choosing their favourite restaurants, resorts, schools, standard of living, etc. But the "life-style" demands of the 1990s have changed all that.

In 1999, The New Rating Guide lists 193 cities of 15,000 - 40,000 population; Places Rated Almanac ranks the "overall quality of life" in 351 North American metropolitan areas.[196] Today's best-place guides combine local data on economic conditions, employment percentages, and crime figures to find their ranking for each area/node. Other categories (in the U.S.) include: [The] Best Places to Earn and Save Money (winner: Minneapolis-St.Paul); Best Small Towns (winner: Elko, Nevada, was the closest to the Pacific Northwest); and Healthiest Place (winner: Boulder, Colorado, with Eugene, Oregon, in second place).[197]

There are still intangible factors which make every small city in North America unique. Trishia Jacobs, a transplanted booster and Peninsula Daily News columnist living in Port Angeles, has answered local sceptics:

"Whereas [Kevin] Heubusch [a New Rating Guide editor] bases his assessments on things like ... practical criteria -- jobs, public transportation -- my list would also include certain aesthetic and sentimental needs.
"What is the town's sunset quotient? Are there mountains and a water view? Is there a Taco Bell, a Dunkin' Donuts and an Olive Garden?"[198]

Port Angeles fills the bill in all the above categories, and surely deserves to rest on national sixth place laurels. (I have often asked myself, why seek a retirement haven in Oregon, Nevada, México, or Ponape?) Throughout FutureFish I have touted the good life in P.A.; written glowingly of the fresh salmon, sweet-corn, yellow Finns (potatoes), and seasonal apple juice-cider that make harvest time so special on the Olympic Peninsula. It was from Port Angeles High School (PAHS) that our three children were graduated to face the U.S. Army, college/university, and the workaday world. It has been in Port Angeles that I have voted in three national elections, served weeks of jury duty, and spent two weekend nights in durance vile. At Port Angeles, I lost a business, won a court case, taught Norwegian, and became an American. It has been a long dozen years since we came to Port Angeles, on 12 January 1987, as Euro-creepy former-Canadian residents from Victoria, B.C., with foreign license plates on the family jalopy.

During these twelve years we have seen many changes in Port Angeles, both good and bad, but mostly bad; mirroring America's cultural, societal, and political decline. I have already dealt at length with the steady encroachment of the U.S. federal government onto the privately-held property of American citizens. More and more, private lands and waters are being taken over as public lands and waters. Under the guise of "protecting endangered species", armed and empowered eco-agencies have slowly but surely extended the boundaries of Olympic National Park...designating contiguous areas as "protected". (More and bigger national parks mean greater environmental "protection".) In 1998, Evergreen State environmentalists -- in all their eco-sophism -- seriously proposed the truly nutty idea of reintroducing timber wolves into Olympic National Park! This cockamamy notion was rejected outright by angry local residents. The Emerald City eco-sophists had evidently forgotten that the Olympic Peninsula (still) has a generally truculent Guns "R" Us independent mentality.

Peninsula Daily News "Forks neighbor" columnist Rayna Abrahams is tiny, female, fiftyish, and Jewish -- hardly a candidate for "dangerous Right-wing crazy" candidacy! But read -- and heed -- her wise words:

...."It is no wonder that such a small percentage of Americans can afford to buy their own home. There is so little privately owned and governed land. This is what will make it impossible for our posterity to flourish. Look at the big picture. More rules and regulations force us to spend more money to carry out these mandated charades.
"Spotted owls, snail darters, marbled murrelets, gray wolves, and the list goes on, are but the 'reason' in disguise for land control. Be careful. Once the land is tightly controlled, so are the people.
"What has happened to the 'land of the free'?"[199]

It might seem a waste that talented columnists like Trishia Jacobs (see note 198) and Rayna Abrahams find themselves writing for a pc/cw McPaper like the Peninsula Daily News, but they have no choice: The PDN is the only daily newspaper published on the North Olympic Peninsula. (Another reason is that both smart women like living here!) Even though the PDN has striven with all its might, during the 1990s, to emulate USA TODAY (the original McPaper), it dutifully printed letters from their readers which reflected all political points of view. But within the last couple of years (1998-1999), the PDN has hired a sleek new editor with a slick new editorial policy. Thus I found my 1998 year-end letter to the PDN butchered beyond recognition. To right (write) a wrong, I hereby include an offendingly-deleted passage below:

...."I find myself in a social ecology of pot-smoking seniors, [grand]mothers for partial-birth abortion, and daffy [pony-tailed] gran'dads taking their [precocious] grandkids to Rolling Stones concerts."

The two Port Angeles radio stations, KONP 1450-AM and KIKN 1290-AM ("Kickin' Country"), offer programming bereft of content and benighted in quality: The former, High School sports and endless trivial chatter; the latter, bad Country music and endless trivial chatter. Churlish and doltish talking heads respectively characterise the local (televised) Northland Cable News (NCN).

It is for too easy (and unfair) for a former big-city boy like me to poke fun at the mini-urban travails of P.A. For the City of Port Angeles is definitely coming of civic age. During the spring of 1997, the City Councils of Port Angeles, Wash., and Victoria, B.C., held two historic meetings to discuss their participation in the Coastal Corridor Scenic Byways Program.[200] Earlier that year, the mayors of Port Angeles and Victoria met and had talks concerning the International Gateway Project, an offshoot of the overall (bureaucratically titled) Coastal Corridor Master Plan.[201] Anyway, the Coastal Corridor Scenic Byways Program was initiated by the State of Washington, with the state's coastal communities, as a way of creating a "scenic" Pacific Coast route alternative to the megalopolitan I-5 corridor. Eventually, the Pacific Coast Byway Program would extend from northern California, through Oregon and Washington, all the way up to Victoria, B.C.[202] The upshot -- by summer 1997 -- was the signing of a Proclamation of Cooperation by the city councils of Port Angeles and Victoria. The bi-nodal hopes visualised joint economic projects, including a shared web-site and coördination of ferry terminus redevelopment on both sides of the Strait of Juan de Fuca.[203] The City of Port Angeles further projected:

"The two cities are now looking ahead to the 21st century. Who knows? Perhaps their common economic interests could create a single, unique international metropolitan area straddling the Strait of Juan de Fuca."[204]

By late 1998/early 1999, the civic phantasies of the City of Port Angeles knew no bounds: "This regional [Pac. N.W./B.C.]economy is known as Cascadia and is based on the mutual interests of citizens doing business across borders that have imposed economic barriers."[205] (Are we already, then, "Slouching Toward Cascadia"?)

But the coming of civic age to Port Angeles has not necessarily meant overall improvement. Indeed, some aspects of late Gay '90s P.A. life are truly terrible. Like the rest of America, Washington State is very much a Cops-and-Guns culture. I have lived all over the world (from Oslo to Honolulu; from Victoria to Cape Town), and I have never borne witness to an omnipresent "police culture" equaling that of Port Angeles, Washington. If British Columbia can be likened to Mainland China, Washington State resembles neighbouring

North Korea; ruled over by Gov. Gary Locke, a stone-cold statist and political Clintonist opportunist. Gov. Locke's dismissive remarks concerning his defeated Republican opponent, Ellen Craswell, after the 1996 gubernatorial race, were absolutely outrageous.[206] The coming Y2K "millennium bug" crisis might test the true mettle of Gary Locke's potential as minimus-dictator. So early as December 1998, Gov. Locke announced his intention to mobilise fully <u>half</u> of the Washington National Guard on 31 December 1999 and 1 January 2000.[207] Thus, for the avowed purpose of assuring essential services in case of computer breakdown, circa 3,000 U.S. Army guardsmen will deploy throughout the State of Washington.[208] In a January 1999 article, the Associated Press reported:

"Brig. Gen. Lee Legowik, assistant adjutant general to the [Washington] [S]tate National Guard, said he didn't expect problems, but the Guard would be available if there was civil <u>disorder</u>....
"We have a society based on <u>order</u>, and I don't expect that <u>order</u> to break down,' he [General Legowik] said.
"The state Military Department, which includes the 8,600-member Army and Air National Guard and the 100-member state Emergency Management Department, will be ready in 39 National Guard armories to assist if needed."[209]

So <u>much</u> governance! Such a riding herd on human beings -- in this scenario U.S. civilians! I can only shake my head and feel a deep sense of foreboding. Even though a non-believer in conspiracy theories, I am highly suspicious of heavily-armed federal government agencies. And they seem to be proliferating in America. A good place to watch training films of Special Weapons and Tactics (SWAT) teams is on the Discovery (DSC) Channel. The very sight -- even on television -- of the black-clad, ski-masked, SWAT-team units mincing along on their little cat-feet, with heavy weapons held at high-port... is blood-chilling. Perhaps yet more frightening, is seeing these same SWAT-team members at ease: Big, strong, top-dog hunters lounging about with a studied insouciance approaching a cocky insolence. They are fellows who really enjoy their work/status. Government-sanctioned, well-armed and very dangerous, the self-confidence of the U.S. federal gunmen is supreme. Heaven help the poor devils who get in their way. (Just ask the remaining Randy Weaver family, formerly of Ruby Ridge, Idaho; or any surviving Branch Davidians, formerly of Waco, Texas.)

An old friend of mine, a "seasoned [Senior] citizen" from nearby Sequim, recently wrote a letter to the <u>PDN</u> editor:

"As the government enacts ever more legislation to take guns away from the citizens [,] more and more guns are being put into the hands of agents of the government.

"Case in point: I visited a national forest yesterday. The forest ranger was packing a very lethal .45 caliber Colt automatic. Last year the same ranger was gunless. Is this a U.S. Ministry of [F]ear?"[210]

If we have to face an incoming Ministry of Fear in Washington State, I would prefer to do so in a small rural city like Port Angeles than in a large urban area like Seattle. And I cannot -- for the life of me -- understand America's fascination with Grunge City.[211] For years, Seattle has been American yuppiedom's favourite metropolis -- variously referred to as Cyber, Cynergy, and Synchronicity City. Microsoft Corp.'s now-hip billionaires, Bill Gates III and Paul Allen ("revenge of the nerds"), have contributed hugely to transforming gray, grungy Seattle into the '90s-chic Emerald City; with adjoining, dowdy Redmond remade as the new-millennial Magic Kingdom. (See `FF 2000, Ch. 3` .) Even at its societal pinnacle, Seattle culture presents a post-modern ambience of vacuous emptiness. Michael Crichton's insightful novel, Disclosure[212], which deals with the Machiavellian machinations of "sexual harassment", is set in Seattle; the perfect stage for 1990s gender-feminist sexual-correctness. It is no cultural coincidence that the mainstream media's heir apparent to Seinfeld's TV crown has been Frasier. Change the venue (from New York to Seattle) and the cultural quotient (from vaguely Jewish -- a parve Yiddishkeit -- to soppy WASPish), and Frasier has fulfilled all yuppie expectations as worthy whitebread successor to Seinfeld:[213] A meaningless sitcom about nothing filled with centreless personae. The latter are strikingly similar to the trendy Ahrimanic/Luciferian lords of the Seattle technocracy, who reign over the Emerald City/Magic Kingdom cultural ecology of aerospace/cyberspace.

Despite living more than twelve years in Port Angeles (mostly happy), I have not let myself become too much of a loyalist home-boy. I was re-taught that valuable lesson, vicariously, by Larry Holmes, the former world heavyweight boxing champion (title holder: 1978-1985). In his autobiography, Against The Odds,[214] Holmes writes of a conversation with Muhammed Ali (a/k/a/ Cassius Clay) in which Ali tells him:

"You know the smartest thing you ever did, Larry?....
"You stayed in your hometown. Man, you were smart. You got a place where you feel you belong, where you have your people. Me, I can go anywhere in the world and people will know me and make a fuss, but I don't really have a hometown no more. I really miss that.'"[215]

True enough, but after the [first]defeat by Michael Spinks (on 21 September 1985) when Larry Holmes returned home to plebeian Easton, Penna., from glittering Las Vegas, Nev., the former world champ received a non-reception for which he was hardly prepared. A crestfallen Holmes relates:

"The next day, I drove from the Allentown airport along Route 22. As I approached the Easton turnoff, I passed the billboard that for seven and a half years had stood there -- the one that said: 'Easton Is the Home of Larry Holmes, World's Heavyweight Champion.' It didn't say that anymore. It hadn't taken them long to change over that billboard."[216]

End Academy/P.A. Fish Notes

In 1999, I attended two good (but sad) lectures at out local institution of higher learning, Peninsula College. The more recent studium, 'Early Scandinavian Immigrant Painters in the Pacific Northwest',[217] was presented during June, the final month of the academic year. I therefore expected the auditorium to be half-empty---and it was. But the lecturer, an earnest young man who proved to be knowledgeable and enamoured of his subject, regarded the auditorium as half-_full_: He actually _praised_ the overwhelmingly half-assed and nit-witted PC students for just being there! (Surely Dr. Magnusson -- the presenter -- was aware that his year-end studium was almost wholly attended by make-up/mandated "kids in the hall" types?) The ensuing lecture was very well done, but I left the auditorium with a sense of sadness. For on turning to leave, I didn't see a single Sons of Norway member in the audience. I gave a New York shrug, and reckoned that local Lodge "brothers and sisters" were too preoccupied working on their _midsommerfest_ costumes to attend this (widely advertised) _gratis_ presentation. It was their Scandinavian cultural loss.

The other sad studium, which I attended during April, was intended to be sad -- and it was. This lecture was titled 'Losing the Last Frontier',[218] presented by Jana Suchy (pron. "Yana Sookhy"), a lovely woman of about my age, whose fine work in photojournalism had long been familiar to me from the pages of _Alaska Fisherman's Journal_ and _Pacific Fishing_ magazine. On an Alaska fishing level, 'Losing the Last Frontier', was sad because of... "the quick progression from 'lifestyle' fishing seasons to the 'deadly derbies' and insanity of twenty-four-hour openings in the crazy boom times, to the sad and bitter reality of the tamed and civilized 'IFQ' days of today -- Individual Fishing Quotas that effectively fenced and divvied up the ocean range in 1995."[219]

And it was sad on a personal level. Jana Suchy came back to the Northwest after an absence of eight years, after she had lived and worked in her native Wisconsin, and then in the Southwest. Her return to the Northwest was to pursue a long planned book/photo-documentary on Alaska fishing, inspired by a fishing friend tragically lost at sea during the mid-1980s.[220] Jana Suchy went Up North to Sitka, her old Alaska home, to complete the non-fiction book proposal -- to expand upon the meaning of the lost fisherman's life and death; to explain the significance of 'Losing the Last Frontier'. But it was (finally) there that Jana Suchy, tough and feisty as she obviously is, ran into the brick wall of her own damaged psyche:

"...[S]he found that commercial fishing had changed so dramatically in recent years -- bitterly dividing the fleet, fisherfolk, and friends she held dear -- that she lost heart and put the project aside."[221]

Jana Suchy's presentation, then, turned out to be a photographic synopsis of a book not written: 'Losing [the book on] the Last Frontier'. Despite her brave mien, irreverent remarks like "pot powered the fleet", and wealth of beautifully-detailed slides showing all aspects of the 1970s-1980s Alaska fisheries[222]... there was a dated, "dear departed", atmosphere to the showing -- reminiscent of looking through old photographic albums, in which deceased relatives and friends are forever smiling and young. You, the beholder, know that "the loved ones" are forever gone, but you irresistibly thumb through the volumes anyway; emotions a-stir from a time that was and won't be again.

Will there now be no more visions of commercial fishing in Alaska for young Port Angelans or, indeed, Washingtonians? And as of early 1999, the very last charter-fishing boat in Port Angeles was retired.[223] The owner/skipper of The Lucky Strike sold the 16-passenger vessel to a whale-watch operator in the San Juan Islands.[224] Ever since the Dino de Laurentiis film "Orca" was released in 1977[225], the killer whale (Orcinus orca) has enjoyed a near-Homo sapiens status; being popularly perceived since then as able to recognise human enemies, and to exact bloody revenge on same. I wish The Lucky Strike well in her new, renovated rôle in searching for grampus/whale-killer pods, but that is of little consolation to area tourists who want to actually go fishing by charter boat.

It does not seem very long ago, when I first visited Port Angeles in 1982, that there were fully four, functioning, fish-processing plants in town. Today there is only one left, and it processes mostly farmed (Atlantic!) salmon. I recall my own hopes and dreams for Pisces Seafoods during the mid-1980s. Across from the old (long-gone) Pisces Seafoods dock on Oak Street, stands the former railway depot, now the Waterfront Art Gallery. I had once dreamt of converting the depot building into a sit-down, white-tablecloth, up-scale seafood restaurant. Farther down West Railroad Avenue is the Waterfront Antique Mall (presently for sale), originally constructed as a cold storage facility. I had once hoped to restore that solid concrete-slab aedifice to a fish refrigeration plant. But all my hopes and dreams of the mid-1980s have disappeared, like the fish, by the late 1990s.

Whale-watching, anyone?

End Politics Notes

> "The men [and women] the American people admire most extravagantly are the most daring liars" --
> -- H.L. Mencken

I have added "and women" in brackets as our First Lady, Hillary Rodham Clinton, appears to prevaricate as daringly as her deceitful and faeculent husband. In these pages, I have likened the Clintons to the cowardly King Ahab and the evil Queen Jezebel of Israel (1 Kings, Holy Bible), and Special Prosecutor Ken Starr to the prophet Elijah. I could also draw a similar parallel with the mongoose, "Rikki-Tikki Tavi" (Ken Starr), and his terrible battle versus the deadly pair of king cobras, Nag (Bill) and Nagaina (HRC), found in Rudyard Kipling's The Jungle Book.[226] But both comparisons -- as to their outcomes -- are wrong. The righteous Elijah prevailed against the weak Ahab and the defiant Jezebel, and Rikki-Tikki-Tavi was victorious over both king cobras. But Ken Starr batted zip at the recent "impeachment"/show-trial of the president regarding Monicagate, and eventually came up with zero-ziltch-nichts-nada evidence regarding Chinagate-Filegate-Travelgate-Whitewatergate (on which to hang the Clintons). POTUS (President of the U.S.), the compulsive liar; is at present taking self-congratulatory credit for gutting the Yugoslavian province of Kosovo. (Fightin' to make the world safe for hypocrisy). FLOTUS (First Lady of the U.S.), a pathological fraud (and Chicagoan), is currently contemplating a run for Pat Moynihan's shortly-to-be-vacated New York senate seat in A.D. 2000. And what of the Special Prosecutor? Citizen Ken Starr, a "class act" all the way, is nonetheless seen (by the current wisdom) as the Mayor of Loserville, and is reputedly headed for sunny Malibu and certain oblivion.

Has outright lying in the White House become the norm, or has it been an American tradition all along? For a quarter century we have been endlessly reminded by the mass media of Richard Nixon's Watergate cover-up. But the truth has been steadily leaking out about those "daring liars", preceding

Presidents LBJ, JFK, FDR. That shameless falsehood in itself has become an integral part of the U. S. political culture is illustrated nightly, via disseminated doublespeak; with revisionist history, specious groupthink, and outright disinformation broadcast by ABC, CBS, CNN, NBC, and PBS. (I have to keep telling myself, It's not a conspiracy -- it's a <u>syndrome!</u>)

<u>End Church Notes</u>

"For the time will come when men will not put up with sound doctrine. Instead ...they will gather around them a great number of teachers to say what their itching ears want to hear".
- - 2 Timothy 4:3, <u>Holy Bible</u>

"In those days men had convictions. We moderns have opinions: -- Felix Mendelssohn (1809 - 1847), German - Jewish composer referring to Martin Luther, on why he wrote the Reformation Symphony. [227]

I have been a happy church-camper ever since leaving the ultra-liberal ELCA (Evangelical Lutheran Church in America) during 1997, and joining the more traditional LCMS (Lutheran Church Missouri Synod) the following year. Before becoming a member of the LCMS in May 1998, I regularly attended divine services at St. Matthew LCMS starting in October 1997. I was soon to be spiritually flabbergasted. After five long years of religiously-correct nutra-sweet coated popcorn at the ELCA, by May 1998 the LCMS pastor, from his pulpit, had: 1) Read aloud from the book of <u>Revelation</u>; 2) discoursed on the Second Coming of Jesus Christ; 3) discussed the horrors of abortion and euthanasia; 4) warned about New Age teachings creeping into the Church; 5) brought up the lack of order and discipline at home and at school; and 6) mentioned <u>positively</u> Dr. James Dobson of "Focus on The Family". None of the above had been vocalised in church during the wretched half-decade I had foolishly wasted at the ELCA. (Sooner or later, of course, the LCMS will go the oecumenically-correct way of the ELCA. But for now I had found a church and pastor I could trust.)

I am not so happy with the irresponsible hysteria surrounding the Y2K "millennium bug", which has been manifested by much of Christian radio and television; by many evangelical books and journals. It has been wild-eyed millennarianism at its worst. Listening to Christian radio throughout the 1990s while penning these lines, has enabled me to understand the real antipathy many North Americians feel towards evangelical Christianity. For example, I used to tune in to KGMW 820-AM (Seattle) for <u>The Bible Answerman</u>, during weekday afternoons; and <u>America Today</u>, on during weekday evenings. The self-styled, Southern California-based "Bible Answerman" knows the Bible thoroughly, and is right about 90% of the time about things theological. But the Bible Answerman is also a domineering, spiritual bully, who verbally browbeats his caller-audience into a stunned silence. And the youngish, tragically-hip host of <u>America Today</u> means well (he is a Christian, after all), but like so many wanna-be Seattle <u>cognoscenti</u>, is a shrill ignoramus who should never venture topically outside the narrow confines of the Emerald City.

(When not listening to Christian or Conservative talk radio, I have been sustained -- during the hundreds of hours of writing and copying by hand -- by hearing the taped recordings of composers/musicians ranging from Richard Wagner to Howlin' Wolf.)

Coda / L'Envoi

"<u>Adjø'</u> (à Dieu) from the All American City at the End of the Road"

Someone once said that if you wait long enough, the world will beat a path to your door. And so it did. On the evening of 13 August 1997, bluesman John Hammond ("Got love if you want it") visited Port Angeles. He was to play a limited engagement at La Casita restaurant ("a taste of Mexico"). I was at his very first show, and thirty-plus years of memories came surging back. John Hammond arose from the folk music revival movement of the early 1960s, and gained public attention with his hit song "No money down".[228] When I first heard that song, so many years ago, I had assumed that John Hammond was a Black Man; I was surprised to find out that he was just a white guy like me. (This was before the British [blues] Invasion.) I have been a loyal John Hammond fan ever since the 1960s, and he was not thirty feet away, singing and playing guitar/harmonica. It's a small world after all!

And here I am at "55 Alive" years, at the end of the road and the second millennium. I hope that this upfront and personal final Part (III) proves conclusively that not all Scandinavian men need to 1) send anonymous e-mail to [hopefully] perfect strangers, or endlessly dance the tango (also with perfect strangers) in silence, for effective emotional communication. And I have learned enough marine biology to know that if an educated young reader (i.e., little Wisenheimer) ever asks me to name the ancient ancestor of modern billfishes, my answer shall be a one-word shout: "ICHTHYOSAURUS!"

C. D. B-H
Port Angeles, Wash.
Midsummer 1999

NOTES

112. Elaine Ellis Stone, 'Norwegian on North American Campuses', The Sons of Norway Viking, Vol.95, No.9, September 1998, p. 18. (See Appendix U.)

113. Author's note: Among the cultural enablers has been Darrell D. Henning, the (1994) director of Vesterheim, the Norwegian-American Museum at Decorah, Iowa; (also) the home of Luther College. I had indirect contact with Vesterheim via The Bay Foundation during 1992-1995, and was subsequently made a Vesterheim Fellow in 1994 as a result of my steering generous foundation funds in the Norwegian-American Museum's direction. Mr. Henning, Vesterheim's once and future curator, was a dead ringer for deceased Canadian New Age magick-practitioner Robertson Davies (1913-1995), author of The Manticore and other Luciferian works. NB: Personal letter to C.D.B-H. from Darrell D. Henning, dated 12 August 1994.

114. Author's note: Norwegian-American cultural enablers and politically-correct fixers have included Alf and Kathleen Knudsen, editors of Western Viking, Seattle, Wash.; and Synnøve Risholt, Christina Pletten, and Marianne Onsrud Jawanda, editrices of Norway Times, Brooklyn, N.Y. The above five harmonious opinions editorialise in prevailing pc lock-step with each other, emanating (respectively) from Ballard and Bay Ridge. Norwegian America -- that which remains of the "scattered community" -- reads and listens, believes and heeds. (Oh, well -- Bay Ridge, Brooklyn, wasn't picked as the site of "Saturday Night Fever" [1977] and "Staying Alive" [1983] for nothing!) NB: My source for Norwegian-language newspaper updates has been Tracy Baumann, 'Hot Off The Press: After more than a century, "Western Viking" and "Norway Times" stay current with the needs of their scattered community', The Sons of Norway Viking, Vol.95, No.8, August 1998, pp.12-13, 50-51.

115. Roger Fitzgerald, 'The Sanitizing of Ballard', Alaska Fisherman's Journal, Vol.20, No.4, April 1997, p.16.

116. For a detailed summary of my (drinking) life in Ballard from 1976-1978, see FF 2000, "Alaskamenn", Fisherpersons, and Hippies with Guns.'

117. Op.cit., Roger Fitzgerald, 'The Sanitizing of Ballard', p.16.

118. Ibid., p.16.

119. 'Kennewick Man has new home', Peninsula Daily News, 30 October 1998, p.A-7.

120. 'Kennewick site protection project debated', Peninsula Daily News, 14 January 1998, p.A-7.

121. 'Pagans angry with care of Kennewick Man', Peninsula Daily News, 31 August 1997, p.A-8.

122. Ibid., p.A-8. 'Kennewick Man leg bones missing', Peninsula Daily News, 11 March 1998, p.A-6.

123. 'Agency to rule on future of Kennewick bones', Peninsula Daily News, 3 April 1998, p.A-8. NB: Brackets this writer's.

124. Ibid., p.A-8. 'Kennewick Man gets memorial', Peninsula Daily News, 8 January 1998, p.A-6.

125. Ibid., p.A-6. 'Separated at birth?', Peninsula Daily News, 11 February 1998, p.A-6. Also, 'Pagans angry with care of Kennewick Man', p.A-8.

126. 'Agency to rule on future of Kennewick bones', p.A-8.

NOTES (cont'd.)

127. Ibid., p.A-8. 'Kennewick site protection project debated', p.A-7.

128. 'Separated at birth?', p.A-6.

129. Ibid., p.A-6.

130. Ibid., p.A-6.

131. 'Don't get Thor: Pagans plan to bless bones', Peninsula Daily News, 20 August 1997, p.A-5.

132. 'Kennewick Man gets memorial', p.A-6.

133. 'Don't get Thor: Pagans plan to bless bones', p.A-5.

134. 'Kennewick Man gets memorial', p.A-6.

135. Ibid., p.A-6. Author's note: The vernal equinox, vårjamdøgn, is a significant day in Norse lore.

136. 'Kennewick Man has new home', p.A-7.

137. Ibid., p.A-7. Author's note: I noticed a proliferation of these pewter mini-mjølner personal ornaments in Oslo while visiting my hometown in 1994. Has Thor's hammer become a symbol of the neo-Gothic movement, Norwegian Black Metal, or other?

138. Ibid., p.A-7.

139. 'If proved to be Norwegian, Indian roles might be reversed!' Letter from J. Nelson of Ocean Shores, Wash. Reprinted in Sons of Norway Brevet (Port Angeles, Wash.), April 1998, pp.2-3.

140. 'Lesbian who fought Army fights new war', Peninsula Daily News, 24 February 1998, p.A-5.

141. 'Grethe's battle', Peninsula Daily News, 26 July 1998, p.A-7. NB: A compliant federal judge later ordered Cammermeyer reinstated. The U.S. government did not appeal.

142. Margarethe Cammermeyer with Chris Fisher, Serving in Silence (New York: Viking-Penguin, 1994). Author's note: Perhaps the title should read Self-Serving out Loud?

143. 'Grethe's battle', p.A-7. Author's note: Glenn Close, along with old pc princess Barbra Streisand, and new pc queen Rosie O'Donnell, would later contribute to Cammermeyer's "political campaign".

144. 'Lesbian who fought Army fights new war', p.A-5. Author's note: Health (their bodies) and education (their minds) are the twin power-nodes of Leftist sociopolitics. Health has declined somewhat as a perennial liberal topic ever since the ClintonCare scare of 1994. NB: Cammermeyer's quoted remarks not listed in order of utterance.

145. 'Grethe's battle', p.A-7.

146. Ibid., p.A-7.

147. Ibid., p.A-7.

NOTES (cont'd.)

148. From Scanorama magazine, July/August 1997, n.p.

149. Liv Dahl, 'Living in a Fairy Tale', S/N Viking, Vol.94, No.6, June 1997, p.18. (Brackets this writer's.)

150. Author's note: Monica Lewinsky, in her bemused musings, often referred to her presidential paramour as "The Big Creep". I hereby dub Jerry Seinfeld "The Little Creep".

151. "Scandinavia" (usually) means Norway, Sweden, and Denmark; the "Nordic countries" (usually) means rump-core Scandinavia plus Finland and Iceland. In my Sons of Norway language classes, I often referred to a "Greater Scandinavia", which includes the tiny Faeroese Islands and gigantic Greenland (both Danish). A case can also be made for formerly-Soviet Estonia; a proud, new (1991) little-sister nation to Finland.

152. There was a brief resurrection of the Scandinavian male in (Swedes) Bengt Forslund and Jan Troell's 1970 film epic, "Utvandrarna" ("The Emigrants"), starring Max von Sydow as "Karl-Oskar". (Who can forget his brilliant performance?) "Utvandrarna" was based on Vilhelm Moberg's four-part series, and broadcast on SR (Swedish) and NRK (Norwegian) state TV networks. There followed a less successful sequel, "Det Nya Landet" ("The New Land").

153. Trygve Lie, In the Cause of Peace (New York: The MacMillan Company, 1954), pp.11, 16-17. NB: From G. Edward Griffin, The Fearful Master: A Second Look at the United Nations (Belmont, Mass.: Western Islands Publishers, 1964), p.114. Author's note: Griffin's is the harshest assessment I have ever read of the UNO, in this increasingly hard-to-find Americanist classic.

154. 'Director Ullmann to Present Film in D.C.', News of Norway, Vol.54, No.7, Oct./Nov. 1997, p.10.

155. Ibid., p.10.

156. Ibid., p.10. Author's note: Bob Hope's comedy vehicle, Global Affair (1964), endeavoured to put an early version of the UN Convention on the Rights of the Child in a funny, friendly light. It didn't work, but Bob Hope didn't disappoint and Lilo Pulver was stunning!

157. Ibid., p.10.

158. See letter of J.W. Christophersen of Burnaby, B.C. to Scandinavian Press (Vancouver, B.C.), Vol.4, Iss.2, Spring 1997, p.8.

159. Ibid., p.25.

160. 'Brundtland Nominated for WHO', News of Norway, Vol.54, No.7, Oct./Nov. 1997, p.2.

161. 'WHO committee votes Brundtland in final election', News of Norway, Vol. 55, No. 1, Jan./Feb. 1998, p.1.

162. 'Scandinavians in the News', Scandinavian Press, Spring 1997, p.25.

163. Ibid., p.25.

164. Ibid., p.25.

NOTES (cont'd.)

165. 'Good Morning Morons Award', <u>Human Events</u>, Vol. 54, No.3, 23 January 1998, p.12. NB: Brackets this writer's.

166. Author's note: Does any thoughtful student of current events <u>really</u> take the Nobel Peace Prize seriously anymore? Some recent recipients have been the mendacious Rigoberta Menchú (1992), and co-winners Leninist Nelson Mandela (1993) and terrorist Yasir Arafat (1994).

167. Gene Edward Veith, 'Feminist imperialism', <u>World</u>, Vol.13, No.44, 14 November 1998, p.27.

168. <u>Ibid.</u>, p.27.

169. <u>Ibid.</u>, p.27.

170. <u>Ibid.</u>, p.27.

171. In Norwegian: "<u>ei rik, utdannet, lett-livet tøs som kjeder seg.</u>" NB: Quoted from a Norwegian female family member.

172. Michael Crichton, <u>Eaters of the Dead: The Manuscripts of Ibn Fadlan, Relating His Experiences with the Northmen in A.D. 922</u> (New York: Ballantine Books, 1988 ed.), p.149. NB: Brackets this writer's.

173. Michael Moynihan and Didrik Søderlind, <u>Lords of Chaos: The Bloody Rise of the Satanic Metal Underground</u> (Venice, Calif.: Feral House, 1998), p.78.

174. <u>Ibid.</u>, pp.78-79.

175. <u>Ibid.</u>, p.79.

176. <u>Ibid.</u>, p.38.

177. <u>Ibid.</u>, pp.84-85.

178. <u>Ibid.</u>, p.197. Quoted is Ihsahn of Black Metal band, Emperor.

179. <u>Ibid.</u>, p.215. NB: Brackets and emphases mine.

180. Op.cit., Moynihan and Søderlind, p.171. NB: Brackets and contents mine.

181. Gene Edward Veith, 'The new state church', <u>World</u>, Vol.13, No.40, 17 October 1998, p.26. (NB: Brackets mine.)
Author's note: A few godly individuals can still make a big difference...even in Norway! One of these has been <u>sogneprest</u> Einar Martinsen (d. 9 July 1998) of Skjaerhalden, Hvaler. ("Well done, thou good and faithful servant".)

182. <u>Ibid.</u>, p.26. (Brackets and contents mine.)

183. Op.cit., Gene Edward Veith, p.26.

NOTES (cont'd.)

184. Bjørn "Swede" Dahlin, For His Glory: From Mercenary to Missionary (P.O. Box 398, McCall, Idaho, n.d.), p.5. Author's note: I had the privilege of having coffee and rolls with Pastor Bjørn Dahlin in Port Angeles, July 1998. (Thanks, John Nutting -- "Semper Fi"!) NB: Amos 8:11, Holy Bible (Revised Standard Version)
  "Behold, the days are coming', says the Lord God, 'when I will send a famine [hunger] on the land; not a famine [hunger] of bread, nor a thirst for water, but of hearing the words of the Lord.'" Amos 8:11, Bibeln (Swedish)
  "Se dagar skall komma, säger Herren, Herren, då jag skall sända hunger i landet: inte en hunger efter bröd, inte en törst efter vatten, utan efter att höra Herrens ord." (Bokförlaget Libris, Örebro, Sweden, 1992.) PS: All brackets mine.

185. 'Hamsun-feber i Europa' ('Hamsun fever in Europe'), Nytt fra Norge (Norway Digest), Nr.44, 4-11 November 1997, p.15.

186. Ibid., p.15. Trans. from the Norwegian: "Knut Hamsun leses og diskuteres som aldri för i Europa."

187. Ibid., p.15. See Appendix C.

188. Ibid., p.15. See Appendix C.

189. Knut Hamsun, Hunger (New York: Farrar, Straus and Giroux, 1996 ed.), pp.123-125. Translated by Robert Bly.

190. 'Put another Lars on the Fire', Esquire, January 1998, p.44.

191. 'PA ranks 7th', Peninsula Daily News, 30 October 1997, p. A-3. Cf. Trishia Jacobs, 'Living the small-town life is heaven', PDN, 16 November 1997, p. C-4.

192. 'PA ranks 7th', PDN, p.A-3. Cf. Trishia Jacobs PDN, p.C-4. Author's note: "Urban proximity" to me means the nearness of sparkling Victoria, B.C.; not the depressing megalopolis snaking along the I-5 corridor (South Tacoma to North Vancouver).

193. Brad Lincoln, 'Port Angeles rated top outdoor sports town', PDN 14 January 1998, p.A-1. Author's note: Sometime during the 1990s, the steelhead trout (S. gairdneri) was upgraded taxonomically to a Pacific salmon (O. mykiss).

194. Ibid., p.A-1. NB: Some Strait of Juan de Fuca sea-lions reach 600 lbs. or more in weight. (And then there are always the whales!)

195. Brad Edmondson, 'Which city is tops? It depends which guidebook you believe', USA WEEKEND, 12-14 March 1999, p.8.

196. Ibid., p.8. Author's note: Puzzling to me, was why Places Rated Almanac awarded Seattle second best area/node in North America, without a single Canadian metro area listed in their Top Ten.

197. Ibid., p.8. Author's note: Minneapolis-St. Paul, as noted, ranked first in Best Places to Earn and Save Money. "Moscow on the Mississippi"? Gopher State public-sector employees? North Star no-smoking law enforcement? Go figure.

NOTES (cont'd.)

198. Trishia Jacobs, 'Living the small-town life is heaven', Peninsula Daily News, 16 November 1997, p.C-4.

199. Rayna Abrahams, 'Preserve "real" endangered species', PDN, 17 July 1997, p.A-6. Author's note: by "real" endangered species, the columnist meant herself -- a member of the human species.

200. 'Two Flags -- One City', City Report (Port Angeles, Wash.), Vol.XI, No.3, Summer 1997, p.1.

201. Ibid., p.201.

202. Ibid. p.1.

203. Ibid. p.1.

204. Op.cit., p.1.

205. 'Partnering with Victoria', City Report, Vol.XII, No.4, Nov.-Jan. 1998-1999, p.6. (NB: My brackets.)

206. Author's note: Mrs. Craswell is a courageous Christian woman stricken with cancer.

207. 'Guard ready for computer chaos', Peninsula Daily News, 7 January 1999, p.A-5.

208. Ibid., p.A-5.

209. Ibid., p.A-5. Author's note: Our state's Emergency Management Department is little-sister clone of the massive Federal Emergency Management Department (FEMA). (NB: Brackets and emphases mine.)

210. George W. Scott, 'Guns and government', Peninsula Daily News, 13 June 1999, p.A-7. Author's note: That ranger is packing a gun because of you, George! "Onward and upward!" (NB: My brackets.)

211. Rated No.2 in 'Best Areas Overall', Places Rated Almanac, 1999. In Brad Edmondson, 'Which city is tops?', USA WEEKEND, 12-14 March 1999, p.8.

212. Ballantine Books, New York, 1994.

213. Author's note: NBC-TV looks to have a patent on the New York-Seattle connection. There's not only Seinfeld/Frasier, there has for years been Saturday Night Live (New York), and its paler imitation, Almost Live! (Seattle). Where are the young Jackie Gleasons (Brooklyn) or Stan Boresons (Ballard)?

214. With Phil Berger, St. Martin's Press, 1998.

215. Ibid., pp.117-118.

216. Ibid., p.237.

NOTES (cont'd.)

217. 'Early Scandinavian Immigrant Painters in the Pacific Northwest', with Dr. Brian B. Magnusson. Studium Generale, 10 June 1999, Peninsula College, Port Angeles, Wash.

218. 'Losing the Last Frontier', with Jana M. Suchy. Studium Generale, 15 April 1999, Peninsula College, Port Angeles, Wash.

219. Studium Generale Program Notes, 15 April 1999, front page. (NB: Notes by Peninsula College)

220. Ibid., front page.

221. Ibid., front page. (NB: Notes by Peninsula College)

222. Author's note: So detailed, in fact, were Jana Suchy's slides, that scenes from an Alaska beach party revealed that Myers's Original Dark Rum was one adult beverage of choice. Dark, "demon" rum was also my favourite liquid refreshment during my "salad days" (1965-1990).

223. Ken Short, 'Last PA charter boat shuts down', PDN, 5 April 1999, p.A-7. Author's note: Surely a cruel April Fools'!

224. Ibid., p.A-1.

225. Co-starring Richard Harris and Charlotte Rampling, but with Orca at top billing.

226. Children's Classics, New York/Avenel, N.J. 1989 Reprint ed.

227. Cited by Pastor Andrew Simcak, Jr., 'Stand Up for Jesus', Portals of Prayer (St. Louis, Mo.: Concordia Publishing House, 1999), 17 June 1999.

228. Author's note: Over the years, John Hammond has recorded on labels ranging from Vanguard to Point Blank Records.

Bibliography for Part III

11. Dublin, Peter. *I Was a Teenage Norwegian*. Press-Tige Books, Catskill, N.Y., 1997.

12. Legwold, Gary. *The Last Word on Lefse*. Adventure Publications, Cambridge, Minn., 1992.

13. Morgan, Murray. *Skid Road: An Informal History of Seattle*. Ballantine Books, New York, 1971. Comstock ed. (Orig. published in 1951)

14. Newell, Gordon and Don Sherwood. *Totem Tales of Old Seattle*. Ballantine Books, New York, 1974. Comstock ed. (Orig. published in 1956)

15. Schwantes, Carlos. *The Pacific Northwest: An Interpretive History*. University of Nebraska Press, Lincoln and London (England), 1989.

Acknowledgements

My thanks to fellow S/N member Loran Olsen for his gift of Peter Dublin's delightful *I Was a Teenage Norwegian.* "Mange takk, Lorang!"

Special Supplemental Bibliography

1. The Alaskans. Time-Life Books, Inc., Alexandria, Va., 1978. (Alaska history)

2. Bernstein, Richard and Ross H. Munro. The Coming Conflict with China. Alfred A. Knopf, New York, 1997. (China/geopolitics)

3. Brown, Lester R. Who Will Feed China? Wake-up Call for a Small Planet. W.W. Norton & Company, New York, 1995. (China/hunger)

4. Buckley, Christopher. Thank You for Smoking. Harper Perennial, New York, 1995 ed. (Nicotine nazis)

5. Caldwell, Francis E. Beyond the Trails: With Herb and Lois Crisler in Olympic National Park. Anchor Publishing, Port Angeles, Wash., 1998. (Pac.N.W. Outdoors)

6. Davis, Wade. The Serpent and the Rainbow. Warner Books, Inc., New York, N.Y., 1987 ed. (Fish pharmacology)

7. Idyll, C.P. Abyss: The Deep Sea and the Creatures That Live in It. Thomas Y. Crowell Company, New York, 1964. (Benthic biology)

8. Lam, Joseph with William Bray. China: The Last Superpower. Leaf Press, Green Forest, Ark., 1997. (China/Christianity)

9. Lawrence, Carl with David Wang. The Coming Influence of China. Vision House Publishing Inc., Gresham, Ore., 1996. (China/Christianity)

10. Lindsey, Hal. Planet Earth 2000 A.D. Western Front Publishing Ltd., Palos Verdes, Calif., 1994. (Christian eschatology)

11. Marrs, Texe. Megaforces: Signs and Wonders of the Coming Chaos. Tyndale House Publishers, Wheaton, Ill., 1987. (Cybertechnology/prophesy)

12. McDonald, Lucile. Swan Among the Indians: The Life of James G. Swan, 1818-1900. Binford's and Mort, Portland, Ore., 1972. (Pac.N.W. Natives)

13. McPhee, John. The Headmaster. Farrar, Straus and Giroux, New York, N.Y., 1992 ed. Orig. published in 1966. (Education/élites)

14. Moynihan, Michael and Didrik Søderlind. Lords of Chaos: The Bloody Rise of the Satanic Metal Underground. Feral House, Venice, Calif., 1998. (Norwegian neo-pagan cults)

15. Nixon, Richard M. Beyond Peace. Random House, New York, 1994. (Pax Americana)

16. Oberdorfer, Don. The Two Koreas: A Contemporary History. Addison-Wesley, Reading, Mass., 1997. (Korea history)

17. O'Brien, Conor Cruise. On the Eve of the Millennium: The Future of Democracy through an Age of Unreason. The Free Press, New York, 1995. (Secular millennarianism)

18. Pacific Northwest Seafood Cookery. Stan Jones Publishing, Seattle, Wash., 1988 ed. (Fish/nutrition)

19. Shea, Nina. In the Lion's Den. Broadman & Holman Publishers, Nashville, Tenn., 1997. (China/South-East Asia/Christianity)

20. Simon, Arthur. Bread for the World. Paulist Press, Ramsey, N.J., 1984 ed. (World hunger)

Chapter 4   "A Modest Proposal" for Micronesia

Part I: The United States, the United Nations, and Micronesia

Section A: A Brief History of U.S. Imperialism
Section B: A Brief History of U.N. Expansionism
Section C: A Brief History of Micronesian Colonialism

Part II: UNCLOS III – The Third United Nations Conference on the Law of the Sea

Part III: Tropical Western and Mid-Pacific Fisheries

Part IV: A Modest Proposal for Micronesia

"Even the names were a tangle. Micronesia, i.e., small islands, was a sort of geographical umbrella, roughly congruent with the area's current political title, The Trust Territory of Pacific Islands, a United Nations trusteeship administered by the United States. We'd captured the place from the Japanese who'd run it under a League of Nations mandate. There were three main groups of islands – the Marshalls, the Marianas, the Carolines – over two thousand islands in all, half the land mass of Rhode Island, scattered over an area the size of the continental United States. A population of less than one hundred thousand that could be comfortably seated in the Pasadena Rose Bowl. Nine mutually unintelligible local languages, plus smatterings of German, Japanese, English, the conquerors' tongues. A shaky, artificial, lopsided economy based on scrap metal (declining), copra (stagnant), fish (caught by foreigners), and government (U.S. subsidized) ... an archipelago of Lilliputs."

-- P.F. Kluge, The Edge of Paradise, 1993

Preface: Millennium Madness

"I am one of those scientists who finds it hard to see how the human race is to bring itself much past the year 2000" -- Dr. George Wald, Nobel Prize – winning scientist (1967, physiology/medicine).

In Port Angeles, Washington, 1 January 2000 A.D. came roaring in briefly like a lion (a sudden early morning rain-squall) – but remained as a lamb all New Year's Day. The sun shone, the weather was mild, and the first day of the dreaded Y2K turned out to be just another pleasant Saturday on the North Olympic Peninsula. Indeed, the final week of 1999 had been so clement that I (wearing sunglasses, yet) had spent part of Christmas Day in my outdoor hot-tub, in salubrious luxury. So far that winter there had been neither snow nor ice. And when I had awakened that special morning, water flowed from the faucets and electricity from the outlets. All in all, except for the historically significant date,[1] 1 January 2000 was merely another warm winter day on the beautiful Olympic Peninsula of Washington State.

But for me, and millions of other North Americans, the gloom and doom of the impending Millennium Bug had hung like a pall over all of 1999. In January of that year I had read Shaunti Christine Feldhahn's unsettling Y2K: The Millennium Bug, and Grant R. Jeffrey's disturbing The Millennium Meltdown. Later, during August 1999, I had struggled through the truly terrifying Facing Millennial Midnight by Hal Lindsey and Cliff Ford. Although all the above are well-intentioned evangelical Christians, they had nonetheless helped fuel Y2K high anxiety and apocalyptic "millennium madness". Sometime in June 1999 I ceased listening to Christian talk-radio altogether. It appeared that every wild-eyed and loose-lipped, air-headed air-wave televangelist was forever teaching Armageddon. As an evangelical Believer myself, I was both shocked and ashamed at such social and spiritual irresponsibility.

The imminent coming of the Y2K collapse was also relentlessly preached by certain economists, some scientists, and all "Guns R Us" survivalists (both from the private and public sectors). To make matters yet worse, the very elements of earth, sea, and sky seemed to conspire with the Y2K doomsday/nay-sayers. On 22 December 1999 there occurred a "confluence of celestial events" during the final full moon of the 20th Century.[2] As a full moon, some say, affects human behaviour, the last full moon of the millennium loomed especially bright – in fact as well as in phantasy. For the final, full winter solstice moon of 1999 was the biggest and brightest (7% brighter than in summer) since 1866; the last year the then-solstice moon (20-21 December 1866) reached perigee and was full all within a 24-hour time-span.[3]

There had also loomed, along with the millennium moon, the pressing threat of geologist/alarmist James Berkland's "Super Syzygy" days of 22-29 December 1999. During that week, according to the information disseminated by one Richard Dobbs of Sequim, Washington, the sun might have triggered the "window of maximum stress", a Super Syzygy time, which could have culminated in a "Megathrust Subduction Earthquake" of up to 9.5 intensity on the Richter scale."[4] Richard Dobbs explained in two autumn newsletters published locally:

" 'The Land Grows Westward!' is the title of one of the 'Boards' on Cascadia, which shows ½ of the states [sic] of Washington and ½ of the state of Oregon has [sic] come out of the sea since the age of Dinosaurs (75 million years ago), 'the Columbia Embayment'.
"....The movement of the earth's plates under us piles up the scrapings off the ocean floor increasing the height of the Olympic Mountains... 2-4 inches per year. As this movement is compounded by the spreading of the Atlantic mid ocean ridge – pushing North America Westward [sic] – there is a distortion caused along this Western edge/a pressure build-up, that when released, is the Megathrust Subduction Earthquake of up to 9.5! A TSUNAMI [sic] is sent across the Pacific Ocean that causes damage to far off lands and heat and steam generated under this subduction zone makes the volcanoes awake with the typical 'blow off the tops' events'...."[5]

But Super Syzygy week came and went without any of the dire predictions having been fulfilled. Thus the first day of A.D. 2000 passed without disaster. And the new millennium dawned first over Micronesia. During the wee hours of 31 December 1999 I watched, in delight, televised scenes of Gilbertese dancers – sturdy and handsome men, women, and children in Tungaru traditional costumes – herald the new century on "Millennium Island" (i.e., Caroline Island, normally uninhabited),[6] the landfall most adjacent to the International Date Line, in the Republic of Kiribati.[7]

And speaking of television, I admit to having tuned to the Arts and Entertainment channel ordered ranking of the Top 100 V.I.P.s of the millennium, broadcast during the waning weeks of 1999. This had been an absolute waste of time on my part, as I already knew: That (1) A&E (and the little sister History Channel) is a PBS wanna-be in drag; (2) that so many of the Top 100 "choices" as possible would be Americans; and (3) that movie director/ manipulator Steven Spielberg would somehow be included on A&E's élite list. Sure enough, A&E's superior minds did choose a preponderance of Americans for their Top 100 V.I.P.s of the millennium (although our Great Republic has existed for a mere 223 years). And, sure enough, with the 1998 film "Saving Private Ryan" (starring omnipresent androgyne Tom Hanks) still fresh in mind, the A&E upper echelons designated cinematic weasel Steven Spielberg 91st most important and influential person of the second millennium!

(Moving pictures have been around for somewhat less than a century, and it's debatable whether Spielberg is even the best director of the last half century! But His Stevenhood keeps on foolin' 'em. And folks have obviously forgotten that along with directing "Saving Private Ryan" in 1998, Steven Spielberg made the truly atrocious "Small Soldiers" – starring Kirsten Dunst and clever animatronics – that very same year. Go figure?)

As for the Time-Life monster's choice of Albert Einstein (1879-1955) for their Man of the Century – all well and good. I'm glad Time, Inc.'s inquiring minds selected Albert Einstein over runners-up Mohandas Gandhi (1869-1948) and Franklin Roosevelt (1882-1945). Einstein, for all his personal flaws, was the genuine article. I strongly feel, however, that both Gandhi and Roosevelt were intrinsically fraudulent each in his own way. But how about Winston Churchill (1874-1965) as the Man of the Century? Churchill arguably straddled the 20th Century "like a colossus", and he affected people and events more than Einstein ever did. Well, conservative talking-head Michael Medved and G.O.P. presidential aspirant George W. Bush have agreed with me!

And so, after having written the preceding thoughts all New Year's Day 2000, I hereby close these preface notes by giving delicious vent to my aching spleen. Here are my own centennial pet-peeves and millennial irritants:

1) Most over-rated and boring land animal: The giant panda (Ailuropoda melanoleuca). It does nothing of note, and can barely reproduce itself.
2) Most over-rated and predictable movie ever made: "Casablanca" (1942). Why are Ted Turner and the American people so obsessed with this film? Humphrey Bogart is (as usual) forever mugging, shrugging, and posturing. And all Ingrid Bergman does (as usual) is make sheep's eyes at her leading man; in this film at "Bogie".
3) Most over-rated and boring comic strip series: "Peanuts", created by Charles M. Schulz (1922-). Snoopy, Woodstock, Charlie Brown, Linus, Lucy, and Schroeder rode especially high during the Sick Sixties and Silly Seventies, but are soon to be retired. (Good grief! How about bringing back to life the Katzenjammer Kids who were really funny?)
4) The decade's worst television programme: USA's Walker, Texas Ranger starring Chuck Norris. Inevitability of outcome, mindless brutality, reverse race-baiting, and kevlar-clad political correctness are all featured in heavy, lumpen measure. KTZZ's New York Undercover is a close second (especially in race-baiting), but it doesn't high-light arch bully Chuck Norris kicking the bad (invariably White) guys around; Walker, Texas Ranger does. Both shows (along with N.Y.P.D.), in just a few years, have undermined all the good police/public relations which Dragnet (and others) had painstakingly built up for decades. Too bad.
5) Most obnoxious new entertainer of the Gay Nineties: Will Smith. The Fresh Prince of Bel-Air thinks he can sing ("Willennium" etc.) as well as act ("Independence Day" et al.), but don't try listening to "Big Willie" being interviewed on pc TV by one of those fawning White talk-show hostesses – unless you have a very strong stomach. Will Smith is definitely in bad need of being taught some humility lessons, which he (unhappily) won't be getting any time soon. A late 1999 rumour had Big Willie – among some other "celebrities", including chief Clinton-contributor Steven Spielberg – spending New Year's Eve with "Slick Willy" at the Hillary White House. (A case of appropriate American justice?)

If the rumour turned out to be true, I can't think of a more fitting male-bonding buddy match-up than Big Willie and the Slickster, yukking it up at the epicentral seat of U.S. power, to end the American Century for good.

January/February 2000

NB: Charles M. Schulz (1922-2000) died on Saturday, 12 February, at San Francisco, Calif., on the eve of his comic strip series finale.

NOTES

1. NB: In actuality the first day of the last year of the 20th century!

2. Paul ("I am Weasel") Gottlieb, 'Rare lunar phenomenon lights the night Wednesday', <u>Peninsula Daily News</u>, 22 December 1999, p. A-1.

3. <u>Ibid.</u>; and 'Moon: Watch from hot-tub', p. A-2.

4. Letters to the Editor, 22 September and 22 October 1999.

5. Op.cit., Richard Dobbs, <u>ibidem</u>. Author's note: My thanks to George W. Scott of Sequim, Wash., for the above newsletters.

6. Phil Wilder, 'The Millennium Comes to Millennium Island', <u>Pacific Magazine</u>, January/February 2000, Vol. 25, No. 1, Iss. 138, p.46.

7. Hunter T. George, 'State poised for Y2K switch', <u>Peninsula Daily News</u>, 24 December 1999, p. A-7.

<u>Bibliography</u>

1. Bloom, Harold. <u>Omens of Millennium: The Gnosis of Angels, Dreams, and Resurrection</u>. Riverhead Books, New York, 1996.

Foreword: Micronesia Notes

"'[Micronesia's] varied and often exciting history has been ignored as if by some tacit agreement.... One wonders that American historians have apparently no contribution to make to the historical study of an American Trust Territory.'"
-- Australian H.E. Maude in The Journal of Pacific History[8]

Pacific historian H.E. Maude expressed the above sentiments some time ago, but his words – academically speaking – are no longer true. Fortunately, the 1990s was a decade during which much fine historiography was done on Micronesia (cf. Bibliography).[9] The reader will note from my Bibliography that the preponderance of updated Micronesian historiography has been accomplished at the University of Hawai'i. It has not, however, always been that way. When I attended U. Hawai'i at Mānoa for graduate studies from 1979-1981, the academic curriculum for Pacific history entailed Australia and New Zealand (what author Paul Theroux has called "Meganesia"),[10] Melanesia (especially Aussie-favourite Fiji), non-Hawaiian Polynesia (Hawai'i Nei rated an entire department of its own), but hardly Micronesia. The only people interested in Micronesia at U.H. Mānoa, it seemed, were cashiered U.S. government civil servants, aging ex-Peace Corps types, and a meagre sprinkling of (mostly) sullen Micronesian students. This writer readily admits that, to him in 1979, "da Paceefic" meant a sort of Anglo-Hispanic/Asian Pacific "Macronesia"; a cross-cultural island/rimland combination of California and Hawai'i, Australia and New Zealand, Japan and the Philippines.

By 1980, I was still wholly ignorant of Micronesia except for Guam (due to a likeable female Guamanian classmate), Truk (because of a platform-shoed male Trukese classmate)[11] and the Marshall Islands.[12] That was it! Although the former Peace Corps ("Piss Corpse" in Micronesian) "headmen" constantly whined about American imperialism in Micronesia,[13] the Australian department head firmly steered the course[s] in the direction of Australia, Papua-New Guinea ("Niugini"), Fiji, and the Solomons; New Zealand, (Western – i.e., non-American) Samoa, the Cooks Islands, the Tokelaus, Tuvalu, and all other areas of Anglo-Australian interest.[14] Indeed, except for a brief examination of Sir Arthur Grimble's beloved Gilbert Islands (today's Kiribati), Micronesia went begging for attention.[15]

It wasn't until fifteen years later that I started to take an interest in American Micronesia. I was teaching a class at Peninsula College (Port Angeles, Washington) during the early months of 1996, when I suddenly realised that the handful of Pacific Islanders attending the college were almost all U.S. nationals from Micronesia. Being American nationals, these Micronesians were not accorded the (usually) high campus social status as foreign exchange students. The same had been true at U. Hawai'i, and whereas Native Hawaiians had been treated like living Anasazi, Micronesians had been regarded somewhat like Puerto Ricans of the Pacific. (Perhaps that explained their apparent surliness.) Indeed, big black Melanesians (exotic) and big dark-brown Polynesians (majestic) had been handled with near-fawning deference at U.H. (especially by the Pacific history department); a respect the smaller, coppery Micronesians (domestic) had scarcely enjoyed.

But Pacific Islanders at (tiny) Peninsula College were fewer and farther between than at U. Hawai'i, and they seemed genuinely happy to be studying in a chilly little corner of "da Mainland." Perhaps it was the times, but the Micronesians at P.C. in 1996 seemed to have none of that spoilt-brat "minority ingratitude" I had encountered at U.H. in 1979-1981. (Or maybe I had mellowed in fifteen years!) And, as pc as P.C. is, the Micronesians were accepted as just other Americans. Not only was that fine with me (as an American), their positive attitude made Micronesian culture far more attractive – and available to me (as an American). So despite my M.A. in Pacific history, lack of Micronesia information suddenly loomed like a large hole in my degree. Since 1996, then, I have tried to fill that hole as thoroughly as possible. With the scattered islets of Micronesia encompassing vast areas of the Central and Western Pacific Ocean – and rich, tropical fishing grounds – my wish for study of Micronesia has been spurred by will for Micronesia fisheries knowledge. I hope that four and a half years of intensive research on my own is manifest in this report.

Before getting down to brass-tacks and the basic facts and figures on Micronesia, I must readily admit that cable T.V. (!) has played a large role in getting fascinated by the four fly-speck but far-flung archipelagoes which comprise Micronesia – the Marshalls and the Gilberts, the Marianas and the Carolines. The only Hollywood feature films (that I know of) supposedly set in Micronesia are (1) "His Majesty O'Keefe" (adventure:1953), starring Burt Lancaster, was all about Yap, but actually filmed on Fiji; and (2) "The Private Navy of Sgt. O'Farrell (comedy:1968), starring Bob Hope, was located on the fictitious island of "Funapee" (Ponape?). As entertainment, that was it for Micronesia.

Otherwise, there was a wealth of information on American Micronesia; from the grim black-and-white graininess of World War II images,[16] to the horror-in-technicolour of post-war atomic/hydrogen/thermonuclear bomb testing.[17] I remember in particular two remarkable programmes on the testing in the Marshall Islands (Eniwetak, Bikini). One of these was "Trinity and Beyond" (1995), rather pruriently advertised on T.V. as "The Atomic Bomb Movie". Film-maker Peter Kuran spared no niceties in displaying the devastation of Operation Crossroads on Bikini during 1946, or the "Ivy Mike" H-Bomb on Eniwetak during the early 1950s. The bombing tests, practiced throughout the Cold War 1950s, sported cute [Pentagon brass-hat] names like "Able", "Baker", "Charlie", "Hardtrack" and "Sandstone" which lent, if anything, an even more sinister tone to the thermonuclear/atmospheric testing. The agony didn't end for the native Micronesians (Marshallese, Gilbertese) until 1963.[18]

The other programme, far more to my liking, was an update on one Marshallese island in 1992 – "Bikini: Forbidden Paradise"; richly produced by David Livingston. (I liked this film so much that I watched it twice in 1997, on DSC's World of Discovery.) Mentally, I dubbed the production "The King of Bikini", as the programme highlighted the return of Bikini's aging [but still reigning] monarch to his homeland. The slender little king was nut-brown, wizened, wore a baseball cap and used a cane. But he walked ashore, proud and standing tall. For he and his loyal Bikinian subjects had lived in exile since 1946 on the island of Kili (nearby but a world away). 1946 had been the terrible year for the start-up of Operation Crossroads. An atomic bomb followed by twenty more bombs had blasted Bikini, several strategically-placed war-ships (including the USS Saratoga) and the surrounding waters (with their marine life) to perdition. But that still hadn't been enough for the powers at the Pentagon: In 1954 for the "Bravo" test, a 15 megaton hydrogen bomb was dropped on the islet.[19]

At the conclusion of "Bikini: Forbidden Paradise", actress Linda Hunt (who narrates) intones: "There is a future on this island, hidden in the[se] waters" – meaning that marine life had returned to Bikini waters, along with the island's king.

As stated, Micronesia consists of four fly-speck but far-flung archipelagoes – the Marianas and the Carolines, the Marshalls and the Gilberts. The coconut-clad islets can be counted in the hundreds, and are sprinkled over a vast expanse of the Western and Central Pacific Ocean; mostly (except for the Gilberts/Kiribati) situated north of the equator.[20] According to Bruce G. Karolle (1993), the total Micronesian cultural sphere encompasses well in excess of three million square miles (greater than 7 million km^2).[21] Other than oceanic area, though, Micronesia really is "micro".[22] For in terrestrial area, all the islands of Micronesia combined make up a mere 1,045.3 sq. mi. (2,707.2 km^2), about the size of Rhode Island.[23] In other words, the entire land "mass" of Micronesia could comfortably fit – several times – into the Big Island of Hawai'i; which is circa 4,000 sq. mi. or around the size of "Little Rhody's" larger (but still small) neighbour, Connecticut.

The single largest Micronesian landspace is Guam, the southernmost island of the Mariana chain, at approximately 214 sq. mi. (554.2 km^2) in area. Guam has been an unincorporated territory of the U.S. since the Spanish-American War of 1898, and has consequently always held a different political status from the other islands of American Micronesia – formerly of the U.N. Trust Territory. All together, U.S. Micronesia adds up to 929.789 sq. mi. (2,408.1 km^2), or approx. 89% of the land area of Micronesia (excluding Kiribati and Nauru).[24]

The Northern Mariana Islands (NMI), with its seat of government at Saipan, are composed of Saipan itself, Tinian (of A-bomb infamy), Rota and islets north to Farallón de Pajaros. Like the rest of American Micronesia (except Guam)[25], the Northern Marianas became a U.N. Trusteeship under U.S. administration after the Pacific War (1941-1945). From 1947-1986, then, American Micronesia held U.N. Trusteeship status under U.S. Administration, officially designated the [U.S.] Trust Territory of the Pacific Islands (TTPI).[26] In 1965, the elected Congress of Micronesia was endowed with legislative powers, and in 1975 a plebiscite held in the NMI resulted in the vote for U.S. Commonwealth status. By 1976, the Northern Marianas were administered separately from the rest of the TTPI.[27]

The extremes of latitude in Micronesia cover approx. 23° -- from 20.33 N. latitude (NMI) all the way [across and down] to 2.39 S. latitude (Kiribati). The greatest north to south latitudinal expanse covers 1,387.2 nautical miles (1,595.28 statute mi.).[28] The east-west distances, of course, are far greater still. To the west, the Marianas act as a virtual Pacific barrier to Japan in the north and to the Philippines in the south. And tiny Tobi in Palau skirts Indonesia at 131.10° E. longitude. But the easternmost atoll of Kiribati is situated at 176.54° E. longitude – a distance of about 2,726 nautical mi. (3,135 statute mi.)[29]. From west to east, this is circa the distance of the U.S.A coast-to-coast!

The other archipelagoes in American Micronesia are the Marshalls and the Carolines. The Marshalls gained self-government in 1986, U.N. trust status ended in 1990, and the archipelago achieved "independence" in 1991 as the Republic of the Marshall Islands. The RMI promptly joined the U.N. on 1

October that same year.[30] The political sequence was basically identical for the Caroline Islands, except for one very notable exception. The Carolines include five districts of the former TTPI : Kosrae, Pohnpei (Ponape), Truk, Yap, and Palau. All the Carolines, but without Palau (Belau), united to form a new political entity: The Federated States of Micronesia (FSM). Like the RMI, the FSM gained self-government in 1986, achieved independence in 1990, and joined the U.N. (also on 1 October) in 1991.[31] Palau, however, remained the Carolinian odd-man-out after 1986; electing to stay outside the FSM, and not becoming independent as the Republic of Palau (Beluu era Bilau) until 1994.[32] (This Palau "story-board" entails an involved Micronesian tale to be told in Part I, Section C.)

The remaining islands of [non-American] Micronesia are the Republic of Kiribati and the Republic of Nauru. Kiribati's territory includes Banaba (Ocean Island). Great Britain ruled the Gilbert and Ellice Islands from 1892 until 1975, when the (Polynesian) Ellice Islands split off from the (Micronesian) Gilbert Islands to become Tuvalu. By 1979, the Gilberts became independent from Great Britain as the Republic of Kiribati. The phosphate-rich (island) Republic of Nauru was formerly administered by Australia under a (post-World War I) League of Nations mandate, and later under a (post-World War II) United Nations Trusteeship.[33] What is of great significance to our study is the fact that scattered, land-poor Kiribati has political jurisdiction over an enormous, marine-rich, section of the Central Pacific Ocean. (Stay tuned!)

Within widely dispersed Micronesia, there are seven major Micronesian cultures : (1) The Chamorro on Guam and the four southern Mariana islands; (2) the Palauan; (3) the Yapese; (4) the Trukese; (5) the Pohnpeian; (6) the Kosraean; and (7) that of Nauru, which is a Polynesian/Melanesian mixture on a Micronesian matrix. Despite the cultural differences, however, the Gilbertese of Kiribati (for instance) and the Marshall Islanders speak a mutually intelligible language.[34]

But every island in Micronesia is physically unique. Travel writer David Stanley has well described the differences:

....."Some are high islands with volcanic peaks, others low islands of sand and coral. All of the Marshalls and Gilberts are coral atolls or islands. In the Northern Marianas, Micronesia's only active volcanoes erupt. Nauru and Banaba [Ocean Island] are uplifted atolls. The Caroline Islands include both volcanic and coral types. Kosrae, Pohnpei, and Chuuk [Truk] are high volcanic islands. Guam and Belau are exposed peaks of an underground sea ridge stretching between Japan and New Guinea, volcanic in origin but partly capped with limestone. Yap is an uplifted section of the Asian continental shelf"....[35]

The coral reefs and silent lagoons of Micronesia abound in clams and oysters, shrimps and lobsters, octopuses and turtles and innumerable species of brightly coloured fishes.[36] But the real riches of marine life are Micronesia's pelagic fishes – the bonito, jack, mahi mahi, swordfish, the many tunas and more.[37] The commercial fin-fish and shellfish of Micronesia will be carefully examined in Part III of this chapter.

My brief foreword has only attempted to introduce wide and varied Micronesia to the North American reader. I hope these Micronesia Notes have somewhat fleshed out those fly-specks on the Pacific wall-map into living, breathing island ecosystems containing ancient cultures of interest and significance. The ancient cultures will become of increasing interest to American history/anthropology, and the island ecosystems will take on increased significance for Micronesian fishing and aquaculture. Micronesia is rich in both culture and fish, and is located almost wholly in the American political sphere. Already, starting in the preceding decade, U.S. historians have been taking Micronesia far more seriously. And it's about time. Just forty years ago, the Encyclopedia of World Travel[38] concluded their Vol. 2 with the following [hardly glowing] description of Micronesia:

"None of the islands, however, can be enthusiastically recommended to tourists. A few still have tribes of cannibals. All are rampant with malaria and other unpleasant diseases. Few have hotel accommodations, and most are inaccessible except by inter-island schooners. Pan American World Airways has five flights a week into Guam and Wake islands. However, these spots cannot be given high itinerary ratings, since the United States has military installations on the islands and you are required to have a special pass from military authorities to wander about."[39]

How the Pacific world has changed in forty years! And how now American attitudes have changed on Micronesia! As I wrote in the preface to this chapter, "…[T]he new millennium dawned first over Micronesia".

NOTES

8. Cited by Robert C. Kiste in 'Editor's Note', Francis X. Hezel (S. J.), <u>Strangers in Their Own Land: A Century of Colonial Rule in the Caroline and Marshall Islands</u> (Honolulu: University of Hawai'I Press, 1995), p. viii.

9. Author's note: I'm proud to announce, as a U. H Alumnus, that U. H. Press has published just about all major recent historiography on Micronesia.

10. In his <u>The Happy Isles of Oceania: Paddling the Pacific</u> (New York: Ballantine Books, 1993 ed.), U. S. citizen Paul Theroux paddled the vast Pacific without <u>once</u> making landfall in widely – scattered Micronesia – an area of American interest. See note 328, p.92, in this volume.

11. The Afro'ed Trukese was simultaneously tacit and truculent and avoided conversation with me, an older White (<u>haole</u>) Guy with a crew-cut. I'm pretty sure that he'd learned most of his reverse-racial prejudice since coming to Honolulu.

12. A neighbour of ours in Hawaii Kai (east of Diamond Head) was one of those sad-faced Mainland Girls who'd had a quickie fling/marriage with an Island Boy, and then been summarily dumped along with a <u>hapa-haole</u> ("half-White") child. The erstwhite lover/husband, who worked as a "Sizzler's" waiter, came from a well-to-do Marshallese family living over in Kalama Valley.

13. The Peace Corps fellows, indoctrinated in '60s thought, regarded Micronesions as an (other) example of the "self-victimization of minorities"; which was (of course) part and parcel of the "Coca-Colonization" of U. S. possessions.

14. Dr. Timothy McNaught of Sydney, N. S. W., headed the U. H. Pacific history department during my attendance, 1979 – 1981. Author's note: Although a White himself, "Tiger Tim" concentrated on Aboriginal Australia and Maori New Zealand. To Dr. McNaught, Aboriginal Australia was an anthropological [Fourth World] extension of Papua-New Guinea (PNG).

15. Sir Arthur Grimble, the Grand Old Man of [Anglo/austrocentric] Pacific historiography who wrote the classic Grimble Papers (1916 – 1926). H. E. Maude edited the papers, and U. Hawai'i published the collection as <u>Tungaru Traditions: Writings on the Atoll Culture of the Gilbert Islands</u> (Honolulu: University of Hawai'i Press, 1989).

16. Excellent, original footage was compiled and produced by Arthur B. Tourtellot (narrated by Westbrook van Voorhees) during the late 1940s / early 1950s, and released as "Crusade in the Pacific". The sources have been the U. S. Armed Forces, and the films were shown weekly, during the 1990s, on the HIST channel.

17. Author's note: The Micronesian Island of Tinian in the Marianas was a busy place in August 1945. This was the drop-off point from which "Little Boy" (the uranium gun-type bomb) was loosed on Hiroshima (6 August 1945), and "Fat Man" (the plutonium implosion-type bomb) on Nagasaki, Japan (9 August 1945).

18. "Trinity and Beyond" (1995) compiled by film-maker Peter Kuran.

19. "Bikini: Forbidden Paradise" (1992) produced by David Livingston. Shown on <u>World of Discovery</u> (DSC) 26 April and 27 July 1997.

NOTES Cont'

20. David Stanley, Micronesia Handbook: Guide to the Caroline, Gilbert, Mariana, and Marshall Islands (Chico, Calif.: Moon Publications, Inc., 1992 ed.), pp. 2-3. Cf. Bruce G. Karolle, Atlas of Micronesia (Honolulu: Bess Press, 1993). p. 1. NB: when in doubt – Stanley and Karolle list differing stats and figures – I have stuck to Karolle, and give kilometres whenever possible.

21. Bruce G. Karolle, p. 1.

22. Micronesia is derived from the Greek micros ("small") and nesos ("island"). The derivation was first used in 1831 by one Domeny de Rienzi in a submission to the Societé de Géographie de Paris.

23. Bruce G. Karolle, p. 1.

24. Ibid., p. 1 and Robert F. Rogers (1995).

25. Guam was ceded to the U. S. by Spain in 1898, but Spain sold her remaining Micronesian possessions to Germany.

26. 'Micropaedia', The New Encyclopaedia Britannica, 15th ed., No. 9 (Chicago, Ill.: Encycl. Brit., Inc., 1990), p. 48.

27. Ibid., p. 48.

28. Bruce G. Karolle, p. 1.

29. Ibid., p. 1.

30. Student Handbook (Desk Reference) Vol. 5 (Nashville, Tenn: The Southwestern Company, 1998 ed.), p. 670.

31. Ibid., p. 671 and Karolle, p.1.

32. 'Micropaedia', Vol. 8, p. 103, ff., Student Handbook, p. 671. For complete details on Palau's difference(s) in political status, see Elizabeth Rechebei (1997).

33. The Student Handbook, pp. 669 – 670, and Karolle, p. 1.

34. 'Micropaedia', Vol. 8, p. 103, ff.

35. Op. cit., David Stanley, p. 3.

36. 'Micropaedia', Vol. 8, p. 103.

37. See David Stanley, p. 12.

38. Edited by Nelson Doubleday and C. Earl Cooley (Garden City, New York: Doubleday & Company, Inc., 1961).

39. Encyclopedia of World Travel, Vol. 2, p. 328. NB: "The Pacific" was the last world-area examined, and "Micronesia" rated one half-page; the very last page (328)!

Bibliography

2. Bugliosi, Vincent with Bruce B. Henderson. <u>And the Sea Will Tell</u>. Ivy Books, New York, N.Y, 1991.

3. Rossi, Mitchell Sam. <u>Truk Lagoon</u>. Pinnacle Books, New York, N. Y., 1988.

4. Theroux, Paul. <u>The Happy Isles of Oceania: Paddling the Pacific</u>. Ballantine Books, New York, N. Y. 1993 ed.

Part I: The United States, The United Nations, and Micronesia

Section A: A Brief History of U. S. Imperialism

"'Who does not see, then, that ... the Pacific Ocean, its shores, its islands, and the vast regions beyond, will become the chief theatre of events in the world's great hereafter?'"
-- William H. Seward, 1852[40]

"The American Century had begun in 1898 at Manila Bay".
-- Patrick J. Buchanan, 1999[41]

"'America is not to be Rome or Britain. It is to be America'".
-- Charles A. Beard, 1939[42]

During the waning days of 1999 I read political columnist/political candidate Pat Buchanan's A Republic Not an Empire. The theme of Pat Buchanan's compelling but controversial interpretation of American history is that the United States has been an empire since 1898... but should never have acceded to her imperial (i.e., overseas colonial) rôle. The second half of Pat Buchanan's book title is: Reclaiming America's Destiny, and the ever-conservative pundit has tried to define what the Republic's destiny was originally meant to be. And Pat Buchanan, despite a couple of glaring inconsistencies (I'll get to them!), has presented his Republican case well. After the U.S. Civil War (the War between the States to Southerners), the Federal Union Army -- almost a million-men strong -- in 1865 stood on the doorstep of British North America to the north, and at the border of Spanish-speaking México to the south. (Or what was left of México after the crippling Treaty of Guadalupe Hidalgo of 1848.) The Mexican War (1846-1848), shortly followed by the U.S. Civil War (1861-1865), convinced the now thoroughly-alarmed British North Americans (above the Great Republic's border) that is was high time (read: Urgent!) to confederate (which they did in 1867) and to consolidate (Manitoba, 1870; British Columbia, 1871) their precarious-looking holdings.[43]

With the U.S. Civil War over with by 1865, and Napoleon III's French troops pulling out of México during 1866, Secretary of State William H. Seward could well boast in 1867: "'[G]ive me fifty, forty, thirty more years of life, and I will engage to give you the possession of the American continent and the control of the world.'"[44] As the foreign policy controller of a powerful new world player, William H. Seward was free to pursue his personal vision of Manifest Destiny; a momentous concept prevalent in Americanist top-dog circles since the mid-1840s. By the late 1860s, culminating with the purchase of Alaska (Russian North America), William H. Seward had become Manifest Destiny's main U.S. proponent.

But, before bowing off the world stage in 1869, William H. Seward had acquired the Midway Islands for the U.S.; had tried but failed to annex Cuba, Puerto Rico, St. Bart's, Canada, Greenland, Iceland, and even a portion of China.[45] In evaluating William Seward's stewardship, Manifest Destiny as an American force, and creeping U.S. imperialism -- Pat Buchanan has assessed that:

"By 1869 America was a completed nation. But a coming generation would contend that Manifest Destiny required not just American rule from Panama to the Pole, but also expansion west to Hawaii, the Philippines, even China -- as the United States assumed its rightful place among the great imperial powers, acquired distant colonies, and undertook the instruction of alien peoples in the superior American way of life. Thus was Seward, that paragon of Manifest Destiny, a transitional figure. Near manic in his acquisition of territory, Seward, like the turn-of-the-century American imperialists, would accept the burden of ruling alien peoples."[46]

Thus, after the Spanish –American War of 1898, Buchanan has concluded that:

"The United States annexed Spain's colonies of Puerto Rico, Guam, and the Philippines, acquiring control of lands we had no intention of converting into states, and of peoples we had no intention of allowing to become U.S. citizens. We also annexed Hawaii. The Republic had become an [E]mpire."[47]

So the Republic is an Empire after all. But it is here that Pat Buchanan displays two glaring inconsistencies completely out of sync with his strong beliefs expressed in A Republic Not an Empire. Very much later in his interesting thesis, Pat Buchanan has written his [truly horrendous] thought that [Danish/Inuit] Greenland "should... eventually be formally annexed to the United States."[48] Yike! Although a proud U.S. citizen, I am of Scandinavian origin and cringe at the very notion of the last Viking outpost becoming just another global/cultural appendage of the greater Anglophone world. (Greenland must remain Grønland/Kalaalit Nunaat and must stay out of North American hands!) Pat Buchanan's second gaffe regarded Canada. North American historians John Herd Thompson (U.S.A.) and Stephen J. Randall (Canada) have reported that:

"Canadians took Patrick Buchanan's Los Angeles Times pipedream of 'a republic that, by the year 2000, encompasses the Maritime and Western provinces of Canada' far more seriously than did any of Buchanan's American readers..."[49]

So, Pat, the Republic really is an Empire after all, as some Asians, most Latin Americans, and just about all Canadians have always known. It didn't take much time, after 1898, for the American spread-eagle to seize even more in her talons. Indeed, the year 1903 loomed as a watershed year for the entire Anglo-American world. First, the British (with direct ANZAC and Canadian involvement) battered (and brutally incarcerated) the hapless Boers into submission in South Africa (1899-1902). Next came the American hegemony over Cuba and Panamá, and the lop-sided settlement of the Alaska/B.C/Yukon boundary dispute by bully-boy President Teddy Roosevelt (1901-1909).[50] (Last but not least, 1903 was also the year during which Henry Ford [1863-1947] founded the Ford Motor Company.) But Pat Buchanan was dead right about the course the Republic would embark on after the Spanish-American War:

"We would eat of the forbidden fruit of imperialism and go out into the world in search of monsters to destroy -- leading to America's involvement in all the great wars of the twentieth century."[51]

The United States did not get involved in World War I (1914-1918) until 1917, two years <u>after</u> the British luxury liner, <u>Lusitania</u> (laden with passengers and armaments), was sunk by a German submarine. Due to a long chain of convoluted and Machiavellian events, fueled by British-only propaganda reaching receptive American ears, U.S. "doughboys" finally went "over there" to make the world safe for democracy.

In January 1918, President Woodrow Wilson (1913-1921) appeared before the U.S. Congress to spell out American war aims, namely his Fourteen Points, the last of which would establish: "'A general association of nations... affording mutual guarantees of political independence and territorial integrity to great and small states alike.'"[52] (It sounded far better than it actually was -- a tragic geopolitical folly.) After the Germans signed the Armistice in November 1918, President Wilson journeyed to Paris to help preside over a supposedly lasting peace. The resultant Treaty of Versailles[53] (extremely punitive) contained the Covenant of the League of Nations, which Woodrow Wilson envisioned as a vehicle for eventual global governance; with himself at the helm.[54]

But November 1918 was also a mid-term election month and year in the U.S.A., and by that time the American people had had ample opportunity to study "Tommy" Wilson's grandiose Fourteen Points. Congressional majority was won by the Republicans, and the Treaty of Versailles failed by seven votes in the Senate.[55] Thus the American people, by refusing to ratify the Treaty of Versailles, had turned thumbs down on Woodrow Wilson's League of Nations and fledgling world government.[56] But Wilsonian concepts of global collectivism would live on after the coming world war in the United Nations Organization (U.N.O.), and be overwhelmingly accepted (this time) by the American people.

Powerful or influential nations such as Great Britain and Sweden, however, were charter members (1920) of the Geneva, Switzerland-based League of Nations. Significant for the Pacific was the awarding of all former German possessions north of the equator to Japan; all those south of the equator to Great

Britain. (Indeed, Japan joined the League in 1920 at the insistence of the United States, who regarded Imperial Japan as a naval threat to her own colonial interests -- Guam and the Philippines.)[57] Consequently, the League of Nations awarded a "class C mandate" to Japan, who established a civil administration in formerly-German Micronesia with its seat at Koror (Oreor), Palau. (German Micronesia had consisted of the Marshalls, Carolines, and Marianas -- except for Guam -- which had been purchased for $6 million U.S. from Spain after her crushing defeat by the Americans in 1898.)[58] Pat Buchanan has also reminded us that strange Pacific bedfellows, the British and the Japanese, had -- so early as 1917 -- concocted a secret naval agreement for dividing German Pacific possessions after the cessation of the Great War.[59]

Despite America's refusal to join the League of Nations (thus assuming an "isolationist" stance), the Republic was a rising global power (and already an Empire) after World War I. The Washington Naval Conference of 1921-1922 concurred on naval armaments ratios among Great Britain, France, Japan, and the United States.[60] Besides forcing Japan to cede back to China her formerly-German Shantung (Shandong) holdings,[61] the Washington Conference insured overall Anglo-American naval dominance of the whole Pacific Ocean; both north and south of the equator. As to colonies and mandates, the French were relegated to their own lonely outposts in Melanesia and Polynesia; but the ambitious Japanese were now in firm control of former German Micronesia -- the Marshalls, the Carolines, and the Northern Marianas.

Meanwhile, after the Roaring 20s and the Great Depression, time and events were swiftly catching up to the ill-starred League of Nations. In 1935, the League failed to stop the Italians from annexing Ethiopia. In 1936, the League failed to stop the Germans from reoccupying the Rheinland. In 1937, the League failed to stop the Japanese from raping Nanking (Nanjing) and razing China's hinterland. In 1938, the League failed to stop the Germans from absorbing Austria and the Munich Pact from transferring Czech Sudetenland to Adolf Hitler. On 1 September 1939, the League failed to stop the Germans from invading Poland, with the complicit Russians standing idly by.[62] Thus the League of Nations failed to stop World War II, and, thankfully, was put out of its misery during April 1946. But for the Pacific islands of Micronesia, the Japanese-administered class C mandate was definitive: For during those years between World War I and World War II, all Micronesia save American Guam, the British Gilberts and [formerly German] Australian Nauru, comprised Nan'yō -- the southern island empire of Imperial Japan.[63]

"The American Century", as indicated earlier in this work, in truth started with the humiliating defeat of Spain in 1898. The "splendid little war", a cocky little phrase uttered by future secretary of state John Hay and echoed by bully-boy Theodore Roosevelt,[64] was "splendid" for the rising American Republic. The United States suffered a mere 4,108[65] casualties during the Spanish-American War (the lowest in U.S. military history until the Persian Gulf War almost a century later), while gaining strategic Guam and the entire Philippine archipelago. American imperialists like Theodore Roosevelt (assistant secretary of the Navy), Henry Cabot Lodge (senator from Massachusetts), and Alfred Thayer Mahan (captain and naval strategist) had gotten their way: Only as a world power could America fulfill her [new] Manifest Destiny; to protect her [new] global trade and her [new] colonial possessions.[66] In the jingoist mind of "Teedy" Roosevelt, war was the highest form of manly endeavour, and William Randolph Hearst (1863-1951), "lord of the yellow press", ably served as the Spanish-American War's willing booster/publicist.[67] Indeed, so early as 1895, Hearst's newspaper, The New York Morning Journal, pushed for armed conflict with Spain and constantly pressured President William McKinley (1897-1901) to declare war.[68] Conservative weekly Human Events commented in a Spanish-American War centennial anniversary issue that:

"Hearst's paper... used phony pictures and fake stories to fire the public rage... He featured trumped-up tales that the Spanish were feeding prisoners to sharks, raping Cuban women, poisoning wells, throwing nuns into prison, and that they roasted 25 Catholic priests alive. When the U.S. battleship Maine blew up in harbor at Havana, Hearst ran the headline: 'WAR! SURE!... Maine destroyed by Spanish...' Modern reinvestigation attributes the explosion to a spontaneous fire that spread to the ship's munitions... but the war was on, and Hearst's circulation soared..."[69]

By the time that the American doughboys had returned from "over there", makin' the world safe for democracy, a new news/warmonger-tycoon was emerging on the U.S. scene. This was Time-Life, Inc's Henry Robinson Luce, a/k/a Harry Luce, born to American Protestant missionaries in Tengchow, China, on 3 April 1898.[70] (The astute reader will notice that this was the same year and very month that the Spanish-American War broke out.) It was Henry Robinson Luce (1898-1967) who coined the fateful phrase, "The American Century", in a Life article by that name on the eve of World War II.[71] W.A. Swanberg, who wrote the master biography (no hagiography!) on H.R. Luce published in 1972, quoted revealing passages from the February 1941 Life article:

"'[T]he complete opportunity of leadership is ours....[72]
[Ie., the United States would not enter the war just to save the British Empire.]

"'Tyrannies may require a large amount of living space. But Freedom requires and will require a far greater living space than Tyranny...'[73]

[Ie., America's Manifest Destiny/Empire is holier/more justifiable than Germany's "Lebensraum"/Drang nach Osten" or Japan's "Nan'yō"/"Greater East Asia Co-prosperity Sphere".]

"'[We must] accept whole-heartedly our duty... as the most powerful and vital nation in the world and... to exert upon the world the full impact of our influence... as we see fit and by such means as we see fit.'"[74]

[Ie., America must be ruthless in carrying out her (new) Manifest Destiny.]

Biographer W.A. Swanberg, although writing from an obviously liberal/Left perspective, summed up Henry Luce's American Century succinctly and well:

"Implicit in this was his [H.R.L.] long impatience with an American constitutionalism he felt in need of overhaul more than ever because of the challenge of world war and revolution. The exact shape of Luce's 'new order' was necessarily vague in its details but not in its fundamental principles. The American Century was a capitalist century. It was a militarist century. It was a century of General Motors, Standard Oil, Pan-Am -- and of Time-Life-Fortune -- entrenched in Asia and Africa with the protection of American military power...."[75]

Had Henry Robinson Luce been a "New Deal" Democrat, he would have loved the imperial presidency of Franklin Delano Roosevelt (1933-1945). But H.R. Luce wasn't and he didn't. Except for Douglas MacArthur's half decade rule of post-War Japan as Supreme Commander Allied Powers (SCAP), Franklin Roosevelt has come closest to reigning as American Caesar at home.[76] And versus a potential Caesar whom he hated (we're back in the pre-War 1930s), Luce turned the full rhetorical force of the (then conservative Republican) Time-Life monster against Roosevelt and his Big Government administration. Swanberg has commented:

"One of [the literary gimmicks], reflecting Luce's fairly continuous effort to depict Roosevelt as a potential dictator, appeared in Life, where careful search of picture files turned up Hitler, Mussolini, Stalin and the President all in similar poses of shaking hands, pointing, petting animals and pinning medals. The headline was 'Speaking of Dictators', and readers who wrote to complain were assured it was only a joke.[77] Another was Time's cover story on the American Communist leader Earl Browder whose color photographs, backed by Communist posters, had a clever guilt-by-association line under it: 'COMRADE EARL BROWDER/For Stalin, for Roosevelt...'"[78]

Even Old Leftist Studs Terkel in his 1970 book, Hard Times, on the Great Depression, reported verbatim man-in-the-street interviews on FDR-as-dictator. In one interview (with a "Doc" Graham), Studs Terkel asked:

Q: "You think Roosevelt had a role to play in this [encroachment of the Federal Government on a citizen's perogative]?"
A: "'Definitely. He [FDR] was perhaps the lowest human being that ever held public office. He, unfortunately, was a despot. I mean, you get an old con man at a point in high office, he begins to believe the platitudes that are expounded by the stupid populace about him.'"[79]

In another man-on-the-street interview, Studs Terkel asked (one Jerome Zerbe):

Q: "What does the phrase 'New Deal' mean to you?"
A: "'It meant absolutely nothing except higher taxes. He [FDR] obviously didn't help the poverty situation in the country... New Deal! God... [the] crap he brought into our country, Jesus!'"[80]

Before turning to FDR and World War II (when the Roosevelt administration truly evolved into an imperial presidency), a word should be said about Herbert Clark Hoover (1874-1964; president 1929-1933), the man FDR trounced in the election of 1932. There was an excellent little bio written on Herbert C. Hoover published in 1960, simply titled Herbert Hoover, Humanitarian.[81] Although Herbert Hoover was a multi-talented man of many achievements, he has been personally blamed for the Great Depression. Repeat a lie often enough and "The [American] People" will believe it. Hoover's culpability has now become part and parcel of American [old] urban legend.[82] The political campaign of 1932 was eerily similar, this writer strongly believes, to the political campaign of 1992. The Republicans were out-hustled and out-muscled by the Democrats, grandiose promises were made for Big Government, and the FDR forces practised the politics of division. Those same dirty tricks were smoothly and effectively employed by the Clintonites sixty years later against another multi-talented man of many achievements, George H.W. Bush.

If Franklin Delano Roosevelt was the Great Faker, then William Jefferson Clinton is the Great Fakir. Bill Clinton, the phony-baloney Great Empath who feels your pain, like the silvery-tongued FDR before him, overwhelmed his less flamboyant opponent, and the American people, with non-stop speechifyin', teachifyin', and preachifyin'. The political knavery and fakery, flummery and mummery, carried the day for the Democrats and won the elections of 1932-1944 and 1992/1996. Even Black blues band, Sacred Steel, recorded a song depicting the great magnanimity of FDR and the mean-spirited smallness of Herbert "Hooverville" Hoover. Things haven't changed much in U.S. politics except that Bill Clinton will be gone, by law, in January 2001, otherwise the easily flummoxed U.S. electorate would re-elect Billy Liar the Lyin' King for life. (The reader will recall that Tom Brokaw's "Greatest Generation" voted for FDR not once, not twice, not thrice, but no less than four times!)

On 7 December 1941, as every American schoolkid should know, the Japanese bombed Pearl Harbor, Hawai'i, and F.D.R.'s fustian phrase, "A day that will live in infamy", went on to characterize an entire era. The self-righteous, high-flown rhetoric of Franklin Roosevelt was matched in style and volume across the Atlantic by that of Winston Churchill. The beleaguered British Prime Minister, representing a still-mighty Empire, was thrilled that his fellow Anglophone despot would now -- at last -- enter World War II (all British plans coming to fruition as in World War I), after the conflagration had raged for more than two years out of control.[83] The proud/noble Brits had been hard-pressed by those dreadful/swinish Huns in the air and at sea. (On the continent, the feared and loathed German dictator had invaded the feared and loathed Russian dictator's territory by land.) So early as 1937, Churchill (a benevolent despot) had written of Roosevelt (a malevolent despot), "Supreme power in the Ruler [FDR], and a clutching anxiety of scores of millions [i.e., the American people] who <u>demanded</u> and <u>awaited</u> orders."[84]

FDR went from being the Great Hornswoggler of the 1930s to being the Great Bamboozler of the 1940s. Alternative historian Dr. Robert N. Crittenden, in his recent <u>Politics of Change</u>, has verily described the wartime situation:

"The authority conferred upon President Roosevelt during World War II was far greater than that granted to President Wilson. Roosevelt received almost unlimited power over natural resources, industry, the rationing of food, and the fixing of prices. There was a universal draft of eligible men, many women entered the work force, and men and women were 'frozen' in their jobs in vital industries. The military took over many of the schools and trained the children, while others, left without parental guidance, roamed the streets..."[85] (Ave Cesare! Hail Caesar!)

National (and then international) wonder-worker Franklin Roosevelt remained unchallenged at the pinnacle of American (and then world) power for the final twelve years of his life (1933-1945); throughout his presidency. Unbeknownst to the American people -- including the "Greatest Generation" risking life and limb for Main Street and Wall Street -- there was a whole lot going on behind the [American] scenes of global significance. There popped up, like poisonous toadstools after a rainstorm, influential (public and private) agencies -- elected by no-one and of which the U.S. electorate remained wholly ignorant -- that would wield inordinate political and extraordinary financial power. These groups and aggregates had been busily planning a New [World] Order <u>long</u> before America entered the War in 1941, and, indeed, before the outbreak of the War itself in 1939.[86]

One of the outcomes of these secretive sessions/meetings culminated in the Bretton Woods Agreement of July 1944. An insider circle of FDR presidential advisors, aided and abetted by John Maynard Keynes (U.K.) and Harry Dexter White (U.S.), convened and then drew up the blueprints for a New [World] political and economic Order. The results of these covert conclaves are still with us: The United Nations, the International Monetary Fund, and the World Bank.[87] As the actual head of the victorious allied powers, and as the unofficial "Commander in Chief of the United Nations" (whose

Declaration the U.S. had signed on 1 January 1942), Franklin Delano Roosevelt briefly ruled the "Free World" in early 1945.[88] But President Roosevelt died on 12 April 1945 -- before Victory-Europe Day (8 May) and Victory-Japan Day (15 August) -- so didn't live to relish the fruits of ultimate (Allied) victory. National Socialist Germany and Imperial Fascist Japan were beaten and lay prostrate. Communist Russia and the European Allies were mostly in ruins; thus in 1945 there remained solely one, real global Empire: The United States. And although America had just lost her Caesar, her world wide hegemony stood undisputed. The essential difference between the Old British Empire and the New American Empire were perceptively noted by historian Amaury de Riencourt in 1957:

"Though nominally a monarchy, Britain was in fact an aristocratic republic, whereas, though nominally a republic, the United States was gradually becoming a Caesarian monarchy under Franklin Roosevelt. Britain and America represent, respectively, the modern equivalents of different phases of Rome's social evolution."[89]

It was now a case of "Post proelium, praemium": "After the victory, the spoils." The United States would, via the stripling United Nations Organization, inherit Japan's Nan'yō -- her southern island empire of Micronesia. And despite all the machinations of Winston Churchill, the post-War grand alliance of the United Nations was now firmly under the suzerainty of the United States. Washington rather than Whitehall would control the inevitable post-War trials in Japan and Germany (1945-1946). The Americans, who held a special grudge against "the sneak-attack Japs" for the Pearl Harbor bombing to the Bataan "death march", took the Pacific War Japanese leaders to swift and deadly account. Seven Japanese leaders were consequently condemned to death, with 18 others imprisoned for "war crimes".[90] Among those hanged were General Tomoyuki Yamashita ("the Tiger of Malaya") in 1946, and Prime Minister Hideki Tōjō ("Kamisori"/"the Razor") in 1948.

But the genuine Hollywood-style "show trial" was to be held at Nuremberg (Nürnberg), Germany. This turned out to be a "show" trial in the truest sense of the word. It was not as if those tried at Nuremberg should have gone unpunished. That is not the issue. Like the Japanese war-lords executed or incarcerated, the absolutely evil Nazi upper echelon richly deserved their fate -- and more. Honest enemy soldiers doing their military duty is one thing; the cruelly calculated, cold-blooded slaughter of millions of innocent men, women and children -- because of ethnic classification or political category -- is something else entirely; something devilish. The Nuremberg War Crimes Trial, however, devolved into an exemplar of self-serving, self-righteous judicial exercise in vast excess. The full, frightful scope of Nazi atrocities somehow lightened the burden of Anglo American arrogance and heavy handedness; somehow lessened the total horror of the Russian rape and vindictive destruction of Berlin and eastern Germany. The film "Judgment at Nuremberg" (1961; starring wartime poster-boy Spencer Tracy)[91] conveyed that terrible, throbbing sense of Allied (read, American) sanctification. To the victors go not only the spoils, but the complete re-writing of history too. (Perhaps that is the greatest spoil of war.)[92]

As a naturalized and proud U.S. citizen, it is especially painful for me to see the style of American jingoistic patriotism which seriously annoys some Asians, most Latin Americans, and just about all Canadians.[93] (Especially by those Americans who have risked virtually nothing but have gotten it all.) Several weeks ago (for instance), on 6 June 2000, CNBC-TV celebrated the 56th anniversary [94] of the D-Day Normandy invasion by having their own Tom Brokaw (The Greatest Generation), along with Steven Spielberg (who directed "Saving Private Ryan", 1998), and Tom Hanks (who starred in "Saving Private Ryan") on camera. There the three sat, with appropriate churchyard faces, moaning and groaning about the sacrifices of the Greatest Generation. Watching the "disconsolate" trio, I uncharitably speculated that not-a-one of those three smug, self-satisfied gentry had served in Vietnam; indeed, most probably were anti-war "Vietniks". Brokaw, Hanks, and Spielberg should spare their crocodile tears for their own lost generation. At least His Stevenhood forebore wearing his usual bombardier jacket. (It's all a game, isn't it, guys? After all, the Greatest Generation is easily snookered…)

With the twin chimaeras of German Nazism and Japanese militarism slain in 1945, the American Empire stood at high tide. And who was best qualified to serve as spokesman for this next phase of The American Century? Why, none other than Henry Robinson Luce, coiner of the phrase, and reigning potentate of Time-Life, Inc! It did not take Luce long (27 May 1945), even before the cessation of hostilities with Japan, to (again) expand America's imperial frontiers:

"'In the last year or two the U.S. Navy has conquered some 20,000,000 miles of ocean (check figure)… The American frontier is no longer Malibu Beach; the American frontier is a line [from] Okinawa-Manila -- and it will never be moved back from there… Americans have a very acute sense of where their "home" is; from now on their home is a continent and the ocean that nearly covers half the globe. This is the political geography for the next round of the human drama.'"[96]

This was H.R. Luce-type spread-eagleism at its most blatant. And Luce, the super-Americanist and premier disseminator of wartime propaganda to the masses, had a personal reason to be extra-patriotic. His son, twenty-two year-old Ensign Hank Luce, was serving in the Pacific theatre on a U.S. Navy destroyer.[97] Luce Senior sincerely believed that God had founded America to be a global beacon of freedom; therefore America was fully justified in spreading this freedom throughout the world.[98] With the Cold War having heated up by the 1950s, America's enemies had changed but not Luce's Americanist message. In a June 1952 speech at Philadelphia, Luce gleefully repeated what Oswaldo Aranha, the Brazilian president of the U.N. General Assembly, had recently said: ""The United States which has disintegrated the atom has the duty now to integrate the world.""[99] In an address to TIME employees at New York City's Union Club, after the 1952 election of General Eisenhower, Luce opined that financial success

> "'... is the cause of America. For me that is the categorical, moral imperative... I mean America in relation to the world, the whole world of mortal men in the 1950s... [British historian Arnold] Toynbee regarded America as simply a peripheral part of European civilization. I regard America as a special dispensation under Providence.'"[100]

By the 1960s, and the final decade of his life, Henry R. Luce's super-patriotic rhetoric had not diminished with advancing middle-age (62/63 years). After reading Herman Kahn's U.S.A.-assuring On Thermonuclear War, Luce ebulliently suggested to a friendly audience at New York City's University Club on 28 January 1961 that:

> "'[Thermonuclear war] could happen. Let's learn to think that terrible thought. Then let's learn to think a novel thought -- that civilization could survive such a war, provided we start now to do the necessary things... I suggest that a general theory covering our activity would contain these two propositions: First, that order must be established in the world -- a global order, a new order. Second, that the major responsibility for bringing about this order rests upon the United States...'"[101]

This was, according to biographer W.A. Swanberg, H.R. Luce's "New Order" speech. But unlike the mostly faceless bureaucrats at the UNO (United Nations Organization), or the one-world movers and shakers behind the American scenes (in Hollywood, Manhattan, or inside the Washington, D.C., "beltway"), Luce was an American First'er (in every sense of the word) without being an "isolationist". The writer must remind the reader that the United Nations, until fairly recently, was almost entirely a creature of the United States. The preceding Luce speech was given in 1961, the Monroe Doctrine was about to be seriously challenged, and the times were soon to become the mini-era of Kennedy, Khrushchëv, and Cuba. Any New [World] Order, in the mind of Luce (and many Americans), would -- and should -- be headed by the United States. If Luce were alive today, he (most probably) would be aghast at the U.S./U.N. rôle reversals (the tottering Empire vis-à-vis the Rising Power). (Luce might be even more shocked, possibly, by the fawning Left/liberal tone of the Time-Life monster, and its lowest common standards of cw/pc group think.)[102] One of Luce's final messages on U.S. global commitment was rendered at the Memphis Bar Association (Tennessee), 3 May 1965, and he told the assembly that, "'the predominant desire of the people and government of the United States is peace.'"[103] But Luce's watchdog-biographer, W.A. Swanberg, would not let this Lucean statement pass without rebuttal:

..."But as [HRL] saw it, America had saved Europe from a military and economic disaster and become the leader of the West. Now it was America's duty to modernize, legalize, democratize and technocratize the East whether the inhabitants like it or not."[104]

Henry Robinson Luce, progenitor of "The American Century" and its main spokesman/cheerleader for more than twenty-five years, died in 1967. Meanwhile, the American Century was already in its first, swirling phase of ebbtide.

After the national trauma of Vietnam (-1973) and Watergate (-1974), the Stars and Stripes appeared to have lost some of its starch. Old Glory still flew all over the world, but it didn't seem to snap so crisply as before. The Jimmy Carter interregnum (1977-1981) of funky music and chicken diplomacy ("Gullible's Travels") didn't much help stiffen U.S. international resolve either. But the advent of the Reagan-Bush period (1981-1993) brought the crackle back to the American flag. Ronald Wilson Reagan, president from 1981-1989, initiated the Strategic Defense Initiative -- a/k/a "Star Wars" -- an ambitious U.S. military defense spending/rearmament programme which the U.S.S.R. simply could not match, rouble for dollar. Simmering ethnic/economic dissatisfaction within the Evil [Soviet] Empire itself, coupled with the failed glasnost/perestroika liberalisation of Communism by General Secretary/President Mikhail S. Gorbachëv (1985-1991), plus four outstanding individuals[105] -- brought the Berlin (Mauer) Wall crashing down in late 1989. U.S. President Ronald Reagan ("Ray-gun" to his enemies; "the Gipper" to his friends) had won the Cold War against the U.S.S.R., a war which had been on-going since 1946.

George Herbert Walker Bush (1989-1993), ascending to U.S. power in January 1989, enjoyed the fruits of President Reagan's victory. He went on to invade General Noriega's Panamá in Operation Just Cause during 1989-1990, and to militarily free Kuwait from Saddam Hussein's Iraq in the Persian Gulf War (Operation Desert Shield/Desert Storm) in 1991. But President George Bush often spoke (alarmingly for some of us) of a "New World Order" throughout 1990, and employed a grand United Nations coalition to help oust the Iraqis from Kuwait. Instead of obtaining U.S. Congressional imprimatur, President Bush looked to the U.N. Security Council to pass a resolution authorising use of U.S./U.N. force against Iraq.[106]

This was a Pax Americana of a different sort. Therefore, Pat Buchanan's assessment of contemporary Bush/Clinton foreign policy rings true:

"The [Bush] school is the Republican-neoconservative establishment that held power under President Bush and believes America should exploit its hour of power to impose a Pax Americana. This elite believes that the great lesson of history was Munich in 1938, that the way to deal with dictators and disturbers of the peace is ruthlessly and early, that while it is best to act in concert with other nations, it is often necessary to act alone."[107]

"In this [Clintonist] vision, the United States should impose a 'benevolent global hegemony' on mankind and stand at the epicenter of a worldwide alliance of democracies, united by trade and investment, whose combined power would dwarf any combination that might arise to challenge it. Theirs [Bush/Clinton] is the vision of America as Rome, with imperial allies, not imperial provinces."[108]

Towards the very end of his refreshingly against-the-U.S.-mainstream study, Pat Buchanan expresses his conviction in A Republic Not an Empire that the 21st. century will (also) be an American Century.[109] It is this writer's contention, to the contrary, that America at the beginning of the 3rd. millennium has all the earmarks of an Empire in decline. From a brutish, dumbed-down popular culture to bombings/shootings in her public schools, to a constrictive pc socio-political tyranny of racial/ethnic, "sexual preference", and trans-gender minorities... America is a society in moral and spiritual decline. Although the legal ground rules in the U.S.A. have been laid down for a basically North European society (tame, cowed), the actual social climate in the U.S.A. is more akin to that of a South American society (violent, dangerous). At Year Zero, the Ayatollah's "Great Satan" is poised to be taken -- from without by the Communist Chinese or nuclear-armed rogue states; from within by America's own "minorities" and her mostly unprotected national borders. It is America's ultimate capitulation to Clintonism.

At the helm of highest power in America sit the nine U.S. Supreme Court justices. Like the tenth member of the clique that really rules America, the Chairman of the Federal Reserve, the nine justices are unelected (viz., archons).[110] (No fear to have Prince Albert Gore, the Manchurian Candidate, hand-pick any of the future Supreme Court justices in a President Gore administration!) These ten -- the nine archons and the Chairman of the Federal Reserve -- remain unelected but nonetheless reign over America; subject, of course, to Caesar (Billy) and Caesar's shrewish wife (Hillary). The latter Twelve Persons, during the past eight years (1993-2001), have wielded far more socio-political authority than both the elected Houses of Congress. Is this the [great] Republic that Pat Buchanan had in mind in 1999, or the Great Republic that Winston Churchill wrote of during 1956-1958? I think not!

No indeed; the United States is an Empire, not a Republic, and has been ever since 1898. And today the American Empire is in a state of serious decline. In <u>Decline and Fall of the Roman Empire</u>, British historian Edward Gibbon (1737-1794) listed five reasons for the collapse of Rome: 1) The increase in divorce, thus undermining the institution of the family; 2) the imposition of higher taxes for bread (sustenance) and circuses (entertainment); 3) the drive for more pleasure, with sport becoming more exciting and brutal; 4) the people losing their religious faith; and, in the words of alternative historian A. Ralph Epperson[111] (paraphrasing Gibbon), 5) "the existence of an internal conspiracy, working to undermine the government from within, all the time that government was proclaiming that Rome's enemy was external."[112] Epperson went on to comment that "Gibbon reported that the conspiracy was building huge armaments for protection against both real and imaginary external enemies, all the while they [the conspirators] were literally destroying the [E]mpire from within."[113]

The February 1946 issue of the <u>New World News</u>, published by Moral Re-Armament of England, made public the plans for a total socio-political revolution, the so-called "Communist Rules for Revolution."[114] The apocryphal tale told was that Allied Forces, directly after World War I in 1919, somehow secured advance copies of the "Communist Rules for Revolution" at Düsseldorf, Germany. Whether the Düsseldorf Rules story has been absolutely verified or not is immaterial. For the Düsseldorf Rules are truly applicable to our sick end-times. The Communist/Düsseldorf Rules of 1919 instructed:

> " A. Corrupt the young; get them away from religion. Get them interested in sex. Make them superficial; destroy their ruggedness.
> " B. Get control of all means of publicity, thereby:
>   1. Get people's minds off their government by focusing their attention on athletics, sexy books and plays and other trivialities.
>   2. Divide the people into hostile groups by constantly harping on controversial matters of no importance.
>   3. Destroy the people's faith in their natural leaders by holding the latter up to contempt, ridicule and disgrace.
>   4. Always preach true democracy, but seize power as fast and ruthlessly as possible.
>   5. By encouraging government extravagence, destroy its credit, produce fear of inflation with rising prices and general discontent.
>   6. Incite unnecessary strikes in vital industries, encourage civil disorders and foster a lenient and soft attitude on the part of the government toward such disorders.
> " C. Cause the registration of all firearms on some pretext, with a view to confiscating and leaving the population helpless."[115]

Although this writer strongly believes that the American Century ended in 1998 (despite presidential impeachment, the Clintonist mid-term victories), the United States still boasts an Empire in

A.D. 2000. These overseas territories include five unincorporated U.S. territories: American Samoa (unorganised), and Guam, the Northern Marianas, Puerto Rico, and the U.S. Virgin Islands (organised). The Northern Mariana Islands (NMI) and Puerto Rico each hold the political status of "Commonwealth."[116] Of greater interest to this report is the "Free Association" political status of the Federated States of Micronesia, the Republic of the Marshall Islands, and the Republic of Palau. (What became American Micronesia was seized from the Japanese during the Pacific War in 1944.)[117] The Philippines, as promised, was granted independence by the United States on 4 July 1946,[118] but the U.S. has held on to her unincorporated territories -- and one possession (Wake Island, 1898) -- in the vast Central Pacific Ocean: Baker and Howland Islands (1856), Jarvis Island (1856), Johnston Atoll (1858), Midway Island (1867), and Kingman Reef (1922).[119] (There is also the ill-fated Palmyra Atoll, annexed with Hawai'i in 1898, which became notorious for the sensational murders of Mac and Muff Graham in 1974.)[120]

The United States of America is still an Empire in the year 2000. But how much longer shall the Great Republic remain an Empire? So long ago as 1942 (with the U.S. at near-zenith), Philip Wylie pointedly asked in his <u>Generation of Vipers</u>:

..."[S]hall we return after this fight [World War II] to a more soul-enslaving merry-go-round of goods and gadgets, nuttier congressmen, viler unions, obscener corporations, stupider soldiers, womanlier men, more manlike women, moms who suck marrow as well as blood, doctors who cure nothing, sappier preachers who belong with undertakers, portrait photographers, and window dressers, to a legion of the clammy-handed that merely wait upon fleshly vanity?"[121]

The answer, Mr. Wylie, is <u>YES</u> to all the above!

Notes to Part I

40. Patrick J. Buchanan, A Republic Not an Empire: Reclaiming America's Destiny (Washington, D.C.: Regnery Publishing, Inc.), p. 127.

41. Ibid., p. 173.

42. Ibid., p. 367.

43. Please see Appendix F, Canada and the United States: Animist Beaver versus the Spread-Eagle, for further details.

44. Quoted by Pat Buchanan, p. 135.

45. Ibid., p. 136. NB: Seward's failed scheme to purchase the Danish Virgin Islands was eventually realised by the U.S. in 1917. Buchanan has written that Seward's designs on a part of China was "according to his enemies."

46. Ibid., p. 141.

47. Ibid., p. 159. Brackets mine.

48. Ibid., p. 370.

49. John Herd Thompson and Stephen J. Randall, Canada and the United States: Ambivalent Allies (Athens, Ga. and London, U.K.: The University of Georgia Press, 1994), p. 4. Author's note: There was much talk during the mid-1990s of a Canadian break-up after a break-away Québec. Pat Buchanan envisioned a voluntary unification of Canada's truncated Anglophone provinces with the United States. See Appendix F.

50. In Thompson and Randall, pp. 68-69. Cf. Appendix F.

51. Op. cit., Buchanan, p. 141.

52. President Wilson cited by Frank Freidel, The Presidents of the United States of America (Washington, D.C.: White House Historical Association, 1987), p. 61.

53. With the Germans. Other Paris suburbs served as sites for the Treaty of St. Germain (with Austria), Trianon (Hungary), Neuilly (Bulgaria), and Sèvres (Ottoman Empire). All the Treaties were harsh and vindictive -- ensuring another world war in the very near future.

54. Robert N. Crittenden, Politics of Change: A Brief History (Carlsborg, Wash.: Hargrave Publishing, 2000), pp. 199-200.

55. See Frank Freidel, ibid., p. 61.

56. See Robert N. Crittenden, ibid., pp. 199-200.

57. Bruce G. Karolle, Atlas of Micronesia (Honolulu: The Bess Press, Inc., 1993), p. 30.

58. Ibid., p. 30.

59. See Pat Buchanan, p. 211.

60. Student Handbook, Vol. 5 (Nashville, Tenn.: The Southwestern Company, Inc., 1998), p. 363.

Notes to Part I (cont'd.)

61. Ibid., p. 647.

62. Ibid., pp. 362-363, 365.

63. Ibid., p. 671 and Bruce G. Karolle, p. 30 ff. Author's note: For a truly superior work on this subject, see Mark R. Peattie, Nan'yō: The Rise and Fall of the Japanese in Micronesia, 1885-1945 (Honolulu: University of Hawai'i Press, 1992 ed.).

64. Stanley Karnow, In Our Image: America's Empire in the Philippines (New York: Ballantine Books, 1990 ed.), p. 79.

65. Bud Bendix et al., '100th Anniversary of the "Grand Little War"', Pacific Magazine, Vol. 23, No. 2, Iss. 128, March/April 1998, p. 43.

66. Stanley Karnow, p. 10.

67. Ibid., p. 10.

68. Supplement to Human Events, Vol. 54, No. 14, 10 April 1998, p. 2.

69. Ibid., p. 2. Author's note: The 1941 film à clef, "Citizen Kane" (starring Orson Welles) vividly portrayed W.R. Hearst as news/warmonger-tycoon. Cf. Student Handbook, Vol. 5, p. 513. (See Appendix H)

70. W.A. Swanberg, Luce and His Empire (New York: Charles Scribner's Sons, 1972), p. 15.

71. 'The American Century', Life, 17 February 1941.

72. Quoted by W.A. Swanberg, p. 180.

73. Ibid., p. 181.

74. Ibid., p. 181.

75. Ibid., p. 182. Brackets and contents mine.

76. I took this idea from William Manchester's superb work, American Caesar (Boston, Mass.: Little, Brown and Company, 1978); a biography of Douglas MacArthur to match W.A. Swanberg's of Henry Luce. But Manchester's is a hagiography!

77. From 'Speaking of Dictators', Life, 18 April 1938. Brackets mine.

78. Cited in Swanberg, pp. 166-167.

79. Studs Terkel, My American Century (New York: The New Press, 1997), p. 118.

80. Ibid., pp. 136-137. Brackets and contents mine.

81. Subtitled: A Biographical Sketch of the Former President of the United States. Written by Mildred Houghton Comfort (Minneapolis, Minn.: T.S. Denison & Company, Inc., 1960). NB: An honest children's book about a great achiever and a good man.

82. See Appendix H, The Late, Great United States: [In] Famous in [Pop] Song, [Wild] History, and [Urban] Legend.

Notes to Part I (cont')

83. Winston Leonard Spencer- Churchill, according to Norman Rose, Churchill: The Unruly Giant (New York: The Free Press, 1995). Cf. Neil Ferrier, ed., Churchill: Man of the Century (London, England: Purnell & Sons, Ltd., 1965). Author's note: Rose's biography of Churchill, originally published in U.K. during 1994, was the less egregious of the two British "Winnie" hagiographies. And the Sir Winston legend lives on. See 'Winston Churchill: The Last Hero' by John Keegan, U.S. News & World Report, 29 May 2000. See Appendix G for revisionist bibliography.

84. Sir Winston Churchill, The Great Republic: A History of America (New York: Random House, 1999), p. 296. From 'Roosevelt from Afar' chapter. NB: Orig. published 1956-1958; ed. by Winston S. Churchill 1999. Author's note: Churchill, my own 20th century "homme du siècle", once called Henry R. Luce the 7th most powerful man in the U.S.A. (All brackets and emphases mine)

85. Robert N. Crittenden, p. 226. Author's note: Dr. Crittenden lives (2000) near me out in Sequim, Washington, but we've never met.

86. See Appendix H and especially Alternative U.S. Bibliography.

87. Ibid., pp. 227-228 and Hal Lindsey, The 1980's: Countdown to Armageddon (King of Prussia, Penna.: Westgate Press Inc., 1980), p. 131. NB: Bretton Woods, New Hampshire.

88. Amaury de Riencourt, The Coming Caesars (New York: Coward-McCann, Inc., 1957), p. 256.

89. Ibid., p. 263.

90. Student Handbook, Vol. 5, p. 652.

91. With all the current American obsession with World War II, "Judgment at Nuremberg" was re-done as a 2000 cable TV movie, starring arch-liberal Alec Baldwin in the Spencer Tracy rôle. To the arch-obnoxious Baldwin I say, "You're no Tracy." (And Kim Basinger is no Katie Hepburn!) But like Tracy in his time, Baldwin is a "social justice celebrity supporter".

92. Thus German soldiers were never "German boys", and Japanese soldiers were never brave -- always "fanatical". And after the fall of the Berlin Wall in 1989, and the subsequent break-up of the Soviet Empire during the early 1990s, not a single former Stasi/Securitate/KGB agent has been brought to justice and placed on trial for murder, mayhem, or the like.

93. U.S. News & World Report voted Uncle Sam "Man of the Century" (January 2000).

94. This writer had his own 56th birthday this year. I was born on Sunday, Mother's Day, 14 May 1944. So 1944 is special to me.

95. Random House, New York, 1998. Author's note: That loud, sucking sound you heard a couple of years ago was Tom Brokaw's hagiography to a generation greater than his own. This is also the generation which (yesterday) voted four times for FDR, and (today) voted twice for Bill Clinton. Go figure. (Guess the Greatest Generation just can't get enough of Big Government!)

96. Cited in W.A. Swanberg, p. 235. Brackets mine.

97. Ibid., p. 234.

98. Ibid., p. 306.

99. Ibid., p. 307.

Notes to Part I (cont'd.)

100. Ibid., p. 332. Brackets and contents mine.

101. Ibid., p. 418. Brackets and contents mine.

102. NB: The Time-Life monster was swallowed, whole-hog, along with Warner/Turner, by America On-Line (AOL) in January 2000. "The times they are a-changin'."

103. Quoted by Swanberg, p. 459.

104. Op. cit., W.A. Swanberg, p. 459. NB: "The East" meaning East and South-East Asia. Author's note: 1965 was the year I first heard words like Vietnik, peacenik, and quitnik. (Brackets mine)

105. Ronald Reagan, Margaret Thatcher, Lech Walesa, and Karol Wojtyla (Pope John Paul II, 1978-).

106. Student Handbook, Vol. 5, p. 628. NB: Despite his total victory in the Persian Gulf, Republican Bush -- like Hoover before him in 1932 -- was outhustled and outmuscled by the Democrats in the election of 1992.

107. In A Republic Not an Empire, p. 359. (Brackets mine)

108. Op. cit., Pat Buchanan, p. 360. (Brackets mine)

109. Ibid., p. 390. (No, Pat, no!)

110. Author's note: There are two meanings of the Greek word archōn:
    1. "[In] Ancient Greece, the chief magistrate[s] in many city-states. The office became prominent in the Archaic Period, when the kings were being superceded by aristocrats...
    "At Athens the list of annual archons begins with 682 B.C. By the middle of the 7th century B.C., executive power was in the hands of the NINE [emphasis mine] archons, who shared the religious, military, and judicial functions once discharged by the king alone" (p. 531)
    2. Archons in Gnosticism: Viewed by Christians as 'maleficent forces'....
    "The recurring image of Archons is that of the jailers imprisoning the divine spark in human souls held captive in material creation".... (p. 532) Sounds like the U.S. Supreme Court to me!
    From 'Micropaedia' (Chicago, Ill.: Encyclopaedia Britannica, Inc., 1998 15th ed.), pp. 531-532.

111. See Appendix H.

112. A Ralph Epperson, The Unseen Hand: An Introduction to the Conspiratorial View of History (Tucson, Ariz.: Publius Press, 1985), p. 323.

113. Ibid., p. 323. Cf. Edward Gibbon, Decline and Fall of the Roman Empire (New York: Harcourt, Brace and Company, 1960). One-volume abridgment, ed. D.M.Low. NB: Brackets mine.

114. In Epperson, p. 399. Cf. Medford Evans, 'The Rules and the New York Times', The Review of the News, 21 October 1970, p. 29.

115. Paraphrased from Epperson, pp. 399-400; and Charles R. Taylor, World War III and the Destiny of America (Nashville, Tenn.: Sceptre Books, 1979), p. 347.

116. Joseph E. Fallon, Deconstructing America: Immigration, Nationality, and Statehood (Washington, D.C.: Council for Social and Economic Studies, 1998), pp. 80-81.

117. Ibid., pp. 82-83. NB: Further details of American Micronesia's political status will be analysed in Sections B and C of this Part.

Notes to Part I (cont'd)

118. The Tydings-McDuffie's Act took full effect on 4 July 1948. Ibid., p. 78.

119. Student Handbook, Vol. 5, p. 680.

120. See Pacific Islands Monthly, June 1986; and Vincent Bugliosi, And the Sea Will Tell (New York: Ivy Books, 1991) for the whole sordid story.

121. Philip Wylie, Generation of Vipers (New York: Pocket Books, Inc., 1955 ed.), p. 263. Originally published in 1942. Brackets mine.

United States Bibliography

5. Buchanan, Patrick J. *A Republic Not an Empire: Reclaiming America's Destiny*. Regnery Publishing, Inc., Washington, D.C., 1999.

6. Churchill, Sir Winston. *The Great Republic: A History of America*. Random House, New York, 1999. (Orig. published 1956-1958; ed. by Winston S. Churchill 1999.)

7. Karnow, Stanley. *In Our Image: America's Empire in the Philippines*. Ballantine Books, New York, 1990 ed.

8. Riencourt, Amaury de. *The Coming Caesars*. Coward-Mc Cann, Inc., New York, 1957.

9. Swanberg, W.A. *Luce and His Empire*. Charles Scribner's Sons, New York, 1972.

10. Terkel, Studs. *My American Century*. The New Press, New York, 1997.

Section B: A Brief History of U.N. Expansionism

"The U.N. [does] not boast a glorious history; only chauvinist nations boast of glorious histories. But the U.N. created Israel, joined the United States in the war against North Korea, helped put an end to colonial empires at Suez, ended the secession at Katanga [!], lobbied for peace in Vietnam, supervised the end of the Iran-Iraq war, authorized the Persian Gulf War and helped keep the peace in Cyprus, Kashmir, El Salvador, Cambodia, South Africa, Mozambique, and a bunch of other trouble spots. It hosted such visionary figures as Dag Hammerskjöld, Adlai Stevenson, and Ralph Bunche. And, through it did not yet know what to do about it, the U.N. learned in the 1990s not to accept that nations can hide behind the sanctity of sovereignty to inflict punishment on their own people."[122]
--Stanley Meisler, The United Nations: The First Fifty Years, 1995

"Some of the elements of a [UNO-governed] New World Order would be: a world tax system...; a world court; a world army; a world central bank; a world welfare state; compulsory worldwide economic planning; abolition of private firearms; mandatory population and environmental control; and centralized control of education. A world government will need a world police, and a [world] military... [is] slated to become the core of... a U.N. police force."[123]
--Donald S. McAlvany, Toward a New World Order: The Countdown to Armageddon, 1992

On the occasion of the fifty-first birthday of the United Nations, I attended a Studium Generale at Port Angeles, Washington's Peninsula College. I remember little of the pro-U.N. presentation, except that the lecturer (rather honestly) admitted that as the United Nations "celebrates its 51st birthday, we find a new set of challenges. Organizational problems, budget deficits, doubts about its evolving mission in a changing world, and continued calls for reform have left the United Nations struggling with its own mid-life crisis. At such times, it is important to take stock and reflect on both the past and the potential for the future."[124]

"To take stock", indeed. And by 1996 I was starting to "reflect" on the U.N. rôle in post-Pacific War Micronesia; namely the U.N. Trusteeship Council (which will be dealt with in detail in Section C). In 1999, United Nations night – 24 October – showed a post-harvest Blood Moon, with Hallowe'en (America's highest holy day) a mere week away. With such a lugubrious tone set, I remember musing that the UNO had been possibly the most dangerous institution conceived by Western Man during the twentieth century.[125]

I hadn't always felt so negative about the United Nations Organization. Growing up on New York City's posh Park Avenue, on the Upper East Side, I hadn't lived far from U.N. Headquarters – located in Manhattan's Turtle Bay district fronting the East River. As a boy of twelve to fifteen years attending the top-drawer Buckley School (1956-1959), this writer had ample opportunity to visit U.N. Headquarters, and did. One of my Buckley School chums was one Tom Urquhart, scion of none other than UNO stalwart and former undersecretary-general, Brian Urquhart.[126] Ah – those innocent early adolescent "wonder years" of political idealism! In 1956 at age twelve, I proudly wore an Adlai E. Stevenson for President button to my overwhelmingly ("I like Ike") pro-Eisenhower private school. I was even prouder of Adlai Stevenson (now a liberal martyr/icon) when he was appointed American ambassador to the United Nations (1960-1965). I remained a staunch U.N. supporter all throughout my teen years, and during my early twenties harboured grand notions of fulfilling my military service under the U.N. (rather than the U. S.) flag. True to my Scandinavian heritage, I sincerely believed that keeping peace (via the U.N. Emergency Force/the U.S. Peace Corps) was far preferable to making war. Vietnam and the United States Selective Service System changed all that for me in February 1966 (but that's another story!).

And what "normal" rich kid, raised in the comfortable heart of the new capital city of global, do-gooder Liberalism, and home-seat of U.N. Headquarters, would not believe in World Peace? Or not strive to become a Citizen of the World? Member nations like Switzerland or Costa Rica might well be better examples of geopolitical neutrality (went the shared sentiment throughout the Cold War years), but only a UNO backed by the U.S.A. would be in a position powerful enough to "promote international peace, security, and cooperation under the terms of the charter signed by 51 founding countries in San Francisco in 1945."[127]

American support – in both money and muscle – is still crucial to UNO military operations whether the august body readily admits that fact or not. Never was American support more crucial to U.N. power and prestige than during the Persian Gulf War of 1991, so successfully prosecuted by U. S. President George Bush. Although, supposedly, a concerted U.N. action, the Persian Gulf War was almost wholly an American military operation waged and won by U.S. Secretary of Defense Dick Cheney, and U.S. Generals Colin Powell and Norman Schwarzkopf. Nevertheless, U.S. President George Bush insisted on referring to Operation Desert Shield/Storm as a victory for the United Nations. Indeed, President Bush spoke so often of a "New World Order" during the early 1990s, that American alternative historians like Donald S. McAlvany understandably sneered:

"[A] global currency, banking system, Environmental Protection Agency, educational system, population control authority, gun control mechanism, and police force under the U.N. are all talked about openly by [George] Bush and his establishment comrades as integral parts of the New World Order."[128]

So much as this writer might agree with Don McAlvany and other anti-UNO American conservatives, I am all too well aware that the U.N. is a U.S. organization... and always has been. The doubting reader has to merely look at any map of Central New York City (Fun City, Midtown Manhattan). United Nations Headquarters, erected on [John D.] Rockefeller-owned land, is a brisk walk from Rockefeller Center. Right across the street stands the [Henry] Ford Foundation, wherein the afore-mentioned Brian Urquhart (U.N. factotum extraordinaire) watched the Brave New World unfold.[129] Not far away, to the near southwest, sits the [J.P.] Morgan Library; and farther up First Avenue to the north, sprawls the Rockefeller University. Go straight up First Avenue from U.N. Headquarters, make a left at East 57th Street, follow 57th Street crosstown to 7th Avenue, and you'll end up at [Andrew] Carnegie Hall. It is no coincidence that U.N. Headquarters is situated, within literal spitting distance, of all these [a]edifices erected for/by/to the Captains (some say Robber Barons) of American Industry. For almost to a man, they were firm proponents of one-world government – with themselves (or their surrogates) in control... from the United States. Historian Amaury de Riencourt, writing at the height of the Cold War, dispassionately analysed the American sponsorship – and essentially American character – of the United Nations with pin-point accuracy:

"It was during the Second World War that America realistically faced the fact that the problem rested on a simple alternative: either one organized world or no world at all. The Declaration of the United Nations was signed by twenty-six countries in Washington on the first day of 1942 – and the name itself, designed to wipe out the unhappy memory of the deceased League's failure, was symbolic. It seemed to ease the transition from United States to United Nations, and under American supervision, implied a promise to apply to the ordering of the world of the future the rich experience learned during the constitutional evolution of the United States. The charter of the UN included a great deal of America's own constitutional past and the initiative itself was largely American. It is highly significant that the Declaration was signed in Washington, that the first Assembly gathered in San Francisco, and that the permanent seat of the United Nations should be in New York – always on American soil and under American auspices. And, to date, the only military success of the UN, highly relative though it may seem, was entirely due to America's dynamic leadership in Korea. But even then, it was made possible by a blunder of the Russians, who could have vetoed UN participation in the Korean War if they had not temporarily withdrawn from the Security Council."[130]

On 27 July 1953, the [U.S./] U.N. Command and the Red Chinese/North Koreans signed the armistice agreement. On 10 November of that year, Trygve Lie resigned as U.N. secretary-general.[131] On 31 March 1954, the U.N. Security Council chose Dag Hammarskjöld of Sweden to succeed (Norwegian) Trygve Lie as secretary-general.[132] It was under the tutelage of self-seen lonely mystic, Dag Hammarskjöld,[133] that the United Nations became a changeling – transmuting from good to bad into ugly.

The UNO in 1960 was a fifteen-year-old tough new kid on the global block, and, during that turning-point year, started acting like a big bully; ever encouraged and egged on by the U.S.A.

It is not the intention of Future Fish 2001 to disparage the work or ethic of the United Nations. Alternative historians and conspiracy theorists have been casting shadows of doubt on UNO geopolitical motives for decades – especially since 1960. Although this writer is in accord with much of that opinion, this report purports only to render a chronology of U.N. expansionism and the effects on U.N. Trusteeship territories in the Pacific. But no report on UNO activities, however brief, would be complete without mention of the systematic and brutal obliteration of Katangan independence (1960-1963).

In the Appendix (I) of his jubilarian history book on the United Nations (subtitle: The First Fifty Years, 1995), apologist Stanley Meisler has written up 'A U.N. Chronology'. His entry for 1960 includes: "June 30 The Belgian Congo becomes independent and then erupts into chaos and bloodshed when the army mutinies against its Belgian officers. Belgian troops seize key points, but the [U.N.] Security Council dispatches Blue Helmets to replace Belgian troops, restore law and order, and end the secession of Katanga province. Peacekeepers put down the secession and remain in the Congo for four years."[134] So neat and clean; so short and simply stated. But if Stanley Meisler's entry under 1960 was short, the entry for the Congo under 1961 was even shorter: "September 18 Dag Hammarskjöld dies in an air crash in Ndola, Northern Rhodesia [today's Zambia] while on a mission [to] the Congo."[135]

But the real shocker to this reader was the empty entry for 1962: For that year was not listed at all by Stanley Meisler in his 'A U.N. Chronology'. Yet 1962 was a year of sheer agony and absolute ruination for the people (both Black and White) and province of Katanga, Congo (later Zaïre). In her engrossing book Who Killed the Congo? published in 1962, former resident Philippa Schuyler directly answers her own question:

" Nine groups of people caused the slow murder of the Congo: the ghost of Leopold II [King of the Belgians, 1865-1909], deeply-imbedded Congolese tribal hatreds, the undisciplined Congolese mob, intriguing Belgian Socialists, American pressures, Russian plotting, Ghanaian conniving, [Patrice] Lumumba's ambition and instability, and the intervention of the United Nations."[137]

As noted, Secretary-General Dag Hammarskjöld had died in a mysterious air crash during September 1961 on his way to the Congo. Philippa Schuyler has projected that Sweden was still enraged in 1962 over the death of her favourite/most famous son, and, blaming Kantagans and Belgian "mercenaries", lent her (Royal/kungliga) jets "unconditionally to bomb Elisabethville [Katanga's metropolis, today's Lubumbashi]. The planes devastated the city. Ill-aimed bombs hit the hospital; the hotels and residential sections were bombarded. U.S. cargo planes ferried equipment and munitions to the U.N. forces." [138] (All this death and destruction was visited on Katanga province during 1962, but was still deemed unworthy of dishonourable mention in Stanley Meisler's U.N. hagiography.)

It's not as if no-one in the United States cared about the terrible tragedy unfolding in break-away Katanga. American political notables such as former President Herbert Hoover and former Vice-President Richard Nixon; various U.S. senators, congressmen and private citizens (who included some famous writers and many professors from eighteen universities) had contributed both their prestige and money, to buy a full page advertisement in The New York Times (14 December [1962]) protesting the bombing of Katanga. [139] The advertisment, in part, declared that the rebellious province had " 'committed no aggression except to want to be free of a Communist controlled central government' ". [140]

But it was not to be. Instead, United States might and money helped support and pay for the brutal United Nations operation which destroyed the rightful government of Katanga (under the truly heroic Moise Tshombe); the most pro-West, most pro-American, most pro-democracy, most pro-free enterprise government (some say ever) in all Black Africa. [141] Philippa Schuyler ended her thought-provoking book with a stern admonition for America:

....."Thus the whole tragic drama of the Congo with all its blunders, massacres, failures and abuses should be seriously studied by every American, for it represents a concentrated example of the mistakes of American foreign policy. America has paid for one-third to one-half of the costs of each highhanded and blundering UN act in the Congo. Yet if the UN can interfere in the internal affairs of the Congo, why have they not invaded Cuba, Hungary, Tibet, where they were really needed, where thousands have been unjustly slaughtered, where not even the pretensions of democracy exist?
"America must bear part of the responsibility for the debacle in the Congo. And every American citizen should ask himself that challenging question, Who killed the Congo?" [142]

Among the seven Principles laid out in the U.N. Charter is Number 7, which unequivocally states: "The U.N. accepts the principle of not interfering in the actions of a member nation within its own borders."[143] And so recently as 1995 (the very year that the UNO so solipsistically celebrated their jubilee), Secretary-General Boutros Boutros-Ghali (1992-1996), referring to U.N. actions in Somalia and Bosnia, pronounced that "three traditional principals must guide peacekeeping operations: U.N. troops must have the consent of the warring parties before entering a country; U.N. troops must be impartial; U.N. troops must not use force except in self-defense."[144]

Like a fat, self-satisfied adult attaining a mature fifty years, the UNO seems to have conveniently "forgotten" the episode of violent crimes committed during its lean and mean youth.

After the débâcle in the Congo (1960-1963), the United Nations has acted in a most predictable fashion – opposing "Rightist" régimes (especially if White), and supporting "Leftist" régimes (especially if non-White) whenever possible, and promoting their neo-Marxist unitary global governance agendum whenever possible. Since the rise of Political Correctness (pc) as an actual New Left socio-political movement/phenomenon succeeding the collapse of [proto-Marxist] Communism (1989-1992), the UNO has wholly embodied the crushing, American-style pc (ethnic "bean-counting" backed by humungous firepower) of a ruling New [Third] World Order. 'A U.N. Chronology' in Stanley Meisler's Appendix to his The United Nations: The First Fifty Years is instructive. My favourite examples are listed below:

1966
"October 27 The General Assembly nullifies the South African mandate over South-West Africa.
"December 16 The Security Council imposes mandatory sanctions on the breakaway British colony of Rhodesia.[145]

1971
"October 25 The General Assembly votes to seat Communist China as a member state in place of Taiwan. The Chinese Communists also replace Taiwan as a member of the Big Five on the Security Council.

1972
"June The first U.N. Environmental Conference takes place in Stockholm.

1975
"June The first U.N. Women's Conference takes place in Mexico City. The conference ends in controversy as Third World delegates push through a resolution denouncing Zionism.
"November 10 The General Assembly passes a resolution declaring that 'Zionism is a form of racism and racial discrimination'. The vote is seventy-two to thirty-five with thirty-two abstentions.

1977
"November 4 The Security Council adopts a mandatory arms embargo against South Africa.

1979
"December 18   The General Assembly adopts the Convention on the Elimination of Discriminations against Women.[146]

1987
"September The Treaty on Protection of the Ozone layer is signed, the first global environment agreement.

1990
"August 2 Troops from Saddam Hussein's Iraq cross the border and seize the emirate of Kuwait. American President George Bush and American Ambassador Thomas R. Pickering galvanize the Security Council into near-unanimous agreement to condemn the invasion and impose sanctions.
"September 30 UNICEF convenes the World Summit for Children.

1991
"January 17 The Bombing of Baghdad begins as the United States leads a coalition in a successful war to oust Iraqi troops from Kuwait.
"February 27 The Persian Gulf War ends. The Security Council imposes a harsh peace on Iraq: Iraq must destroy all its programs of building weapons of mass destruction, ending its nuclear,

chemical, and biological warfare programs. Sanctions will remain in place until the Security Council is satisfied that Iraq is a peaceful country that no longer harbors aggressive intents against its neighbors.[147]

1992

"June The Earth Summit – the U.N. Conference on Environment and Development – is held in Rio de Janeiro and attended by 104 heads of state and government. [The notorious "Carnival of Dunces". See Future Fish 2000 for details.][148]

1995

"January Boutros-Ghali issues a supplement to his [17 June 1992] 'Agenda for Peace' that retreats somewhat from some of his original ideas on peacekeeping. He proposes that future missions be much more limited than the intervention in Somalia. [Too bad for the wretched Kantangans that Boutros-Ghali wasn't in the Congo, circa 1961. Would – could – the Egyptian (and African) secretary-general have made a difference? Naeh!][149]

After World War II ended in 1945, the newly-born United Nations Organization took up where the dying (April 1946) League of Nations had left off (essentially in mid-conflict). In the Western Pacific, (Melanesian) Papua New Guinea had been a League-mandated territory from 1921 to 1942, administered by Australia.[150] Occupied by the Japanese 1942-1945, Papua New Guinea was then again administered by Australia but as a U.N. Trust Territory. In 1975 PNG became fully independent as a member of the (British) Commonwealth of Nations.[151] In the South Pacific, (Polynesian) Western Samoa had been a New Zealand-administered League mandate from 1920 until the Pacific War's end in 1945. Western Samoa became a U.N. Trust Territory – again administered by new Zealand – from 1945 until achieving complete independence as "Sāmoa" in 1962.[152] In the Central Pacific, the tiny (phosphate-rich, Micronesian) island of Nauru was jointly administered as a League mandate by Great Britain, Australia, and New Zealand from 1920 to 1942, when invaded and occupied by Japan until 1945. In 1945, Nauru again became jointly administered by the British-[mainly] Australians-NZ'ers as a U.N. Trust Territory until 1968, when "independence" was granted, with Australia providing defense needs.[153]

The three Micronesian island groups (excepting the formerly British Gilberts/Kiribati) of the Marianas, the Marshalls, and the Carolines have had a uniquely complex colonial history, which will be examined in the next Section (C) of this Part (I). Suffice it for now to say that all three archipelagoes were administered by Japan under the League of Nations mandate (1921-1944), until captured by the advancing Americans during the Pacific War. As a result, the Northern Marianas, the Marshalls, and the Carolines in 1947 all become American Micronesia; that is, the United Nations Trust Territory of the Pacific Islands (TTPI) administered by the United States.[154] Also during 1947, the Northern Mariana Islands (NMI) elected to stay close within the American fold, and the U.S. fully implemented NMI's Commonwealth status in 1986. In 1990, the Marshalls became independent as the Republic of the Marshall Islands (RMI), and joined the U.N. in 1991. The Carolines (except for Palau) also became independent in 1990 as the Federated States of Micronesia (FSM), and also joined the U.N. in 1991.

But Palau, in the Western Carolines, remained the odd-man-out. Due to Palauan problems with the Compact of Free Association with the U.S. ("no nukes" etc. to be explained in Section C), Palau didn't become the Republic of Palau until 1994.[155] Because of Palau's unique political status, the entire modus operandi (and raison d'être) of the U.N. Trusteeship Council has come under question. One instance of naysaying was on the CBS-TV programme, 60 Minutes (19 September 1993), where Mike Wallace interviewed Charles Lichtenstein, a former U.S. representative to the U.N. (1981-1984). Conservative author Cliff Kincaid has commented:

"One such program is the U.N. Trusteeship Council, one of the six major organs of the U.N., established back in 1946 to oversee the governing of some eleven territories around the world. Today, [Mike] Wallace noted, there is just one territory left, Palau, with fifteen thousand citizens. And yet, the U.N. Trusteeship Council keeps meeting to discuss what Lichtenstein laughing called 'urgent issues'."[156]

As noted, Palau became the Republic of Palau (Beluu era Bilau) on 1 October 1994 – the very same day that the (U.N.) Trusteeship was terminated for Palau, and the (U.S.) Compact of Free Association was implemented. On 15 December 1994, the Republic of Palau was accepted as the 185th member of the United Nations.[157] On 17 September 1999 the Republic of Kiribati and Nauru (both Micronesian but neither American) were admitted to the United Nations as fully independent states.[158] As I watched the preceding proceedings on CNN, this writer wondered wh y there always appeared to be such a rush to U.N. membership by newly-emerged nations... and such eagerness by the UNO to embrace them? Of late, I have found no better answer (from an Americanist perspective) than that given by Christian authors/broadcasters Hal Lindsey and Cliff Ford:

"UNITED NATIONS

"Power Grab

"1. The United Nations, aghast at the concept of '17 remaining non-self governing territories' has set up a committee charged with stripping those territories from the United States and the United Kingdom.
"2. 'As non-self governing' territories, like the Virgin Islands, Guam and Samoa they are outside the reach of the globalists at the United Nations.
"3. The UN would like to see Guam and American Samoa 'join the family of nations' because they would bring two new votes to the assembly – votes of equal weight on many issues as that of the US.

"4. The UN's push to divest America of its possessions is unilateral and without the consent of the US Congress. At the time, UN-sanctioned organizations promoting independence for the State of Hawaii have been seen handing out literature in Honolulu at various high-profile events.

"5. What effect the loss of these territories would have on US national security and that of its allies is unclear and the issue is certain, eventually, to provoke an intense debate in both Houses. Not that it will make any difference, unless you think the administration will really go up against the existing de-facto global authority.

"Assessment: At the heart of the effort one can see the UN's real agenda, exposed and naked to the world. <u>The death of the nation-state and its replacement by a centralized global authority</u>...."[159]

Are the proud new states of Micronesia, with their ancient cultures and beautiful sea/[is]landscapes, condemned to change into merely a part of a United Nations racial (First Peoples), spoils system, rather than one of the United States? Let us hope for the sake of a truly free Micronesia – for all her component parts of whatever political status – that these Pacific Islanders scrupulously avoid either one of these materially-tempting, but independence-deadening, U.S./U.N. geopolitical traps. And that is exactly what this Chapter attempts to address.

<u>Post Millennium Summit 2000</u>

For several days, during early September 2000, midtown Manhattan was all a-whirl with the goings-on at UNO Headquarters. This was the year 2000, and U.N. Hq. served as the site of the grandly-monickered Millennium Summit. On Tuesday 5 September, Secretary-General Kofi Annan kicked off the Great Event by beseeching (in his usual dulcet tones) the mostly-motley aggregate of kings, presidents, generals, and tin-pot dictators, to utilise the unprecedented meeting for establishing peace and ending poverty in the 21st century.[160] By the following day, Wednesday 6 September, the globalist body had enunciated its New Age visions for the third millennium at the extraordinary convocation of world leaders – who ranged from the celebrated to the notorious to the downright obscure.[161]

The American president was there; the silly-billy Boy from Hope, Ark., desperately making a final bid to seal his failed "legacy", his "place in history", with a false and elusive peace in the Middle East.[162] Fidel Castro also attended the U.N. Summit, and thank God, was limited to a five minute speech (rather than his usual five hour tirade). I watched that week on C-SPAN the agitated expostulations of the Western Hemisphere's last Leninist, now a pathetic and toothless Lion-in-Winter. And at second glance, Boy Clinton looked none the worse for wear in trying (his level best) to sell [eretz] Israel down the [Jordan] river....

But wait! Who <u>was</u> that oafish-looking, sheep-faced man shown shmoozing with the U.S. Prez on CNN's <u>Diplomatic License</u>?[163] Why, it appeared to be none other than the ovine king of Norway, His Majesty Harald V! I knew then why the UNO has been possibly the most disastrous institution conceived by Western Man during the twentieth century.[164] And nothing, I repeat, <u>nothing</u> was achieved at the grandiloquent U.N. Millennium 2000. Turtle Bay's "Goombay Summer" went bust.[165]

Notes to Part I (cont'd.)

122. Stanley Meisler, United Nations: The First Fifty Years (New York: The Atlantic Monthly Press, 1995), p.339. Brackets mine.

123. Donald S. McAlvany, Toward a New World Order: The Countdown to Armageddon (Phoenix, Ariz.: Western Pacific Publishing Co., 1992 ed.), p. 287. Brackets mine.

124. Dr. Daniel Stengel, ' The United Nations at Mid-life Crisis', Studium Generale Program Notes (Peninsula College, Port Angeles, Wash.), 24 October 1996, p. 1.

125. Ibid., p. 2. Author's note: Principal organs of the UNO number six – General Assembly, Secretariat, Security Council, Trusteeship Council, Economic and Social Council, International Court of Justice – with their forty satellite agencies.

126. NB: Brian Urquhart, I was to learn decades later, also served as personal assistant to first U.N. Secretary-General (and Norwegian), Trygve Lie (1946-1953). Urquhart's overall appraisal of Lie was harsh:

" 'As Secretary-General of the new world organization... he was out of his depth. Lie was an unsophisticated man who relied more on peasant shrewdness and what he called his 'political nose' than on intellectual effort or hard diplomatic work. He was, in public life at any rate, a naturally suspicious man with a hair-trigger temper.
" 'Lie had the massive physique of an athlete who had run to seed in middle age.... He was apt to go dark red in the face with rage and to utter, jowls quivering, complex and ominous Norwegian oaths.... I felt that he was confused, temperamental, and insecure in his new and demanding job and was often carried away by the emotions of the moment.'" In Stanley Meisler, p. 23. (Author's note: "How are the mighty fallen!")

127. Webster's Encyclopedic Unabridged Dictionary of the English Language (New York: Gramercy Books, 1996), p. 2076. NB: There were fifty signers at San Francisco in 1945; the 51st. U.N. member-state was Poland, added in 1946. See Stanley Meisler, p. 358.

128. Op.cit., Donald S. McAlvany, p. 293.

129. Cf. Stanley Meisler, p. 334.

130. Amaury de Riencourt, The Coming Caesars (New York: Coward-McCann, Inc., 1957), pp. 317-318. (NB: Riencourt was/is probably not an American historian.)

131. In Stanley Meisler, p. 359.

132. Ibid., p. 359. NB: Meisler has commented: "Dag Hammarskjöld never married, and rumors hovered for years that he was homosexual, a state of life that might have brought him down if proven and publicized in those days. Brian Urquhart, his biographer and a junior member of the Hammarskjöld U.N. team, denied the rumors emphatically and accused Trygvie Lie of spreading them." In Meisler, pp. 78-79. (Author's note: "Another one bites the dust!")

133. Ibid., p. 79. NB: Meisler has reported that an English edition of Dag Hammarskjöld's posthumously published notebook/diary, Markings, underwent 38 reprintings by 1994.

134. Ibid., p. 360. (Brackets mine)

135. Ibid., p. 360. (Brackets mine)

136. The Devin-Adair Company, New York.

Notes to Part I (cont'd)

137. Ibid., p. 258.

138. Op.cit., Philippa Schuyler, p. 292. (Brackets mine)

139. Ibid., p. 293. (NB: Year of Times ad is unclear; writer has assumed 1962.)

140. Quoted by Philippa Schuyler, p. 293.

141. Ibid., p. 272.

142. Op.cit., Philippa Schuyler, pp. 272-273. Author's note: In 1965, General Joseph Mobutu takes over the Congo; renames the Congo "Zaïre" in 1971, renames himself "Mobutu Sese Seko", and "Africanizes" Zaïre in his own image. In 1997, General Kabila's rebel troops enter Kinshasa, the capital. Mobutu dies, and Zaïre is back to being "Congo"... and to Square One. (See Student Handbook, Vol. 5. p. 730.)

143. Student Handbook, Vol. 5, p. 244.

144. Op.cit., Stanley Meisler, p. 335.

145. Ibid., p. 361.

146. Ibid., p. 362.

147. Ibid., p. 364.

148. Ibid., p. 365. (Brackets and contents mine)

149. Ibid., p. 366. (Brackets and contents mine) Author's note: For an excellent account of the agony in Katanga, see G. Edward Griffin (1964), passim.

150. Student Handbook, Vol. 5, p. 665.

151. Reilly Ridgell, Pacific Nations and Territories: The Islands of Micronesia, Melanesia, and Polynesia (Honolulu: The Bess Press, Inc., 1995 3rd ed.), p. 104.

152. Student Handbook, Vol 5. p. 668; Reilly Ridgell, p. 130.

153. Ibid., p. 669; ibid., pp. 92-93.

154. Ibid., pp. 670-671; ibid., pp. 64-69.

155. Ibid., pp. 670-671; ibid., pp. 64-69.

156. Cliff Kincaid, Global Bondage: The UN Plan to Rule the World (Lafayette, La.: Huntington House Publishers, 1995), pp. 125-126.

157. Elizabeth Diaz Rechebei and Samuel F. McPhetres, History of Palau: Heritage of an Emerging Nation (Koror, Republic of Palau: The Ministry of Education, 1997), p. 356.

158. Seen on CNN's Diplomatic License, 18 September 1999.

159. International Intelligence Briefing/www.iib-report.com, June 2000, pp. 3-4.

160. 'World leaders bring own agendas to U.N. summit', Peninsula Daily News, 6 September 2000, p. C-1.

161. 'Clinton asks world leaders to help in talks', Peninsula Daily News, 7 September 2000, p. C-1.

162. See 'Arafat rejects proposal', PDN, 8 September 2000, p. C-1; and 'Clinton, Barak try to salvage Mideast talks,' PDN, 10 September 2000, p. C-1.

163. Shown on CNN's Diplomatic License, 9 September 2000. Cf. Mindy Belz, 'UNdistinguished', WORLD, 16 September 2000, p. 8.

164. Since the Korean War (1950-1953), the only U.N. action I can think of really deserving merit is that recently in East Timor. Eg., 'U.N. workers killed, Peninsula Daily News, 7 September 2000, p. C-1.

165. Taken from the Bahamian slogan, "Goombay Summer Goes Boom", 1973.

United Nations Bibliography

11. Ferencz, Benjamin B. and Ken Keyes, Jr. PlanetHood: The Key to Your Future. Love Line Books, Coos Bay, Ore., 1991.

12. Griffin, G. Edward. The Fearful Master: A Second Look at the United Nations. Western Islands Publishers, Belmont, Mass., 1964.

13. Kincaid, Cliff. Global Bondage: The UN Plan to Rule the World. Huntington House Publishers, Lafayette, La., 1995.

14. Meisler, Stanley. United Nations: The First Fifty Years. The Atlantic Monthly Press, New York, 1995.

15. Schuyler, Philippa. Who Killed the Congo?. The Devin-Adair Company, New York, 1962.

Section C: A Brief History of Micronesian Colonialism

" 'The Spanish came for God, the Germans came for glory, the Japanese came for gold, the Americans came for good.'"
--Quoted by P.F. Kluge in The Edge of Paradise, 1993

By now the reader should have a somewhat clearer picture of Micronesia than when this addended Chapter started. I have generously handed out facts and figures on all Micronesia in Sections A and B. Rereading these sections and reviewing my notes, however, I seem to have forgotten to include that the Micronesian languages are of the Austronesian (Malayo-Polynesian) family; that Western Micronesia (Palau and the Marianas) were settled from Indonesia and the Philippines, while Eastern Micronesia was settled from Eastern Melanesia.[166] Somewhere I did mention that Micronesia, as a whole, has had the most complicated colonial history of the entire Pacific. (Dealing with the Spanish, Germans, and Japanese; rather than just the British or French.) The Pacific War (1941-1945), of course, changed all that, and since 1945 the Pacific Ocean has been so much an American lake as it had been a Spanish lake prior to 1898. (For the purposes of this report, commercial fishing history in Micronesia begins with the Japanese Nan'yō; officially the Japanese-administered "class C mandate" as awarded by the League of Nations in 1920.)[167]

To simplify the complex colonial history of Micronesia, let me divide the islands into their current geopolitical groupings:

I. U.S. Territory of Guam
    1668-1898   Spanish era
    1898-1941   Rule by U.S. Navy
    1941-1944   Japanese Occupation
    1944-1950   Organic Act of 1950. Rule transferred from U.S. Navy to U. S. Department of Interior. Guamanians granted U.S. citizenship; Guam an unincorporated U.S. territory.

II. Commonwealth of Northern Mariana Islands (NMI)
    1668-1899   Spanish Era. Spain sells NMI to Germany.
    1899-1914   German period until outbreak of World War I
    1921-1944   Japanese administration under League of Nations class C mandate.
    1947-1975   U.N. trusteeship administered by U.S. (TTPI)
       -1975   NMI votes for U.S. Commonwealth status.
    1976-1978   NMI administered separately from the rest of the TTPI.
       -1978   NMI a U.S. Commonwealth

III. Republic of Palau
    1885-1899   Spanish era. Spain sells Carolines to Germany.
    1899-1914   German period until outbreak of World War I
    1921-1944   Japanese administration under League of Nations class C mandate
    1947-1994   U.N. trusteeship administered by U.S. (TTPI)
       -1994   Independence, but under U.S. Compact of Free Association (1993)

IV.   Federated States of Micronesia (FSM)
        [Yap (Waab), Truk (Chuuk), Ponape (Pohnpei), Kosrae]
        1886-1899   Spanish era. Spain sells Carolines to Germany.
        1899-1914   German period until outbreak of World War I
        1921-1944   Japanese administration under League of Nations class C mandate
        1947-1986   U.N. trusteeship administered by U.S. (TTPI)
        -1990   Independence, but under U.S. Compact of Free Association (1986)

V.   Republic of the Marshall Islands (RMI)
        1885-1899   Spanish era. Spain sells Marshalls to Germany.[168]
        1885-1914   German period to outbreak of World War I
        1921-1944   Japanese administration under League of Nations class C mandate
        1947-1986   U.N. trusteeship administered by U.S. (TTPI)
        -1990   Independence, but under U.S. Compact of Free Association (1986)

VI.   Republic of Kiribati (including Ocean Island/Banaba).
        1892-1941   Gilbert and Ellice Islands under British protectorate
        1941-1943   Japanese occupation
        1943-1975   British rule loosens – (Polynesian) Ellice Islands separate from
                     (Micronesian) Gilberts to become independent Tuvalu.
        -1979   Independence of Kiribati (Gilberts)

VII.   Republic of Nauru (Naoero)
        1888-1914   German period (as part of Marshalls) until outbreak of World War I
        1920-1942   Administered by Great Britain/Australia/New Zealand as League of
                     Nations mandate
        1942-1945   Japanese occupation
        1947-1968   U.N. trust territory administered by Great Britain/Australia/New
                     Zealand[169]
        -1968   Nauru independent under Aussie wings

NOTE: Pohnpei served as the centre of both Spanish (1886-1899) and German administrations
        (1899-1914) in the Carolines. Palau, with Koror (Oreor) as capital, was the bustling heart
        of the Japanese Nan'yō (1914-1944).
        Guam, then Saipan, were (and are) American Micronesia's respective metropolitan and
        administrative centers (1898-1944, 1944 --).

American Micronesia since 1945

Significant dates since the Pacific War:

1947:   Micronesia a U.N. Trusteeship under U.S. administration
1952:   Legal basis for government laid down in Code of the Trust Territory. U.S. government
        responsible to U.N. Trusteeship Council for civil administration of Micronesia.
1965:   Congress of Micronesia elected by Territory, endowing it with legislative powers.
1975:   Plebiscite in Northern Marianas which voted for U.S. Commonwealth status.
1976:   Northern Marianas to be administered separately from the rest of the Territory.
1978:   Remaining island groups reorganised (again) into 6 districts, and voted on proposed
        constitution for the Federated States of Micronesia. Four of these six districts – Ponape
        (Pohnpei), Kusaie (Kosrae), Yap (Waab), Truk (Chuuk) – approved new constitution.
1979:   The 4 districts – all in the Carolines – were established as the FSM. The 2 dissenting
        districts, the Marshall Islands and Palau (Belau, in the Carolines) formed republics.
1981:   The FSM (the 4) plus the 2 republics voted in referenda for the Compact of Free
1982:   Association with the United States; i.e., internal self-government, independent foreign
1983:   diplomacy, but U.S. defense responsibility.

1986: U.S. government dissolved Trust Territory agreements (under Soviet anti-colonialisation pressure at U.N.). The Federated States of Micronesia (FSM) and the Republic of the Marshall Islands (RMI) become sovereign, self-governing states, with U.S. responsible for their security and defense. The Northern Mariana Islands (NMI) formally declared U.S. Commonwealth.

1994: Belau becomes a republic, but under (U.S.) Compact of Free Association.

The year 1945 spelled the end of the Second World War and the almost exact mid-point of the American Century. The United States stood astride a shattered planet at her zenith of power, and nowhere was this more apparent than in the Pacific basin.

The Micronesian island of Guam, a U.S. flag territory since 1898, is the logical place to start our examination. Guam has served as the western bastion of the overseas American Empire ever since its seizure from Spain. So soon as 1904 the Navy was requesting Congressional funds to fortify Guam; by 1906 the Navy had started formulating secret contingency war plans – the infamous "War Plan Orange".[170] As the U.S.A. is invariably "blue" in American war plans, it was Japan (cast as the future naval threat after the Russo-Japanese War of 1904-1905) designated as enemy "orange". The Washington military/naval mind-set regarding the Pacific was two-fold: 1) The U.S. Army would defend the Philippines until the U.S. Navy could cross the Pacific (from Hawai'i) carrying men and materiel; (2) meanwhile, the U.S. Navy's mission success was dependent on holding Hawai'i, and Guam if possible, to secure the line of communications to the Philippines.[171] By 1907-1908, the U.S. Army War College started playing Blue-Orange war games, and in 1911 the American War Plan Orange had become a combined Army/Navy affair. Getting wind of all the American activities, the Japanese commenced drafting their own battle plans in case of war. The Japanese military/naval "brass hats" matched perfectly their American counterparts in war mongering paranoia. Guam historian Robert F. Rogers has commented:

"The American Congress, however, assumed that war in the Pacific was unlikely and repeatedly refused to provide money for any large-scale fortifications on the American islands west of Hawai'i. A corollary of the American Orange [War] Plans and of [N]avy policies in general was that any interference in military authority on Guam by civilian concerns was to be opposed in the interests of U.S. national security. The highly secret Orange [War] Plans were not known, of course, to the people of Guam and would be kept secret until after World War II."[172]

To mollify the U.S. Navy and to maintain Guam's security at low federal cost, President William F. ("Big Bill") Taft (1909-1913), declared on 23 September 1912 that Subic Bay (Philippines), all of Guam, Pearl Harbor (Hawai'i), and Guantanamo Bay (Cuba) were off limits to foreign vessels of commerce.[173] Thus did the entire island of Guam become a closed U.S. Navy reserve, shutting out all

American civilian or foreign exchange for the next near-half century. That which routinely came to and went from Guam...would be a rotating succession of U.S. naval governors. And thus Guam would remain somnolent and severed from the outside world, until the rude Japanese wake-up invasion of 1941 and occupation until 1944.

The Great Pacific War was finally over and delegates of forty-six nations signed the U.N. Charter at San Francisco on 26 June 1945.[174] Then was created the U.N. Trusteeship Council, as we have seen, to monitor the promises of self-determination for eleven trusteeships the world over. The formerly Japanese League of Nations class C mandate of Micronesia was to constitute the eleventh trusteeship; now under United States administration. But the American "brass hats" at the Pentagon demanded the <u>direct</u> retention of Micronesia by the United States for reasons of national defense. U.S. military/naval leaders and their Congressional backers contended that (1) Micronesia was too strategically valuable, plus (2) too much American blood had been shed for (3) the islands to be given up for idealistic notions of decolonisation.[175]

Indeed, the U.S. Navy supported the passage of Bill H.R. 7044, introduced to the 79th Congress (1945-1946), to annex Micronesia making Guam the capital of a new American territory; including an organic act so that the territory could eventually achieve U.S. statehood.[176] H.R. 7044 failed due to State Department opposition. In lieu of annexation, the U.S. requested the U.N. for authorization to fortify the trusteeship islands – and restrict them to solely U.S. personnel. To ameliorate liberal proponents of decolonialisation, President Truman pledged to facilitate the future self-determination of (American) Micronesia. But, in a "snake move", the Micronesian trusteeship was placed under the U.N. <u>Security</u> Council, where the U.S. wields veto power, rather than under the U.N. <u>Trusteeship</u> Council, where a majority vote rules.[177] (Just who was to gainsay the U.S.A. during the late 1940s? And the U.S.A. had already commenced atomic testing at Bikini Atoll on 1 July 1946!)

Sly old political fox that Harry S. Truman was, the American president acquiesced to the Soviet Union's ongoing occupation of the four southern Kuril Islands north of Japan – snatched by the U.S.S.R. during the waning days of World War II – in return for Russian recognition of absolute U.S. suzerainty over Micronesia.[178] The U.N. Security Council then unanimously approved the strange-but-truly strategic accord with the U.S. on 2 April 1947, and the Congress ratified the agreement on 15 July 1947 by a joint Senate/House resolution. The title of the new dependency, as we all know by now, was the Trust Territory of the Pacific Islands (TTPI). President Truman turned control of the TTPI over to the tender ministrations of the U.S. Navy, and appointed CINCPAC (to administer from Hawai'i) as the first high commissioner. Guam, of course, as a U.S. flag territory and not a part of the TTPI, continued to be administered by the U.S. Navy.[179]

Also in 1947, the National Security Act created the U.S. Air Force (USAF) – spawning, too, the Central Intelligence Agency (CIA) and the National Security Council (NSC) – in a humungous reorganisation of the U.S. Armed Forces into a tripartite system master-minded by the new (and first) Secretary of Defense, James V. Forrestal (1947-1949).[180] Purportedly, the National Security Act would unify the U.S. Armed Forces, but instead engendered three massive, semi-independent military services with the Forrestal-favoured Navy emerging with even greater power and influence than ever before. Guam historian Robert F. Rogers has wryly noted:

> "With a free hand, the U.S. Navy implemented what critics called a 'zoo policy' of keeping the TTPI isolated and underdeveloped like an exotic anthropological park. This militarized Pax Americana would transform [non-TTPI] Guam into a small but highly valuable U.S. Gibraltar of the Pacific second only to Hawai'i, just as navy strategists had advocated ever since the first Orange Plans.
>
> "For the indigenous people of the region, however, it would be a different story. In the judgment of an analyst writing in 1975, American administration of Micronesia [TTPI] failed 'dismally...to meet the diverse needs which exist among the islands'. The militarized U.S. regional policy constricted Guam's development even though the island enjoyed an unprecedented economic boom the first five years after World War II."[181]

Meanwhile, (still) during 1947, a socio-political firestorm over Guam's status was being fueled by liberals in the U.S. Congress and in the American media. And John Q. Public's awareness of Guam's plight was further inflamed by the (timely) appearance in 1947 of Laura Thompson's third edition of her 1941 book, Guam and Its People. The revised, post-War edition of Laura Thompson's study had been the only serious analysis of Guam since the turn of the century.[182] Unfortunately for the U.S. Navy, Laura Thompson was married to a Mr. John Collier, president of the Institute of Ethnic Affairs (of which Laura Thompson had been a founding member) in Washington. D.C., a veritable warehouse of anti-Navy sentiment. Not only this, John Collier had served as commissioner of Indian affairs in the Harold Ickes Department of the Interior.[183] Thus the Institute put out a virulently contra-Navy newsletter, the Guam Echo (1947-1950), supported by and subscribed to by both Guamanians and Mainland liberals. John Collier, along with Guam Echo's editrix, Dolores Coulter, splattered America's newspapers with poison-pen letters highly derogatory of U.S. Navy rule in Guam, the TTPI, and American Samoa.[184] And so it went for three years.

Eventually, to break a quickly-rusting deadlock, President Truman signed H.R. 7273. the Guam Organic Act, on 1 August 1950 at the White House. (The presidential signature made the Act a federal law codified as "48 U.S.C.∫1421, et.seq.").[185] (Historian Robert F. Rogers has informed us that the only Guamanian present at the signing ceremony was a certain Carlos P. Taitano, later speaker of the Eighth Guam Legislature.) Guam has been an unincorporated territory ever since (1950) – belonging to, but not part of, the United States. Like Puerto Rico and the U.S. Virgin Islands, Guam is in the ambivalent position of being a mere geopolitical appurtenance of Mothership U.S.A. In 1970, Guamanians elected their own governor for the first time; since 1972 an elected but non-voting congressman has represented Guam in Washington, D.C. In 1979, the voters of Guam overwhelmingly rejected (by 81.7%) a draft constitution (to replace the outmoded Organic Act of 1950 with an NMI-type Commonwealth status) as it didn't, Guamanians believed, fully protect (indigenous) Chamorro rights.[186] By 1987, Guam voters had approved a draft Guam Commonwealth Act which (at last look) has languished in the U.S. Congress since 1988; stuck in the mud of a provision granting Chamorros the exclusive right of self-determination.[187] In A.D. 2000, Guam ("Where America's Day Begins") remains a U.S. colony in all but name.

The hip/youth-oriented travel publication, <u>Lonely Planet</u>, recently posted a decidedly uncharitable review on their website about Guam which has caused some controversy on the American Gibraltar, to wit:

> "'This strategic U.S. outpost is Gilligan's Island on steroids, a tropical dirty-weekend diorama and a real nice place for thousands of military personnel to get a tan....Health risks: Sunburn, fungal infections and gut trouble.'"[188]

And so we come to the colonial history of the political and legal status of the (U.S.) Trust Territory of the Pacific Islands (TTPI). I have earlier warned readers that the colonial history of American Micronesia is complicated – but the political and legal status of each island –chain sub-unit is possibly more complicated still. As travel writer David Stanley has correctly told us, "Guamanians cannot vote in U.S. presidential elections, yet any U.S. citizen can vote in a Guam election as soon as he/she gets off the plane."[189] On the other hand, though, citizens of the Federated States of Micronesia have <u>unrestricted</u> entry to the American Mainland and her territories (eg. Guam), but Mainland American citizens <u>do not</u> enjoy such full access to the FSM[190]...or to other (U.S.) Pacific territories. Conservative writer Joseph E. Fallon has studied extensively this surely one-sided sociopolitical situation, and has observed:

> "Recognizing that immigration can have an adverse impact on the demographics of a land, Congress has given American Samoa, the Federated States of Micronesia, [the Republic of] the Marshall Islands, [the Commonwealth of] the Northern Marianas, and [the Republic of] Palau the right to control immigration to their territories so those [I]slanders can preserve their ethnic, racial, and cultural identities."[191]

In other words, Joseph Fallon has explained plainly, U.S. territories – unlike U.S. states – have become "de facto ethnic-based polities that exercise political powers denied to the states."[192] Ergo, following Fallon's thesis to its conclusion, the various U.S. commonwealths and territories..."each possesses power greater than the [50 individual] states."[193] Not only that, Fallon has reminded us, but the U. S. Congress has gone so far as to allow U.S. territories to enact land alienation laws, id est:

> "To place race-based restrictions on the ownership of land. In order to own land in the Federated States of Micronesia, [the Republic of] the Marshall Islands, or [the Republic of] Palau, one [that's you!] must be a citizen of that particular 'republic', which, in effect, means one must belong to a specific ethnic group."[194]

Good gad!, the American historian or regular U.S. citizen might well exclaim in consternation. But in the total colonial sum story of the TTPI and American Micronesia, the horrific effects of the Pacific War and of the post-War bomb-testing on the Islanders – the dedicated Pacific historian has only to do a cursory study of the previous Japanese, German, and Spanish régimes to land...smack-down, squarely, on the side of the Micronesians. This position might (somewhat) surprise my readers, as this writer is a late middle-aged White Guy, evangelical Christian, political conservative, very tribal, and who regards post-modern "correctness" – whether religious, political, racial or ethnic – with a deep and abiding loathing. But...Joseph E. Fallon, whose views I respect, has his points. Therefore we must find out how these apparently unbalanced (from both sides) conditions came to be.

We have examined the unincorporated Territory of Guam, therefore the next logical step would be to look at the Commonwealth of the Northern Mariana Islands (NMI). When the Northern Marianas are mentioned, one immediately thinks of the islands of Saipan, Tinian, and Rota (Luta). And these are the three main islands of the NMI. But the elongated island chain stretches north from Guam and includes the islets of Aguijan, Farallon de Mendinilla, Anatahan, Sarigan, Guguan, Alamagan, Pagan, Agrihan, Asuncion, Maug Is., and Uracas.[195] As noted, the NMI were administered by the Japanese as a "class C mandate" from 1921 until the Pacific War. So, unlike American Guam to the south, the NMI were essentially under the aegis of Japan for thirty years, from 1914 until the U.S. invasion of Saipan on 15 June 1944.[196] By 9 July 1944, the U.S. Armed Forces had swept the island clean of all remaining Japanese opposition, and Old Glory was officially raised over Saipan the following day.[197]

But at what a price – especially for the Japanese! From the ignominious "Great Marianas Turkey Shoot", to the desperate banzai charge across the Tanapag plains, and the terrible tragedy of the mass-suicide [of Japanese soldiers and civilians, women and children] over the steep cliffs at Marpi Point; with over 3,000 Americans killed, another 11,000 wounded, and the death of 419 Saipanese…Saipan endured the worst devastation of any Pacific island.[198] In quick time, the U.S. Armed Forces had landed on Guam by 21 July 1944 and on Tinian three days later.[199] From these islands, American bombers started leveling the cities of Japan. Indeed, tiny Tinian became the homebase of North Field, which in the words of David Stanley, "was [1944-1945] the busiest airfield in the world, with two B-29s taking off abreast every 45 seconds for the seven hour ride to Japan. The nuclear age began here when the Enola Gay left Tinian for Japan."[200]

As my readers know by now, the Northern Marianas became part of the American-administered TTPI, along with the rest of formerly Japanese Micronesia, in 1947. Until 1962, the NMI were placed under U.S. Navy rule, and utilised as a Central Intelligence Agency (CIA) training ground.[201] However on 4 July 1962 (Kennedy administration), little Saipan became the capital, the administrative centre, of the entire TTPI.[202] Serious political-status negotiations commenced in 1969 (Nixon administration), and in February 1975 (Ford administration) the Marianas Covenant was co-signed, which separated the NMI from the remainder of the TTPI. A mere four months later, in June of 1975, the NMI voted: 79% of the circa 5,000 voters approved the Covenant and their annexation by the U.S.A. Thus the NMI were by 1978 – with full implementation plus U. S. citizenship in 1986 – an American Commonwealth like Puerto Rico.[203] David Stanley, hip travel author and no conservative, has nonetheless wondered out loud:

"A factor in the favorable vote was a desire to become eligible for a federal food stamp program, which feeds [1992] a third of the population. There's also Medicaid, school lunches, education, and old-age assistance. Federally subsidized housing starts at $8 a month. No other area under U.S. sovereignty gets services like these![204]
"….The Marianas government is very good at asking for federal handouts but the money often disappears into private pockets. To qualify for extra disaster relief benefits, [the local] people have gone to the extreme of bulldozing their own homes after typhoons. The Japanese are buying back the island [Saipan] they lost during the war, piece by piece. They've gained control of large areas through Chamorro

{the indigenous people] fronts who buy up land for them. Many locals appear naïve and greedy, unwilling to learn the lessons of the Hawaiians."[205]

In the Foreword to this Chapter, I wrote a brief paragraph on the post-War American bombing tests in the Marshall Islands (1946-1963). During the months following the Pacific War and into mid-1946, a major portion of Los Alamos (N.M.) scientific brain-power was channeled into Operation Crossroads – the Bikini tests.[206] With U.S. General Leslie R. Groves in logistical charge, Los Alamos "Manhattan Project" technicians produced, assembled, and tested the weapons components; controlling the timing of Test Able and detonating the bomb for Test Baker.[207] In his 1979 book, <u>World War III and the Destiny of America</u>, Christian writer Charles R. Taylor quoted a lengthy – but pertinent – paragraph from a 26 March 1953 article from <u>The New York Times</u>:

"'In the climax of the blast a whole island disappeared, transmuted into deadly vapor and ash. Since then, however, this explosion has been dwarfed by the even larger thermonuclear blast of March 1, 1953. The latest blast surprised even the controlling scientists, according to President Eisenhower. Its effect flared beyond the control boundaries and the "fallout" of radioactive ash burned Americans, natives of Pacific islands and Japanese fishermen many miles away.'"[208]

Charles R. Taylor, a conservative evangelical, struck me as hardly "a bleeding-heart liberal". Indeed, he seemed somewhat hawkish and "gung-ho" military. But from the nature of the passage cited from <u>The New York Times</u>, Mainland Americans of all political persuasions – including this writer – can sympathise with the Bikinian victims of the Marshall Islands. However, even the sad plight of the Marshallese is able to be twisted politically by evil minds – both White (liberal lawyers) and Brown ("activist" leaders). By the time the U.N. trusteeship (TTPI) had ended in 1986 (Reagan administration), the 1300 Bikinians and other Marshallese affected by the American atomic/hydrogen bomb-testing programmes (1946-1963), had been promised full monetary compensation for all nuclear claims…plus Compact money.[209] The United States agreed to invest a $150 million trust fund to create a <u>permanent</u> endowment for all health payments and nuclear claims.[210] Much negative political savvy ("victimization", entitlement groupthink, "playing the race card" etc.) has been perfected in the American Empire since 1986. Even New Leftistish Pacific historian, David Hanlon, in 1998 allowed (that):

"Witness the disturbance and suffering caused by nuclear testing in the Marshall Islands and the more recent efforts of the Marshallese government to capitalize on them… A history of colonialism in American Micronesia must account for these sad conflicts as well; there is no simple story that is only about external efforts at domination and local means of resistance."[211]

As noted, the U.S.-administered [U.N.] Trust Territory of the Pacific Islands comprised hegemony over the Marshalls and Carolines (except for Palau) from 1947 to 1986. Also, as noted, American control over the TTPI passed from the U.S. Navy to the Department of the Interior in 1951. Finally, as noted, in 1954 the Office of the High Commissioner moved from Honolulu to Agana (Hagatna, Guam) and then, in 1962, the TTPI Hq. shifted to Saipan (NMI). There were seven [at first six] districts: 1)Saipan District (NMI), 2) Yap District (Waab), 3) Palau District (Belau), 4) Marshalls Districts, 5) Truk District (Chuuk), 6) Ponape District (Pohnpei), and 7) Kusaie District (Kosrae).[212] There were fully eleven (formerly colonial possessions) areas of the post-War world which had been designated as subject to the U.N. trusteeship system, but in 1947 (now) American Micronesia became a "strategic" trust territory – the only "strategic" trust territory of the eleven U.N. trusteeships.[213] Although the U.S. was given responsibility to administer the TTPI by the U.N., the Micronesian islands were not annexed outright. Indeed, under Article 76 of the U.N. Charter, the U.S. was sworn "'to promote the political, economic, social, and educational advancement of the inhabitants of the trust territories, and their progressive development towards self-government or independence as may be appropriate to the particular circumstances of each territory and its peoples and the freely expressed wishes of the peoples concerned.'"[214]

The TTPI HQ. relocation from Guam to Saipan coincided with John Fitzgerald Kennedy's "New Frontier" administration (1961-1963). In1962, JFK appointed economist Anthony Solomon to head up an investigative commission. The Solomon Report, which resulted from a year's work of the imperially-minded commission, recommended that if Micronesians were to remain within the American sphere of influence (with all the material benefits that promised), substantial [infrastructural] improvements would have to be made.[215] Therefore in 1963, President Kennedy issued National Security Action Memorandum 145, a policy which would further absorb Micronesia into the American body politic. To accomplish this (according to numerous Pacific writers and historians and I concur), the Micronesians would have to be lured away from their tradition Island way of life, and then made dependent on U.S. government assistance, subsidies, and hand-outs.[216] David Stanley (no Right-winger he) has related that:

"Despite the flood of money essential services…remained as bad as ever. It seemed most of the money went into government salaries for people who performed no specific task[!] From the start politics in Micronesia was based mostly on family ties and the immediate benefits [local] politicians could deliver to voters."[217]

But President John F. Kennedy's NSAM 145 would be of negligible influence on American Micronesia compared to what was about to come – Lyndon Baines Johnson's "Great Society" along with 900 (at highest tide) Peace Corps volunteers arriving on (less than) 100 inhabited Micronesian isles from the U.S. Mainland.[218] A veritable tsunami of young, [mostly] White, American idealists who wanted "to make a difference" and to "change the world." In Tides of History: The Pacific Islands in the Twentieth Century (1994), historian Robert C. Kiste has drawn a sharp critique of "da Piss Corpse" – even going so far as to describe its advent to American Micronesia as having… "events in the trust territory [run]amok".[219] Pacific historian Kiste (an academian, remember!) has built a solid case against the Peace Corps – hence the U.S. federal government being yet more culpable – acting as main vector/agent in the "Coca-Colonization" of American Micronesia; that U.S. government programmes have literally destroyed social mores and cultural structures of the Islands, and the net economic result has been to create a massive welfare state.[220] Micronesians of the former TTPI are, of course, still heavily dependent on the United States (perhaps even more so after "independence".)

Robert C. Kiste has presented his argument in chronological order. In 1966 (an "All the Way with LBJ" year), a large contingent (delegation, really) of Peace Corps volunteers was sent to American Micronesia by well-meaning but ill-informed members of the U.S. Congress. Social legislation had been amended… "to make overseas territories eligible for welfare measures originally designed for the [Mainland] American poor and disadvantaged."[221] By the late 1970s (during the "Jimmy" Carter administration), the U.S. federal government was operating 166 different (and unrelated) programmes in the Trust Territory. These programmes ranged from giant-scale food subsidies for the impoverished (inappropriate in subsistence-based Island societies), to a plethora of assistance programmes for the aged (unnecessary in a traditional culture which honoured and provided for elder members). Kiste has also mentioned U.S. federal employment-training programmes that were more suited to Urban America than to American Micronesia.[222] In all, Robert C. Kiste's analysis of U.S. federal paternalism and largesse was devastating – especially from a man of letters.

Of a different tone, but equally devastating, was the assessment of Peace Corps volunteers given by a certain Brian Orme, described as "South Seas Beachcomber, Irishman" in Thurston Clarke's Equator: A Journey (1990). Travel writer Thruston Clarke interviewed Brian Orme, after the latter had rented his

Abemama, (formerly British) Kiribati, pub to the Peace Corps for a conference. The Irishman's hilarious remarks, with expletives undeleted, are reproduced here in their entirety:

"They [the volunteers] wanted to sleep on pandanus mats, but I said, 'Fuck you, I'm not pulling bloody beds out of the rooms for you lot. The people here sleep on mats because they can't afford a bloody bed. They'd have beds if they could, and they think anyone who doesn't is a bloody idiot.' All the volunteers had this lank hair, and one was a nutrition expert. Ha! Bloody nutrition expert. This poor son of a bitch couldn't even feed himself – looked like a dead rat caught in a drainpipe. 'What can you do?' I asked. 'You can't mechanic, [you] can't fish, [you] can't carpenter, fix a car, build a house. You don't shave. What the bloody fuck can you teach the people here? The Peace Corps ought to send us a bicycle repairman, a hotel manager, a retired plumber, not these useless assholes. The people here, they keep Peace Corps volunteers as pets.'"[223]

As stated, the Carolines made up five districts of the TTPI: Palau, Yap, Truk, Pohnpei, and Kosrae. The Carolines, except Palau, confederated to form the Federated States of Micronesia (FSM) in 1990, after attaining self-government in 1986. The political fortunes of the Marshalls ran concurrently – self-ruled granted in 1986, and independence achieved as the Republic of the Marshall Islands (RMI) in 1990. Both the FSM and RMI joined the United Nations on the very same day, 1 October 1991.[224] But as noted, Palau elected to remain in the TTPI after 1986, after choosing not to be united to the Federated States of Micronesia.[225] Now we ask the question: Why didn't Palau join the FSM in 1990? The answer is simple but not easy: Palau's constitution clashed with the U.N. concocted/U.S. engineered Compact of Free Association. Exactly what is the Compact of Free Association? Conservative political analyst Joseph E. Fallon has described Free Association clearly and succinctly:

"'Free Association' is recognized by the United Nations as an alternative to independence for a former trust territory. It enables the local population to enjoy a maximum degree of self-government – including representation in international organizations and the right to negotiate and sign treaties – while insuring the former administrating power continues to finance and defend that territory."[226] (Such a deal!)

But Free Association seems to fly in the face of the Micronesian draft constitution of 1975, which in part (rather grandly) reads:

"'To make one nation of many islands, we respect the diversity of our cultures....Our differences enrich us. The seas bring us together, they do not separate us. Our islands sustain us, our island nation enlarges us and makes us stronger....With this Constitution we, who have been the wards of other nations, become the proud guardian of our own islands, now and forever."[227]

Despite the high aspirations of the Constitution, Micronesians basically got everything they wanted from the [U.S.] Compact of Free Association (1982, 1986): To be entirely self-governing, relying on the Americans for "financial assistance"; conceding to the United States the onus of their defense and security.[228] The United States, Pacific historian Francis X. Hezel, S.J., has posited, is the long-term winner by keeping all foreign nations out of the islands...and with Micronesia "within the American [cultural/]military sphere for a very long time."[229] Meanwhile, bit by bit, the wily Island negotiators loosened their political dependence on the United States utilising – according to F.X. Hezel, S.J. – a "distinctively Micronesian strategy involving a combination of attrition and pragmatism."[230]

The result was threefold: (1) The drafting of Micronesian constitutions which accorded the islands out-and-out sovereignty; (2) the filibustering of U.S. objections to gaps in these constitutions and the Compact of Free Association; and (3) the real achieving of a political autonomy that was, in theory, unacceptable to the U.S.[231] For permanent U.S. denial rights in all three Micronesian TTPI entities, the FSM and RMI today have flags, overseas embassies, and UNO membership...and unending economic assistance from the U.S.A. under the Compact.

(Since the Compact of Free Association went in effect for the FSM and the RMI in 1986 for 15 years, the time is almost up for renegotiation – 2001. Kwajalein in the RMI has Free Association security provisions for yet another 15 years – until 2016 – due to the atoll's important U.S. military facilities.[232] Before getting to the unique case of Palau, let it be known that the major island of Pohpei, the heart and soul of the FSM, voted against the terms of the Compact (viz., against further "Coca-Colonization"![233])

So why didn't wild card Palau vote to join the FSM in 1978, or implement the compact of Free Association – signed in 1986 – until 1993? (After eight plebiscites!) This is indeed Pacific political history to be told on a painted Palauan story-board! The hesitation can be traced back to [the Palauan island of] Peleliu (Beleliou), September 1944. During the interwar period under the Japanese, Koror (Oreor) became the capital of all Micronesia, with a large and thriving Japanese population. A streetcar line ran through downtown, and during the late 1930s the Japanese erected fortified bases and Palau was closed to all foreigners ("permanent denial rights" which continued under the Americans until 1962).[234]

The U.S. Armed Forces attacked Peleliu in September 1944, to secure a springboard from which to launch an all-out invasion of the Philippines. The Japanese defenders entrenched themselves in caves on a limestone ridge, rather than challenge the Americans on the beaches.[235] After two and a half months of intense engagement, 11,000 Japanese lay dead, with 2,000 Americans killed and 8,000 wounded.[236] The U.S. Armed Forces bombed and "neutralized" the main island of Babeldaob (Babelthuap) and Koror proper but never invaded – leaving the 25,000 Japanese troops to sit out the action. But prior to the landings on

Peleliu, the Japanese herded the Palauan civilian population onto central Babeldaop where 526 persons (almost 10% of Palauans at the time) lost their lives. David Stanley, travel writer, has aptly stated the Palauan case: "Memories of this period are still [1992] vivid in Belau, and the [I]slanders don't want to get caught between two crocodiles [Japan and the United States] again."[237]

As noted, it wasn't until 1962 that the U.S. authorities lifted the ban on outsiders entering Palau. By 1972, Palauans were again struggling to lift the yoke of American military domination; this time as the Pentagon brass-hats had announced their intention of erecting military bases on the islands (as the Japanese had during the 1930s). In July 1978 Palauans voted to separate from the rest of (American) Micronesia, and by January 1979 the first Belau Constitutional Convention gathered to draft a constitution…for an independent republic! The Palauan delegates – to the dismay of U.S. officialdom and the Pentagon brass-hats – incorporated provisions in their Constitution which: (1) Banned all nuclear materials from Palau; (2) prevented the [Palauan] government from using "eminent domain" powers to benefit a "foreign [i.e., American, Japanese] entity"; and (3) declared a 200-MEZ around the entire Palauan archipelago. Plus, it would require a 75% referendum vote to override or change the anti-nuclear provisions in the Constitution. And so it went; the people of Palau believing strongly in consensus.[238]

That high 75% consensus foreordained at least four different votes on the nuclear-free Constitution, and fully eight separate referenda held on the Compact of Free Association.[239] The terms of the Compact would grant Palau self-government, relegate security rights to the United States for 50 years, and allocate $1 billion U.S. to Palau for 15 years.[240] Meanwhile, during the 1980s – a decade of real turmoil in the Islands – there were shootings, bombings, and multiple cases of arson. First Palauan President Haruo Remeliik was assassinated (1985); his successor, Lazarus Salii committed suicide (1988); and more than twenty feisty Palauan women, with Ms. Gabriela Ngirmang in their forefront, fought the pro-nuclear Compact tooth and nail.[241] (It's a stirring tale of the Western Pacific to be etched onto a Palauan story-board!) But by 1993, Machiavellian political machinations by both Palauans and U.S. Mainlanders had insured the passage of the Compact of Free Association; "free", that is, of local Island court challenges (i.e., the Compact passed by only 68% rather than the required 75%).[242] On 1 October 1994 Palau became a republic, but under the terms of the Compact of Free Association. In various ways, both Palauans and Americans were, simultaneously, winners and losers.²⁴³ P.F. Kluge, who knew Palauan President Lazarus Salii personally, has sadly surmised:

> "The old Trust Territory of [the] Pacific Islands was a perfect arena for Palauans. They didn't know it at the time. They thought they wanted to be on their own. Do their thing. Palau for the Palauans. A separate Palau, separate Compact of Free Association. Independence, maybe. Independence of all the other districts, that's for sure,.... But when that happened...their world shrank. They lost all the other islands where they could start businesses, take jobs, pour drinks. They lost that porous, comfy headquarters on Saipan, that safety net of a Trust Territory government that offered a home – and a job – away from home."²⁴⁴

As stated, the Compact of Free Association is due for renegotiation next [this] year, 2001, in both the Federated States of Micronesia and the Republic of the Marshall Islands. By July 2000, a four-day session had already taken place between the U.S.A. and the FSM at the East-West Center (U. Hawai'i at Mānoa) in Honolulu. Representing the U.S. federal government was Washington, D.C., special negotiator Allen Stayman; his FSM opposite number was Compact Joint Committee executive-director Asterio Takesy.²⁴⁵ The main reason for the meeting at the East-West Center had been the FSM's request for $84-million in renewed (i.e., <u>continued</u>) annual funding, once the current 15 year provisions for Compact aid dry up in 2001. The FSM delegation, under Mr. Takesy, arrived with detailed specifics regarding future use of incoming U.S. funds in five basic areas: Health, education, infrastructure, development of the private sector, and building (better) local government.²⁴⁶

And Mr. Stayman arrived in Honolulu with an agenda of his (viz., the State Department's) own. The lack of accountability of funding [seemingly] <u>transferred en masse</u> (rather than <u>provided as grants</u>) under the existing Compact had been a contentious issue before a U.S. congressional committee in June 2000. Allen Stayman, clearly a bureaucrat of another stripe, is determined to rectify the present situation: That the Micronesian government (viz., the FSM) <u>must</u> improve its own capacity to plan, manage, and account for the use of resources.²⁴⁷ Thus, Mr. Stayman had announced to the FSM delegates:

> "'We [the U.S.] will support specific areas of governance and hold you to performance (measurements) in those areas....I believe that the U.S. Congress should direct the [S]ecretary of the [I]nterior (who is responsible for administering Compact support) to assure that there is progress toward mutually agreed objectives each year.
> "'What we don't want is to get into the problem we have been in the past where not only has there been no tracking of performance, there have been no clear objectives and no identification of performance indicators. That all has to be built in now.'"²⁴⁸

After the July 2000 mini-summit at Honolulu with the FSM delegation, Stayman immediately quit Hawai'i for the Republic of the Marshall Islands. In the RMI, Stayman would be confronted by a new government at Majuro which would pursue <u>additional</u> nuclear claims compensation – more money – from the U.S.A.[249] The evils of "Coca-Colonization" aside, there is a growing no-nonsense mood on the U.S. Mainland (quite outside the D.C. "beltway boys" and our priapic President Clinton) regarding the one-way government gravy train to the (fraying) outer edges of the American Empire. <u>Pacific Magazine</u>, in September/October 2000, recorded the wise words of former U.S. ambassador to the Marshall Islands (and other Pacific nations), William Bodde, Jr.:

"'I would expect there's going to be continued funding. But I would expect it will be both more specifically targeted and more accountability will be built into it....
"'I think congress is willing to put up money, but it will want it to be for things that are specific....I also personally believe that money should be targeted wherever it can to encourage the private sector. Without that, you won't get economic development. You have to do that.'"[250]

How the United States and American Micronesia can act as co-equal partners in the Pacific, justly and honestly trading quid[s]pro quo[s], is the central thesis of this Chapter, <u>"A Modest Proposal" for Micronesia</u>. It can be done! The people of Micronesia will be able to proudly walk tall and to live truly free again, and there shall be a full stop to the "'deplorable, outrageous fleecing of the American taxpayer'"[251]...forever!

Non-American Micronesia since 1945

There are but two remaining areas of Micronesia, and these were never under U.S. control. They are the tiny (8 sq. mi./21 km^2) phosphate-rich island republic of Nauru, and the land-poor (281 sq. mi./728 km^2) but water-rich (more than 3.5 million km^2) Republic of Kiribati.[252] As stated, Nauru was taken by the Japanese in 1942, and occupied until the end of the Pacific War in 1945. From 1947 to 1968, Nauru (Naoero) was administered by (chiefly) Australia, New Zealand, and Great Britain as a United Nations Trust territory.[253] In 1968, Nauru became "independent" under Australian auspices. Until the 1990s the Republic of Nauru exported two million tons of phosphate per annum to Australia, New Zealand, South Korea, and the Philippines, bringing in $100-million Australian – a tidy sum for a population of circa 10,500.[254] But starting in 1990, phosphate shipments plummeted, due to both economic recession and new agricultural methods, but mainly because of phosphate reserve depletion. In 1992, travel writer David Stanley vividly described Nauruan monetary policy at home and overseas:

"The Nauru Government takes half the revenue from phosphate sales; the rest is split between the Local Government Council, the Nauru Phosphate Royalties Trust, and landowners. The trust fund is now worth over A $1 billion, designed to provide the inhabitants with a future income. Investments include flashy office buildings in Melbourne ("Birdshit Tower"), Honolulu, Manila, and Saipan, two Sheraton hotels in New Zealand, Fiji's Grand Pacific Hotel, the highly successful Pacific Star Hotel on Guam, the disastrous Eastern Gateway Hotel on Majuro, housing developments in Oregon and Texas, and a seven-stor[e]y 'Pacific House' in Washington, D.C."[255]

Meanwhile, little Nauru's limited topography could physically end as an empty husk; as an arid lunar landscape of total industrial devastation. That's her colonial legacy.

If anything, Kiribati (Tungaru) – the former Gilberts – has been even more (utterly) boned and gutted by colonialism in Micronesia than Nauru. (There were no rich, generous, and well-meaning Yanks here!) As noted, the Gilbert and Ellice Islands were a British protectorate from 1892 to 1941, when the Japanese invaded and occupied the islands until 1943. Japanese rule was rudely interrupted by the U.S. Marine Corps' amphibious landing at Tarawa, which commenced 20 November 1943.[256] Thus Great Britain regained control of the Gilbert and Ellice Islands in 1943, and remained until 1975. During that year the Polynesian Ellice Islands seceded, via plebiscite, from the Micronesian Gilberts (Tungaru) to become the separate entity of Tuvalu.[257] Ocean Island (Banaba), another miniscule but phosphate-rich islet like Nauru, attempted (like the Ellice Islands) to break away from the Gilberts during 1975, but without success. Great Britain (mini-England now) granted self-governance to the Gilbert Islands in 1977, and official independence followed two years later. The brand-new Republic of Kiribati included the 16 Gilbert Islands, the 8 Phoenix Islands, and the 8 Line Islands.[258] It all sounds like grand new-nation building, but British timing clearly manifested the absolute truth of the designation "Perfidious Albion". And David Stanley has hit the [colonial] nail squarely on the [British] head. Forsooth:

"Official independence followed on 12 July 1979. The British had timed this carefully to coincide with the playing out of the phosphate deposits on Banaba (the mine closed in 1980). With no other easy resources to exploit, and a growing, impoverished population, the British were more than happy to get out. Probably no other former British colony got such a lousy deal at independence; even Tuvalu had seceded from the Gilberts."[259]

However, terrestrial resource-poor but marine resource-rich Kiribati might still have the last laugh. For no political unit on the planet can boast a greater sea-to-land ratio. Kiribati is also the largest atoll state on earth; hugging the equator for 3,218 km from Ocean Island (Banaba) to Christmas Island (Kirimati), and crossing the International Date Line.[260] That's a whole lot of Central Pacific Ocean, plus Kiribati's 200-mile Exclusive Economic Zone! And recently, widely-scattered Kiribati has been trying to gain control over three fly-speck/islet U.S. possessions in the Central Pacific.[261] Though tiny and unpopulated, the Phoenix islands of Howland and Baker, and the Line island of Jarvis, loom significantly in the mind of

Teburoro Tito, President of the Republic of Kiribati. President Tito has expressed the belief (quite understandable) that Howland-Baker-Jarvis comprise an integral/geographic part of Kiribati.[262] A high American source has insisted that the islands were "'never an issue...never were on the table'" during the 1979 and 1982 negotiations, and U.S. ownership of Howland-Baker-Jarvis dates back to the passage of the Guano Mining Act of 1856.[263]

So Kiribati shall have to wait, bide her time, as will all Micronesia...at the pleasure of the American Empire, to see how the geopolitical fate of Howland-Baker-Jarvis eventually plays out. This, too, is a legacy of Coca-Colonialism in the Pacific. But time and space are on the side of the Micronesians.

Notes for Part I

166. 'Micropaedia', No. 8, The New Encyclopaedia Britannica (Chicago, Ill.: Encycl. Brit., Inc., 1990, 15th ed.), p. 103.

167. Effective in 1921. See Bruce G. Karolle, Atlas of Micronesia (Honolulu: Bess Press, '93), P.30. NB: But the Japanese seized all of the formerly-German Micronesia at the outbreak of World War I.

168. In 1874, the Holy Father at the Vatican in Rome mediated a Spanish-German dispute over the Marshall Islands in favour of Spain. But Germany was granted trading rights and, in essence, annexed the Marshalls. See David Stanley, Micronesia Handbook (Chico, Calif.: Moon Publications, Inc., 1992 ed.), p. 17.

169. Author's note: David Stanley (among others) has Australia solely administering Nauru; other sources give Great Britain and New Zealand, with Australia, as jointly administering Nauru after both wars – first as a League of Nations mandate, then as a U.N. Trusteeship.

170. Robert F. Rogers, Destiny's Landfall: A History of Guam (Honolulu: University of Hawai'i, 1995), p. 131.

171. Ibid., p. 131.

172. Op.cit., Robert F. Rogers, p. 131. (My brackets and emphases)

173. Ibid., p. 131.

174. Ibid., p. 206. NB: Robert F. Rogers lists forty-six nations signing the U.N. Charter; Stanley Meisler (1995) has fifty, p. 358. (See Section B)

175. Ibid., p. 206.

176. Ibid., p. 206.

177. Ibid., p. 206.

178. Ibid., p. 207. Author's note: Of intense interest to the Japanese, especially with the demise of the "Evil Empire", is the fate of those four southern Kurils – Habbomai, Etorofu, Kunashiri, and Shikotan. There are certain Japanese irredentists who also want all of Sakhalin Island (Karafuto) back from the Russians. (NB: Kuril or Kurile Islands/Chishima in Jpnse.)

179. Ibid., p. 207.

180. Ibid., p. 207. Author's note: Will a conspiracy theorist please tell me the actual circumstances surrounding Secretary Forrestal's (1892-1949) untimely death?

181. Op.cit., Robert F. Rogers, p. 207. (Brackets mine)

182. Ibid., p. 212

183. Yup; that Harold Ickes.

184. Ibid., p. 212. Author's note: Although crusading for Guamarian social-political rights is a platform worthy of full support, the now-familiar theme of White, well-heeled, liberal lobbyists fighting for "Indigenous Peoples" is a sorry old story. But at least somebodies, in Washington, D.C., were standing watch on the U.S. Navy.

Notes for part I (cont'd.)

185. David Stanley, <u>Micronesia Handbook</u> (Chico, Calif.: Moon Publications, Inc., 1992 ed.), p. 206.

186. <u>Ibid.</u>, p. 206.

187. <u>Ibid.</u>, p. 206.

188. Cited in <u>Pacific Magazine</u>, Vol. 25, No. 3, Iss. 140, May/June 2000, p. 20.

189. David Stanley, p. 206.

190. <u>Ibid.</u>, p. 99. Author's note: There go my retirement plans!

191. Joseph E. Fallon, 'Territorial Bliss', <u>Chronicles</u>, Vol. 22, No. 11, November 1998, p. 46. (My brackets)

192. <u>Ibid.</u>, p. 45. (My emphasis)

193. <u>Ibid.</u>, p. 46. (My brackets and contents)

194. <u>Ibid.</u>, p. 46. (My brackets and contents)

195. Bruce G. Karolle, <u>Atlas of Micronesia</u> (Honolulu: Bess Press, 1993), p. 92.

196. Earl Hinz, <u>Pacific Island Battlegrounds of World War II: Then and Now</u> (Honolulu: Bess Press, 1995), p. 83.

197. <u>Ibid.</u>, p. 84.

198. David Stanley, p. 233.

199. <u>Ibid.</u>, p. 234.

200. Op.cit., David Stanley, p. 234. (My brackets)

201. <u>Ibid.</u>, p. 234.

202. Bruce G. Karolle, p. 92.

203. David Stanley, p. 234.

204. <u>Ibid.</u>, p. 234. (My brackets) See also Reilly Ridgell, <u>Pacific Nations and Territories: The Islands of Micronesia, Melanesia, and Polynesia</u> (Honolulu: Bess Press, 1995 ed.), pp. 72-73.

205. Op.cit., David Stanley, p. 238.

206. Leslie R. Groves, <u>Now it Can Be Told: The Story of the Manhattan Project</u> (New York: Harper, 1962), p. 384.

207. <u>Ibid.</u>, p. 384. NB: There were two atomic tests at Bikini: Test Able with an aerial explosion, and Test Baker with a submarine explosion.

208. Charles R. Taylor, <u>World War III and the Destiny of America</u> (Nashville, Tenn.: Sceptre Books, 1979), p. 41.

Notes for Part I (cont'd.)

209. The Asia & Pacific Review 1987 (Lincolnwood, Ill.: NTC Business Books, 1986), p. 187.

210. Ibid., p. 187.

211. David Hanlon, Remaking Micronesia: Discourses over Development in a Pacific Territory, 1944-1982 (Honolulu: University of Hawai'i Press, 1998), p. 13. Author's note: Are you insinuating, Doctor, that some smart Marshallese leaders – born since the U.S. nuclear testing – are cynically milking the guilt-ridden Mainlanders, for all they can get?!

212. Ibid., p. 3. Cf. Robert C. Kiste, 'United States', Tides of History: The Pacific Islands in the Twentieth Century (Honolulu: University of Hawai'i Press, 1994), p. 230.

213. Ibid., pp. 229. (My emphasis)

214. Article 76 of U.N. Charter cited by Robert C. Kiste, pp. 229-230. NB: Kiste has drily stated, "The Department of War had won the day."

215. Robert C. Kiste, p. 231. NB: The Solomon Report was shelved after JFK's assassination on 22 November 1963.

216. David Stanley, p. 26.

217. Ibid., p. 26. (My brackets and emphasis)

218. Ibid., p. 26.

219. Robert C. Kiste, P. 231. (My brackets)

220. Ibid., p. 232.

221. Op.cit., p. 231. (Brackets and contents mine)

222. Ibid., pp. 231-232.

223. Thurston Clarke, Equator: A Journey (New York: Avon Books, 1990), pp.369-370. Author's note: One of the best travelogues ever!

224. Student Handbook, Vol. 5 (Nashville, Tenn.: The Southwestern Company, 1998), p. 671. Cf. Bruce G. Karolle, p. 1.

225. 'Micropae dia', Vol. 8, p. 103. ff.

226. Joseph E. Fallon, Deconstructing America: Immigration, Nationality and Statehood (Washington, D.C.: Council for Social and Economic Studies, 1998), p. 79. (My emphasis)

227. Francis X. Hezel, S.J., Strangers in Their Own Land: A Century of Colonial Rule in the Caroline and Marshall Islands (Honolulu: University of Hawai'i Press, 1995), p. 350. (My emphasis)

228. Ibid., p. 365.

229. Op.cit., p. 365.

230. Op.cit., p. 365.

Notes for Part I (cont'd.)

231. Ibid., p. 365.

232. Joseph E. Fallon, Deconstructing America, pp. 81, 83.

233. David Stanley, p. 97; cf. David Hanlon (1988).

234. Ibid., p. 170.

235. See Mark R. Peattie, Nan'yō: The Rise and Fall of the Japanese in Micronesia, 1885-1945 (Honolulu: University of Hawai'i Press, 1992 ed.), passim. Author's note: Peattie vividly describes the gallant Japanese defense of "Bloody-Nose Ridge", led by the formidable Col. Kunio Nakagawa. For the American view I consult my friend Ed Moriarty, formerly of the U.S. Marine Corps (now of Port Angeles), who fought at Peleliu. "Semper Fi," Ed!

236. David Stanley, p. 170.

237. Op.cit., p. 170. (Brackets and contents mine)

238. Elizabeth D. Rechebei and Samuel F. McPhetres, History of Palau: Heritage of an Emerging Nation (Koror, Republic of Palau: The Ministry of Education, 1997), p. 273, ff.

239. Ibid., p. 291, ff. Cf. David Stanley, p. 170.

240. P.F. Kluge, The Edge of Paradise: America in Micronesia (Honolulu: University of Hawai'i Press, 1993), p. 7.

241. Ibid., passim; Cf. Elizabeth Rechebei and Samuel McPhetres, p. 320, ff.

242. Rechebei and McPhetres, pp. 344-348.

243. Joseph E. Fallon, Deconstructing America, pp. 83, 100.

244. Op.cit., P.F. Kluge, p. 212.

245. Al Hulsen, 'Different Method of Compact Funding Seen', Pacific Magazine, Vol. 25, No. 5, Iss. 143, September/October 2000, p. 24.

246. Ibid., p. 24.

247. Ibid., p. 24.

248. Ibid., p. 24. Cited is Allen Stayman, special negotiator for Compact of Free Association, Bureau of East Asian and Pacific Affairs, Department of State. (Brackets and contents mine)

249. Ibid., p. 25.

250. Ibid., p. 25. Quoted is former U.S. Ambassador William Bodde, Jr.

251. Ibid., p. 25. From remarks made by Rep. Doug Bereuter (R—Nebr.)

252. Student Handbook, Vol. 5, pp. 669-670. Cf. David Stanley, pp. 273-274.

253. See note 169 in this Chapter.

Notes for Part I (cont'd.)

254. Student Handbook, Vol. 5, p. 669; David Stanley, p. 263.

255. Op.cit., David Stanley, p. 263.

256. See Earl Hinz, pp. 25-29 for U.S. Operation GALVANIC.

257. Student Handbook, Vol. 5, p. 670.

258. Earl Hinz, pp. 29-30. Author's note: The U.S., in the 1979 Treaty of Friendship, released all claims to the Phoenix and Line Islands to Kiribati. In 1982, Kiribati bought Fanning and Washington atolls back from transnational corporation Burns Philp.

259. Op.cit., David Stanley, p. 277.

260. Ibid., p. 274.

261. Al Hulsen, 'U.S. Not Likely to Part With Central Pacific Islands', Pacific Magazine, Vol. 24, No. 5, Iss. 137, September/October 1999, p. 36.

262. Ibid., p. 37.

263. Ibid., p. 36. Quoted is former U.S. Ambassador William Bodde, Jr. (See note 250)

Micronesia Bibliography

16. Fallon, Joseph E. Deconstructing America: Immigration, Nationality, and Statehood. Council for Social and Economic Studies, Washington, D.C., 1998.

17. Hanlon, David. Upon a Stone Altar: A History of the Island of Pohnpei to 1890. University of Hawai'i Press, Honolulu, 1988.

18. _____. Remaking Micronesia: Discourses over Development in a Pacific Territory, 1944-1982. University of Hawai'i Press, Honolulu, 1998.

19. Hezel, Francis X., S.J. The First Taint of Civilization: A History of the Caroline and Marshall Islands, 1521-1885. University of Hawai'i Press, Honolulu, 1994 ed.

20. _____. Strangers in Their Own Land: A Century of Colonial Rule in the Caroline and Marshall Islands. University of Hawai'i Press, Honolulu, 1995.

21. Karolle, Bruce G. Atlas of Micronesia. Bess Press, Honolulu, 1993.

22. Kluge, P.F. The Edge of Paradise: America in Micronesia. University of Hawai'i Press, Honolulu, 1993.

23. Peattie, Mark R. Nan'yō: The Rise and Fall of the Japanese in Micronesia, 1885-1945. University of Hawai'i Press, Honolulu, 1992 ed.

24. Rechebei, Elizabeth D. and Samuel F. McPhetres. History of Palau: Heritage of an Emerging Nation. The Ministry of Education,, Koror, Republic of Palau, 1997.

25. Rogers, Robert F. Destiny's Landfall: A History of Guam. University of Hawai'i Press, Honolulu, 1995.

Part II: UNCLOS III – The Third United Nations Conference on the Law of the Sea

"International law, as reflected in the United Nations Convention on the Law of the Sea [UNCLOS], offers the most appropriate and comprehensive framework as a legal instrument for the management of the worlds' oceans."
--From the UNO's AGENDA 21, adopted at the Rio Earth Summit in 1992[264]

...."Other proposed rules, if adopted, would establish a world government for the seabed beyond national jurisdiction, a regime that states have been unwilling to accept anywhere else on earth."[265]
--Robert E. Riggs and Jack C. Plano, The United Nations, 1994

"With its new Law of the Sea Treaty ... the United Nations is reaching control of – and a right to royalties on – the resources on the seabeds of the world's oceans...."[266]
"[T]he United States should reject the UN Law of the Sea Treaty."[267]
--Patrick J. Buchanan, A Republic Not an Empire, 1999

Prelude

The Law of the Sea has a centuries-old history. The deeply interested student of legal maritime history might well look up Latin phrases such as mare liberum ("free sea") and mare clausum ("closed sea"); research personages like Elizabeth I of England (1558-1603) and Holland's Hugo Grotius for the "common-sea" concept.[268] During modern times, in 1930 an international conference at the Hague (Netherlands) attempted to codify the Law of the Sea in a coherent treaty form, but failed to achieve accord.[269] Then came the terrible holocaust of World War II and its aftermath, where "freedom of the seas" lost all meaning. On 28 September 1945, U.S. President Harry S. Truman was decidedly quick to issue two policy directives on oceanic affairs:

(1) Proclamation No. 2667: "'[T]he government of the United States regards the natural resources of the subsoil and the seabed of the continental shelf beneath the high seas but contiguous to the coasts of the United States as appertaining to the United States, subject to its jurisdiction and control.'"[270] (For American oil interests)
(2) Proclamation No. 2668, according to Edward Wenk, Jr., of U. Washington, "dealt with coastal fisheries in relation to the high seas and served to carefully differentiate jurisdiction over conservation zones in the water column from that over resources on the seabed beneath."[271] (For the American fishing industry)

At this juncture (latter 1940s), certain West Coast nations of South America, having narrow continental shelves, declared their own 200-mile "fishing limit". The South American rationale was logically based on the 200-mile neutrality-zone mandated by FDR for the "protection" of western South America during the Pacific War.[272] So, largely motivated by the Truman Proclamations of 1945 (which had so much – if not more – to do with undersea drilling/mining as with fishing), the Convention on the Continental Shelf was framed at the (first) United Nations Conference on the Law of the Sea.[273] UNCLOS

I convened at Geneva, Switzerland, on 24 February 1958 and concluded on 29 April 1958; with 86 nations represented. UNCLOS I adopted four landmark Conventions, with three Conventions emerging from that initial Conference, plus a concluding Conference (UNCLOS II) held in 1960 (which concerned the territorial sea and contiguous zone, the high seas, fishing and the conservation of "living resources").[274] Strange to relate, in light of contemporary American resistance (in some quarters) to UNCLOS III, Edward Wenk, Jr., of U.W., wrote in 1972 that the U.S. Senate had "<u>casually</u> ratified the sea law treaty (UNCLOS II) on May 26, 1960, with scarcely a thought about ambiguities in text or other implications. Discussion on the Senate floor took less than half an hour."[275]

UNCLOS I was fully ratified on 30 October 1958. The four Conventions adopted were: 1. The Territorial Sea [and the Contiguous Zone], 2. The High Seas, 3. Fishing and Conservation of the Living Resources [of the High Seas], 4. The Continental Shelf.[276] Convention 1, Article 14, contained six Rules on the Right of Innocent Passage. The fifth Rule stated:

> "'Passage of foreign fishing vessels shall not be considered innocent if they do not observe such laws and regulations as the coastal State may make and publish in order to prevent these vessels from fishing in the territorial sea.'"[277]

Under Convention 2 (The High Seas), Article 2 declared – <u>inter alia</u> – the unequivocal "'Freedom of fishing'."[278] According to Ronald B. Kirkemo, who wrote <u>An Introduction to International Law</u> in 1974, <u>only</u> the Convention (2) on the High Seas satisfied the general principles of international law.[279] Was <u>that</u> one reason for the failure for the agreement on 1) the breadth of the territorial sea and fishing jurisdiction, b) the breadth of sovereign rights over the sea-bed resources of the Continental Shelf, or c) the dangers of marine pollution? Whatever the reasons for lack of accord at UNCLOS I, UNCLOS II (held a mere two years later in 1960) did no better to resolve the disagreements outstanding from UNCLOS I. In 1974, marine legal expert Ronald B. Kirkemo concluded that:

> ...."In the years since the 1958 and 1960 conferences supertankers have increased the danger of pollution from oil spills; methods of fishing have become more sophisticated, thereby increasing the danger of overfishing; technology for extracting the mineral resources of the Continental Shelf has increased; and states have enlarged their claims of jurisdiction over the seas, some claiming jurisdiction up to 200 nautical miles from shore. In response to these developments, another Law of the Sea Conference is scheduled for 1974.[280]

Much would transpire, however, before the nine-year-in-the making UNCLOS III (the Third U.N. Convention on the Law of the Sea) was signed and sealed by 117 nations on 10 December 1982.[281] As noted, the Second Conference (UNCLOS II) of 1960 had been unable to resolve the disagreements held over from UNCLOS I of 1958, and, during the 1960s, national jurisdictions – and claims – proliferated over the oceans. A number of states, most of them Latin American (no doubt following the earlier example of some South American countries), went so far as to claim a "territorial sea" (rather than a "fishing limit") of two hundred nautical miles.[282] If established – and enforced – these jurisdictions would control navigation and ocean-floor/sea-bed resources.

Enter now – 17 August 1967 – one Arvid Pardo, U.N. ambassador from Malta. For that was the day that Ambassador Pardo personally electrified the usually somnolent U.N. General Assembly. Tiny Malta had requested the inclusion, at the General Assembly's 22nd Session, of an item ponderously entitled: "'Declaration and Treaty Concerning the Reservation Exclusively for Peaceful Purposes of the Sea-Bed and of the Ocean Floor, Underlying the Seas Beyond the Limits of Present National Jurisdiction, and the Use of their Resources in the Interests of Mankind.'"[283]

Ah, but the oration was the thing! And Arvid Pardo's double-barreled proposal had them (the General Assembly) sitting up and paying attention: The UNO ought to (1) "internationalize" the sea-bed beyond some arbitrary limit of national jurisdiction set by a "particular interpretation" of the Convention on the Continental Shelf, or, if required, by its amendment; and that the UNO ought to (2) create a new agency to administer the internationalised sea-bed.[284] Ambassador Pardo climaxed his presentation by declaring that, once fixed limits had been drawn for national jurisdiction, the resources of the sea-bed beyond those limits would become "'the common heritage of mankind'".[285] These were words, and ideas, guaranteed to warm the cold hearts of those listening that summer day at the 22nd Session of the U.N. General Assembly. There would be six long years to 1973, but preparations were already underway for UNCLOS III ... with the initial impetus for the Conference directly attributable to Ambassador Arvid Pardo of Malta's seminal speech.[286]

UNCLOS III (1973-1982-1994)

"Beyond customary law, if the treaty should take effect as it stands, with a seabed authority empowered to govern and exploit an international undersea territory, it would be a large step toward international government in that limited area".
--Robert E. Riggs and Jack C. Plano[287]

On 16 November 1994, the Third United Nations Convention on the Law of the Sea became "enforceable" when Guyana became the 60th nation to ratify the Treaty.[288] This writer puts "enforceable" in quotes due to the unyielding opposition of American sea-bed mining interests, for UNCLOS III, (Convention 4) remains a paper-tiger without U.S. might and muscle behind the Treaty. But, with the U.S. Armed Forces behind it or not, most Convention policies have been integrated into the corpus of international law since 10 December 1982.[289] (And, although UNCLOS III established the 200-nautical mile [371-km] EEZ for "coastal States", about two-thirds of the earth's oceans are not under [any] national jurisdiction. This fact – "jurisdictional gap" – is especially significant for the "fishing States" of the globe.)[290]

UNCLOS III essentially kept, but also changed and expanded, the four Conventions written up in 1958 at UNCLOS I; then added comprehensive new provisions on: 1) An exclusive economic zone (EEZ), 2) navigation, 3) access to the seas for land-locked nations, 4) regimes for both national and trans-national ("common") areas, 5) scientific marine research, 6) protection of the marine environment, 7) exploitation of [the] living resources, 8) sea-bed/ocean-floor drilling and mining, and 9) the vehicle for settling international disputes.[291] Notable changes from previous law, written into the new Treaty, included: Establishment of a fixed "territorial sea" of up to 12 nautical miles (22 kilometres), over which a coastal State exercises national sovereignty (subject to the right of "innocent passage); with an additional 12 nautical miles of contiguous zone over which a coastal State has control necessary for maintaining (national) law, order, and safety. The Convention also provides a 200-nautical mile EEZ (as we have seen) within which coastal States control the economic resources of the sea – and the subsoil beneath – but not jurisdiction over navigation.[292]

There was more, much more; mostly to do with marine mineral rights and royalties, and undersea drilling and mining. As FutureFish 2001 is a book on Pacific fishing, marine "living resources" are of greater interest to us, writer and reader alike, than sea-bed mineral deposits. However, as stated, circa two-thirds of the earth's oceans are outside all national jurisdictions. Ergo, the most troublesome provisions of the new Treaty would be those clauses setting aside the sea-bed/ocean-floor beyond (any) national jurisdiction (i.e., "jurisdictional gap") as… "'the common heritage of mankind'" – in the exact words of the Maltese ambassador to the UNO, Arvid Pardo. Not only that, but Ambassador Pardo had called for a (U.N.-sponsored) International Sea-Bed Authority to administer the "'Area'."[293]

As alluded to earlier, the U.N. ocean-floor provisions were strongly opposed by American sea-bed mining interests; so the United States, under the (Republican) Reagan administration (1981-1989), voiced qualms about an "International Sea-Bed Authority"... and subsequently refused to sign the Treaty. By November 1992 (and the final full year of the Bush administration), 53 nations had ratified – still seven short of the 60 necessary for the Convention to take effect.[294] By November 1994, Guyana became the 60th – and the last – nation to ratify the Convention, which has officially been in effect since then. But November 1994 was also the mid-term election of an anti-Clinton/anti-UNO 104th U.S. Congress.[295] As conservative columnist Cliff Kincaid has commented, "U.N. proposals such as the Convention on Biological Diversity and the Law of the Sea Treaty were clearly going to have a difficult if not impossible time moving through he Senate."[296]

And thus has it gone at the U.S. Congress, but the beat goes on at the U.N. Organization....

UNCLOS III: Effects on Mid-Pacific Tuna Fisheries

"All nations should act to ensure that marine living resources under national jurisdiction are conserved and managed in accordance with the provisions of the United Nations Convention on the Law of the Sea."
--From the UNO's AGENDA 21, adopted at the Rio Earth Summit in 1992[297]

Living in Honolulu, Hawai'i, as I did for four years (1978-1982), this writer (a U.H. student then) became acquainted with the Hawaiian tuna-fishing industry. I quickly discovered that the tunas were as significant a marine resource to the Central Pacific Islands as the salmons are to the Pacific Northwest Coast. And tombo (albacore), aku (skipjack), ahi (bigeye) and yellowfin ahi were all tunas found fresh and readily available at local fish markets and seafood stores. Other local pelagic fish of high quality were mahimahi (dolphinfish) and ono (wahoo). Off Kailua-Kona, on the Big Island of Hawai'i (my favourite place in the Aloha State), Mainland and Japanese sport anglers sought yet other pelagic giants as game fish: Hebi (shortbill spearfish), kajiki (Pacific blue marlin), nairagi (striped marlin), and sometimes even the mighty shutome (broadbill swordfish).[298] Instead of Mainland meat and potatoes, the Island diet was [said to be] fish and rice (not to mention sushi and sashimi). I will identify and examine the Central Pacific tuna fisheries in Part III of this Chapter.

UNCLOS III, U.S. Congressional approval or not, has been part and parcel of international law since 10 December 1982, and had been signed and sealed (but not yet delivered) on that date by fully 177 states of the United Nations. After the text of the New Convention had been adopted by UNCLOS III, the focus turned from the Conference negotiations to the problems of interpretation and implementation of the Treaty once it became ratified (which, of course, UNCLOS III has been officially since 1994). Therefore, during 1982, the U.N. in the form of the Food and Agriculture Organization (FAO), commissioned a legal paper under the FAO Fisheries Law Advisory Programme as a background paper for the South Pacific Forum Fisheries Agency (FFA). The FFA held a Workshop on Fisheries Access Rights Negotiations at Port Vila (Vanuatu/New Hebrides) in September 1982, and the FAO a Regional Seminar in Monitoring, Control, and Surveillance of Fisheries in Exclusive Economic Zones at Mahé (Seychelles); also in September 1982.[299] The paper commissioned by the FAO was written by one W.T. Burke, an American Professor of Law at the University of Washington (Seattle), and also an FAO consultant in 1982. The FAO decided to republish Professor Burke's paper in their Legislative Series (no.26) ... "in view of its general interest and applicability to tuna fisheries throughout the world."[300]

W.T. Burke, in his legal paper, 'Impacts of the UN Convention on the Law of the Sea on Tuna Regulation', starts right off claiming all views therein as his own and not necessarily those of the FAO. (A good start!) After this auspicious beginning, Burke gets right to his main thesis:

"The Treaty includes additional provisions dealing with certain specific species: highly migratory, anadromous (salmon being the most important), catadromous, sedentary, and marine mammals....
"Highly migratory species (HMS) are singled out for special treatment for obvious reason: they move considerable distances in the ocean, sometimes spanning several zones of national jurisdiction as well as crossing vast expanses of high seas. The [T]reaty thus has a separate provision not only for the area within national jurisdiction but also for high seas beyond. The treatment of HMS within the EEZ is widely discussed, <u>but it is seldom noticed that the [T]reaty also provides a special high seas regime as well, one which appears to work a radical change in the traditional right of freedom to fish.</u>"[301]

Burke then proceeds to analyse the convention EEZ fisheries provisions pertaining to coastal fisheries rights, with the "specific perspective" of HMS (highly migratory species) in mind. This writer shall attempt to cut through all Burke's academic "legalese"; trying then to make <u>fair</u> sense of his "specific perspective" on UNCLOS III and its impact on tuna regulation.

According to the FAO, the greatest possible effects on tuna management are contained in Article 64 of the Convention. The implications of Article 64 are many-fold. Among them are: 1) The requirement that coastal and fishing States <u>co-operate</u> regarding conservation, and 2) the optimum utilization of specified HMS both within the region and beyond the EEZ (Burke: 1982). I have underlined the word "co-operate" as it a buzz-word in any binding treaty – especially a U.N.-sponsored treaty – that's fraught with global ramifications concerning the independence and sovereignty of States. As to Article 64 of UNCLOS III, Burke questions the exact meaning of "co-operation" and the scope of its subject matter: When the duty for co-operation is fulfilled; the consequences of failure (or termination) of co-operation; the forms and nature of co-operation; and (of course) the applying of co-operation to fishing in high seas enclaves. On the "duty" for co-operation, Burke writes:

"The coastal State has final decision making authority in the exercise of its sovereign rights over HMS but instead of exercising that authority and employing discretion, it has the duty of cooperating with other coastal States and with distant water fishing nations (DWFN) to achieve conservation and optimum utilization..."[302]

Under Article 64, the scope of this co-operation involves the marine resources and fishing activities for HMS both within the 200-nautical mile EEZ and beyond – that is, the fish stocks of the entire "region" relevant to the coastal States involved: "In some instances this may include a relatively large number of States, many coastal and some DWFN, and an enormous geographic area, as in the South and West Pacific and Indian Ocean" (Burke: 1982).

The reverse side of the duty for co-operation is the failure (or termination) of co-operation. Coastal States have to decide – make choices – in establishing a TAC (total allowable catch), adopting conservation measures and management to ensure against over-exploitation, determining optimum yield (OY), and promoting the proper "allocation of resources". According to Burke, if co-operation between coastal States fails to produce the desired results, "Article 64 allows coastal States of a region to join in making decisions regarding a stock of common interest that is available for harvest in each of their zones. Fishing States who seek access to this stock must operate in accordance with the decisions of the coastal States."[303]

Halfway through his legal paper for the U.N. ("[a]n FAO EEZ Programme Activity"), W.T. Burke finally arrives at the knotty/thorny problem of Treaty enforcement. The American legislative scholar's sub-heading, although carefully worded, contains the first hint of a fist in the hitherto velvety glove; thus:

"(iii) Manner and form of cooperation: direct interaction and international organizations (IO).
"Article 64 poses important questions about the manner and form of cooperation including (a) whether it permits a choice between direct cooperation and use of an international organization (IO) or whether an IO must be employed even if direct cooperation is also preferred, and (b) what characteristics or elements of an IO are required by Art. 64, if any."[304]

If there had been a fist (viz., U.N. agendum) hidden in the velvety glove, it wasn't that of W.T. Burke! Indeed, FAO legal consultant or not, Burke first disassembles Article 64 generally and then analyses two clauses in particular. Burke strongly criticizes the very first two sentences of Article 64 (1) as mutually excluding each other. Concerning how States – both "coastal" and "fishing" (from afar, DWFN) – are to co-operate respecting HMS, the first sentence of Art. 64 (1) states:

"'[T]he coastal State and other States whose nationals fish in the region shall cooperate directly or through appropriate international organizations.'"

Burke quite rightly translates this as meaning that the States involved (both coastal and fishing) may deal directly with each other in one-on-one interactions, or in a manner involving one group interacting with another State or group of States. But the second clause of Art. 64 (1) directs that:

"'In regions for which no appropriate international organization exists, the coastal State and other States whose nationals harvest these [highly migratory] species in the region shall cooperate to establish such an organization and participate in its work.'"

Burke contends that the latter sentence seems to contradict the former, in that the first sentence of Art. 64 (1) provides that States co-operate directly OR through "'appropriate international organizations [IOs]'."[305] And Burke takes his critique even further:

"An interpretation that Article 64 mandates cooperation exclusively through an IO is not credible for several reasons. First, the use of the disjunctive in the first sentence of 64 (1) is consistent with leaving the final choice of the form of cooperation to the States concerned. Second, the notion that only an IO may be employed for cooperation is inconsistent with all previous and current practice. The pattern on the several decades since World War II has been to employ simultaneously direct cooperation between States and international organizations. It would be surprising if the drafters of Article 64 believed that in the future all States would find that cooperation were possible only if they cr[e]ated a separate institution and refrained from direct interaction. Such an expectation by the drafters is simply not plausible."[306]

On Article 64, Burke concludes that while Art. 64 ought not be seen as requiring coastal and fishing (in far waters, DWFN) States to cooperate solely via IOs, the second clause in 64 (1) surely appears to envision just that. Burke gets frustrated at this juncture: "The difficulty is that Article 64 does no more than that – it offers no details, no direction, no concrete guidance into the operational significance of cooperation through IOs." The confusing wording of the convention, surmises Burke, probably reflects the truly fundamental differences between coastal and fishing States in the UNCLOS negotiations. Burke clinches his FAO Legislative Study no. 26 of Article 64 with a Treaty interpretation in favour of national sovereignty in respective EEZs:

"The outcome seems heavily tilted toward the coastal State position because Art. 64 does include HMS within the sovereign rights of the coastal State in the EEZ, but barely mentions IOs or give[s] them any power or responsibility. Since <u>nothing</u> is agreed about the operational details of the IOs, <u>it is up to the States directly concerned</u> (coastal and fishing States) to develop such international bodies as they can agree on .....

...."There is <u>no single</u> 'appropriate' IO."[307]

However, the reader might immediately ask, what about the FAO/UNO <u>itself</u> as "single" IO? This writer can only reply that the major sections of UNCLOS III, although fully ratified only since 16 November 1994, have been embedded in international law since 10 December 1982. (And W.T. Burke's FAO legislative study was published in 1982.) <u>That</u> year was Year Two of the Reagan administration (1981-1982), and the political atmosphere in the United States was very different from that of the Carter administration (1977-1981); four years of disco, chicken diplomacy, and "Gullible's Travels". The new ("Ray-Gun") American truculence in foreign affairs extended to the United Nations: President Reagan's like-minded political cohorts bridled at the U.N.'s call for an "International Sea-Bed Authority". The upshot was that the United States didn't sign UNCLOS III in 1982, and American political conservatives (plus corporate ocean-floor mining/drilling interests) have subsequently kept the U.S. from signing both UNCLOS III and the Convention on Biological Diversity.[308]

The second query the reader might have is, Doesn't Article 64 of UNCLOS III <u>support</u> the sovereign rights of "coastal States" in their EEZ? My answer would be: Unreservedly <u>Yes</u>! Professor Burke's very thorough examination of UNCLOS III, especially of Article 64, leaves no doubt as to the intent of the Treaty. So for the tiny but widely-scattered islands of Micronesia straddling the vast Central Pacific (teeming with fish), the sovereign rights of coastal States are crucial indeed – especially regarding highly migratory species (HMS). For "fishing States" (distant water fishing nations, DWFN) – like Japan, South Korea, and Taiwan – however, UNCLOS III acts as a geopolitical bridle to their hitherto unrestricted harvesting (and often poaching) of fish in the world's oceans (mostly the wide Pacific Ocean). The U.N. further restrained the DWFNs of Japan, South Korea, and Taiwan in late 1991, by enacting a global ban on high-seas driftnet fishing (again, mostly in the Pacific).[309]

Thus, for the fishing States (small nations with large fleets – egs., Japan, South Korea, Taiwan, Spain with her enormous Vigo-based deepwater fleet), UNCLOS III is decidedly <u>bad</u>. But for coastal States (big or scattered island countries with large EEZs – egs., Canada, U.S.A., formerly-American Micronesia, Kiribati), UNCLOS III is good, very good (<u>for now</u>!). But fishing and coastal States alike should remain forever vigilant to the ever-pressing UNO ambitions for global governance and "internationalizing" the planet's seaways. Blue-helmeted and armed planeteers can just so easily be mustered at sea as deployed on land, even though (at last look) this writer has seen no signs (yet) of a United Nations navy! If the U.N. Ambassador from Malta in 1967 could call for <u>setting aside</u> the sea-bed/ocean-floor <u>beyond</u> any national jurisdiction (<u>viz</u>., "jurisdictional gap") as "the common heritage of mankind" (to be administered, naturally, by the UNO), the same might happen – any day now – to "jurisdictional gap" high seas fishing grounds.[310] With their post-Pacific War American connections (except Nauru, Kiribati) and half-century know-how of United Nations methodology, the small land-poor coastal States of Micronesia are in a prime position to shore up a near-monopoly of HMS fishing in the Central Pacific Ocean. But the Islanders must act soon, very soon (while there is still time), or their golden opportunity will be lost to others, forever. And <u>that</u> is the thesis of this Chapter!

Notes for Part II

264. See Daniel Sitarz, ed., Agenda 21..., p. 145. (Brackets mine)

265. Robert E. Riggs and Jack C. Plano, The United Nations..., p. 219.

266. Patrick J. Buchanan, A Republic Not an Empire: Reclaiming America's Destiny (Washington, D.C.: Regnery Publishing, Inc., 1999), P. 387.

267. Ibid., p. 389.

268. Wesley Marx, The Frail Ocean (New York: Ballantine Books, 1970), p. 190. NB: Ch. 15, 'The Queen, the Dutchman, and Mare Liberum.'

269. Robert E. Riggs and Jack C. Plano, p. 220.

270. Cited by Edward Wenk, Jr., The Politics of the Ocean, p. 253-254.

271. Op.cit., Edward Wenk, Jr., p. 254.

272. Cf. Wesley Marx, Ch. 16, 'Flag-Flying Fish', passim.

273. See Edward Wenk, Jr., p. 255.

274. Ibid., p. 255; cf. Benjamin B. Ferencz and Ken Keyes, Jr., PlanetHood: The Key to Your Future (Coos Bay, Ore.: Love Line Books, 1991), p. 165.

275. Op.cit., Edward Wenk, Jr., p. 257. (My emphasis and brackets)

276. For the United States Convention 4, the Continental Shelf, was (and continues to be) the most troublesome. At UNCLOS I, no agreement had been reached on the breadth of sovereign rights over the seabed resources of the Continental Shelf (Kirkemo: 1974). Cf. Riggs and Plano, p. 220.

277. Cited in Ronald B. Kirkemo, An Introduction to International Law, p. 41.

278. Ibid., pp. 49-50.

279. Ibid., p. 71.

280. Op.cit., Ronald B. Kirkemo, pp. 71-72. NB: The first brief session of UNCLOS III kicked off during December 1973 – with the final session closing in December 1982. Cf. Benjamin B. Ferencz and Ken Keyes, Jr., p. 168.

281. Stanley Meisler, 'Appendix I', United Nations: The First Fifty Years (New York: The Atlantic Monthly Press, 1995), p. 363.

282. Robert E. Riggs and Jack C. Plano, pp. 220.

283. Cited by Edward Wenk, Jr., p. 260. Author's note: Whew! Let's hope Ambassador Pardo trimmed some Maltese "logorrhoea" while speechifying!

284. Ibid., p. 260. Author's note: Was Arvid Pardo's mother Swedish? Only in Sweden has this writer heard the name "Arvid". (Those Scandinavian women sure do get around!)

285. Robert E. Riggs and Jack C. Plano, p. 220.

Notes for Part II (cont'd.)

286. Wesley Marx, pp. 248-249.

287. The United Nations…, pp. 221-222.

288. C.D. Bay-Hansen, FutureFish 2000 (Victoria, B.C.: Trafford Publishing, 2000), p. 130. (Cf. note 259.)

289. Ibid., p. 130. (See note 258 in FutureFish 2000.)

290. Ibid., p. 141. (See note 292 in FutureFish 2000 for "jurisdictional gap")

291. Robert E. Riggs and Jack C. Plano, p. 220. Cf. Elliot A. Norse ed., Global Marine Biological Diversity (Washington, D.C. and Covelo, Calif.: Island Press, 1993), passim.

292. Ibid., p. 220-221; cf. Elliot A. Norse, passim.

293. Ibid., p. 221.

294. Ibid., p. 221.

295. Cliff Kincaid, Global Bondage: The U.N. Plan to Rule the World (Lafayette, La.,: Huntington House Publishers, 1995), pp. 99.

296. Op.cit., pp. 99-100. Author's note: The Cato Institute, a libertarian think-tank, even prepared a handbook for the 104th Congress. In the Cato handbook were four anti-U.N. steps, one of which advised "'…refuse to ratify the Law of the Sea Treaty and reject similar schemes if they arise….'"(p. 100)

297. Daniel Sitarz, ed., AGENDA 21…, p. 159.

298. Poster of "Hawaii Seafood". State of Hawaii, Dept. of Business, Economic Development & Tourism, Ocean Resources Branch, Honolulu, 1993.

299. Dante A. Caponera, 'Foreword', p. iii; in W.T. Burke, Impacts of Tuna Regulation….

300. Ibid., p. iii. Author's note: The FAO's decision has been a Godsend for this writer as Burke's paper has been the sole resource I could find on this complicated but important topic!

301. Op.cit., W.T. Burke, Impacts on Tuna Regulation…, p. 1. (My emphasis)

302. Ibid., p. 5. (Brackets and emphasis mine) NB: "Optimum utilization" is an FAO-type term used to refer to allocation of resources. See measures identified as "conservation" under Article 61 and "optimum utilization" by Article 62. (Burke, p. 6)

303. Ibid., p. 7. NB: Under Article 64, the coastal State "'shall determine the allowable catch' of fish within its EEZ". (Burke, p. 2) Author's note: So far so good!

304. Ibid., p. 10. Author's note: Article 64 in the Convention is the main Article having to do with coastal States' EEZs, high seas fishing, and highly migratory species (HMS). Burke's focus is centred almost entirely on Article 64, but related Articles 61 and 62 are briefly highlighted; and mentioned are Articles 56, 69, 70, 73, 87, 116-119, and 297 as pertinent. (My emphasis)

305. Ibid., p. 11. (My brackets and contents)

Notes for Part II (concl'd)

306. Op.cit., W.T. Burke, p. 11. (His emphases)

307. W.T. Burke, pp. 11-12. (My emphases)

308. NB: Despite two pro-U.N. U.S. administrations since 1989, this writer has neither seen nor heard any news of the United States signing UNCLOS III or the Convention on Biological Diversity. (Author's note: I could well be wrong on the latter!)

309. See FutureFish 2000, Ch. 2, Pt. III, Sec. C for further details. Author's note: In 1994, our own Professor W.T. Burke (along with two other Seattleites) co-authored a discordant study on the U.N. ban, 'United Nations Resolutions on Driftnet Fishing: An Unsustainable Precedent for High-Seas and Coastal Fisheries Management.' (This took guts in 1994!) Cf. 'Study Criticizes U.N. Driftnet Ban', Pacific Fishing, September 1994.

310. An example of "jurisdictional gap" familiar to the Northwest Coast reader is that area known as "the Donut Hole" in Bering Sea – surrounded by American and Russian EEZs but still outside either jurisdiction. NB: Approximately two thirds of the earth's oceans are not under any national jurisdiction.

Main References for UNCLOS III

A. Burke, W.T. 'Impacts of the U.N. Convention on the Law of the Sea on Tuna Regulation'. Food and Agriculture Organization of the United Nations, Rome, 1982.

B. Kirkemo, Ronald B. 'The Sea', An Introduction to International Law. Nelson-Hall, Chicago, 1974.

C. Riggs, Robert E. and Jack C. Plano. 'The Law of the Sea/UNCLOS III', The United Nations: International Organization and World Politics. Wadsworth Publishing Co., Belmont, Calif., 1994, 2nd ed.

D. Sitarz, Daniel, ed. 'The Oceans', AGENDA 21: The Earth Summit Strategy to Save Our Planet. Earthpress, Boulder, Colo., 1993.

E. Wenk, Edward, Jr, Ch. 6, 'One Sea and One World', The Politics of the Ocean. University of Washington Press, Seattle and London, 1972.

Part III: Tropical Western and Mid-Pacific Fisheries

"A great fishery requires an abundant stock, and the rapid growth and turnover of the yellowfin meets this requirement. The fish must also be nutritious and appealing. The protein content of tuna matches that of red meat without the accompanying fat calories. It contains iodine to prevent goiter, fluorine to thwart tooth decay, phosphorous to build bone, and vitamin B_{12} to enrich red blood cells. The yellowfin perfectly exemplifies the nutritious richness of the ocean, a richness many food authorities see as the key to man's survival on a congested planet."
-- Wesley Marx, The Frail Ocean, 1970 [311]

" 'Over the next fifty years, the world's population is expected to double [and]...rational management of the world's tuna ... in an even more crowded and competitive world will not be easy' ".
-- Fisheries expert Dr. James Joseph, head of the Inter-American Tropical Tuna Commission, quoted in 1998 [312]

Introduction: The Wide Pacific And Scattered Micronesia

The reader ought now to spread a map of the Pacific Ocean in front of him/her. Starting at Hawai'i, let the eye of the reader scan directly southward from the Aloha State (between ca. 20° N. and the Tropic of Cancer) until reaching the Line Group, approx. 5° North of the equator. Excepting the U.S.-held Kingman Reef, Palmyra Atoll, and Jarvis Island, the Line Group is a geopolitical part of the Republic of Kiribati. Directly south of Hawai'i nei, the Line Group lies between about 150° and 160° West Longitude. As our eye travels due west just south of the Equator, we reach 170° West Longitude and the Phoenix Islands. Except for the American Howland and Baker Island[s], we are still in Kiribati. Indeed, to attain the western extremity of the Republic of Kiribati, we must cross the International Dateline and journey beyond 170° East Longitude to finally arrive at Ocean Island (Banaba)!

The vast, watery breadth of Kiribati, added to the huge oceanic swathes cut by the neighbouring Republic of the Marshall Islands (RMI) and the Commonwealth of the Northern Mariana Islands (CNMI) next-door, all contain the major sector of the Central Pacific Ocean just north of the Equator. Add to these Micronesian borders (to the west of Kiribati, towards the Asian mainland) the Nauruan EEZ, the Federated States of Micronesia (FSM), and the Republic of Palau (directly north of western New Guinea) ... and the Micronesian areas, with their aggregated archipelagoes, encompass a goodly portion of the Western Pacific Ocean too. All in all, the geographic evidence printed on my "[The] New Pacific" [313] wall-map is impressive: Like a giant, maritime crescent stretching from south of Hawai'i (Teraina islet in Kiribati) to directly north of Irian Jaya (Tobi islet off Palau), Indonesia, Micronesia – area EEZs easily straddle 150° West Longitude (westward) to 140° East Longitude. That is truly mind-boggling!

The Pacific itself is a mind-boggling ocean. Easily the planet's largest body of water, the Pacific is also (mostly) the deepest. The great oceanic trenches are all here -- the Kermadec, Kurile-Kamchatka, Japan, Marianas, Tonga, and the Philippines (Trench). Until the middle of the twentieth century, the Philippines Trench was thought to be the world's deepest submarine canyon. Then it was discovered that the Philippines Trench was exceeded by three other "vast furrows" -- the Marianas, Kurile-Kamchatka, and Tonga Trenches -- all in the Pacific! In fact, the Marinas Trench, or Challenger Deep, is the absolute abyss in all the oceans, with its actual (recorded) depth altered several times since (so recently as) 1951.[314] (In Micronesian waters no less!)

As for the circulation system in the North Pacific, the North Equatorial Current is jump-started by the trade winds and maintained in a permanent east-west flow. The North Equatorial Current heads past the Philippines, flows north past Taiwan and then east of Japan. Here the North Equatorial Current becomes the kuroshio ("black current"), the "Asiatic Gulf Stream". Carried across the wide Pacific to the shores of North America, the kuroshio warms the coasts of Southeast Alaska -- which are in sharp contrast to the frigid interior of the Last Frontier. We of the Pacific Northwest and British Columbia live in an "Evergreen Playground" due to the temperate influence of the kuroshio. Flowing southward along the American West Coast as the California Current, the living waters turn west (again) to join the North Equatorial Current off the rocky littoral of Baja California, Mexico.[315]

The Tunas: Tribe Thunnini, Family Scombridae

" 'A shortage of tuna raw material continue[s] to overshadow the market' " -- The United Nations' FAO, quoted in 1998 [316]

The "highly migratory species" (HMS), constantly referred to in UNCLOS III and by W.T. Burke in his legal paper for the FAO, are (of course) none other than those pelagic species of commercially – fished tunas. (See Appendix I.) The four species of tunas important to this study are: 1) Albacore/tombo (Thunnus alalunga), 2) bigeye/ahi (Thunnus obesus), 3) skipjack/aku (Katsuwonus pelamis), and 4) yellowfin/ahi (Thunnus albacares).[317] But there are nine other true tunas -- black skipjack, blackfin, bluefin, bullet and frigate mackerels, kawakawa, little tunny, longtail, and southern bluefin. (See taxonomy and local Pacific names in Appendix I.) All the true tunas share certain features; the most noticeable being the nearly round body in cross-section, rather than compressed, and being elongate and fusiform in profile.

The first dorsal fin is high in front, and occasionally sweeps down in a concave curve posteriorly, rather than merely tapering off. The second dorsal fin and anal fin are alike in size and shape, and each is similarly curved. (The yellowfin possesses extremely elongated lobes.) [318]

The scombroid, bullet-like body of the warm-blooded tuna is driven by its powerful, deeply-forked tail -- the long-lobed caudal fins. On both sides of the caudal peduncle are strong lateral keels which are supported by prominent bony extensions of the vertebrae. Where the caudal peduncle ends, each lateral keel is followed by two smaller ridges. The almost hot-blooded tunas possess a kind of "heat exchanger" that permits the fish to maintain a body temperature sometimes 18° F (10°c) higher than that of surrounding waters. According to ichthyologists at the Woods Hole Oceanographic Institution at Cape Cod, Mass., a 10° F rise in body temperature translates into a three-fold power and response of muscle-mass.[319] As the International Game Fish Association reported in 1982 on the giant bluefin tuna:

"Bluefin tuna (which grow to weights of 1,000 pounds -- 454 kg --- and more) have been clocked at speeds of up to 43.4 miles per hour in bursts of 10 to 20 seconds duration. It is little wonder that these fish are among the most popular species that anglers encounter!" [320]

Of the Pacific Ocean commercially-harvested tunas, the albacore/tombo is easily distinguished by its very long pectoral fins (vid., alalunga), which extend beyond the anal fin. Thus alternate names for the albacore are longfin tuna and long-finned tunny. The fins are dark-yellowish in colour, except for a white, trailing edge on the tail. The anal finlets are dark. The albacore feeds on anchovies, mullet, sardines, sauries, squid and other small fish. Trolling with feathered jigs, lures, and spoons is the favoured method of fishing; whole-bait fishing is also employed.[321]

The bigeye/ahi does appear to have a big eye! The first dorsal fin is a deep yellow, the second dorsal fin and anal fin are blackish-brown or yellow (and may be edged with black), and both dorsal and anal finlets are bright yellow with narrow black edges. The bigeye feeds on crustaceans, mullet, sardines, small mackerels, squids, and some deep-water fish species. Ways of fishing include deep-water trolling with mullet, squid, and other small baits; or with artificial lures and live-bait fishing. The mass-market importance of bigeye translates into a huge Pacific commercial fishery utilising methods of longlining and purse-seining.[322]

The gregarious skipjack/aku may form schools of 50,000 or more fish. The skipjack is recognised by the lack of markings on its back, but the lower flanks are silvery with four to six prominent, dark lateral stripes running from about the pectoral fins towards the caudal peduncle. The "surface-loving little skipjack" feeds high up in the water column on clupeiods (herring-like species), crustaceans, euphasiid shrimps, lanternfish, small scombroids (mackerel-like fishes), and squids. Skipjack will strike trolled feathers, plugs, spoons, strip-baits, and small whole-baits. Commercially significant, skipjack are also taken individually via casting, jigging, or live-bait fishing offshore.[323]

And so we come to our fourth and final tuna under study, the yellowfin/ahi. The "bright yellowfin" is, according to the IGFA, arguably the most colourful (thus most beautiful) of all the tunas:

"The back is blue-black, fading to silver on the lower flanks and belly. A golden-yellow or iridescent blue stripe ... runs from the eye to the tail. All the fins and finlets are golden yellow, though in some very large specimens the elongated dorsal and anal fins may be silver edged with yellow. The finlets have black edges. The belly frequently shows as many as 20 vertical rows of whitish spots." [324]

The yellowfin feeds on locally-abundant small fishes and on crustaceans, flying fish, and squids. Preferred fishing methods are trolling with small fishes, squids, and other troll-baits including artificial lures and strip-baits. Chumming and live-bait fishing methods are also used. With flesh almost so light in colour and texture as that of albacore, yellowfin is an extra-valuable commercial tuna, with hundreds of thousands of tons harvested yearly by longliners and purse-seiners.[325]

The latter three tuna species described above are the significant highly migratory species (HMS) of the tropical Central-Western Pacific (CWP), and represent the economic future -- and hope -- of the Micronesian commercial fisheries. Of the four species of tunas under our study, albacore is included as it is 1) important to U.S. fishermen, and 2) marketed as the "white meat tuna" highly regarded by North American consumers. Albacore/tombo also seems to head any list of preferred tunas sought in Hawaiian waters ... and often off the Pacific Northwest Coast. Otherwise, the three tunas of real commercial significance in the warm-temperate sectors of the Central-Western Pacific are, as stated, bigeye/ahi, skipjack/aku, and yellowfin/ahi.[326] All three tunas are similarly schooling, pelagic, and seasonally-migratory species. In fact, the bigeye was once not recognised as a separate tuna species, but thought to be a variation of yellowfin.[327] And the far-ranging, small but-swift skipjack often schools in the Pacific with yellowfin. Hence the three tunas appear to be irrevocably intertwined -- both as fish and as food. In the United States, bigeye, skipjack, and yellowfin are all canned and sold together as "chuck light tuna".[328] So the three tunas plus albacore, debateably the Pacific's most harvested food-fish, have (like the salmon) sustained enormous commercial fishing pressure for many years. In 1998, concerned fisheries expert Carl Safina wrote:

> "By the late 1980s, about four tons of tunas were yanked from the oceans each minute. In 2000 it will be twice that. The bright yellowfin and the surface-loving little skipjack tuna bear the brunt of this -- about 75 percent, by weight, of the kill. Bigeye tuna, a sushi favorite, and albacore, beloved by American moms preparing sandwiches, absorb 20 percent or so...."[329]

Tuna Fisheries of the Central-Western Pacific Ocean

> "The proud California fishermen in their streamlined clippers were forced to rely on a grueling, primitive method of fishing. Once the seabirds had led the clipper to a tuna 'boil', a crew member would 'chum' (toss out live anchovetas or other bait fish scooped inshore and kept in bait wells). Other crew members unhinged ramps on the side of the clipper. They stepped down onto these ramps, their legs awash in the Pacific, their hands gripping bamboo poles that dangled a barbless hook shrouded in chicken feathers. The feather lures were flicked into the boil.. The frenzied tuna failed to discriminate between the lures and the anchovetas. Once he felt a wrench, the fisherman, like a human crane, would arc a forty-pound tuna overhead. At the top of the arc, the fisherman would release the barbless hook by simply letting up on the tension. The world's fastest fish plopped onto the clipper deck as the released hook was whipped back into the tuna boil. This feat was accomplished in one sweeping motion and repeated for hours on end by a dozen men knee-deep in the Pacific. In schools of large-sized tuna, as many as five fishermen would link their poles to a common lure, hook a tropical tuna, pull together, and arc an iridescent, hundred-pound torso over their backs to the clipper deck. The tuna, their mouths bleeding, their iridescence fading to gray, drummed on the deck with their tail fins -- now just cargo in the brine tanks, awaiting a quick freeze...."[330]
> - - Wesley Marx, The Frail Ocean, 1970 (From a description of traditional California commercial tuna-fishing methods)

> "At present a purse seine fishery is developing in the [CWP] area and this fishery is largely dependent on floating objects for its catches. Tuna schools can be caught with greater ease in the eastern Pacific because the fish will not dive to escape the encircling net; a shallow and abrupt vertical temperature gradient (thermocline) through which the tuna are reluctant to pass restrict them to shallow water. In the western Pacific, however, this thermocline is deeper and allows the tuna to dive beneath the net to escape.
> "In the western Pacific, consequently, tuna seiners set their nets mainly around schools associated with floating objects. Once an object with an associated school of tuna is spotted, a radio beacon is attached to it. Its position is then monitored electronically. Only around dawn or dusk is a net set : the object is to make a set when there is just enough light for the fishermen to see what they are doing, but not enough for the fish to see the encircling net and dive in time to escape it."[331]
> - - R.E. Johannes, Words of the Lagoon, 1992 (From a description of modern Palauan commercial tuna-fishing methods)

Before journeying across the ocean to the Central-Western Pacific, the reader must indulge me in a last mention of albacore in Pacific Northwest waters. After the red-hot Asian economy tanked in 1997, the South Korean (R.O.K.) and Taiwanese (R.O.C.) distant-water fishing fleets...switched over from supplying bigeye, yellowfin, and bluefin tuna for the Japanese sashimi trade, to harvesting albacore for the U.S.-owned canneries in American Samoa.[332] By late summer 1998, the U.S.-owned canneries -- Starkist, Bumblebee, and Chicken of the Sea -- in American Samoa and Puerto Rico, had all the tunafish they could handle to satiate the $459 - million U.S. albacore market. Literally hundreds of American West Coast albacore fishermen during 1998 were wondering where to unload their catch. By early September, the U.S. fleet had only sold about 7,000 tons of albacore mostly to Spanish buyers, while circa 3,000 tons of the fish lay frozen in the hold of vessels tied to the dock; with 4,000 additional tons of product aboard boats headed to port.[333] The final paragraph of an article in our local Port Angeles (Wash.) newspaper showing the effects of the tuna war at the Newport (Ore.) waterfront, ended on a tentative note:

> "To sell fish in the future, U.S. tuna fishermen hope to persuade [American!] canners to switch from the larger, whiter, and less fatty albacore caught at mid-water depths by the Asian long-liners to the smaller fish loaded with more of the heart-healthy Omega-3 fatty acids that the American fleet catches with jigs trolled on the surface."[334]

It must have been locally-fished light-fleshed albacore which initially drove and inspired the California tuna-fishermen (I would contend), and they were soon ranging farther a-field, into the tropical Pacific, to catch yellowfin. Meanwhile, the stateside onshore California canners kept a-breast of the fishermen; perfecting the technology of tuna preparation and packaging; enabling them to steam, bone, gut, and slice the yellowfin; finally wrapping the tuna in sanitary steel to produce for, and present the fish to, a vast nation of meat-eaters.[335] The humungous American advertising industry, which had grown exponentially by the late 1950s, did the rest. And, as luck would have it, back in the early 1950s the stocks of North America's chief canned fish, the Pacific salmon, were already being severely taxed. American, Canadian, and Japanese fishermen had exerted enormous fishing pressure on the Pacific salmon (off-and-on) throughout the century, utilising fish-traps, river nets, gillnets, and purse-seines.[336] With salmon prices rising after World War II, fishing intensity changed over to the tropical yellowfin tuna. The slow pace of traditional pole-fishing methods would not, and could not, keep up with the new demand. As with so much fishing innovation since 1945, the Japanese were in the forefront of finding a new way to catch tuna. In 1970, Wesley Marx described the maximum efficacy of Japanese long-lining:

> "Across the Pacific other high-seas fishermen were exploiting tuna stocks with an absurdly simple but highly successful strategy. Instead of pursuing feeding seabirds, the Japanese tuna fleet merely drops stationary lines into the ocean. These lines, which can be set at any depth, are attached to the beautiful purple glass globes that occasionally wash up on California beaches. The lines contain hundreds of hooks baited with frozen saury. The success of 'long lining' depends on the extent of coverage, and extensive coverage, in turn, depends on a ready supply of inexpensive labor. The Japanese began stringing the Western Pacific with long lines ...."[337]

In our own day, 1997 – 2001 A.D., the tunas, fishermen, vessels, and methods are all roughly the same except geometrically multiplied. The albacore, bigeye, skipjack, and yellowfin are still being landed mostly by East Asians and North Americans. Long-liners harvest deep-water tunas such as bigeye and large yellowfin; purse-seiners (and pole-and-line boats) target surface-swimming tunas like skipjack and small yellowfin.[338] The [South Pacific] Forum Fisheries Agency (FFA) has estimated that in 1996 well over a thousand vessels fish offshore for tuna in the Central-Western Pacific (CWP). Of this formidable flotilla catching tuna, the FFA calculated that 950 vessels were longliners, 170 purse-seiners, and 80 pole-and-line boats.[339] Virtually all of this fleet is owned/operated by nationals of Pacific Rim nations; the Pacific Island countries, nonetheless, partake in a tuna fishery that produces over 1 million tonnes of product per year, valued at more than $U.S 1.5 billion.[340] Pacific Island canneries process circa 20% of the CWP tuna catch, the majority in American Samoa. About a further 35% of the regional tuna harvest (by tonnage) is air-freighted to the high-value fresh-fish markets of Japan and South Korea.[341] By the year 2000, tuna-processing in American Samoa was doing very well; supplying more than $40-million to the overall U.S. tunafish industry, and with the Territory's two canneries (Star Kist Samoa and COS Samoa Packing) working full-time six-day weeks to keep up with landed product.[342]

In 1996, the Forum Fisheries Agency (FFA) had surmised, rather gloomily, that Pacific Island tuna canneries survived (hardly thrived) due soley to preferential access to European and North American markets.[343] As quota protections and tariff barriers disappear under global free trade open-doors, Pacific Island nation canneries will sink or swim in an increasingly Darwinian/libertarian economic atmosphere. Canneries in American Samoa, though under the Stars and Stripes, will face far stiffer competition in the future U.S. market from Central and South America.[344] The only hidden flaw in the bright Y2K tuna picture on American Samoa, was the problem of the independent U.S. purse-seiners based in the Territory: The independent fishermen were not receiving (they believed) a fair price for tuna product, and thus were idling their vessels in harbour.[345] And according to the Samoa News, the independent U.S. fishing boats inject greater than 20% of the wealth into the local economy with their purchase of fuel, food, goods, and services.[346] (Ah, price wars! Mainly because there was a glut of tuna product landed at Pago Pago harbour during spring 2000!)

The Functions and Work of the Forum Fisheries Agency

"The functions of the Agency...are to provide scientific, commercial and technical information and advice to member countries in relation to the living marine resources of the region and in particular the highly migratory species [HMS]....

"[T]he Agency's work has concerned the negotiation and implementation of related agreements among its members and with distant water fishing nations [DWFN] ...."[347]
-- The South Pacific Forum Fisheries Agency, 1997

The Forum Fisheries Agency (FFA) was founded in 1979 as a response to two connected global marine developments: 1) During the 1960s and 1970s, paradigm shifts took place in the international Law of the Sea; and 2) the level of commercial tuna fishing climbed steeply in the Central-Western Pacific (CWP). The Forum Fisheries Agency (FFA), which arose from the independent Oceanic states comprising the South Pacific Forum, is made up of a Committee (FFC), a Secretariat, and is headquartered at Honiara, Solomon Islands. The FFC convenes as the Agency's governing body, twice or three times a year. At last count (1997) there were sixteen member Pacific nations, among them the Micronesian states of the FSM, Kiribati, the Marshall Islands, Nauru, and Palau.[348] In the FAA's own words regarding its raison d'être

"Through [the] FFC, member countries determine the priorities and direct the work programme of the Agency and seek to mobilise the resources needed for its operations .... Within that broad scope, under the direction of [the] FFC, the Agency has in practice concentrated on assisting member countries in the management and development of their tuna resources."[349]

As faithful readers know from my two previous books,[350] distant water fishing nations (DWFN) such as Japan, South Korea, and Taiwan were abruptly stopped from further driftnetting in the Pacific (or anywhere else) by United Nations decree during late 1991.[351] The Forum Fisheries Agency had been there with member nations, in a determined solidarity to draw back the dreaded "curtains of death" for good.[352] And the FFA (plus a body known as the Nauru Group) had won its spurs early in the 1980s by leading the coalition of Pacific Island nations to rightfully gain a fair share of economic benefits derived from the huge tuna harvest, via co-operative enforcement and paid-for fishing permits. The two Pacific "crocodiles" (to Palauan eyes), the United States and Japan, and other encroaching DWFN, had been fishing in the enormous EEZs of the Pacific coastal States with impunity -- and without offering any financial compensation whatsoever. For certain land-poor countries like Kiribati, large EEZs and the marine resources they encompassed was all that was available to them. The FFA leadership had demonstrated how exercising a common Pacific will could, and would, benefit all the nations in the region.[353]

Readers of FutureFish 2001 have already become acquainted with the [South Pacific] Forum Fisheries Agency in Part II, 'UNCLOS III -- the Third United Nations Conference on the Law of the Sea.' It was the FFA that held two workshops at Port Vila, Vanuatu, and at Mahé, Seychelles, both during September 1982. The workshops had, variously, to do with fisheries access rights and monitoring, control, and surveillance of fisheries within EEZs.[354] Indeed, W.T. Burke's legal paper on UNCLOS III and tuna regulation was commissioned by the UN's Food and Agriculture Organization (FAO), under the FAO's Fisheries Law Advisory Programme, as a background paper for the Forum Fisheries Agency.[355] Professor Burke's paper, which strongly supported the sovereign rights of "coastal States" (viz., the Pacific Island nations) in their EEZs (as opposed to DWFN, i.e., the "fishing States" of East Asia and the U.S.A.), lent truth, logic and weight to the avowed mission of the FFA: "The Agency's mission is to enable its [16] member countries to obtain maximum sustained benefit from the conservation and sustainable use of their fisheries resources."[356]

Thus the arm of the Forum Fisheries Agency is as far-reaching as it is well-meaning. From its inception in 1979 to the present, the Agency has proven itself -- time and again -- to be the driving force behind co-operation and co-ordination of fisheries management and conservation in the South and Central-West Pacific; especially regarding tuna and other highly migratory species (HMS).[357] Although tuna management and conservation are priorities for the FFA, the Agency is helping member countries obtain "maximum sustained" development of the tunafish resource. And rather than having the members just demand fishing access fees, the Agency encourages the 16 to look for a more active involvement in tuna fishing and processing. This would include the provision of crews, services, and supplies; and the establishment of local onshore infrastructure of fishing bases and canneries. Lastly, the FFA seeks policy reform for local governments which often hinder, rather than facilitate, new fishing ventures (including JVs) or local boat/plant ownership.[358] As the FFA itself has forecast:

> "Important realignments in the tuna fishery are likely in the next few years .... New patterns of ownership of vessels and processing plant[s] are sure to emerge from the interaction of liberalised international trade and closer management of tuna resources. Meanwhile FFA's members are moving to put in place harmonised management regimes that use regional solidarity to protect their national interest."[359]

Certainly, the achievements of the Forum Fisheries Agency show what can be accomplished by [even] small nations working together to further their common interests. As noted, the Agency is affiliated with the United Nations in that the FFA holds observer status in the Food and Agriculture Organization (under a "memorandum of understanding"), and attends other U.N.-sponsored meetings as an advisor to its members' national delegates.[360] Foreseeing the future, the FFA envisioned in 1997 a Corporate Plan for 1998-2001: A corporate plan as the Agency, composed of sixteen investor nations, acts very much like a service board answerable to its shareholders. The member countries, bound together by their shared commitment to sound resource management, all [are supposed to] play a fully equal rôle in its direction. (But Australia and New Zealand, as the two predominant members with First World living standards and financial clout, are the natural Oceanic leaders politically and economically.) The Agency has, moreover, kept its far-flung membership remarkably cohesive since 1979, and the FFA's hitherto-successful corporate experience promises solid, long-term results in tropical Pacific tuna management, conservation, and development.[361]

During the Forum Fisheries Agency's four-year Corporate Plan, particular emphasis would be placed on: "[1] Implementation of UNCLOS [III] and the establishment of compatible in-zone and high seas regimes...; [2] strengthening the scientific basis of resource management; and [3] the strengthening of FFA's finances by cost recovery mechanisms."[362] And, finally, the basic financial framework of the Agency, overseen by the Forum Fisheries Committee (FFC), is composed of two groups - - A and B. Group A, in turn, is made up of Australia and New Zealand, who together contribute 74%; and then Group B, consisting of the rest of the fourteen Pacific Island nations, who allocate the remaining 26%.[363] Donor groups aiding the Forum Fisheries Agency financially, are the U.K., Commonwealth Funds for Technical Coöperation (CFTC), FAO, EU, U.N. Development Programme (UNDP), Asian Development Bank (ADB), Taiwan, and South Korea. Interestingly (and presciently), the ADB, Taiwan, and South Korea have indicated <u>increased</u> support for the FFA, while all the others have <u>reduced</u> their giving in recent years.[364]

It is not surprising, perhaps, that East Asian distant water fishing nations (DWFN) are more committed to the South Pacific Forum Fisheries Agency than Canada, the European Union, and the United Nations. But the former are also the very "fishing States" (along with the U.S.A.) from whom UNCLOS III and the FAA -- in effect -- are <u>protecting</u> the "coastal States"! That East Asian fishing States actually <u>support</u> the mission and the work of the FFA should be interpreted as the best possible news for the coastal States of the Pacific Islands (including Micronesia), and for the tuna resource of the South Pacific and CWP. Surely co-operation between the fishing States and the coastal States, and the common commitment to the management, conservation, and development of highly migratory species, are in the best interests of all user groups in the region?

The Forum Fisheries Agency evolved into the South Pacific Forum Fisheries Agency largely as a result of the U.N. – sponsored Straddling Fish Stocks and Highly Migratory Fish Stocks Agreement. The Agreement was initially signed by the United States and 27 other countries and territories in December 1995, and was to have an enormous -- and almost immediate -- impact on the U.S. Atlantic and Pacific tuna and swordfish fleets (not to mention the Bering Sea pollock fishery).[365] Once the U.N. treaty was ratified and in place, the (then) 28 signatories would have to obey a far-ranging and nit-picking set of detailed rules and regs...meaning an <u>end</u> to the first-come, first-served, fishing-as-usual <u>outside</u> the 200-mile EEZs. The new, legally-binding Agreement, wrote <u>National Fisherman</u> field-editor Michael Rivlin in March 1996, "call[ed] for fisheries to be managed <u>globally</u> under a strong and <u>enforceable</u> framework of <u>international law</u>."[366]

(The U.N. Agreement sends up a red flag of warning to this writer, anyway, about UNO encroachment on "jurisdictional gaps" and those two-thirds of the earth's oceans <u>not</u> part of any national EEZs. See Part II in this Chapter.)

The U.N. Agreement would have significant effect in the South and Central-Western Pacific, where a (ca.) 50-vessel U.S. skipjack/yellowfin tuna purse-seine fleet home-ported (then as now) at Guam and American Samoa. In early 1996, the United States was the sole DWFN to have seriously discussed fishing rights with the 16-member Pacific-nation FFA. It was around this time, spinning off from the U.S.A./FFA negotiations, that the Forum Fisheries Agency decided to merge with the South Pacific Commission, a scientific body which had assessed fish stocks and made [fish] management recommendations.[367] <u>National Fisherman's</u> Michael Rivlin prognosticated that the newly-fused organisation would become the Pacific regional entity which DWFNs such as the East Asian countries (China, Japan, South Korea, Taiwan) and others (eg., U.S.A.) will belong. But <u>National Fisherman's</u> field-editor had preceded his March 1996 piece with a small, and chilling warning:

"It was once true that -- torturing an old adage -- all fishing was local. Not any more. Today, it's global and ecosystem-wide. If any U.S. fishermen hadn't yet woken up to this [B]rave [N]ew [W]orld by the end of 1995, the conclusion in December of a <u>global</u> treaty designed to regulate high-seas fisheries was a very powerful shot across the bow."[368]

A wee bit frightening, too, had been the physical presence, at the formulation of the Agreement, of "non-governmental organizations" (NGOs) with activist environmentalist agenda; namely Greenpeace and the World Wildlife Fund (<u>that</u> WWF!). And <u>National Fisherman</u> reported that attorneys-at-law from the

Audubon Society and the Natural Resources Defense Council <u>had actually sat</u> with the U.S. delegation. (<u>Yike!</u>) Because of the American eco-legalists having been present, buzz-phrases like "paradigm shift" were bandied about, and the very language of the U.N. treaty was skewed by ecospeak. As a consequence, "maximum sustainable yield" (MSY) was rejected in favour of "conservative management".[369] (Which can mean whatever the treaty framers and their eco-cohorts want it to mean.... Whatever happened to "optimum yield" [OY] for a CWP tuna resource described by the FFA in 1997 as having... "not so far been over fished"?)[370]

Meanwhile, in the early morning of the new millennium, the FFA is doing its job of <u>sensible</u> "conservative [tuna] management" and protecting the rights of coastal States and territories in the South/Central-Western Pacific. During the spring of 2000, <u>Pacific Magazine</u> told a tale of a Spanish armada of 15 super-size purse-seiners trying to get fishing access to the tuna-rich waters around Kiribati and Vanuatu.[371] Not only could the Spanish vessels handle up to 2,500 tons of fish (twice the amount of an American purse-seiner), but their gaining access would undermine the FFA regional treaty/license agreement; an agreement which had -- and has -- benefited all 16 Pacific Island nation members. At the time, the governor of American Samoa expressed his intention to consult with the prime minister of [independent, formerly Western] Samoa in order to jointly influence other Pacific leaders to uphold the FFA treaty/license agreement.[372]

Thus, so far, the FFA has been as equitable as it has been effective. The powerful U.S. State Department negotiates on behalf of American flag fishing vessels, and the FFA established the maximum number of boats that a DWFN -- or fishing State -- may enter in the South/Central-Western Pacific region. The mighty U.S.A., the global super-power, is limited by FFA agreement to exactly 42 boats.[373] Also during the spring of 2000, the Taiwan Tuna Fishing Association signed a three-year fishing license agreement with Nauru. The tiny Micronesian island-nation (via the FFA) granted permits to ... 42 (yes! <u>exactly</u> 42!) Taiwanese tuna boats to operate in her 200-mile EEZ.[374] How's <u>that</u> for even-handedness?!

Let us hope during the years ahead, of the twenty-first century, that the South Pacific Forum Fisheries Agency continues to delegate such fair economic opportunity and to ever remember its mission:

"The Agency's mission is to enable its member countries to obtain maximum sustained benefit from the conservation and sustainable use of their fisheries resources."[375]

The alternative path is for the FFA to merely descend into becoming yet another [feared and loathed] globalist tentacle of the UNO/FAO. That would surely be a most ignominious end to the both equitable and effective South Pacific Forum Fisheries Agency.

A Frigate Bird's-eye Overview of HMS Fishing by DWFN in the CWP, 1986-2001

"The central and Western Pacific tuna fishery is the only one of the world's major fish resources that has not so far been overfished. As other stocks decline or are brought under management, and world demand for wild-caught fish increases, pressure on the resources of the central and western Pacific will intensify."
-- South Pacific Forum Fisheries Agency, 1997[376]

Although Part III so far has dealt mainly with the commercially-important tropical tunas (bigeye, skipjack, yellowfin) and albacore, the highly migratory species (HMS) in the Central-Western Pacific (CWP) include various billfishes, swordfish, sharks, and large pelagic fishes.[377] Among these are the sought gourmet favorites mahimahi or dolphinfish (Coryphaena hippurus), and the broadbill swordfish or shutome (Xiphias gladius). And sport-caught billfishes such as Pacific blue marlin/kajiki (Makaira nigricans), black (or silver) marlin/a'u (Makaira indica), and striped marlin/nairagi (Tetrapturus audax), support significant recreational fisheries (as well).[378] Otherwise, American commercial billfish fisheries -- excepting swordfish -- had been virtually supplanted, by the early 1990s, by foreign (i.e., East Asian) long-line and drift-gillnet fleets. The U.S. management authority for billfish, and for tuna, in the mid-Pacific and CWP resides with the Western Pacific [Regional] Fishery Management Council (WPFMC) for Hawaiian and Western Pacific waters.[379]

From 1970 to 1980, the American tuna fishing fleet in the Eastern Tropical Pacific (ETP) predominated, but tuna harvests in the ETP became less lucrative during the 1980s (porpoise mortality concerns etc. etc.). Consequently, many U.S. tuna fishermen left the industry or ventured into the Central-Western Pacific to try their luck; homeporting at Guam or American Samoa as we have seen. So instead of competing with Mexicans, Central and South Americans in the ETP, U.S. tuna fishermen now (late 1980s/early 1990s) had the Japanese, Taiwanese, and the South Koreans as rivals in the CWP. As noted, long-line gear is used to catch deep-water tunas such as bigeye/ahi and yellowfin/ahi, and purse-seining to harvest small yellowfin and skipjack/aku in the CWP -- these same methods had been employed by U.S. tuna fishermen in the ETP. Besides challenging the Americans at long-lining and purse-seining, the East Asians utilised the traditional tried-and-true methods of ring-net, handline, pole-and-line. According to the

National Marine Fisheries Service (NMFS), in 1989 the total number of purse-seiners (taking 30-50% of yellowfin landings) amounted to more than 120 vessels; during 1990-1992, circa 42 U.S. purse-seine boats operated in the CWP.[380]

In The Asia & Pacific Review, 1987 edition, little emphasis was made on commercial fisheries in Micronesian waters. Under 'Belau', A&PR staff writers commented that "It has potentially rich marine resources, but these are still largely untapped"...and "A dormant tuna processing facility in Belau...may reopen in 1987. The government is promoting investments...in...fisheries development."[381]

On the Federated States of Micronesia, the A&PR reported that: "The main potential wealth of the FSM lies with the sea, for fish resources now.... The export of fresh fish and shellfish...[is] receiving attention in all four states in the Federation."[382]

The last entry of The Asia & Pacific Review for American Micronesia was the Marshall Islands: "Fishing is mainstay of economy, supplying principal source of protein and export revenues.... Majuro [the capital] has [1986] one of the two currently operating katsuobushi processing plants in the Pacific islands, under a joint venture with Japan, for producing smoked tuna for the Japanese market."[383] (See Part IV for history of katsuobushi processing in Micronesia.)

But there was a Big Fish Story of the 1980s, and it didn't originate in American Micronesia but from the formerly – British Gilberts; since 1979 the Republic of Kiribati. After all was said diplomatically and done politically to what remained of Kiribati (as related), the young republic's EEZ still encompassed 3,550,000 square km of the richest fishing grounds in the CWP. Her marine resource – rich EEZ was Kiribati's main -- and only -- asset for earning badly-needed [Australian] export dollars. For many years Kiribati, like comedian Rodney Dangerfield, had "not been gettin' any respect at all" -- especially from American (i.e., California) purse-seiners fishing in Kiribati waters, who considered tuna "migratory" ... hence not subject to licensing fees.[384] So the tiny Pacific Island country, with the huge "exclusive" economic zone, in October 1985 signed a licensing agreement with the Russian fishing company, Sovryflot.[385] The licensing agreement permitted 16 Russian vessels to fish for tuna within the 200-mile EEZ, but staying outside Kiribati's 12-mile territorial waters, for 12 months (that year) at $2.4 million Australian.[386]

The enigmatic and pragmatic head of state responsible for the controversial Soviet fishing pact was (first) Kiribati president Ieremia Tabai. That worthy was determined that his new nation would earn her own way in the world and not live on Great Powers/UNO aid handouts. According to The Asia & Pacific Review, President Tabai had been [willing but] unable to negotiate a [fair] fishing deal with the American Tunaboat Association (ATA) in Kiribati's [best] interest. The U.S. fishermen, facing a massive slump in the tuna industry and with expensive purse-seiners tied to the dock, refused to meet President Tabai's asking price. Thus, "dissed" and bullied little Kiribati, turned to the big and bad Soviet Union.[387] Kiribati's fishing treaty with the U.S.S.R. most assuredly got the concern of her Pacific neighbors - - Australia, New Zealand, and the United States. They were sitting up and paying attention now. Indeed, Pacific travel writer David Stanley in 1992 remarked that "the mere specter of their [Red] presence led to a dramatic increase in the U.S. and New Zealand aid to their country [Kiribati]."[388]

The Soviet-Kiribati fishing licensing agreement was not renewed for 1986, and by April 1987 the U.S. government had at last stepped back; recognising the EEZs of the 16 (FFA – mentored) Pacific Island nations ("coastal States") and agreeing to pay regular fishing license fees.[389] No retreat was sounded, however, before the resolute Kiribati president, Ieremia Tabai, had caused a Red Scare among the "brass hats" at the Pentagon, Washington, D.C. (and the Aussie Cold War warriors at Canberra, A.C.T.), plus engendering a whole new opposition party at home; the Christian Democratic Party! But the two-term Kiribati president, who was scheduled to leave office in 1987, had proven his point ... and become the leading voice of true economic [and real political] independence among the proud new, but previously powerless, Pacific Island nations. And even U.S.N. Rear-Admiral Edward Baker, Pacific regional director of the Pentagon's office of international affairs, was compelled to eventually admit that Kiribati's Russian fish affair had been for money -- and nothing more.[390]

In the U.S. Commerce Department's December 1993 edition of Our Living Oceans[:Report on the Status of U.S. Living Resources, 1993], the NOAA/NMFS has made all kinds of projections for tropical tunas -- but mostly in the "Unknown" category. And the projections targeted only two tuna species, skipjack/aku and yellowfin/ahi.[391] This fish-glitch notwithstanding, recent average yield (RAY) for CWP skipjack tuna taken by U.S. and foreign fleets (all DWFN) for 1989-1991 was 752, 200 tons (t), and the NMFS allowed that skipjack in the CWP was " underutilized." The conservation - minded NMFS, however, referred to skipjack tuna's long-term potential yield (LTPY) as merely "Unknown." For the entire Pacific region, the skipjack harvest (1989-1991) amounted to about $590 - million U.S.[392]

According to the NMFS, the recent average yield (RAY) of yellowfin tuna in the CWP for 1989-1991 was 320, 233 tons (t), with earnings well in excess of $410 - million in the Pacific as a whole. The NMFS characterised yellowfin stocks as "Unknown", and the tuna's long-term potential yield (LTPY) as also "Unknown."[393] The RAY for bigeye/ahi in the entire Pacific region, in the 1993 NMFS report, was 134,000 t with a [then] current potential yield (CPY) and LTPY of 160,000 t (1988-1990). These figures might appear modest compared to those of skipjack and yellowfin, but bigeye is the sashimi (raw, sliced for Japan) tuna, and thus generated (ca.1990) ex-vessel revenues of approx. $1 billion U.S.! The bigeye/ahi brought (and doubtless brings today) the highest dockside price of any of the tropical tunas, anywhere: In 1990, close to $7000/t. In the 1993 NMFS report, the bigeye tuna's status of utilisation was listed as "Full", and its status of stock level was categorised as "Near" (just above "Below").[394]

(There were no figures for albacore/tombo harvests in the 1993 NMFS report pertaining to the Central-Western Pacific, and the Table [Figure 18-3] listed landings solely for the North Pacific Transition Zone [NPTZ], 1970-1991. But the information regarding this most favoured of tunas was instructive:

"Albacore is fished from the northern limits of the North Pacific Transition Zone (NPTZ) to about lat. 15°N, and from Japan to North America....
"In the North Pacific, the total catch, catch rates, and fishing effort in the U.S. troll fishery and the Japanese pole-and-line fishery have all been declining....")[395]

But the declining albacore/tombo stocks in the North and South Pacific during the early 1990's, have been more than made-up by the abundance of tropical tunas during the late 1990s. Fili Sagapolutele reporting for Pacific Magazine in the summer of 1999, wrote that:

"More than $400 – million worth of canned tuna from American Samoa was shipped to the U.S. Mainland in 1998, showing a continuing annual increase in production for the territory's two canneries".[396]

(NB: While slogging through Part III, 'Tropical Western and Mid-Pacific Fisheries', the reader ought always bear in mind that those harvesting these rich fishing grounds -- off the tiny islands of the enormous archipelago which comprises Micronesia -- are overwhelmingly from DWFN: [the "fishing States" of] U.S.A., Japan, South Korea (R.O.K.) and Taiwan (R.O.C.). Micronesians --and other nationals residing in Micronesia --plying their own waters (today and yesterday) will be fully examined, with their fisheries, in Part IV.)

NOTES for PART III

311. Wesley Marx, The Frail Ocean (New York: Ballantine Books, 1970), p. 196.

312. Carl Safina, Song for the Blue Ocean (New York: Henry Holt and Company, 1998), p. 54.

313. TM of the Pacific Magazine Corporation, 1986. My wall-map is the 1996 Thirteenth Edition of "Pacific Geographic Maps" (also TM!)

314. C.P. Idyll, Abyss: The Deep Sea and the Creatures That Live in It(New York:Thomas Y. Crowell Company, 1964), p. 21.

315. Ibid., p. 62. (NB: All quotes those of C.P. Idyll.)

316. Cited by Carl Safina, p. 395. (My brackets)

317. The eds., 1982 World Record Game Fishes (Fort Lauderdale, Fla.: The International Game Fish Association, 1982), p. 222. NB: For a taxonomical table of all the tunas, see Appendix I: A Pacific Fish And Shellfish Gazetteer.

318. Ibid., p. 223.

319. Ibid., p. 223.

320. Op.cit., International Game Fish Association, p. 223.

321. IGFA, p. 231.

322. Ibid., pp. 287-288.

323. Ibid., p. 290.

324. Op.cit., IGFA, p. 291.

325. Ibid., p. 291.

326. As listed by the National Fisheries Corporation, Kolonia, Pohnpei, FSM. (See Appendix I)

327. See IGFA, pp. 287-288.

328. Ibid., pp. 290-291; Carl Safina, p. 85. (My emphasis)

329. Op., cit., Carl Safina, p. 54. NB: Carl Safina was featured on a National Geographic Explorer programme highlighting giant bluefin tuna. Presented on Sunday, 11 March 2001, by CNBC-TV.

330. Op. cit., Wesley Marx, p. 198.

331. R.E. Johannes, Words of the Lagoon: Fishing and Marine Lore in the Palau District of Micronesia (Berkeley, Calif.: University of California Press, 1992), p. 97.

332.. 'Tuna fishermen angle for markets', Peninsula Daily News, 4 September 1998, p. A-8.

333. Ibid., p. A-8.

334. Ibid., p. A-8. (My brackets and emphases)

NOTES for PART III (cont'd.)

335. Wesley Marx, p. 199.

336. Ibid., p. 199.

337. Op. cit., p. 199.

338. National Fisheries Corporation, Kolonia, Pohnpei, FSM.

339. 'South Pacific Forum Fisheries Agency: Corporate Plan 1998-2001' (Honiara, Solomon Islands: FFA, 1997), p. 4.

340. Ibid., p. 4.

341. Ibid., p. 5.

342. Fili Sagapolutele, 'American Samoa's Tuna Picture Looking Rosy', Pacific Magazine, Vol. 25, No. 2, Iss. 139, March/April 2000, p. 17.

343. FFA (1997), p. 5. NB: Pacific Island nations such as the Solomons, PNG, Fiji.

344. Ibid., p. 5.

345. Fili Sagapolutele, 'American Samoa's ….', p. 17.

346. Ibid., p. 17. NB: Pacific Magazine reported that Star Kist Samoa had ownership in 8 of the 30-40 purse-seiners operating out of Pago Pago, thus these boats were unaffected by the lower price; unlike the 18 U.S. independents who were.

347. "The Evolution of FFA", FFA (1997), pp. 2-3. (Brackets mine)

348. Ibid., p. 2. NB: The other members are: Australia, Cook Islands, Fiji, New Zealand, Niue, Papua New Guinea, Samoa, Solomon Islands, Tonga, Tuvalu, and Vanuatu.

349. Op. cit., FFA, p. 2. (My brackets)

350. Fisheries of the Pacific Northwest Coast, Vantage Press, Inc., N.Y. Vol. I (1991) and Vol. II (1994).

351. See FutureFish 2000, Ch. 2, Pt. III, Sec. C for details.

352. Elliott A. Norse, ed., Global Marine Biological Diversity: A Strategy for Building Conservation into Decision Making (Redmond, Wash.: Center for Marine Conservation, n.d.), p. 174. NB: Final draft copy was published by Island Press in August 1993.

353. Ibid., p. 174. Author's note: Global Marine Biological Diversity, despite its ominous title, was written before Mr. Elliott A. Norse devolved into a great big eco-nuisance!

354. See note no. 299 in this Chapter.

355. See note no. 300 in this Chapter.

356. Op. cit., FFA, p. 10: "The Mission of FFA". (My brackets)

357. Ibid., pp. i-ii : "Statement by the FFC Chairman", Moses Amos, May 1997.

NOTES for PART III (cont'd.)

358. Ibid., pp. iii-iv : "Director's Contribution", Victorio Uherbelau, June 1998.

359. Op. cit., FFA, pp. 5-6: "Overview of Tuna Industry in Central and Western Pacific."

360. Ibid., p. 12: "Institutional Environment."

361. Ibid., p. 14: "Relations with Stakeholders."

362. Op. cit., p. 16: "Response to Members' Needs."

363. Ibid., p. 21: "Financial Framework."

364. Ibid., p. 22.

365. Michael Rivlin, 'A U.N. treaty takes on the world's fisheries', National Fisherman, Vol. 76, No. 11, March 1996, p. 28.

366. Op. cit., p. 28. (My brackets and emphases)

367. Ibid., p. 28 cf. Student Handbook, Desk Reference Vol. 5 (Nashville, Tenn.: The Southwestern Company, Inc., 1998), p. 660. See FFA (1997), p. 34. Author's note: The South Pacific Commission, formed at Canberra (Australia) in 1947 was made up of old colonial masters U.K., U.S.A., Australia, New Zealand, and France. According to the above sources, the old SPC (South Pacific Commission) morphosed into the new SPC (Secretariat for the Pacific Community), and is headquarted at Noumea, (French) New Caledonia. This writer can only conclude that the FFA gained the old SPC's fish scientists/managers; thus renaming itself the South Pacific Forum Fisheries Agency. (Then again I could be wrong!)

368. Op. cit., p. 28. (My brackets and emphasis)

369. Ibid., p. 84.

370. FFA (1997), p. 4.

371. Fili Sagapolutele, 'American Samoa's…', p. 17.

372. Ibid., pp. 17-18.

373. Ibid., p. 17.

374. Ibid., p. 17. Radio Australia, 'Nauru Signs Fishing Agreement with Taiwan.'

375. FFA (1997), p. 10: "The Mission of FFA."

376. FFA (1997), p. 4.

377. Our Living Oceans. NOAA Tech. Memo., NMFS – F/SPO-15 (Silver Spring, Md.: U. S. Dept. of Comm., 1993), p. 97.

378. Ibid., p. 97. Cf. IGFA, passim; "Hawaii Seafood", State of Hawaii, Ocean Resources Branch, Honolulu, 1993.

NOTES for PART III (cont'd.)

379. Ibid., p. 97. NB: Billfish and tuna are also managed by the Pacific Fishery Management Council (PFMC) for North American waters, although the PFMC has delegated authority to the State of California over some sharks, striped marlin, and swordfish.

380. Ibid., pp. 97-98: "Tropical Tunas."

381. The Asia & Pacific Review, 1987 ed. (Lincolnwood, Ill. : NTC Business Books, 1986), p. 174: 'Belau'.

382. Ibid., p. 178: 'The Federated States of Micronesia.'

383. Op. cit., p. 187: 'Marshall Islands.' (Brackets and emphasis mine)

384. David Stanley, Micronesia Handbook (Chico, Calif.: Moon Publications, Inc., 1992 ed.), p. 279. NB: The Americans also refused to recognise Kiribati's 200-mile EEZ, although the U.S. claimed a like zone about her own shores.

385. Ibid., p. 39.

386. Ibid., p. 279.

387. The Asia & Pacific Review, 1987 ed., p. 185.

388. Op. cit., David Stanley, p. 279. (Brackets mine)

389. Ibid., p. 279.

390. A&PR, 1987 ed., pp. 185-186. Cf. David Stanley, pp. 39, 279. See also Thurston Clarke, Equator: A Journey (New York: Avon Books, 1990), p. 370.

391. NMFS (1993), pp. 97-98: Table 18-1

392. Ibid., pp. 98-99: Figure 18-1

393. Ibid., pp. 98-99: Figure 18-2

394. Ibid., p. 98: Table 18-1

395. Ibid., pp. 99-100: "Albacore".

396. Fili Sagapolutele, 'American Samoa Shipped $400-Million Worth of Canned Tuna to U.S. in '98', Pacific Magazine, Vol. 24, No. 4, Iss. 136, July/August 1999, p. 13.

Bibliography for Part III

26. Fielding, Ann and Ed Robinson. <u>An Underwater Guide to Hawai'i</u>. University of Hawai'i Press, Honolulu, 1993.

27. Hviding, Edvard. <u>Guardians of Marovo Lagoon: Practice, Place and Politics in Maritime Melanesia</u>. University of Hawai'i Press, Honolulu, 1996.

28. Johannes, R.E. <u>Words of the Lagoon: Fishing and Marine Lore in the Palau District of Micronesia</u>. University of California Press, Berkeley, Calif., 1992 ed.

29. Knop, Daniel. <u>Giant Clams: A Comprehensive Guide to the Identification and Care of Tridacnid Clams</u>. Dähne Verlag GmbH, Ettlingen, Germany, 1996. (Translated by Dr. Eva Hert and Dr. Sebastian Holzberg)

30. Safina, Carl. <u>Song For the Blue Ocean: Encounters Along the World's Coasts and Beneath the Seas</u>. Henry Holt and Company, New York, 1998.

Main References

F. Mangrove Action Project. <u>Mangroves at the Root of the Sea</u>. Puffin Graphics, Port Angeles, Wash., 1994.

G. 'South Pacific Forum Fisheries Agency: Corporate Plan 1998-2001', FFA, Honiara, Solomon Islands, 1997.

H. <u>1982 World Record Game Fishes</u>. The International Game Fish Association, Fort Lauderdale, Fla., 1982.

Part IV: "A Modest Proposal" for Micronesia

"Pacific Islanders stand roughly today where Alaska stood 10 years ago. Their tuna industry has room to grow; some estimates suggest the harvest could safely increase by 40%. Their challenge is to make sure [I]slanders receive a greater share of the benefits from the fish that -- so far -- have been hauled out of their waters mostly by foreign, distant – water fleets."
-- Brad Warren, ed., Pacific Fishing, December 1997[397]

(A) A Brief History of Commercial Fishing in Micronesia

"The true history of commercial fishing in Micronesia started with Okinawan immigrants after World War I."
-- C.D. Bay-Hansen, 2001

(1) The Spanish/German period, 1790-1914

According to Bruce G. Karolle, Professor of Geography at the Micronesian Area Research Center (MARC), University of Guam, the indigenous Island subsistence economy began to subtly shift as a result of first [commercial] contacts with Europeans (i.e., the Spanish) during the 1500s.[398] Professor Karolle has asserted that the history of economic impact on Micronesia by external forces can be roughly divided into three periods: 1) The China Trade (1790-1850), 2) The Whaling-ship Trade (1840-1860), and 3) The Copra Trade (1850 to the present).[399] During the China Trade, overseas merchantmen traded Western goods such as cloth and ironware for dried sea cucumber (bêche-de-mer/trepang),shark fins, and turtle shells. In the brief but sea-change years of the Whaling-ship Trade (1840-1860), certain Micronesian islands (Palau, Pohnpei et al.) served as regular sources of supply (and rest and recreation; R & R) for the [mostly British and American] whalers - - food, water, timber, tobacco…and native women, too. The Copra Trade (commercial planting and processing of coconuts) was initiated by the Germans during their years of Micronesian colonialism (1899-1914; 1885-1914 in the Marshalls; never in the Gilberts or Guam), was continued by the Japanese during their years of Micronesian colonialism (official mandate in 1921, but actual control from 1914-1944) -- and the Copra Trade continues on to this day.[400]

The earlier years of the China Trade (1790 to ca.1810) were dominated by the proud Spanish -- based at Manila, Philippines -- and the lowly (and unlikely) sea cucumber. The trepang (or in Guam, the balate) is, of course, the common echinoderm found throughout the shallow waters of the tropical Pacific. Abundant, easily-gathered and processed, the marine animal's smoked and shriveled skins were bagged and shipped by the long ton from all over the tropical Pacific to China -- where they would sell for " ' 15

Spanish dollars the Hundred pounds.' "[401] In China proper, the much sought and valued bêche-de-mer was simmered in soup and believed to be an aphrodisiac. On Guam, the Spanish governor held a tight monopoly on the balate trade.[402] Not only that, but the Spanish in the Philippines had learned the method of curing the sea slug, and marketed the finished product to Chinese buyers long before the British and Americans comprehended the process in 1829.[403] Francis X. Hezel, S.J., has described some of the Spanish explorations and efforts in the China Trade around Micronesia during the early years:

> "Not only British ships were making discoveries in the area. Spanish trading vessels from the Philippines were, from the late 1700s on, sailing into the Carolines -- some of them en route to Lima and other ports on the coast of South America, others to search for cargoes of shell and bêche-de-mer for the Manila-based China trade....
> "The crews of some of these ships put ashore at islands to collect bêche-de-mer, the sea slug found in warm tropical waters that is a gourmet dish to Chinese wealthy enough to afford such a delicacy. The master of a British merchantman lying off Palau in 1802 learned that four Europeans had been left ashore there by a trading vessel to gather bêche-de-mer, tortoise shell, and shark fins for sale to China... A few years later, the Spanish ship Modesto also spent time at Palau, after visiting Fais, in search of bêche-de-mer....
> "At first, Spanish trade vessels roamed rather freely throughout the Carolines in search of the precious holothurian, but in later years -- after 1810 or so -- they confined their activities largely to the vicinity of Palau and Yap."[404]

Moreover, the trade in sea cucumbers, like that of copra, continues to this day in the tropical Pacific. But the days of the once-vast Spanish Empire were numbered. During the later years of the 19th century, the up-and-coming British and Germans had all but driven the steeply-declining Spaniards from the major Pacific trading sea-lanes. In 1898, a young and strong America booted an old and decrepit Spain out of Guam, out of the Philippines, and indeed, out of the Pacific entirely...for good. The defeat of proud Spain in the Spanish American War had been utter and devastating. But the Pacific Ocean, the former "Spanish lake", was not to become an all- "American lake" just yet.

(2) The Japanese/Okinawan period, 1914-1945

As stated in the epigraph by this writer, the real history of commercial fishing in Micronesia started with the Okinawan Japanese during the inter-World War years. Bruce G. Karolle of MARC has quite rightly credited the Japanese with great and good economic deeds for Micronesia, however selfishly motivated:

> "Under the Japanese...Micronesia attained a measure of self-support in the 1930s and by 1936 generated an annual surplus of one million yen. The sectors mainly responsible for this high level of production were sugar, phosphate fertilizers, copra, and marine resources (mainly dried tuna). All the

products were exported to Japan.... Training, technology, and research programs in agriculture and marine resources were provided for the benefit of Japanese commercial undertakings.

"The Micronesians derived scant direct benefits from these aggressive development programs, which favored only the Japanese national economy and Japanese enterprizes."[405]

The quick-witted and acquisitive Japanese got off to an early commercial start in Micronesia. Even prior to the Spanish fire sale of the Northern Marianas, Carolines (Palau and the present FSM), and Marshalls to Germany in 1899, Japanese merchants were busily trading copra and coconut oil during the mid-late 1880s.[406] Indeed, by 1906, 80% of the entire trade in [now] German Micronesia was in Japanese hands. Two years later, in 1908, the two main Japanese commercial rivals in the Northern Marianas and the Carolines merged to form the Nan'yō Bōeki Kabushikigaisha (South Seas Trading Company); to be known as the eventually formidable Nambō.[407]

By 1914 Nambō predominated in the commercial life of [still] German Micronesia: The company ran the copra and trepang trade, commercial fishing operations, inter-island postal service, as well as the freight and passenger-ferrying business. When World War I broke out in 1914, Nambō held a near monopoly over all commerce in central and western Micronesia. Thus Japan's seizure of German Micronesia merely strengthened Nambō's already vice-like grip on Island trade. The occupation by Japan of the Nan'yō (South Seas) -- which the League of Nations sanctioned in 1921 -- spurred Japanese investment in agriculture and fishing in addition to trading. Through the founding of a development company to be known locally as Nankō,[408] enterprising Matsue Haruji was able to grow sugar on Saipan (in the Northern Marianas) into the most profitable of all Japanese agricultural efforts in the Island mandate.

Matsue's Nankō was so successful that other Japanese entrepreneurs were motivated to follow Matsue-san's shining example. The colonial authorities (Nan'yō-chō) knew a good thing when they saw it, so the government launched especial efforts throughout the 1930s to integrate the Island mandate both rapidly and extensively into the greater imperial economy.[409]

The end result was a special commission appointed by the government to speed up Japanese migration into the Nan'yō, to accelerate the commercial use of Island and marine resources, and to hasten the promotion of tropical industries generally. The formation of Nantaku[410] followed a-pace; an agency of Nan'yō-chō -- complete with a board of directors selected from both public and private sectors. Nantaku quickly assumed direct control of the phosphate mines in the Western Carolines (i.e., Palau) and then established several affiliated companies overseeing electrical energy, refrigeration, bauxite mining, pearl diving, and market agriculture. University of Hawai'i historian of Micronesia, David Hanlon, has accurately summarised the 1930s situation in the Nan'yō: "The activities of Nantaku and its subsidiary enterprises graphically demonstrated the mutually beneficial reinforcing alliance between Japanese government and business in Micronesia."[411]

In fact, business was booming, life was good -- very good --- in the pre-Pacific War Micronesia... for everybody, but especially for the Japanese living in the Island mandate.

In the Federated States of Micronesia is the State of Chuuk (Truk). Chuuk is probably best known for its famous lagoon, which, semi-encircled by atolls from one to two hundred miles distant, is filled with high but small basaltic islets. The enormous lagoon served during the Japanese era as the headquarters for the Imperial Navy. The presence of the Japanese went mostly unremarked by the native Chuukese until about the late 1920s.[412] Then, slowly but surely, the district of Truk underwent a degree of "Japanization." In 1928, a determined Okinawan immigrant named Tamashiro responded to Japan's call for settlement of the Nan'yō; made the overseas journey to the Caroline Islands, and settled in Wonei on the western side of the Chuuk Lagoon. Here Tamashiro-san built a fishing boat, a 14-ton dory on which he installed a 20-horsepower inboard engine. Tamashiro and his new motor-vessel, the Kongo Maru, were primed and ready to go fishing.[413]

Along with a crew of ten colonials, Tamashiro started a commercial tuna fishing operation, which would soon harvest several hundred bonito per day. Tamashiro-san, helped by the members of his family and a few Chuukese workers, cleaned and cooked the catch, laid the fish out in a shed to be dried and smoked, and then packed the finished product for export as katsuobushi -- the flaked bonito greatly prized by the Japanese as a soup flavour-enhancer.[414] The Kongo Maru and Tamashiro's processing plant, both subsidised by the Nan'yō-chō, set a precedent. Tamashiro's fellow Okinawans, getting wind of their homeboy's good fortune, literally swarmed into Chuuk. Bait was plentiful, and, assisted by government aid, the Okinawan immigrants constructed small fishing boats to start-up their own family businesses. The actual "Japanization" of Chuuk came about in a natural and (surely) benign way, and has even been warmly portrayed by Fr. Francis X. Hezel:

"When their work [of fishing, for hours or sometimes days at sea] was finally done, the Okinawans would unwind as often as not with a raucous evening of drinking and singing, merriment in which they usually invited some of their Chuukese friends to join them. There was nothing aloof or pretentious about these simple fisherfolk. Unlike the Japanese civil servants who were hidden away in their enclaves, the Okinawans made their homes in the villages and shared in the daily life of the Chuukese to a degree that few foreigners had up till then. Many of them married local women and became important figures in their local communities."[415]

A further wave of venturesome Okinawan immigrants descended on Palau and Pohnpei (Ponape) as well as Chuuk, and production of katsuobushi soon became a major industry in the Nan'yō. As usually happens in free-market/capitalist economies when a particular industry suddenly "takes off", the more powerful firms don't wait long to take control. The same held true in the Japanese Island mandate. By 1934, Nankō was building katsuobushi processing plants at Palau and Saipan, with Nambō planning to construct a facility at Pohnpei.[416] During the mid-late 1930s, over 300 medium-sized tuna/bonito vessels were operating in the Carolines; with more than 50 boats fishing out of Chuuk, and most of the rest working out of Palau. By 1937, close to 6,000 tons of katsuobushi were exported yearly to earn 5.5 million yen (¥). For dried-smoked tuna/bonito, katsuobushi, had become the second largest money-maker in the Nan'yō --ahead of phosphate and even that old stand-by, copra. Only the sugar plantations of Saipan (NMI) brought more financial income from the Island mandate back to the Home Islands.[417]

"Micronesian waters were inhabited by nearly two thousand kinds of fish, one tenth of the world's known species. Among them were varieties prized by fishermen around the globe: the strong, fast-moving Thunnidae -- tuna, bonito, skipjacks, and albacores -- as well as mackerel, mullet, herring, sardines, sea bass, and snappers. Prized above all by the Japanese were the bonito, which, dried to rock-like hardness became katsuobushi, used for flavoring soup, and the yellowfin tuna, which yielded the rich red flesh sought by sushi shops and canneries alike."[418]

The above was written by one Mark R. Peattie, who, with Nan'yō: The Rise and Fall of the Japanese in Micronesia, 1885-1945, has arguably produced the finest work in English ever done on Micronesia during the mandate period. Although surprisingly sympathetic to the Japanese (especially for an Anglophone author), Mark Peattie nonetheless has always kept in mind (like Fr. Francis Hezel) that native Islanders were indeed "strangers in their own land." (And not just under Japanese rule!) In Nan'yō, Peattie has noted that Micronesians possessed neither the capital nor the technical skills to start a commercial fishing venture -- an economic activity quickly dominated by Okinawans, as we have seen. Small industry victories were won, however, by the Islanders: "The Mortlockese harvested and smoked trepang for export to China and Japan, and some Trukese were able to fish commercially for bonito."[419]

But the mercantile omnipresence of the few Japanese mega-companies permeated the Micronesian land -- and sea -- scape. For instance, once every three months (every four months on the outer islands), the Nambō steamship would anchor off the beach to disgorge a fresh load of supplies for the many, tiny Japanese emporia -- biscuits, cigarettes, clothing, dishes, fishing gear, flour, hardware, kerosene lamps, miso paste, pans, pots, rice, sewing machines, soy sauce, sugar and tea, tobacco, towels, and watches. These were some of the goods that would mould and shape the taste and lifestyle of the Island villagers whose palm-thatched huts crowded close by. In return, the Nambō steamer would on-load the cargo that awaited -- dried tuna/bonito, trepang (namako to the Japanese), and turtleshell -- a cycle of trade which changed little during the mandate years.[420]

And Nambō (or Nankō or Nantaku) seemed to be everywhere all-the-time in the mandate; it managed fishing fleets and processing plants on Chuuk and Pohnpei (in the Eastern Carolines) and on Saipan (in the Northern Marianas). Nambō itself was based on Koror, Palau, in the Western Carolines, the modest but bustling capital city of the Japanese Nan'yō.[421] (A portrait of the busy little Japanese metropole will follow shortly.)

With typical Japanese thoroughness and attention to detail, the Nan'yō-chō (colonial administration) remained true to [national] character throughout the 1930s. As noted, the Imperial Government launched especial efforts during the 1930s to integrate the Island mandate into the greater Japanese economy. To assist the already-flourishing (mostly Okinawan-run) tuna/bonito fishery, the Nan'yō-chō established the Marine Products Experimental Station (Suisangyō Shinkenjō) at Koror on Palau in 1931. But this decision was not made until the Nan'yō-chō had (very) carefully researched the potentialities of marine resources in Micronesia... for eight years, no less (since 1923)! And not before, after carefully considered planning, the colonial administration had encouraged the nascent fishing/processing industry by distributing direct subsidies for the purchasing of boats, gear, and equipment.[422]

The Marine Products Experimental Station at Koror had been founded to complement the impressive Japanese achievements in colonial agricultural research throughout the Island mandate (eg., Pohnpei); to study further marine resources in the Nan'yō; and to perfect the catching and processing of fish and shellfish for export. Excited (and elated) by the prospect of harvesting the rich and virtually untouched Micronesian fishing grounds, and solidly backed by the Nan'yō-chō itself, the Okinawan fishermen -- and Japanese fishing companies -- poured into the Island mandate in greater numbers than ever, during the early 1930s. These proto-professional fishermen in Micronesia would be working out of Garapan on Saipan, Malakal on Palau, Kolonia on Pohnpei, and Dublon Town on Chuuk.[423]

As mentioned, the local corporate giants in the Japanese Island mandate, Nambō and Nankō (plus Nantaku, a creation of the Nan'yō-chō), came to control -- wholly -- the trading/mercantile life of the Nan'yō. (Nambō was formed so far back as 1908 during the German era and was commercially predominant by 1914.) By the mid-1930s, Nambō and Nankō were also running -- entirely -- the fishing/processing industry in Micronesia, and employed the greatest number of fishermen and processors. Due to its powerful subsidiary company, Nankō Marine Products Company (Nankō Suisan), Nankō was ichiban (#1) in fishing fleets, refrigerator plants, drying (katsuobushi) sheds, and canneries on Saipan, Palau, Chuuk and Pohnpei. Nambō was nīban (#2); followed by smaller, Island-based firms such as the South Seas Products Company (Nan'yō Bussan) which operated out of the Northern Marianas. So rewarding was commercial fishing in the Nan'yō, in fact, that some fishing outfits based in the southern Home Islands would send 60-100 ton vessels to harvest Micronesian waters...and then return to Japan without making landfall at any of the mandate Islands.[424] Indeed, the Japanese economic interest in the marine resources of the Nan'yō, coupled with a geopolitical stake in Micronesia, was deep, abiding, and total. Pacific historian Mark R. Peattie has observed:

"A host of other commercially valuable marine products from the island waters had been sought by Japanese fishermen long before the mandate period: the slippery trepang (bêche-de-mer), which was dried, packed, and shipped to the Chinese market; shark fins and shark-liver oil; the great hawksbill turtle, sought for its flesh and beautiful translucent shell; and a whole series of commercially valuable shells -- trochus shells, white oyster shells, and mother-of-pearl -- all gathered by Japanese divers....
"The fishing industry, especially pearl fishing, and the demands it made for goods and services, transformed Koror from a drowsy colonial into a thriving small city."[425]

The reader will recall the creation of the Marine Products Experimental Station (Suisangyō Shinkenjō) by the Nan'yō-chō in 1931. The Station was housed at Koror, on Palau, in a brand new building at the west end of town, with two small fishing vessels and a pearl oyster nursery under its supervision. The Station had been established to provide scientific information and technical assistance to both large companies and small concerns (mostly Okinawan) with interests/investments in the fishing/processing industry. The Station also researched and developed new marine products -- like pearls. The pearl oyster nursery, supervised by the Station was, as Father Hezel has related,

"…located in the shallow waters between the tip of Koror and Malakal… [in] racks of caged trays with thousands of mother-of-pearl shells that had been seeded [spat-on-cultch] to produce cultured pearls. The warm waters of Palau, the Japanese discovered, produced pearls in just three years."[426]

As in the case of big money-maker katsuobushi, the larger Japanese companies moved in to muscle aside the smaller firms; one of these giants being the subsidiary of Nankō, Nankō Suisan (Nankō Marine Products Company). But the pearl industry (both diving and cultured), despite all the Japanese effort and expertise, never showed more than slim profits -- exports never reached figures greater than 20-30,000 yen (¥) per annum. The ever-energetic Japanese even attempted farming sponges, but with negligible results. Ultimately, then, besides the tried-and-true marine products like turtleshell, shark fins, and sea slug (namako to the Japanese; chesobel to the Palauans), the most valuable export items taken from Micronesian waters proved to be the various shells -- trochus, mother-of-pearl, and white oyster shell. But the most reliable marine export product in value found throughout the Island mandate was tuna, always tuna (and bonito). [427]

In 1935, the Marine Products Experimental Station was replaced by a larger one placed under the aegis of the government-created entity, Nantaku. The new station's raison d'être was to envision a broader view of the wide Pacific than the mere limits of the Nan'yō, and then how better to exploit and utilise its rich oceanic resources. Indeed, the Nan'yō-chō, staying in step with the ambitious weltanschauung of Tokyo during the mid-1930s, looked far beyond the borders of the Island mandate…and the singularly successful tuna/bonito fishing industry stood as primary exemplar. Tuna/bonito fishing and katsuobushi processing earned close to 3-million yen (¥) a year by 1937, with nearly treble that amount by 1940. Roughly three thousand Japanese were employed in the fishing/processing industry by the end of the decade, and the colonial authorities sought to expand the Nan'yō's fishing ventures into new seas -- especially the grounds north and southwest of the Marshall Islands. Canneries had been built in the Marshalls by 1940, and the Nan'yō-chō -- always on the alert for new marine products and expanded seafood markets -- was considering (during those final years of peace in the Pacific) salting fish for the Dutch East Indies and elsewhere in South-East Asia (the Greater East Asia Co-Prosperity Sphere). The new station was even experimenting on possible commercial uses for poisonous fish! [428]

We must return to Koror, Palau's main settlement, and since 1920 the administrative headquarters of the Nan'yō-chō. During the next twenty years, Koror made the transition from dusty village to colonial capital and industrial centre. At first (in the 1920s), the Japanese and Okinawans in both the public and private sectors increased Koror's population to nearly two thousand. By the early 1930s, Okinawans were crowding into nearby Malakal to start-up their small fishing operations and ancillary marine activities; which included pearl-culturing. A large cannery with accompanying refrigeration plant was also constructed. No less than six Japanese contracting firms were kept busy at double-time: Building homes, offices, schools, and stores. And in 1935, an amphibious aëroplane touched down in the waters off Koror for the first time -- signaling the regular flight service that would soon commence linking Yokohama to Palau, in April 1939.[429] Even by 1935, however, Koror had not peaked yet as overseas Japanese entrepôt. Father F.X. Hezel has vividly portrayed boom-town Koror, ca. 1935 – 1940:

"Koror…still had not reached its zenith. It experienced another surge of growth after 1936, when it became a resort for Japanese pearl divers fishing for months at a time in the Arafura Sea, southwest of New Guinea. A fleet of ships, numbering 170 at the height of the pearling boom, left Palau in April each year and returned in October to winter during their off season. For several months each year, more than a thousand Japanese divers descended on Koror with money to spend and a thirst for excitement. The town could barely build enough bars, geisha houses, cafés, and brothels to keep up with the demand of these vacationing divers, who constituted the beginnings of the leisure industry in the territory. Beneath the respectable face of Koror in 1940 existed its more disreputable underside, honky-tonk quarters with 56 liquor stores and 42 cafés served by 77 geishas and 155 barmaids. Many Japanese who had originally settled on Saipan, but had been a little too late to cash in on the boom years there, moved to Koror to make another try. While the Japanese population of the Marianas declined during these years, Palau's grew more rapidly than ever; nearly 20,000 Japanese were living in Palau by the end of the 1930s. By then Koror had become a genuine city…the largest and most developed urban area the Carolines has ever seen."[430]

Koror, and especially Malakal Island, owed their phenomenal growth and expansion (for tiny, scattered Micronesia) almost entirely to the tuna/bonito fishing and pearl diving/culturing industries. Fishing concerns built their own facilities at the east end of the island, including a concrete wharf; which, along with the large cannery and refrigeration plant referred to earlier, were all constructed by Nankō (Suisan). With the continuing influx of Okinawan fishermen to Palau, to live and work in greater Koror, a model (commercial) Japanese fishing village had been created on Malakal Island -- the only one in Micronesia.[431]

Japanese mandate historian Mark R. Peattie has noted that although Koror and Malakal owed their existence to fishing, processing, and aquaculture, these industries in turn owed their development to improvements in fishing equipment and harbour facilities begun in the late 1920s.[432] For example, land was reclaimed for wharfage on Malakal Island, and a concrete pier was constructed which thrust one thousand feet out into the deeper waters of Koror harbour. Concrete causeways were also built to connect Malakal and Arakabesan Island[s] with Koror proper. Mark Peattie has summarised:

> "By the opening of the Pacific War, Koror was the mandate's unrivaled entrepôt for general commerce, it's waters sheltering passenger liners, freighters, and fishing craft -- almost all, of course, Japanese."[433]

But with the hot pursuit and terrible ending of the Pacific War, neither the tuna-fishing Japanese nor the skilled-at-sea Okinawans would return to the South Seas for twenty long years (1944-1964). In his epilogue, Peattie has rather wistfully concluded that today "Japanese tuna boats ply the waters of all major island groups as they did years ago, though by and large, the rich loads that fill their holds are destined for the fish markets of Japan rather than for katsuobushi factories in Micronesia."[434]

(3) The American/Micronesian period, 1945-present

> "Failed attempts to revive the once vibrant Japanese-manned fishing industries in Chuuk and Palau quickly demonstrated that Micronesians still viewed fishing largely in terms of subsistence."[435]
> -- David Hanlon on post-War commercial fishing in Micronesia

With proverbial heavy-handedness, the victorious new American administration didn't take long to round up and send all enemy/"Jap" civilians packing back to the Home Islands. This was accomplished without regard to the considerable numbers of Japanese men who had lived in the Island mandate for many years, had married local women, had fathered children and headed families; had even become community leaders.[436] For those Japanese and Okinawan men, Micronesia had become much more than Nan'yō the colony: The islands were work-place, home, and the familiar loci in which they thrived as self-sustaining patresfamilias. No matter -- the arbitrary U.S. authorities ordered these men "back" to the bombed-out husk that was post-War Japan. (It is this proudly American writer's opinion that, except for Operation Keelhaul which sent thousands of Russian refugees back to certain death in the Soviet Union, the United States had never committed a more stupidly arrogant or foolishly chauvinistic blunder following World War II. The European blunder was deadly for the Russians seeking freedom; the Asia/Pacific blunder was thoroughly destructive to Micronesian family life and to the Island commercial infrastructure.)

And the all-powerful and tunnel-visioned U.S. administration would pay a steep price for official Mainland ignorance. Pacific historian David Hanlon has recounted a telling tale: Assessing the U.S. Navy's efforts at economic development in post-War Micronesia, one USCC employee had written, " ' We have destroyed a twentieth-century economy, and we are now trying to put it back [together] with bailing wire and splintered boards.'"[437]

The cocky and self-confident American conquerors soon discovered the seemingly insurmountable problems of exploiting fish stocks and other marine resources such as trochus shell. Post-War markets were afflicted by fluctuating prices, trochus-shell buttons were being replaced by plastic, and the logistical difficulties of doing business in tiny but far-flung Micronesia were enormous.[438] To make matters worse, the huge U.S. investment in technical studies and training programmes weren't working. The Islanders simply were not buying into the idea of commercial fishing for profit, rather than food-fishing for subsistence. David Hanlon has recounted another telling tale. One crusty American observer, a Lieutenant-Colonel Dorothy Richard, who wondered why marine resource development was failing in (resource-rich) Micronesia, concluded that

..."the waters around Micronesia had little value for Micronesians except as a source of subsistence food and a pleasant place to sail: 'The greatest number of sea creatures in the world might inhabit those waters but they were safe from molestation by the natives'".[439]

Finally, in 1964, the Japanese and Okinawans returned to Micronesia by sea; however, this time they came back not as sailors or marines in the Imperial Navy but as the expert tuna and bonito fishermen they were. And 1964 marked the first time Micronesia had received a real influx of outside capital since the Pacific War ended. The Van Camp Seafood Company constructed a million-dollar freezer/processing plant on Palau, started up a tuna fishery by importing a fishing fleet manned by Okinawans, and commenced tuna-fishing/processing operations.[440] Van Camp encouraged three Palauans to enter the fishery, using Okinawan boats crewed by Okinawan fishermen. The Van Camp fishing fleet expanded from six to sixteen vessels, but the company never could find the Islanders willing to take over on the boats from the Okinawans -- although they were perfectly content to work ashore.[441]

Despite the Palauan aversion to work aboard the American company boats, Van Camp was landing - - in the space of a few short years - - 20-million lbs. of tuna, valued at $3-million U.S., per annum. This fishing/processing activity was pumping circa $230,000 a year into the local economy; a goodly portion of that sum representing the pay for Island plant and dock labour. Due to some notable early success, Van Camp and now Starkist too, seriously mulled the feasibility of establishing a tuna fishery on Chuuk. But this was one fish tale with a sad ending. Fr. Francis Hezel has revealed the unwanted outcome:

"Micronesia needed more Van Camp-type operations, medium to large industries that could turn its unexploited resources into capital. All the studies had called for outside investment in the Trust Territory, but American companies had never shown much interest in business ventures in the islands, and it would be a few more years yet before the ban on foreign investment was finally lifted. In the end, Van Camp and Starkist decided against beginning an operation in Chuuk, and even the Palau fishing enterprise was closed down in the early 1980s. The hope of attracting new and more lucrative business operations was proving futile."[442]

While the American companies Van Camp and Starkist strove mightily to get (and keep) the seafood industry going in Palau, the U.S. federal government - - in the form of the Trust Territory administration - - had been equally busy during the mid-1960s. Governmental authorities responsible for the USTTPI hired cracker-jack Mainland consultants, Robert R. Nathan Associates, to study Micronesia's industrial potential (agriculture/fisheries/tourism) and then to write-up a long term prospectus for the Territory. The Nathan Report, issued in December 1966, recommended that in order to make Micronesian agriculture, fisheries, and tourism profitable, that: 1) A major relocation of outer-islanders to inner-island work places be effected; 2) a large number of foreign workers (about 20-30,000 Asian aliens) be brought in to alleviate the [Native] labour shortage; and 3) all the islands of the Territory be opened to foreign investment (at least a capital outlay of $150-million U.S. be made available).[443]

The American high commissioner, one William R. Norwood, ordered a formal review of the Nathan Plan, which generally agreed with the more than 280 separate suggestions for economic development of the USTTPI. A Mr. Peter Wilson, a Territorial government fisheries officer, approved the Nathan Plan's idea of bringing foreign workers into the Islands; indeed, he concurred, local tuna fisheries and boat-building operations could hardly be expected to expand without imported labour. Furthermore, wrote Peter Wilson cited by David Hanlon,

"Micronesians...needed to be shown 'What it means to put in a full day's labor.' Wilson felt that Micronesian laborers loafed on the job, used their potential and their ethnic identity as excuses, and were secure in the knowledge that it would be almost impossible to replace them. Real economic development argued Wilson, would depend on Micronesians' ability to compete effectively against Japanese, Okinawan, and Taiwanese fishermen..."[444]

Luckily (?!) for the Micronesians and the Americans, the Nathan Plan's ways and means for solving the USTTPI's dearth of labour and shortage of capital proved unacceptable to the powers that were…and the Nathan Report languished along with many others in that great maw of forgotten U.S. files.[445]

All was not lost, however, for the cause of commercial fishing in Micronesia. Despite the eventual shutdown of Mainland fishing and processing enterprises on Palau during the early 1980s, Islanders showed their commitment to marine resource development and aquaculture, by establishing in 1973 the Micronesian Mariculture Demonstration Center in Palau.[446] Thus a flurry of maritime-oriented activities followed: Local boat-building projects, stepped-up marine infrastructure development; with various surveys and workshops run by international agencies such as the Japan International Cooperation Agency, the South Pacific Commission (SPC), and the U.N.'s Food and Agriculture Organization.[447] In 1974, Japan donated seven fishing vessels (some have charged to ward off accusations from that nation's reluctance to redress Micronesian war claims); and an American corporate-sponsored organization, the Pacific Tuna Development Corporation, offered assistance to Island fishermen (some have charged to facilitate its own access to the tuna-rich waters of Micronesia). But, as David Hanlon has analysed:

"All of these efforts seemed not to make a great deal of difference. Although the reported total value of marine exports increased from $27,000 in 1963 to roughly $3-million by 1976, commercial fisheries in Micronesia never generated the kind of revenues required to make fishing a viable, major contributor to the [T]erritory's economy…. Only with the establishment of the Micronesian Maritime Authority in 1978 and the subsequent licensing of foreign fishing vessels within the territorial waters of the three self-governing entities to emerge from the Trust Territory would commercial fishing provide any consistent form of revenue."[448]

A potentially big economic breakthrough came at last for the Islands in 1974. For 1974 was the year that U.S. Secretary of the Interior Rogers Morton (who was head honcho for the TTPI) announced the lifting of restrictions on all foreign investment in (American-controlled) Micronesia. Outside American investment, although permitted since 1962, had simply not been enough to fire-up the barely-sputtering Island economies. The Congress of Micronesia (COM), which had been voted into place by a Territory-wide election in January 1965 (replacing the more U.S. – compliant Council of Micronesia), firmly spelled out the regulations and laid down the conditions for the Foreign Investment Act. Thus the U.S. administration in 1974 had opened the door to investors other than American (i.e., the Japanese) to contract business and to sell goods in the hitherto untouchable U.S. Trust Territory of the Pacific Islands.[449]

For Micronesia, the suddenly-opened portal to the world set quite a precedent. The Congress of Micronesia had waited for this juncture for nearly a decade (and more), and wasted no time in passing legislation that established district fishery (and other) offices; sponsored developmental projects in agriculture, fisheries, tourism, and vocational education; and granted tax exemptions to needy local businesses, while providing tax relief for local harvesters of copra and trochus shell (Trochus niloticus).[450] And there was more. The Micronesian Development Bank (MDB) was set up and empowered to locally manage the funds appropriated by the U.S. Congress to the Economic Development Loan Fund (EDLF). The Congress of Micronesia (COM) also created foreign investment boards in each Island district, levied taxes on "luxury items" to curb consumption, encouraged "import substitution", and bolstered local entrepreneurs. But the crowning achievement (and of greatest interest to us) was the passage of a 1977 law which provided for a 200-mile [Island] EEZ - - with local jurisdiction over all commercially exploitable "living" marine resources within it.[451] Micronesia's place in the sun seemed to have finally arrived.

Francis X. Hezel, S.J., who has lived in Chuuk and Pohnpei for an accumulated thirty-five years, has made a cogent statement on the overall effect of the Law of the Sea (UNCLOS III, 1973-1982 phase) on the Island geopolitical psyche:

"The soundest economic hope for Micronesia seemed to be in its abundant but largely unexploited marine resources. The Congress of Micronesia therefore showed a keen interest in the series of UN-sponsored international conferences to draw up the Law of the Sea, a treaty governing the global distribution and use of ocean resources. In 1972 the congress created a joint committee that met with the United States delegation to promote its position; when it discovered that its interests were not necessarily those of the United States, Micronesia began sending its own representatives to the international conferences. After years of such meetings, the Law of the Sea was finally completed and signed by representatives of nations around the world. The law gave each nation exclusive control of a two-hundred-mile economic resource zone in its offshore waters, along with all fishing rights and undersea mineral rights. Meanwhile, in October 1977, the Congress of Micronesia passed its own legislation establishing a two-hundred-mile zone around its islands, giving Micronesia fishing rights to nearly two million square miles of sea. Land-poor Micronesia had suddenly become ocean-rich. The Micronesian Maritime Authority was created to regulate this economic zone, and the new Micronesian nation-states were able to begin negotiating treaties with foreign nations that soon brought in yearly payments of about $5-million in exchange for fishing rights in their waters...."[452]

But there was an additional impact from Micronesian involvement in the Law of the Sea - - UNCLOS III manifested to Islanders the pragmatic significance (as well as local pride) of their retaining sovereignty in foreign affairs if they were to take responsibility for their own future economy. The United States agreed in 1978 (Carter administration) to amend the Compact, thus granting the new Micronesian governments control over their own foreign affairs.[453]

Micronesia was on her way, but with increasingly less rationale to blame all ills, real and imagined, on the American crocodile. The onus since independence has been on Micronesians to show the world what they can do.[454] The fish are there....

(B) "A Modest Proposal" for Micronesia

"The central and western Pacific tuna fishery is the only one of the world's major fish resources that has not so far been overfished."
- - Forum Fisheries Agency, 1997 [455]

...."Coral atolls in Micronesia might be converted into tuna pastures that fatten yellowfin and restrict their foraging on anchovies and other valuable herbivores."[456]
- - John Isaacs of the Scripps Institution of Oceanography (LaJolla, Calif.) on solving world hunger in International Science and Technology, April 1967, cited by Wesley Marx (1970)

(1) Past: The Pacific War and the American post-War nuclear bombing tests, 1941-1963

After the horror and destruction of the Pacific War (1941-1945), followed by the American post-War nuclear bombing tests (1946-1963), the waters around the Marshall Islands most assuredly did not appear a likely marine area for reviving commercial fisheries (of any kind, any time soon). In his fashionably pessimistic The Frail Ocean published in 1970, eco-Cassandra Wesley Marx pointed with accusatory finger at the "hot clams" of the Marshall Islands.[457] Marx indicated that although the Marshalls had not been exposed to radioactive fallout for two years, scientists (remember, during the mid-1960's) had been puzzled to find high concentrations of the radioisotope Cobalt-60 in clams near the Islands. Furthermore, despite the two-year absence of radiation, scientists knew that Cobalt-60 is not a by-product of atomic fission. So where had the radioactivity present in the clams come from? For a possible explanation, Marx quoted Dr. Lamont Cole of Cornell University (Ithaca, N.Y.):

" 'It [Cobalt-60] must have been produced by the action of radiation on some chemical, presumably a stable cobalt isotope, naturally present in the water.' "[458]

In other words, pursued Marx, the Ocean has an ill-defined and poorly understood capacity to react to nuclear irradiation. Marx went on to gloomily guesstimate, in the enviro-alarmist tradition of Rachel Carson and Paul Ehrlich, that within a few decades (from 1970...that's now!) the Earth would be faced with the disposal of 1,000 tons of high level nuclear waste-fission products per annum. And what "better" repository for these highly-toxic nuclear wastes would do than the very bottom of the vast, deep Pacific Ocean - - preferably the Central-Western portion; around and about sparsely populated by (polluting) Man? Indeed, claimed Marx, the 1954 A-bomb tests mixed fission by-products - - up to a half-ton in volume - - into the upper water column of the CWP within a short time, to wit:

"'That this was near the limit of safety [was] evidenced by the capture in adjacent areas of specimens of tunas and other fishes with sufficient radioactivity to be doubtful for human consumption.'"[459]

Marx re-quoted "'adjacent areas'" to make a salient point - - namely, that 1-million square miles of Micronesian waters were affected by/infected with highly-toxic nuclear wastes. A gripping fear of their own environment kept Marshallese Rongelap Islanders from feeding on coconut crabs (Birgus latro), a staple. For five years thereafter, reported Marx dolefully, U.S. federal government food shipments subsidised the Rongelap diet.[460] Despite his typical 1960s moaning and groaning, Wesley Marx did make his point. And this writer basically concurs, as the reader has discovered from my earlier "Micronesia Notes" (Foreword to this Chapter).

After the horror and destruction of the Pacific War and the post-War American nuclear tests in the Marshall Islands, the waters around Micronesia did not appear a likely marine area for reviving commercial fisheries of any kind; at any time soon.

(2) Present: The Micronesian Commercial Fisheries, 1990-2000

In Part III we thoroughly examined the pelagic fishes and tuna fisheries of the Central-Western Pacific (CWP). Also, in former parts of this concluding chapter, we have analysed some of the international ramifications that influence fishery [geo]policies of the small "coastal States" which comprise the Pacific Island nations. This has included the mission, work, and protective role of the 16-member South Pacific Forum Fisheries Agency. As we have seen, the FFA handles all fishing agreements between the tiny Pacific Island members (the coastal States) and the powerful distant water fishing nations (DWFN) of the United States, Japan, South Korea, and Taiwan (the "fishing States").

In 1990, the U.S. Congress voted ("overwhelmingly") to manage federally the tuna stocks within the American EEZ; thus settling once and forever the Pacific Island claims to all pelagic fishes inside their own 200-mile limits.[461] (The fish once scoffingly referred to by American tuna fishermen as "migratory", now had owners.) Today, the Tuna Treaty provides U.S. help to FFA member-nations in establishing fishing industries of their own.[462] The United States has also sufficiently eased (too) tight U.S. Customs rules and regulations to permit the Northern Marianas, Palau, FSM, and the Marshalls - - as a group - - the option of supplying up to 10% of the American canned tuna market, without being penalised. Meanwhile the Japanese, in order to maintain their interests in the Micronesian region, have sponsored fisheries development and constructed Made-in-Japan freezer-plants throughout the former Nan'yō/TTPI.[463]

The Americans and the Japanese, as expected, take the [sea] lion's share of the tuna from the CWP fishing grounds; with (ca. 1992) 32 of the 110 high-tech purse-seiners in Japanese hands, and a further 65 affiliated with the American Tunaboat Association (ATA). East Asian longliners supply the enormous Japanese sashimi market. Some Island nations have tried to catch up, to compete, with their old-style pole-and-line boats, but the American, Japanese, and other DWFN big-money/high-tech purse-seine fleets leave the Micronesians far behind in their wake. Travel author David Stanley, in his Micronesia Handbook, observed during the early 1990s:

"For the foreseeable future, licensing fees will be the [I]sland governments' only share of this [marine] resource. Meanwhile fisheries researchers in Belau have been successful in breeding giant clams. The clams, which produce meat faster than any terrestrial animal, could become the basis of a major seafood industry serving the $100-million-a-year Asian clam-meat market."[464]

But a united, co-operating Micronesia - - all seven political entities - - can have her [fish] cake and eat it too. Together the seven Micronesian areas (five presently American or formerly U.S.-administered, and two formerly British or Commonwealth-administered) might only occupy a land area of 3,227 sq. km, but their combined EEZs encompass 11,649,000 sq. km of Central-Western Pacific Ocean.[465] And therein lies the key to Micronesia's power and future. Before spelling out the specifics of a "modest proposal" for inter-Island geopolitical co-operation however, a brief updated review of Micronesian commercial fisheries is in order.

I. U.S. Territory of Guam

U.S. federal regulations (stupidly?) prohibit foreign tuna-fishing vessels from off-loading in American ports. The (obviously!) bad results are two-fold: Nearly all seafood consumed on [the island of] Guam is processed and purchased overseas, and the enormous potential of fishing-related industries - - canneries and freezer-plants - - has never been developed. Fortunately for the U.S. Territory, much chilled product is unloaded from American tuna boats at Guam (where many ATA fishing vessels also homeport) and air-freighted to Japan (Tokyo's Tsukiji Central Wholesale Market) for the sashimi trade. Less fortunately (perhaps) for Guamanians, control of all marine resources and undersea minerals within the Micronesian island's 200-nautical mile EEZ remains firmly in federal hands at Washington, D.C.[466]

II. Commonwealth of Northern Mariana Islands (CNMI)

In David Stanley's comprehensive Micronesia Handbook (1992 ed.), there is only a single snippet of information on commercial fishing in the CNMI (and I quote): "U.S. and [East] Asian fishing boats use Tinian's spacious harbor to transfer tuna caught in FSM and PNG [Papua New Guinea] waters to refrigerated freighters bound for U.S. canneries...."[467]

As the NMI enjoy U.S. commonwealth status rather than territorial, I surmise that this fact/difference explains the presence of Asian vessels off-loading at Tinian Harbor. On the Micronesia Handbook map of San Jose (city), I spy a freezer-plant overlooking the water-front.[468] I would wager that both fishermen and shore workers are grateful for it.

III. The Republic of Palau

David Stanley has opined that "[d]espite lacking a beach, Koror is by far the most scenic town in Micronesia."[469] Maybe location (as well as geography) was a major reason why Koror (today's Oreor) was made capital city - - to become main entrepôt - - of the Japanese Nan'yō. And today, as under the Japanese Nan'yō-chō, nearby Malakal Island serves Koror - - indeed, all Palau - - as industrial suburb and commercial port. Malakal boasts a cold-storage plant for tuna exported to Japan, the next door Fisheries Co-op vends fresh fish daily, and close by is situated the Belau Boat Yard. Down at the road's end on Malakal Island proudly stands the Micronesian Mariculture Demonstration Center (MMDC), known locally as "da ice box" - - the aedifice was once a Japanese ice-making plant.[470]

The MMDC building/complex, which started its new life under the Americans as a Trust Territory marine research facility in 1974, currently produces commercially the largest volume of tridacnid (giant) clams on the globe. On tiny Palau, the seedlings of Tridacnidae are sown to replenish and repopulate the earth's tropical seas with this fascinating (and increasing rare-in-the-wild) giant clam. Not only has Tridacna recruitment been regenerated, the MMDC has accomplished minor miracles in hatching endangered hawksbill sea-turtle (Eretmochelys imbricata) eggs, and raising the young in holding tanks until ready for release into the ocean. Other marine life such as trochus (shell) and reef fish are also farmed commercially in outdoor tanks at the MMDC on Malakal Island.[471]

If diminutive Palau's herculean (and fruitful) efforts in mariculture were not sufficient, the Palauan marine environment is itself absolutely astonishing; for the scuba-diving tourist on eco-vacation as well as the more serious-minded marine biologist. It was the classic book on "ethnofishing", <u>Words of the Lagoon</u> by R.E. Johannes first published twenty years ago (1981), which brought Palau's marine environment to the attention of interested readers world-wide.[472] In his updated and excellent review of Palau's marine environment, in <u>Song for the Blue Ocean</u> (1998), Carl Safina has described in detail the uniqueness of life beneath Palauan seas:

> "The richest and most diverse collection of marine life in the world lives among the tens and thousands of islands of Indonesia and the Philippines. Standing closest to this biological hot spot, tiny Palau harbors a richer species assemblage than any other single oceanic island group. In the 1970's, two of Bob [R.E.] Johannes's colleagues found thirteen species of fish that were new to science during a two-hour dive. [!] Another colleague later recorded 163 species of corals during a four-hundred-yard swim in water less than twenty feet deep: twice as many species as are known from the entire Caribbean to a depth of three hundred feet.[!!] That survey was done in Palau's main harbor, slated at the time for dredging to construct a U.S. submarine base during the Cold War.
>
> "Few places in the world have more than two species of sea grass. Palau has nine. Hawaii has thirteen species of damselfish. Palau, about eighty. One tiny cove in Palau holds all seven of the world's species of giant clams.[!!!] Palau has fringing reefs, lagoons, patch reefs, barrier reefs, mangrove communities, and more than twenty marine lakes holding delicate forms of life not seen elsewhere. A greater variety of marine habitats pack Palau's waters than in any area of similar size anywhere else in the world, and more than most places regardless of size.[!!!!]"[473]

Besides Palau's great natural beauty and exciting dive-sites making the republic an eco-tourism mecca, it is the wide (and deep) diversity of the archipelago's marine life that makes the waters of Palau very special. For example, a field station of the Coral Reef Research Foundation (est. 1994), partially funded by the National Cancer Institute (a division of the National Institutes of Health in the U.S.A.), is located at Palau. It is here, run by field station manager Larry Sharron, that various marine invertebrates such as sea sponges and sea squirts (ascidians) are collected for experimentation as potential pharmaceuticals to combat AIDS, cancer and other fatal human maladies. Centres of high bio-diversity such as tropical rain forests and coral reefs have unlocked a multiplicity of priceless bio-medical treasures.[474]

<u>Blue Ocean</u> author Carl Safina noted, while visiting the CRRF field station on Palau, that the aquarium tanks featured toxin-laden fishes and venomous sea urchins. When asked why by Carl Safina, station-manager Larry Sharron answered (in effect): The more toxic, the more promising! Which meant that some of those sea creatures' poisonous defense-chemicals make good compounds for drugs. For instance, Sharron continued, certain sponges are soft-bodied, brightly-coloured, and appear to live

dangerously in the open. Yet these sponges are rarely (if ever) preyed upon due to their poisonous defense-chemicals. These chemicals in many organisms kill cells indiscriminately, but the National Cancer Institute (NCI) seeks chemicals not harmful to cells generally; just those which might kill cancer cells or malignant viruses. With an almost incredible range of bio-diversity in Palauan waters, concluded Larry Sharron to Carl Safina, this [Micronesian] region is a great place to start searching![475]

Palau field manager Sharron, Safina reported, humbly regards himself as a mere "collecting agent." The Coral Reef Research Foundation sends specimens on to the NCI, which grinds up the samples, and, using a computer, tests extracts of the sample specimen against sixty types of cancer cells and the AIDS virus. In the case of the CRRF finding a marine substance that turns out to be the sure-cure for AIDS, the selfless CCRF would receive no monetary compensation…but the Republic of Palau would. Indeed, the CRRF has entered into a "letter of intent" agreement between the host government[s] and the National Cancer Institute. As the Coral Reef Research Foundation is federally funded by the U.S.A., bio-medical science, the Republic of Palau, and the human AIDS or cancer patient all win.[476] (And, for once, the American tax-payer!)

During his sojourn on Palau, Carl Safina also "bumped into" Gerald Heslinga, the renowned aquaculturist who devised the successful method for farming giant tridacnid clams (discussed previously). Heslinga, a non-Islander haole, accomplished that great bio-scientific feat while [very gainfully!] employed by Palau's governmental Division of Marine Resources at the Mariculture Center. Because of Heslinga's truly brilliant work, cultured Tridacnidae have replenished the severely depleted stocks of over-harvested giant clams in the wild. And for every Tridacna sold commercially, Heslinga and project colleagues would transfer giant clams back to the ocean or donate seedlings to the breeding programmes of other Pacific Island nations.[477]

Carl Safina summed up Gerald Heslinga's giant clam farming project as… "a model program, filling a real conservation need both locally and over a large region, while making its own way in the commercial market, attracting investment, providing food, and turning a profit."[478]

At the bottom of the last page of Daniel Knop's fine (German-published) book on tridacnid clams, is a lithograph by a W. Schmettkamp.[479] The engraving depicts, in minute detail, a block of (1996) stamps from Palau. The four postage stamps portray the giant clams being farmed - - and in the wild. Above the block of four the heading proudly states:

"CONSERVATION AND CULTIVATION OF GIANT CLAMS IN PALAU
"The Micronesian Mariculture Demonstration Center in Palau has pioneered in low cost, low technology methods of raising giant clams. MMDC has produced millions of seed clams for distribution to all coastal areas of Palau and to more than 17 foreign countries."[480]

As for commercial fishing for pelagic species in the seas off Palau, the former Van Camp freezer-plant (1964-1982) is now utilised by the tuna-boats of three major fishing companies - - two are foreign with one owned by Palauan and "outside interests."[481] They are (respectively) Palau International Trading Inc., Micronesian Industries Corp., and Kuniyoshi Fishing. The high-quality Island tuna, considered top-grade sashimi by the Japanese, is flown fresh to [Tsukiji Central Wholesale?] market from Palau on two company-owned planes.[482] All is not lost for local fishing, though. There are thriving commercial and subsistence (food) fisheries, which mainly target reef and offshore species; with brisk local sales and consumption to match. Elizabeth Rechebei and Samuel F. McPhetres (1997) have written in History of Palau: "The Palau fisheries cooperative, based on Ngemelachel, has provided fresh fish for local consumption and shipment to overseas relatives for many years."[483]

IV. Federal States of Micronesia (FSM)

"But fishing spurred the first significant population increase on the island [of Pohnpei]. The first Okinawans to enter Ponape Harbor were independent fishermen and their families, but they were soon out numbered by those working for the Nankō Suisan, whose katsuobushi factory continued to draw in fishermen and laborers, thereby setting off a booming industry along the waterfront. As on the other islands, the needs of a fast-growing fishing community stimulated a demand for a variety of goods, services, and housing."[484]

The above passage is taken from Mark R. Peattie's Nan'yō, the outstanding work on the rise and fall of the Japanese in Micronesia (see note 418). As expressed and reiterated throughout Part IV, the Japanese - - the Okinawans - - started up the first real commercial fisheries in Micronesia during the mandate years. The reader will notice that I have written "fisheries"...plural. Recently, the Japanese presence regarding fisheries - - in this case mariculture - - surfaced again after more than half a century. In an interesting sidebar of the March/April 1998 edition of Pacific Magazine, columnist Gene Ashby wrote that in 1937 the Nan'yō-chō had had 12,000 trochus shell (Trochus niloticus) seedlings planted in Pohnpei's lagoon.[485] Then, Ashby followed up, "The relatives of these progenitors are much coveted for the manufacture of buttons by the Japanese."[486]

The upshot of Ashby's 1998 report was that a foreign investment business permit application had, of late, been submitted to the FSM government by the Tomei Shoji Company Corp. of Japan. For in 1998, Tomei Shoji seriously considered buying trochus shell from all four States of the FSM (Pohnpei, Kosrae, Chuuk, and Yap); with a shell-button plant to be established at Weno (Moen) on Chuuk. Tomei Shoji's up-front investment was said to amount to $U.S. 100,000 - - with the trochus shell-buttons planned for local use but mostly to earn (badly needed) export dollars. Ashby concluded his sidebar with the wry notice that, "comments from the four state governments are required before the permit can be granted."[487] (Since then, although a religious reader of/subscriber to Pacific Magazine, I have neither seen nor heard of any further developments. Hopefully the deal went through…for Micronesia's sake.)

The only other mention of a trochus shell industry in Pohnpei I found was in David Stanley's Micronesia Handbook, 1992 ed.,

"A branch of Island Traders behind the Chinese Embassy nearby [in Kolonia] makes buttons from trochus shells and mother-of-pearl…"[488]

Some might sneer at these small, so-called "cottage industries", but there exists no doubt concerning the crucial significance and future potential of the FSM commercial fishing industry. In 1988, fish was the primary product of the FSM, accounting for (a modest) U.S. $1,524,000 in earned income. The next year, 1989, saw (a mere) $10-million in fees paid by circa 200 foreign (DWFN) vessels to fish within the FSM 200-mile EEZ…the irony being that approximately $200-million of valuable tunafish would be hauled from Micronesian waters by non-Islanders. (Albeit $1.2-million was collected in penalty fines from boats caught fishing illegally).[489] Anyway, by 1990 the FSM authorities took a pro-active step on their own: The government-owned National Fisheries Corp. (NFC) ordered a fleet of long-liners, and combined in a joint-venture with a Franco-Australian company to form the Caroline Fisheries Corp., with three purse-seiners.[490]

The Compact of Free Association, as noted, permits the duty-free entry of FSM canned tuna product to the American market - - with up to 10% of Mainland consumption. To take advantage of this largesse in U.S. federal laws (during the early 1990s), canneries were planned for some islands, a fish-processing plant was constructed near Pohnpei Airport, and an aquaculture centre was opened on Kosrae. In 1991 a magnanimous Australia grandly presented the FSM with a single patrol boat, the FSS Palikir, with which to police her vast fishing grounds.[491](!Where had been vastly richer, former mistress, America?)

And by 1992 the governmental Economic Developmental Authority had built the brand-new Pohnpei Fisheries Complex, which included the fish-processing plant near Pohnpei Airport...supplied with tuna provided by DWFN fishermen.[492] (Another sad irony!)

As stated at the outset, there are no doubts concerning the crucial significance and future potential of the FSM commercial fishing industry. For the FSM represents, more than any other Island group, the very idea of Micronesia - - and Micronesian pride. In fact, the Federated States of Micronesia is the sole Micronesian political unit with "Micronesia" in her name. As the reader well knows by now, the FSM is composed of the four States of Yap, Chuuk (Truk), Pohnpei (Ponape) and Kosrae (Kusaie). The administrative capital, Palikir, is situated in Pohnpei State. From west (Yap/Waab) to east (Kosrae), the FSM's 200-mile EEZ encircles 1.3 –million sq. mi. (greater than 3-million sq. km.) of Central-West Pacific Ocean.[493] The view - - the agenda - - of commercial fishing in the collective mind of the FSM can best be understood through the institution of the National Fisheries Corporation. The NFC is a public corporation established by the Federation Government in 1984. (The reader will recall that the Compact of Free Association with the U.S.A. was approved in 1986, and independence under the Compact as the FSM followed in 1990.) The stated aim of the NFC is "to develop and promote a profitable and long-term commercial fishery within the FSM."[494] In addition to pursuing its own industry development programmes, the NFC works closely with the individual States in joint fishery projects.

As examined previously, the fisheries of highly-migratory species (HMS, the tunas) - - of all Micronesia - - are the most important commercially in the CWP. (See Appendix I.) It is no different in the rich waters of the FSM, which, during the mid-1990s, yielded in excess of 150,000 tonnes of tunafish annually.[495] The commercial tuna fishery of the FSM, like others in the CWP, is two-fold: (1) the purse-seine fishery, which targets surface-schooling fish (skipjack/aku and small yellowfin/ahi); and (2) the long-line fishery, that targets deepwater species (bigeye/ahi and large yellowfin/ahi). A commercial groundfish (bottomfish) fishery has also been underway in the FSM, with targeted species being deepwater snappers (Family Lutjanidae) and deepwater groupers (Family Serranidae). According to the National Fisheries Corp., more than 50 locally-operated "smaller-vessels" commercially fish within the FSM 200-mile EEZ, utilising long-line, pole-and-line and bottomfishing methods. These "smaller-vessels" vary in size from 1.5 up to 30 tonnes, and are fitted with modern equipment.[496] Meanwhile (however), over 100 purse-seiners and 300 long-liners from DWFNs continue to haul valuable tuna from the rich waters of the FSM.[497]

Regarding opportunities for foreign investment and co-operation, the FSM (via the NFC) has tentatively offered:

"One of the national goals in the FSM is to become largely involved in the development and management of its resources. To achieve this goal the National Fisheries Corporation is seeking joint ventures and closer co-operation with other countries and their fishing industries.
"The National Fisheries Corporation is also the conduit through which transshipment by foreign fishing vessels maybe [sic] conducted in the FSM, offering reasonable and uniform arrangements from FSM States."[498]

This writer has employed the word "tentatively", as the FSM - - like the other Island nations of the CWP - - sorely needs financial help from, and joint-ventures with, the richer and more powerful "fishing States" such as the U.S.A. and Japan. But the FSM (similar to the other "coastal States") is fiercely independent and jealously protective of her EEZ sovereignty and rights. The determined national will is expressed in a no-nonsense Statement of Purpose, Section 101 (the Preamble), Title 24 of the Code of the Federated States of Micronesia:

"The resources of the sea around the Federated States of Micronesia are a finite but renewable part of the physical heritage of our people. As the Federated States of Micronesia has only limited land-based resources, the sea provides the primary means for the development of economic viability which is necessary to provide the foundation for political stability. The resources of the sea must be managed, concerned, and developed for the benefit of the people living today and for the generations of citizens to come. For this reason the harvesting of this resource, both domestic and foreign, must be monitored, and when necessary, controlled. The purpose of this title [24] is to promote conservation, management, and development of the marine resources of the Federated States of Micronesia, generate the maximum benefit for the Nation from foreign fishing, and to promote the development of a domestic fishing industry."[499]

To back up the Nation's Code (i.e., give it some teeth), the FSM established the Micronesian Maritime Authority (MMA) composed of five members appointed by the President; one at-large member plus a member from each of the four States. All appointments to the MMA would be for a term of two years, and none (interestingly) could serve on the Board of Directors of the National Fisheries Corporation (NFC).[500] The establishment of the Micronesian Maritime Authority, this writer strongly believes, was a firm first step in Micronesian economic independence, and historic in its implications.

The MMA didn't have long to wait before getting right to work. Although not able to effectively develop a draft National Fisheries Policy by 1992,[501] the MMA nonetheless did some rough negotiating with tough and troublesome DWFNs such as the United States, Japan, South Korea (ROK), and Taiwan (ROC) during 1992-1993.[502] In 1993, the MMA even entered into talks with two long-liner companies from the People's Republic of China (PRC) - - marking the FSM's first fisheries access agreement with the PRC.[503] And in June 1993, the MMA, on behalf of the FSM, proudly hosted the Sixth Standing Committee Meeting on Tuna and Billfish plus the Third Western Pacific Yellowfin Research Group at Palikir, Pohnpei. (Both meetings were high-tech and attended by marine scientists, resources managers, and policy makers from the South Pacific region.)[504]

Besides continued wrangling with the potent and always-intrusive "fishing States", the MMA saw the demise in 1994 of the Caroline Fisheries Corporation; the failed joint-venture referred to earlier.[505] The other joint-venture in the FSM, between the National Fisheries Corp. and Japan Air-Freighter JV, appeared headed for fruition in 1994. Under the proposed NFC/Japanese joint-venture, approximately ¥ 700-million (yen) would be made available to the NFC (via the Japanese government) under a special grant from the Japan Overseas Fisheries Cooperative Foundation to the JV for purchase/lease of an aëroplane...to serve/support Japanese long-liners based in the FSM (!). But the JV was also set up to spawn new fishing outfits in the FSM...whose various affairs/activities would be handled by the Japanese(!!).[506]

If the above JV sounds beneficial only to the Japanese, the reader should recall the essential tiny-ness, the micro-ness, of Micronesia in comparison to the wealth and sea-power of the "fishing States." The FSM government and the individual State authorities have ambitiously developed fisheries projects. In addition to the National Fisheries Corp. (NFC) with five long-liners in 1994, the Pohnpei Fisheries Corp. (PFC), a State-owned/operated fish-processing facility, was constructed during late 1992. And in 1994 the FSM secured a U.S. $6.5-million loan from the Asian Development Bank (ADB) that would create a private-sector, all-Island, long-line fishing company; the Micronesian Longline Fishing Company (MLFC). Under the ADB loan package, six brand-new longliners will (1994) be purchased, and then managed/operated by Micronesian citizens.[507]

To complement the State of Pohnpei's efforts to develop the Micronesian fishing/processing industry, the State of Yap in 1994 operated the Yap Purse Seine Fishing Corp. (YPSFC), with four purse-seiners working the 1993/1994 season - - with a fifth purse-seiner being refitted for the 1994/1995 season. In 1992 the Chuuk State government registered two longliners, and by 1993 a third boat was registered and being managed/operated by the NFC. And two longline vessels were registered by Kosrae State for domestic fishing permits in 1992 and 1993. Last but not least, again through an Asia Development Bank

(ADB) financial loan, and with instructive hands-on help (as well as donating badly-needed fishing equipment) from the Japanese Overseas Fisheries Cooperative Foundation, the Micronesian Maritime and Fisheries Academy (MMFA) has been sustained by both. The MMFA continues to operate, currently (1994) funded by the FSM and the State of Yap, with the Pacific Missionary Aviation Corporation overseeing the daily management of the Academy. The MMFA's longline fishing programme has successfully emphasised (on-the-job) training and testing of young Micronesians at sea. With fisheries development in the FSM at high tide, the Micronesian Maritime and Fisheries Academy will play an even more vital rôle in the future success of the domestic fishing industry in the Islands.[508]

Meanwhile, vigilantly watching all the various national bodies and taking notes on the international events, has been the MMA - - the Micronesian Maritime Authority indeed. The MMA, as stated, hosted two Pacific Rim fisheries meetings at Pohnpei in June 1993,[509] and represented the FSM at the U.N. Conference on Straddling Fish Stocks and Highly Migratory Fish Stocks held at UNO HQ at New York City, July 199[4].[510] As important an occurrence in 1994, this writer believes, was the FSM/Kiribati Reciprocal Licensing Arrangement. In effect, the two Pacific Island nations agreed to permit one purse-seiner, each from the other's fleet, to fish in the other's EEZ. This no-small-feat, too, was accomplished by the MMA.[511] The MMA concluded both their 1992-1993 and 1994 Report[s] with these thoughts:

"There is evidently a major shift in FSM fisheries policy over the last two years. The traditional payment of fishing rights fees to the FSM by DWFN's, to access the tuna resources of the FSM, is being replaced by new emphasis on developing the local capability to harvest the tuna resources and to build up the necessary infrastructures to support an FSM tuna industry. These sentiments were reiterated and continue to be the major issues at the National and State Leadership Conferences held over the last two years. The Authority [MMA] will play a major role in the development of the FSM tuna industry for the years to come, and to this effect, will rely heavily on the continued cooperation and support of the National and State Governments and the various development entities throughout the FSM.

"As the demand for access to the FSM EEZ by new entrants into the fishery expands and the exploitation of fisheries resources increases, the Authority predicts that fisheries issues will inevitably bring countries of the region together more often to discuss issues of mutual interest. This will result in increased future cooperation on fisheries with our neighboring countries and DWFN's."[512]

If there is ever to be an all-Micronesia combined fishing effort, the four united states of the FSM will lead the way; they are already well and ably represented by the Micronesian Maritime Authority. (We will return to the MMA at the conclusion of Part IV.)

V. Republic of the Marshall Islands (RMI)

Although the RMI is still heavily dependent on financial aid from the United States, the young Pacific Island nation appears to be doing all she can to develop her own industries - - and that naturally includes fishing. During the early 1990s, two hundred or so Japanese vessels continued to pay close to $1-million U.S. annually to haul $150-million worth of expensive tuna from Marshallese waters.[513] And the early 1990s saw the highly "Americanized" centre of Majuro becoming a transshipment point for chilled skipjack/aku air-freighted to Japan via Hawai'i. One project underway has been the culturing (as on Palau) of giant Tridacna clams on the islands of Arno, Mili, and Likiep. Another project has been the initiation of Japanese-funded "black pearl" (Pinctada margaritafera) farms on Arno and Namorik islands. A further marine-related industry has been the export of live tropical fish to aquaria on the U.S. Mainland.[514]

By the late 1990s, the fishing picture in the Marshall Islands had undergone a sea-change. First, the Marshallese equivalent of the Micronesian Maritime Authority (MMA), the Marshall Islands Marine Resources Authority (MIMRA), looked to be in the good hands of one Danny Wase, MIMRA director.[515] Second, the MIMRA sponsored a [black-lipped] pearl oyster-spawning project that successfully reproduced spat locally in land-based tanks. The pearl oyster-spawning project, done in conjunction with Black Pearl, Inc. of Hawai'i, is being supported by funds donated by the U.S. Dept. of the Interior. Successful spawning locally means no more sending brood stock to the Big Island of Hawai'i; a profitable aqua-business in the RMI can be achieved by building a large enough, full-scale, hatchery to eventually produce pearl oysters in commercially viable quantities.[516]

A third change in the RMI fisheries scenario of the late 1990s was the temporary but drastic reduction of Japanese vessels fishing Marshallese waters. (Both Asians and Islanders learned a valuable lesson.) In fact, the number of Japanese tuna boats licensed to fish had dropped to less than thirty in 1997. But in 1998 the Japanese were returning; the increase mainly due to new applications by purse-seiners, along with the usual longliners and the pole-and-line vessels normally used. During the 1990s, the RMI had earned on average of U.S. $1.2-million per annum in fishing revenue from the Japanese, or 5% of the landed value of the catch. The RMI hoped in 1998 to top the high set during 1994 of $1.75-million.[517] However, inferred Giff Johnson of Pacific Magazine, a new respect has been instilled in the Japanese for the Marshalls which had partially caused the initial drop-off:

"The Japanese concerns were over their fear that their vessels were being targeted for arrest by the Marshalls Sea Patrol after one of their boats was boarded for alleged illegal fishing. Marshalls officials said this wasn't the case and promised better coordination in the future to avoid misunderstandings."[518]

As hoped, for revenue from fishing fees in the RMI not only jumped; it made a quantum leap in 1998! Indeed, the Marshalls gained more than $U.S. 4-million for that year. But MIMRA officials also attributed the greatly increased income to the results of a bold new fisheries policy - - that is filling local coffers. For example, during five months of 1997, the sale of purse-seiner licenses doubled; and, for instance, since September 1997 numerous purse-seiners and large transshipment carriers have been off-loading at Majuro's lagoon for export to canneries in Thailand and points elsewhere.[519]

The new RMI fisheries policy - - the fourth major change in Island fisheries - - has evinced a renewed interest in Majuro by the Asian DWFNs; thereby eliciting a revitalised financial backing by the Asian Development Bank (ADB). This fortuitous chain of events has motivated Danny Wase to redirect MIMRA from less direct [government] involvement in the fishing industry, to shifting MIMRA into becoming more of a regulatory agency. In 1998, said RMI fisheries advisor Simon Tiller, MIMRA has been...

..." 'converting policy into practice to make money from fisheries…. The Marshall Islands [Majuro] has considerable competitive advantage above other ports….Good docks, air and shipping service, banks, communications, refueling and other services are all available here.' "[520]

The fifth major change in the RMI commercial fisheries was a pro-active, Pacific-wide, move requiring a vessel monitoring system (VMS) installation on every DWFN tuna boat before obtaining license to fish in the South and Central-Western Pacific.[521] Representatives of eight Island nations met at Majuro in late 1998 (all parties to the Nauru Agreement), and reached accord that the VMS requirement be implemented a.s.a.p. Announced MIMRA director Danny Wase, " 'We're sending a message to the distant water fishing nations…. The VMS will become a condition of licenses.' "[522] Also discussed at the Majuro meeting was the problem of "enclaves" - - high seas areas bordered on all sides by EEZs of Pacific "coastal States" - - and how to manage and control fishing within the enclaves. Once again Danny Wase of MIMRA spoke for the Island nations: "While acknowledging that everyone has 'equal rights' on the high seas, [Wase said] island countries are in agreement that islands bordering these enclaves will monitor fishing actions in those areas and use their licensing procedures to prevent illegal activities."[523] (So long as the

Island EEZ monitors won't get *too* carried away and rambunctious! Those high-seas enclaves, are, after all, "No man's [sea]."]

The sixth, and final, sea-change for the Marshallese fishing industry was the RMI's link-up with Taiwan (R.O.C.) in 1999. In the Business section of Pacific Magazine of March/April 2000, a sidebar reported:

"Although diplomatic ties between the Marshall Islands and Taiwan are barely a year old, Taiwan's fishing fleet is already showing its preference for these North Pacific Islands....

"A record of more than 150 Asian purse-seiners used Majuro's harbor for transshipping tuna and taking on supplies in 1999, with Taiwanese vessels accounting for more than 65 percent of the total."[524]

Indeed. During most of the 1990s, the RMI took in about $3-million all told from [mostly Asian] DWFNs - - a combination of fishing license fees plus spin-off revenues to local businesses; supplying the vessels and crews with food, fuel, ice, and sundries. For just the year 1999, the MIMRA projected that Marshallese income from fisheries should approach $8-million, and presented a list of facts to back their estimate:[525]

(1) Fishing license fees increased from $U.S. 1.6-million during fiscal 1997 to $3.8-million in 1998, with license applications jumping from 132 during 1996 to 279 in 1998.
(2) Majuro became the centre for tuna transshipment in 1998, as noted, taking in $78,000 for 133 transshipments for East Asian, Pacific, and American fishing companies. The new RMI fisheries policy is business-friendly (fourth in my order of mention but first in reason for the sea-change), has slashed government fees, and removed bureaucratic red-tape. The tuna boats now started making Majuro a port of call; previously the Marshallese capital had been given a wide berth!
(3) In 1998, for the first time, South Korean (R.O.K.) purse-seiners bought licenses and transshipped tuna, while the Japanese fishing fleet was back after a (dissatisfied) hiatus of three years.
(4) And, as noted, the Taiwanese are now (2000) in the RMI commercial fishing picture. A pact signed with the RMI permitted 43 Taiwanese purse-seiners to fish Marshall Islands waters.[526]

With a pro-private sector government in charge at Majuro, and with the competent Danny Wase at the helm of the MIMRA, the commercial fisheries - - both foreign and domestic - - in the Marshall Islands are ready to tackle the 21st century.

VI. Republic of Kiribati

In Part III of this chapter we discussed the Big Fish Story of the 1980s, which was the star-crossed Kiribati/Sovryflot fishing agreement of 1985/1986.[527] Although the Russian licensing arrangement was not renewed after that first turbulent year, Kiribati president Ieremia Tabai had effectively served notice to the Great Powers and DWFNs that his tiny Island nation was not to be bullied. Land poor but marine-rich with a vast (3,500,000 sq. km.) expanse of EEZ in the Central Tropical Pacific, the Republic of Kiribati is the formerly-British Gilbert Islands. Although truncated by the runaway Ellice Islands (as the Polynesian Tuvalu) in 1975, the Micronesian Gilberts (Tungaru) achieved independence in 1979, and has managed to hang on to Ocean Island (Banaba), the [eight] Phoenix and [eight] Line Islands.[528] As stated, the Republic of Kiribati is the largest atoll state on earth, and, with her humungous portion of CWP Ocean, no fisheries co-operation between the various political entities of Micronesia is possible without the participation of Kiribati…period!

During the early and mid-1980s, the young Republic did better at cultivating seaweeds than the ill-fated Sovryflot deal. In 1984, a trial seaweed farm on South Tarawa was deemed successful enough to be expanded to 200 hectares.[529] Four years later, 70 metric tons of seaweed were grown, and the project extended farther to Abaiang and other atolls. Seaweed culture has remained popular among the Gilbertese (I-Kiribati) as this economic activity makes more money for less work compared to the onerous production of copra. By 1990, Kiribati had exported $500,000 worth of dried seaweed to Asia and Europe, where it is utilised in food emulsifiers and pharmaceuticals.[530]

During the late 1980s eight Gilbertese pole-and-line tuna boats, based at the harbour town of Betio (pron. "Bedjo") and belonging to the government-owned Te Mautari Ltd., would unload their frozen catch onto a mother ship, which in turn delivered the tunafish to canneries in Fiji or American Samoa. But, less happily (than the seaweed venture) for Kiribati, Te Mautari Ltd. declared bankruptcy in 1990 with a debt of A $7-million (Australian) having accumulated over three years. ("A situation," according to David Stanley, "blamed on poor management.")[531]

During the early 1990s, however, Japan provided millions of yen to Kiribati in fisheries and development - - in return for allowing her 400-vessel fleet ensured access to the Island nation's tuna-rich EEZ. And fish ponds, like the ones next to Tarawa Airport, were being developed to [aqua] culture milkfish (Chanos chanos) for big-fish bait.[532] Risible as these seemingly-puny efforts at building a fishing industry might seem to DWFNs, Kiribati is so deadly serious about her economic independence as political sovereignty. At last look in early 2000, Spain was about to bring twelve large fishing "factory ships" into the CWP to operate out of the Gilberts.[533] An alarmed Radio Australia reported:

> "There were deep concerns expressed by Island leaders at the Palau summit over the Spanish plan, which is larger than any previous fisheries operation in the region and would be a serious threat to tuna stocks. Pacific tuna stocks are the last great fishery in the world that hasn't been severely depleted by over fishing."[534]

Is this great draught of fishes a boon or a bane for Kiribati? That all depends on the Republic's marine resource management and 200 nautical-mile EEZ control. As for the last, the small-but-scattered and sparsely-populated atolls that comprise the Gilberts shall have to seek help/support from a larger, more powerful, Pacific national (or international) body. Kiribati can do no better than to move - - soon - - in co-ordinated action with other Micronesian Island nations plagued with correspondingly similar fishery problems (although possessing larger populations but with smaller total EEZ areas to police.) Part of the Kiribati problem - - and for all Micronesia - - of geopolitical empowerment, has already been solved by the mission and work of the South Pacific Forum Fisheries Agency (FFA).[535] There is no better place from which to start a Central-Western Pacific all-Micronesia fisheries watch-dog group than at the FFA. It is time for Micronesia to get going!

VII. Republic of Nauru

Except for being a micro-island in Micronesia, and never having been in American hands, the islet Republic of Nauru has absolutely nothing in common with far-flung Kiribati save, perhaps, the latter's minuscule, phosphate-rich possession of Banaba. Like the formerly-U.S. Micronesia (TTPI), Nauru was once a League of Nations mandate after World War I and a ward of the United Nations after World War II.[536] (For both post-War periods, Nauru languished under a tripartite British, Australian, and New Zealander administration.) In 1968, Nauru became independent under Australian auspices; in 1994, so diminutive a geopolitical entity as Nauru is, UNCLOS III awarded the Republic a 200-nautical mile EEZ backed by international law. For the [phosphate] mined-out husk of a far isle, the greatly-expanded marine borders gave revitalised economic breathing room to Nauru - - which meant new oceanic responsibilities. Thus, in mid-November 1994, little Nauru with a population of 8,000 and an area of 21 sq. km., now ruled over 320,000 sq. km. of CWP waves[537]...a rôle for which she was hardly prepared.

In 1998, Nauru sought the neighbouring Marshalls (RMI) for help in protecting the pelagic fish stocks in her EEZ. Nauru owned no patrol boat, and had good reason to suspect that certain DWFNs were illegally fishing in Nauruan waters.[538] The update of Nauru's request of the RMI led to a proposed treaty between the two Micronesian Island nations, with the Marshalls also having participated in combined training exercises and joint patrols with another Micronesian neighbour - - the Republic of Kiribati.[539] Giff

Johnson, reporting in Pacific Magazine during the summer of 1999, commented on intra-Micronesian co-operation:

"Micronesian countries are moving into a new era of cooperation to protect the vast marine resources of their Pacific area.

"Pacts among the nations are on the verge of being signed that for the first time will provide for joint policing of the 200-mile Exclusive Economic Zones that extend over what Asian and U.S. fishermen consider to be the prime fishing grounds in the Pacific.

"Palau and the Federated States of Micronesia have approved, in principle, an agreement that calls for sharing fisheries information and conducting joint surveillance and law-enforcement patrols of their waters. Officials in the Marshall Islands said they hope their nation will join the pact."[540]

"A Modest Proposal for Micronesia": Finale

While conducting intensive research on the Islands throughout the years 1996-2000, I have watched "my" Modest Proposal for Micronesia become a self-fulfilling wish. For me, the idea of Micronesian fisheries co-operation germinated (and grew) after a November 1995, 22-nation, conference on tuna development was held at the Maui Pacific Center in Hawai'i.[541] The conference was attended by government heads, fisheries officials, and economists (both public and private sector) from all over the Central-Western and South Pacific. Little wonder: The tropical Pacific's tuna fisheries in late 1995/early 1996 accounted for "at least [U.S.] $1.7 billion at the dock by one estimate - - offer[ing] the region's best hope for economic self-sufficiency."[542] The three main recurring themes voiced at the Maui conference may be briefly recapitulated:

(1) The Pacific Island nations (the "coastal States") must enact political reforms to attract private-sector investment in commercial fisheries, and learn how to prosper in a free-market economy;
(2) the Pacific Island nations must develop regional systems to better manage their fisheries and to better market their tuna harvests;
(3) the Pacific Island nations must act together, in conjunction, to negotiate with DWFNs from a position of unified strength. Together, the "coastal States" will more effectively manage their living marine resources and fisheries, market their tuna harvests, while maintaining sovereignty over their 200-nautical mile EEZs.[543]

The real pan-Pacific celebrity "attending" the Maui conference was unsmiling Hawai'i Democrat Senator Daniel K. Inouye. In a videotaped address to the delegates, Senator Inouye observed that some large Asian nations, especially China (PRC), are buying and building naval and fishing fleets which might presage a frightful new era of " 'conquest and exploitation' " in the Islands during the next century.... " 'People must be fed, and these huge entities will do just about anything to feed their people.' "[544] The Senator advocated that the Pacific Island nations confederate, form a consortium, to defend (and promote) their interests at international councils and global forums. The Hawai'i Democrat warned darkly that should push come to shove, " 'the Law of the Sea would not be sufficient to protect your [Island] interests.' "[545]

Daniel K. Inouye's words apply especially to the tiny Island nations of Micronesia.

There was a series of seven Multilevel High-Level Conferences (on Highly Migratory Fish Stocks in the Western and Central Pacific, or MHLC) held, throughout the 1990s and Y2K, on co-operative tuna management in the CWP. The aim of the MHLCs is an eventual common treaty agreement, with global imprimatur, reflecting conference-discussion results. As the reader might guess, the MHLCs are an outgrowth of UNCLOS III, the Third U.N. Convention on the Law of the Sea, which went into effect in mid-November 1994. Another MHLC was held at Honolulu for ten days, during September 1999, with delegates from "28 Pacific Islands" (sic) attending.[546] The MHLC chairman, Ambassador Satya N. Nandan of Fiji, expressed satisfaction at the " 'good progress' " made towards the pending treaty; with many detailed decisions reached, but all basically designed to protect tuna stocks in the CWP. Follow-up MHLCs were scheduled for March and June 2000, after which a fishery commission would be formed. A chief function planned for the commission was to set a total allowable catch (TAC) for bigeye/ahi, yellowfin/ahi, skipjack/aku, and southern albacore/tombo.[547]

Yet another MHLC was held at Honolulu, with 250 delegates from 28 Pacific nations attending during summer 2000.[548] But this time around, representatives of distant-water fishing nations (DWFNs) were present; high-tech big-boat guys such as the United States and Japan. But the Big Boys would later split on voting for a regulatory fisheries commission - - the United States, Australia, the Philippines, and several other nations voting for creation of the commission; Japan and South Korea opposing any restrictive new regulations, while China (PRC), France, and Tonga would abstain from the preliminary vote.[549] Japan disagreed with certain aspects of the treaty; including concerns regarding a nay-saying bloc of voting members, instigated by Australia and New Zealand, that (Japan believed) was hostile to East Asian fishing nations. Japanese delegate, Masayuki Komatsu, threatened that his country might " 'continue [our] fishing in the area outside of the proposed convention.' "[550]

The Japanese and the South Koreans wouldn't be the only parties dissatisfied with the MHLC proceedings. Our old friend, William T. Burke, professor emeritus of law at the University of Washington

("Uddub"), remarked afterward: " 'In all my years in the field, I have never seen a proposed fisheries convention as complicated and unworkable as this.' "[551]

One could rightly ask Professor Burke, even if in partial or indeed total agreement, what did he expect from a UNO-inspired series of international conferences? He (of all people) should know the answer to that question, having written a legal report on tuna regulation commissioned by the FAO in 1982![552] But W.T. Burke is more aware than most that, although Micronesia has a far smaller land-mass and population than either Polynesia or Melanesia (hence wields far less geopolitical power), since November 1994, [Micronesia] has exercised control over the most generous and valuable (i.e., tuna-rich) portions of the Central-Western Pacific Ocean.[553] Thus Micronesia - - tiny, scattered, and geopolitically powerless - - needs all the helping friends she can get…whether they come from the U.N., the U.S.A., or the South Pacific FFA (Forum Fisheries Agency). Micronesia has had the FFA working for and representing her fisheries best interest since 1979; five Micronesian entities have been - - or are - - under direct American rule or influence (except Kiribati and Nauru); and all Micronesia (save Kiribati and Guam) has been a ward of the League of Nations (after World War I) and the United Nations (post World War II). Surely Micronesia can utilise these geopolitical associations, past and present, to her future advantage?

And Micronesia has a lot of fish to fight for. The CWP yields two-thirds of the tuna harvested worldwide, with the fish stocks in the region accounting for more than 85% of the globe's supply of the best sashimi-quality bigeye/ahi and yellowfin/ahi; and of the skipjack/aku and southern albacore/tombo targeted for high-grade canning product. The total landed value of the fishery is estimated at between $U.S. 1.7-2 billion.[554] So Micronesia's interest in this fishery is paramount - - indeed, her entire future depends on it. MHLC chairman Satya Nandan concluded the sixth conference (on Highly Migratory Fish Stocks in the Western and Central Pacific) with these pronouncements:

" 'Joint management of the stocks is being pursued to ensure their long-term sustainable use….
" 'The management regime will result in one of the world's largest fishery management commissions….' "[555]

Indeed it would. The commission is slated to include a secretariat, a scientific committee, a technical and compliance committee, a panel to review disputes, and an observer programme.[556] (Whew! Maybe Professor Burke is right!)

The seventh and final session of the MHLC was held at Honolulu rather than on Fiji as originally planned. As noted, DWFNs Japan and South Korea voted in opposition to adoption of the convention, with China (PRC), France and Tonga abstaining. The two-thirds majority vote required for adoption, however, cleared with 19 (out of the 24) nations voting in favour. Those voting in favour of the convention included the FSM, Kiribati, the Marshall Islands (RMI), Nauru, Palau, and the United States.[557] Among the issues resolved was one regarding participation of territories - - the last vestiges of Pacific colonialism. Thus American Samoa, French Polynesia, Guam (U.S.), New Caledonia (France), Northern Marianas (U.S.), Tokelau[s] (N.Z.), Wallis and Futuna (France) are now entitled to be present, and to speak, at future meetings of the fisheries commission and of its subsidiary bodies.[558] So "burdensome and impractical" (W.T. Burke) or not, a favourable fisheries agenda has been set for all Micronesia backed by international law. It is now up to Micronesia to act independently and to move co-operatively.

My finale two queries, at the close of this exhaustive (not exhausting for the reader I trust!) chapter on the commercial fisheries in the tropical mid-Pacific, are simply these: Question One - - What does Micronesia really want? And Question Two - - How can Micronesia achieve these goals?

For the answers to these questions we must narrow our scope to a specific Micronesia, the Federated States of Micronesia. The States of Pohnpei, Kosrae, Chuuk, and Yap bonded together in union partially because they were the "have not" entities of Micronesia. The Marshalls and Palau, with their strategic locations in the Western Pacific and other logistical advantages, were favoured geopolitically by the United States. Palau negotiated her own Compact of Free Association arrangement with the U.S. in 1994, and the Marshalls voted for the Compact in 1986 and achieved independence in 1990 - - the very same years as the FSM. (After 15 years, the Compact expires and is therefore up for renewal this year, 2001.) The Cold War is now over, but an expansive ChiCom threat is a cause for concern in some quarters of Washington, D.C. The Red Chinese presence is perceived, however, as an opportunity by FSM president Leo Falcam. President Falcam has held out the possibility of granting exclusive long-term fishing rights to one country (the PRC?) to harvest tuna in the FSM's greater than 1-million sq. mile EEZ.[559] The president's intentions are extremely significant as no geopolitical moves regarding the Micronesian fisheries are possible without the FSM. In late 2000, Pacific Magazine's Floyd Takeuchi interviewed the FSM president. On fisheries, President Falcam had much to say about DWFN fees but nothing on local fishing:

FALCAM: " 'I would like to explore the possibility of saying to the distant [sic] fishing nations - - such as the United States, Japan, and Korea - - that, "Hey, you fish in our waters for the next 50 years, and the fee is this. And only you fish." It would be an exclusive franchise to fish our waters. They would put up a fee equal to what we charge today, or perhaps a bit more. It would mean that instead of signing a (fishing) agreement for a fee every year, take it for 50 years and pay us a fee up front. We can use it toward our proposed trust fund. The idea intrigues me. We sign fishing licenses every year with many nations on a year-to-year basis. I think it would be useful for us to say, "OK, China, Japan or the U.S., you have exclusive rights to fish in our EEZ for a fee." The fee would be put up front, and put in our trust fund for a rainy day.' "[560]

All well and good, Mr. President, but what about developing a <u>local</u> fishing industry? The FSM's Micronesian Maritime Authority (MMA) appears far ahead of their chief executive on national fisheries policy. (See note 512 for MMA conclusions.)

When Floyd Takeuchi of <u>Pacific Magazine</u> asked President Falcam if the U.S. is obligated to pursue an active relationship with Micronesia, the politico answered:

FALCAM: " ' I think so, very much so. If you look at beyond just purely the special relationship between the United States and the FSM, you run into a triangle. You run into Korea, Japan, and other governments in the Far East.... We are still, I think and hope, an essential part of the U.S. policy in the Pacific. They aren't going to let Guam go. Their relationship with Japan is still there. Talk about North Korea. Many obligations. And we are a part of that region.' "[561]

For Micronesia sake, let's hope that the conclusions of the MMA prevail, along with the dynamic leadership of men like Danny Wase, the director of MIMRA, the Marshall Islands Marine Resources Authority. Rather than MMA's five members (one from each FSM State), have a membership of seven - - one for each Micronesian political entity in a new, all-Micronesia organisation. A single at-large special member, empowered by a united Micronesian front, could represent Micronesian fisheries policy at the South Pacific FFA and other international bodies. Meetings could alternate annually at the Micronesian capitals of Oreor, Hagatna, Palikir, Ba iriki et al. I have titled my chapter 'A Modest Proposal for Micronesia' because my proposal <u>is</u> modest: All the building materials are there; they just need to be assembled.

Parting Shot

And the Americans are back in Micronesia!

Throughout this chapter on the Micronesian commercial fisheries, I have asked/appealed, "Where were (or are) the Americans?" Well, this naturalised U.S. citizen is proud to announce that the Yanks are back in Micronesia…as <u>locally-based</u> tuna fishermen. I'm doubly proud as Mr. Timon Tran, the president of Mid Pacific (a longliner fishing company), is a naturalised U.S. citizen but with much harder-earned credentials than mine. For Timon Tran came to America in 1975 as a refugee from South Vietnam (RVN), settling in Hawai'i.[562] Once in the Aloha State, Timon Tran started-up several enterprises; one of these being a longlining outfit, with fellow Vietnamese, Tan Nguyen. Tran and Nguyen went on to expand their longliner operations to the U.S. Gulf Coast.[563]

In spring 1999, president Timon Tran, fishing supervisor Tan Nguyen, and vice-president Duc Ngo headed a delegation of top Mid Pacific executives to Majuro, Marshall Islands (RMI) and Colonia, Yap (FSM). Mid Pacific had journeyed to Micronesia to implement agreements with government officials of both nations to fish the waters of their EEZs. Mid Pacific plans called for nine longline vessels to initially move to Yap, and in the Marshalls to take over the fishing-base previously leased to the Ting Hong fishing company. Summed up president Timon Tran:

" ' I've been traveling, studying and learning about the potential of the fishing industry in Micronesia for the past two years.... I can see the governments slowly beginning to move in the right direction, which is attractive to investors. I feel we have a good team of technical, management, marketing and community relations people.' "[565]

Yes, the Americans are back in Micronesia!

NOTES for PART IV

397. "Pacific Islands Mirror Alaska," Pacific Fishing, December 1997, p.5.

398. Atlas of Micronesia (Honolulu: The Bess Press, Inc. 1993 ed.), p.53.

399. Ibid., p.53. Cf. Hanlon (1988) and Hezel (1994).

400. Ibid., p.53. Cf. Hanlon (1988) and Hezel (1994).

401. Robert F. Rogers, Destiny's Landfall: A History of Guam (Honolulu: University of Hawai'i Press, 1995), p. 89. Rogers quotes Haswell (1917), and describes the sea-slug:
"Trepang: Echinoderms, mostly Holothuroidea, of which H. aculeata or Microthele nobilis (teat fish or mammy fish), fat and whitish, is the most edible." (Rogers, p.313)

402. Ibid., p.89 Cf. C.D. Bay-Hansen, Fisheries of the Pacific Northwest Coast, Vol.2 (New York: Vantage Press, Inc., 1994), passim.

403. Francis X. Hezel, S.J., The First Taint of Civilization: A History of the Caroline and Marshall Islands in Pre-Colonial Days, 1521-1885 (Honolulu: University of Hawai'i Press, 1994 ed.), p.83.

404. Op.cit., p.83. He cites Hockin (1803) and Kotzebue (1821).

405. Bruce G. Karolle, Atlas of Micronesia, p.54.

406. David Hanlon, 'Colonisation', p.106. In K.R. Howe et al., Tides of History: The Pacific Islands in the Twentieth Century (Honolulu: University of Hawai'i Press, 1994.)

407. Ibid., p.106. The former rival Japanese companies were the Nan'yō Bōeki Hiki Gōshigaisha and Nan'yō Bōeki Maruyama Gōemeigasha.

408. Ibid., p.106. Matsue's Nankō was officially the Nan'yō Kōhatsu Kaisha; The South Seas Development Company.

409. Ibid., p.106. NB: And into the Japanese polity - - kokutai.

410. Ibid., p.106. Nantaku was the local designation for Nan'yō Takushoku Kaisha - - The South Seas Colonisation Corporation.

411. Op.cit., David Hanlon, p.106.

412. Francis X. Hezel, S.J., Strangers in Their Own Land: A Century of Colonial Rule in the Caroline and Marshall Islands (Honolulu: University of Hawai'i Press, 1995), p.186.

413. Ibid., p.186. Francis X. Hezel doesn't mention Tamashiro's given name.

414. Ibid., p.186 This bonito (Sarda orientalis) is, of course, not a true tuna. The Japanese for bonito is hagatsuo or kitsunegatsuo (IGFA, p.227). Then again, "bonito" is often used as a colloquialism for several scombroid fishes.

415. Op.cit., Francis X. Hezel, p.187. (Brackets and contents mine). Author's note: These migrant Okinawans can rightly be considered the first truly commercial fishermen in all Micronesia.

416. Ibid., pp.187-188.

NOTES for PART IV (cont'd.)

417. Ibid., p. 188.

418. Mark R. Peattie, Nan'yō: The Rise and Fall of the Japanese in Micronesia, 1885-1945 (Honolulu: University of Hawai'i Press, 1992 ed.) p.138.

419. Op.cit., p.102. The Mortlocks are located between Chuuk and Pohnpei; today part of the FSM.

420. Ibid., p.122.

421. Ibid., p.122.

422. Ibid., p.138. Cf. Francis X. Hezel (1995), p.194. According to Fr. Hezel, Gov. Hayashi Hisao said in the mid-1930s: " 'The possibilities of the islands are limited; of the sea, unlimited' ", (p.194).

423. Ibid., p.139. Dublon Town is today's Tonoas, Chuuk, FSM.

424. Ibid., pp.139-140.

425. Op.cit., Mark R. Peattie, pp.140-141. NB: Both Peattie (pp.140-141, 333, passim) and Hezel (pp.194-195,202-203) have discussed the Japanese pearl-diving industry, which flourished not only in Palauan waters, but in the Timor and Arafura Sea[s] off northern Australia. That, however, is another story....

426. Op.cit., Francis X. Hezel, pp.194-195. (Brackets and contents mine)

427. Ibid., p.195.

428. Ibid., p.195. Author's note: Fugu, too?!

429. Ibid., p.202. NB: There were as many Japanese living on Palau in 1940 as there were Palauans living on Palau in 2000. (Progress?!)

430. Op.cit., Francis X. Hezel., pp.202-203.

431. Mark R. Peattie, p.144.

432. Ibid., p.152.

433. Op.cit., Mark R. Peattie, p.144.

434. Ibid., p.319.

435. David Hanlon, Remaking Micronesia: Discourses over Development in a Pacific Territory, 1944-1982 (Honolulu: University of Hawai'i Press, 1998), p.37.

436. Eg. Koben Mori and the case of the Mori and Suzuki families. See Pattie (1992 ed.) and Hezel (1995), passim. Also see Kiste et al. (1999), pp.203-204.

437. Cited by David Hanlon (1998), p.37. (Brackets mine)

438. Ibid., p.55.

NOTES for PART IV (cont'd.)

439. Ibid., p.55. NB: Lt. Col. Richard penned a three-volume work on the U.S. Navy's administration of Micronesia (published in 1957).

440. Ibid., p.99; cf. Hezel (1995), p.319.

441. Francis X. Hezel, pp.319-320.

442. Op.cit., p.320.

443. Ibid., p.319. Author's note: "Foreign investment", of course, meant that by the present occupying power, the United States. In some Palauan minds, one "crocodile" (Japan) had merely been replaced by another "crocodile" (the U.S.A.).

444. Op.cit., David Hanlon (1998), pp.97-98.

445. Francis X. Hezel (1995), p.319.

446. Hanlon (1998), p.111. NB: Stanley (1992 ed.) gave a 1974 date, and Hanlon has mistakenly substituted "Maritime" for "Mariculture."

447. Ibid., pp.111-112. NB: Surely Hanlon meant the U.N.s FAO - - the Food and Agriculture Organization?! (He wrote "Fish"!)

448. Op.cit., p.112. (My brackets and contents) NB: the "three self-governing entities", of course, are Palau, the RMI, and the FSM. Author's note: Hanlon has referred, however, to a November 1971 report on fishing in Chuuk, written by one Ronald Powell, a Trust Territory fishing gear and methods specialist. In his 'Progress Report: Inshore Fisheries', Mr. Powell had carefully noted that best results occurred when Chuukese fishermen worked (1) in familiar areas, (2) during [personally] preferred times to fish, (3) with gear they had chosen, (4) under leaders with stature in the local Island community. See Hanlon (1998), pp.112-113,253.

449. Ibid., p.142; Hezel (1995), p.359.

450. Hanlon (1998), p.142.

451. Ibid., p.143. NB: COM Seventh Congress, first regular session, January-February 1977. In Hanlon (1998), p.257.

452. Op.cit., Hezel (1995), p.360.

453. Ibid., p.360. For an explanation of the Compact of Free Association, see Part I, Section C in this Chapter.

454. Of the "three self-governing entities," the RMI and the FSM have been independent under the Compact since 1990; Palau only since 1994.

455. 'South Pacific Forum Fisheries Agency: Corporate Plan 1998-2001' (Honiara, Solomon Islands: FFA, 1997), p.4.

456. Wesley Marx, The Frail Ocean (New York: Ballantine Books, 1970), p.217.

457. Ibid., p.68.

NOTES for PART IV (cont'd.)

458. Ibid., p.68. (My brackets)

459. Ibid., p.68. Cited by Marx is one Dr. Schaeffer. (My brackets)

460. Ibid., pp.68-69.

461. David Stanley, Micronesia Handbook (Chico, Calif.: Moon Publications, Inc., 1992 ed.), p.34.

462. Ibid., p.35. NB: The Tuna Treaty was negotiated by the FFA.

463. Ibid., p.35. Stanley writes darkly of the Japanese "buying influence."

464. Op.cit., David Stanley, pp.35-36. (My brackets)

465. Ibid., p.34.

466. Ibid., p.207. ATA: American Tunaboat Association.

467. Op.cit., p.253 (Brackets mine)

468. Ibid., p.252.

469. Op.cit., p.181

470. Ibid., pp.181-183.

471. Ibid., p.183. Cf. Elizabeth Rechebei and Samuel F. McPhetres, History of Palau: Heritage of an Emerging Nation (Koror: The Ministry of Education, 1997), pp.343-344. See also Daniel Knop, Giant Clams: A Comprehensive Guide to the Identification and Care of Tridacnid Clams (Ettlingen, Germany: Dähne Verlag GmbH, 1996), pp.208-209.

472. R. E. Johannes, Words of the Lagoon: Fishing and Marine Lore in the Palau District of Micronesia (Berkeley, Calif.: U. Cal. Press, 1992 ed.).

473. Carl Safina, Song for the Blue Ocean: Encounters Along the World's Coasts and Beneath the Seas (New York: Henry Holt and Co., 1998), p.312. (My brackets)

474. Ibid., p.327.

475. Ibid., p.328.

476. Ibid., p.335.

477. Ibid., pp.335-336.

478. Op.cit., Carl Safina, p.336.

479. See note No. 471 for details, and Appendix I for Giant Clams of Kiribati.

480. From Daniel Knop, Giant Clams..., p. 236.

481. Elizabeth Rechebei and Samuel F. McPhetres, p.343.

482. Ibid., p.343.

NOTES for PART IV (cont'd.)

483. Op.cit., p.343.

484. Op.cit., Mark R. Peattie, p.177. (My brackets)

485. Pacific Magazine, Vol. 23, No. 2, Iss.. 128, March/April 1998, p.27.

486. Op.cit., Gene Ashby, p.27.

487. Ibid., p.27.

488. Op.cit., David Stanley, p.114. (Brackets mine)

489. Ibid., p.98. NB: All figures those of Stanley.

490. Author's note: Unhappily for FSM commercial fishing, the Caroline Fisheries Corporation (CFC) went into receivership during 1994 by court order. The CFC, also undergoing management review in 1994, is/was a JV of the National Fisheries Corp. (NFC), the State of Pohnpei, and the Quails and France PTA. Ltd. Of Australia. See '1994 MMA Annual Report', p.16.

491. David Stanley, p.98. Stanley has accused the Japanese of self-interest regarding Micronesia, but not the Australians. Strange!

492. Ibid., p.114.

493. National Fisheries Corporation, Kolonia, Pohnpei (FSM), p.2.

494. Op.cit., NFC, p.2.

495. Ibid., p.2.

496. Ibid., p.2. See Appendix I for FSM groundfish species.

497. Ibid., p.3.

498. Op.cit., NFC, p.3.

499. 'Laws of the Federated States of Micronesia Concerning Fishing and the Use of the Sea as of August 1989', Territory of Micronesia, p.6.

500. Ibid., p.21. (Section 301)

501. ' Two Year Report of the Micronesian Maritime Authority, 1992-1993.' Government of the Federated States of Micronesia,' Palikir, Pohnpei, p.2.

502. Ibid., pp.8-12. For details, cf. 'MMA Two Year Report 1992-1993.'

503. Ibid., p.13. The UN-hated Taiwanese arranged not only to manage the ChiCom boats and train their crews, but also to provide proper fishing gear to the clueless Red Chinese. One rarely reads true stories like this in the world press!

504. Ibid., p.3. Author's note: I will assume that this would include the entire Central-Western Pacific region as well.

NOTES for PART IV (cont'd.)

505. 'Micronesian Maritime Authority 1994 Annual Report.' MMA, Palikir, Pohnpei, p.16. NB: See note 490 in this Part.

506. Ibid., p.16. Again, where were the Americans?

507. Ibid., p.17.

508. Ibid., p.18.

509. Cf. note 504; 'MMA Two Year Report 1992-1993,' p.3.

510. '1994 MMA Annual Report,' pp.5,41. NB: The Law of the Sea (UNCLOS) commenced to be fully implemented in November 1994.

511. Ibid., p.4.

512. Op.cit., MMA (1994), p.41. (Brackets mine)

513. David Stanley, p.69.

514. Ibid., pp.69-70.

515. Giff Johnson, 'Pearl Oyster Spawning Succeeding in Marshalls,' Pacific Magazine, Vol.23, No.2, Iss.128, March/April 1998, p.24.

516. Ibid., p.24. Once seeded, pearl oysters are grown as strings in the lagoon; taking 12-18 months to mature.

517. 'Japanese Fishing Boats Returning to Majuro,' ibid., p.25.

518. Op.cit., Giff Johnson, p.25.

519. Giff Johnson, 'Fishing License Fees Produce Big Bucks,' Pacific Magazine, Vol.24, No.2, Iss.134, March/April 1999, p.14.

520. Ibid., p.14. Quoted is Simon Tiller, RMI fisheries advisor. (My brackets)

521. Giff Johnson, 'Island Nations Moving to More Fisheries Control,' Pacific Magazine, ibid., p.[15].

522. Ibid., p.[15]. Quoted is MIMRA director, Danny Wase.

523. Ibid., p.[15]. Paraphrased is Danny Wase of MIMRA.

524. 'Taiwanese Fishing Fleet Dominant in Marshalls,' Pacific Magazine, Vol.25, No.2, Iss.139, March/April 2000, p.16.

525. Giff Johnson, 'Fishing Revenues Expected to Triple in Marshalls,' Pacific Magazine, Vol.24, No.4, Iss.136, July/August 1999, p.15.

526. Ibid., p.15. All figures are from the MIMRA.

527. Cf. Chapter 4, Part III, pp.350-351.

NOTES for PART IV (cont'd.)

528. See Ch. 4, Pt. I, Sec. C, pp.315-316.

529. David Stanley, p.279.

530. Ibid., p.279.

531. Op.cit., p.279.

532. Ibid., p.279.

533. Radio Australia, 'Pacific Tuna Catch Increases by 25 Percent,' Pacific Magazine, Vol.25, No.1, Iss.138, Jan/Feb. 2000, p.17.

534. Op.cit., p.17.

535. See Ch. 4, Pt. III, pp.343-349.

536. Cf. Ch. 4, Pt. I, Sec. C, pp.300, 314-315.

537. K.R. Howe et al., Tides of History: The Pacific Islands in the Twentieth Century (Honolulu: University of Hawai'i Press, 1994), p.327. NB: Pop. figures are from 1987.

538. Giff Johnson, 'Micronesians Cooperating to Keep Fishermen in Line,' Pacific Magazine, Vol.24, No.4, Iss.136, July/August 1999, p.15.

539. Ibid., p.15.

540. Op.cit., pp.14-15.

541. Brad Warren, 'Pacific Islands Poised For Tuna Growth,' Pacific Fishing, Vol.XIX, No.1, January 1996, p.18.

542. Op.cit., p.18. (My brackets)

543. Ibid., pp.18-19.

544. Op.cit., p.19. Cited is Senator Daniel K. Inouye (D-Hi). Author's note: It's been many years, Senator, since you got all that media face-time during the Watergate hearings!

545. Brad Warren, 'Pacific Islands Mirror Alaska,' Pacific Fishing, December 1997, p.5. (My brackets)

546. 'Island Nations Nearing Tuna Managing Treaty,' Pacific Magazine, Vol.24, No.6, Iss.138, Nov./Dec. 1999, p.15.

547. Ibid., p.15.

548. 'Pacific Tuna Treaty Near Completion,' Pacific Magazine, Vol.25, No.4, Iss.141, July/Aug. 2000, p.19.

549. 'Japan may ignore tuna agreement,' Peninsula Daily News, 6 September 2000, p.C-1.

550. Ibid., p.C-1. Quoted is Japanese delegate Masayuki Komatsu.

NOTES for PART IV (cont'd.)

551. Ibid., p. C-1. Quoted is U.W.'s William T. Burke. (See Part II, pp.327-332, for Prof. Burke's analysis of UNCLOS III on tuna regulation)

552. W. T. Burke, 'Impacts of the U.N. Convention on the Law of the Sea on Tuna Regulation' (Rome: Food and Agriculture Organization, 1982).

553. See map '200 Mile Exclusive Economic Zones of the Pacific Islands,' in K.R. Howe et al., Tides of History. Center for Pacific Islands Studies, U. Hawai'i at Manoa, 1993.

554. 'Pacific Tuna Treaty Near Completion,' p.19.

555. Ibid., p.19. Cited is Ambassador Satya N. Nandan of Fiji, MHLC chairman. (Emphasis mine)

556. Ibid., p.19.

557. 'Pacific Nations Agree on Treaty to Manage Tuna,' Pacific Magazine, Vol.25, No.6, Iss.144, Nov./Dec. 2000, p.11.

558. Ibid., p.11. Author's note: Participation of Guam and the CNMI means that all Micronesia is included, independent or not.

559. Floyd K. Takeuchi, 'What Micronesia Really Wants,' ibid., p.30.

560. Ibid., p.31. Quoted is FSM President Leo Falcam.

561. Op.cit. p.31.

562. 'U.S. Firm to Begin Fishing Operations in Micronesia,' Pacific Magazine, Vol.24, No.3, Iss.135, May/June 1999, p.14.

563. Ibid., p.14.

564. Ibid., p.14.

565. Op.cit., p.14. Quoted is Timon Tran of Mid Pacific.

Bibliography for Part IV

    31. Anderson, Alan, Jr. The Blue Reef: A Report from Beneath the Sea. (The adventures and observations of WALTER STARCK, marine biologist and authority on sharks, at Enewetak Atoll, a coral reef in the Pacific Ocean.) Alfred A. Knopf, New York, 1979.

Main References

    I. Howe, K.R., et al. Tides of History: The Pacific Islands in the Twentieth Century. University of Hawai'i Press, Honolulu, 1994.

    J. Stanley, David. Micronesia Handbook: Guide to the Caroline, Gilbert, Mariana, and Marshall Islands. Moon Publications, Inc., Chico, Calif., 1992 ed.

FSM Government Documents

    a) 'Laws of the Federated States of Micronesia Concerning Fishing and Use of the Sea as of August 1989.' (Territory of Micronesia)

    b) 'Two Year Report of the Micronesian Maritime Authority, 1992-1993.' Government of the Federated States of Micronesia, Palikir, Pohnpei.

    c) 'Micronesian Maritime Authority 1994 Annual Report.' MMA, Palikir, Pohnpei, FSM.

    d) 'Public Law No.9-047, Congressional Bill No.9-122, C.D.1, C.D.2' Ninth Congress of the Federated States of Micronesia, Second Regular Session., 1995.

Addendum:[566]

"Just Another Shitty Day in Paradise"? The Writers' View of Post-Pacific War Micronesia, 1945-2001

"Economists teach that the market is the fundamental social phenomenon, and its culmination is money. Anthropologists teach that culture is the fundamental social phenomenon, and its culmination is the sacred.[567]
"Anthropology is the only social science discipline still exercising the charm of possible wholeness, with its idea of culture, which appears more really complete than does the economists' idea of the market."[568]
--Allan Bloom, The Closing of the American Mind, 1987

With the wise words of Allan Bloom in mind, I have approached this final addendum through mostly the eyes of U.S. anthropologists rather than from just the Pacific historical viewpoint. Although I was trained as a Pacific historian at U. Hawai'i at Mānoa (1979-1981), my interest in Micronesia didn't awaken until fifteen years after my academic graduation—five years ago in 1996, at the relatively late age of fifty-two. Subsequently I have never been to Micronesia, and the only intimate portraits I have of Micronesians is from the four years I spent in Honolulu (1978-1982). Thus I must rely on the works, anecdotes, and impressions of others. For these, up front and very personal, I've stuck mainly to American anthropologists—who know Micronesia best; having lived with and studied the natives of the USTTPI since 1945. Also, naturally, I have carefully perused the records of my fellow Pacific historians, and, indeed, some of the writings of former Peace Corps volunteers (yup; the "Piss Corpse"). Although the volunteers' view is nearly always tinted by the psychedelic colours of the countercultural 1960s, their testimony is undeniably valuable to understanding the various peoples and problems of Micronesia. So without further ado, I'll start from the beginning... with the view from a former Peace Corps volunteer.

Prior to mid-1999, David Hanlon edited The Contemporary Pacific: A Journal of Island Affairs, and since then has been general editor of the Pacific Islands Monograph Series (co-published by the University of Hawai'i Press and the Center for Pacific Island Studies).[569] David Hanlon went to the District of Ponape in the TTPI's East Carolines (today Pohnpei State, FSM) as a Peace Corps volunteer in 1970, and served there for four years (1973). Later in the 1970s and during the 1980s, Hanlon attended the University of Hawai'i at Mānoa earning both M.A. and Ph.D. degrees.[570] Hanlon's years of Pacific Island studies were paid off with a fine (and unique) history of Pohnpei, published in 1988 by U.H. Press.[571] Despite some of his predictably Leftish conclusions, Hanlon's book about his old stomping grounds (Upon a Stone Altar: A History of the Island of Pohnpei to 1890) was affectionately written in addition to being diligently researched. Of great interest to me was the depiction of one Lucy M. Ingersoll, a U.S. medical doctor who arrived on Pohnpei in 1888. Ms. Ingersoll didn't think much of either native Islanders or Christian missionaries. On the futility of more than thirty-five years of American missionary efforts on Pohnpei, the crusty Lady Doc was recorded as uttering:

"'All seems to be for nothing; the people will never by anything more than what they are now. The best thing for them would be to be left entirely by themselves, and not depend upon foreigners for everything.' "[572]

As if Dr. Ingersoll's remarks had not also contained sufficient dismissal of American missionaries, David Hanlon (as ex-Peace Corps volunteer) provided it; while defending his Noble Savages:

"The missionaries saw Pohnpeians as enslaved by their beliefs in gods and spirits. Perhaps they were, but no more so than people who held that the purpose and work of life was constant penance for the basic sinfulness of human nature. In a sense, the missionaries offered the people of the island an exchange of one form of enslavement for another..."[573]

However, in his summing up of modern Pohnpei's problems, David Hanlon (as thoughtful Pacific historian) concluded:

"Known in the middle of the nineteenth century as the vice capital of the Pacific, Pohnpei now sees itself defined by outside observers in terms of a host of social ills ranging from acute economic dependency on the United States to a spiraling birth rate, alcoholism, suicide, juvenile delinquency, and environmental pollution. Against the norms and values of a culture that defines itself largely in terms of productive economic activity, Pohnpeians are called passive, indifferent, lazy, irresponsible, and unreliable...."[574]

In his final assessment, albeit ever defensive of his former charges, Hanlon has expressed pessimism about Islanders' social and economic adaptation to the closely encroaching outside world. But Hanlon, like the majority of objective Pacific historians and anthropologists, is well aware of the Micronesian penchant for trickery and manipulation.[575] It was Hanlon, after all, who cited U.S. anthropologist William A. Lessa (1950) as having described the Yapese as "'a [historically] difficult, uncooperative, dishonest, quarrelsome, and arrogant people.' "[576]

So even U.S.-ruled Noble Savages have flaws! That Micronesians are smart, plenty akamai folks, was quickly discovered by idealistic American anthropologists after the creation of the post-Pacific War USTTPI. A manual prepared for the World Federation for Mental Health, and edited by the [in]famous Margaret Mead, contains a chapter on Palau and Palauans. A forgotten early-era Territorial anthropologist had commented:

"Most human affairs are subject to manipulation, and those [Islanders] who display virtuosity in the handling of persons and processes in social situations are greatly admired. Deceit is both legitimate and expected, and even honoured if the outcome is successful." [577]

On a lighter note was that same American anthropologist's amusing description of early-era Territorial Palauan dining out—an eclectic milieu that seems to have changed but little a half-century later:

....."A meal may be comprised of a mixture of American canned and South-East Asian derived foods, prepared in Spanish iron pots, placed in Japanese dishes on a table made of Philippine driftwood, eaten with German silverware in a commercial restaurant housed in an American Quonset hut which has been financed by a Japanese credit pattern and is operated by a native clan as a family enterprise, while a Japanese version of an American phonograph plays African-derived dance music. These are merely illustrations of the interconnexions between the traditional and newer forms of economic patterns in contemporary Palau."[578]

But some things have changed in the blended cultural atmosphere of Micronesia since those formative TTPI years. An Island repast today usually consists of much more "American canned" than "South-East Asian derived foods." In a very recent (September 2001) article in Pacific Magazine, correspondent Robert Spegal has written of the many ills plaguing modern-day Micronesians; caused mostly by behavioural patterns and eating/drinking habits. Robert Spegal largely blames "the westernization of the Pacific" for poor Islander diet and lack of exercise that cause diabetes, heart disease, and stroke. And, of course, Spegal correctly identifies cancer, liver conditions, and injuries due to violence... as all related to the use (and abuse) of tobacco, alcohol, and other drugs.[579] We in the West have heard it many times before, but Spegal does put it in an Island context:

"Populations that used to engage in physical labor to obtain the daily bread are now largely sedentary. They take the car to work, sit most of the day at a desk or machine, eat convenience and junk foods, and then spend evenings watching TV. They smoke cigarettes, chew betel nut, and drink alcohol and/or kava to excess. First white rice, and now more recently instant noodles, have replaced nutritional local foods like breadfruit, bananas, yams, and taro."[580]

Age-old cultural beliefs that have been accepted unquestioningly for generations are today obstacles to preventive-medicine/health practices. An example is Bigness (i.e., a large body/persona), which traditionally personified high status in Island societies, is now a high-risk factor for many non-communicable diseases throughout Micronesia. Correspondent Spegal cites that obesity in some Island populations is as high as 75%, and that

"[c]onversely, infants and young children are often borderline to severely malnourished, subsisting on a diet mostly of white rice, bread, instant noodles and occasional canned meats.
"Nauru, one of the Pacific's richest countries, has the highest rates of obesity and diabetes. Kiribati, one of the region's poorest countries, has some of the lowest rates of non-communicable diseases."[581]

In the very same 'Health' issue of Pacific Magazine referred to above, was a truly assinine editorial by Robert Keith-Reid, publisher, who has loftily concluded:

"Perhaps Pacific Island governments should go much further than that by closing down McDonald's and Kentucky Fried Chicken outlets, and banning Coca-Cola and so-called 'snack food' factories that they greet now as symbols of development."[582]

No, Mr. Keith-Reid, what to eat or drink is for parents, family, or the individual adult person to decide! Islanders must possess those same life-style choices that we wrestle with over on the Mainland. If a future Orwellian Ministry of Culture in Micronesia decides to prohibit American fast-food and soft drinks from the Islands, how about a strict censoring of the American "cultural" fluff and garbage emanating from Hollywood and Manhattan via video-cassettes and TV programming? I'll bet, though, that Robert Keith-Reid, like all good Leftish totalitarians, eschews the evils of "censorship". But I would wager, too, that publisher Keith-Reid assiduously snoops about and sniffs around his magazine's parking lot for errant smokers! (His kind usually does.)

Having vented that, this writer (nonetheless) entertains no doubt that Islander behavioural patterns and eating/drinking habits have led to the rise—over the last fifty years—of diabetes, heart and liver disease, stroke, and cancer. And yes, injuries from violence have been due, overwhelmingly, to use and abuse of drugs and alcohol. All those writing on Micronesia today and yesterday—journalists, historians, and anthropologists—have concurred on these conclusions; a veritable litany-list of dead-end hopelessness. Even hip, sunny-travel author David Stanley (Micronesia Handbook, 1992 ed.) appears profoundly pessimistic about the sociocultural climate of contemporary Micronesia:

....."Alcoholism is also a serious problem. Culture change in Micronesia is farthest advanced in the Marianas, where Spanish and American influence was strongest. The large numbers of police one sees in the towns of Micronesia reflect the lack of social control by village elders; on the outer islands relatively few police are present."[583]

Commenting on Palau (Belau), David Stanley has observed that Palauans "appear to be the most Americanized of Micronesians" (his emphasis)—but beneath their Western-style-clad exteriors are very tradition-oriented Islanders. And that means that Palauans, like other Micronesians, have "learned the subtle art of seeming to comply while outlasting those who would control them."[584] Nevertheless, Palau has been a favoured tourist destination since the Peace Corps years of the 1960s for young Mainland Americans... but for SCUBA-diving or recreational drugs? The modish Stanley has reported:

"In recent years Belau became a transshipment point for... Southeast Asian heroin, being smuggled into the U.S. Top Belauan officials have been involved in the heroin trade. Out of a population of 15,000, there are 400 heroin addicts in Belau. Marijuana (udel) grown illegally on Peleliu and Angaur remains the country's only cash crop. Custom inspectors on Guam know all about Belau's drug problems, and thorough spot checks are carried out."[585]

Do corrupting young (or youthful "baby-boomer") Americans—along with other affluent trippers from around the Pacific Basin—exercise a malign influence on the Islanders, or do Palauans have an insatiable craving for pakalolo and other drugs? David Stanley's factual report doesn't tell us, but he does accuse "Westernization" as largely culpable for the wrack and ruin of the Marshallese people:

"Westernization, the breakdown of the traditional family structure on the outer islands, and the experience of colonialism have led to alcoholism, hypertension, and high rates of suicide among young males and pregnancy out of wedlock among young females. The consumption of imported refined foods and sugar has led to widespread obesity, diabetes, dental problems, and infant malnutrition."[586]

Fortunately for the Yanks, travel writer Stanley (a Canadian, I think) has spared his lowliest assessment of Micronesia for non-"Americanized" Nauru. As my readers know by now, tiny Nauru and scattered Kiribati are the administrative results (rejects, really) of the wisdom, justice, and enlightened rule of the British, Australians, and New Zealanders. David Stanley has cluckingly recited:

"Nauruans have the world's largest incidence of diabetes—a shocking 30.3%. Medical journals cite Nauru's epidemic as a classic case of junk-food-induced, as opposed to hereditary, diabetes. The special diabetic section in the supermarket says it all. High blood pressure, heart disease, alcoholism, and obesity are rampant. The heaps of jettisoned Foster's beer cans have led to suggestions that Nauru be renamed 'Blue Can Island' or 'Foster's Island'. Traffic accidents on the one road (!) and the high mortality rate make a 50th birthday a big event for Nauruans. Few males live much beyond this age and only 1.2% of the population is over 65—the lowest such ratio in the Pacific islands. Visit a cemetery and note the ages on the tombstones."[587]

As Foster's Lager is proudly brewed in Melbourne, Australia, America—for once—can hardly be held solely accountable for such human misery and degradation. (But "Westernization" is still under indictment rather than bad individual choices.)

It is not only the politically-correct journalist, travel writer, or academician who blames "Westernization" for the socio-cultural maladies of Micronesia. Earl Hinz is (1995) a former U.S. Marine, World War II veteran, and Pearl Harbor survivor... scarcely the picture of an aging, pony-tailed, "Piss Corpse" hippy. In his informative book on the Pacific War and Island battle-sites (<u>Pacific Island Battlegrounds of World War II: Then and Now</u>, The Bess Press, 1995), Earl Hinz has pulled no punches in blaming "[a]ll the ills of western civilization" for the present plight of Micronesians.[588] But notice that Hinz has fairly indicated "the <u>ills</u>" of Westernization (my emphasis). Remarking on the modern-day Marshall Islands (RMI), Hinz's passage reads:

"Life today in the Marshall Islands is a mixture of high tech at Kwajalein, traditional life in the outer islands and a muddled western culture in Majuro [the capital/centre]. <u>All the ills of western civilization</u> are present, such as alcoholism, TV, fast foods and unemployment. Much of the Marshallese problem has to do with overpopulation brought on by a high birthrate, almost the highest in the world.
"The Marshallese, who have lived a traditional subsistence life heretofore, are having a hard time assimilating the trappings of the Western world. <u>Maybe their lot was better before the Germans, Japanese and Americans arrived.</u>"[589]

Does ex-"leatherneck" Earl Hinz sound a whole lot like crusty Lady Doc Lucy Ingersoll, who expressed pretty much the same thoughts about the Pohnpeians circa 1888 (see note 572)? Earl Hinz appears to be no more sanguine about the present State of Chuuk, FSM:

"Today's Truk has a different name; it is now Chuuk and the people are 'Chuukese'. It is still an agrarian society with strong overtones of western civilization wreaking havoc with its people. Beer, cars, tobacco and boom boxes use up much of the people's resources.

"Underemployed men have tended to violence in both the home and on the streets, keeping all but the most determined of tourists away."[590]

Another Bess Press book from Honolulu—this one an excellent Pacific Studies textbook for young people—was written by Reilly Ridgell, a schoolteacher, and (also) published in 1995. But it was the true-to-life photographs of one Carlos Viti, in <u>Pacific Nations and Territories: The Islands of Micronesia, Melanesia, and Polynesia</u>, that impressed me deeply. The photos, brief portraitures of the Islanders, spoke volumes. In the Micronesia section, on page 67, is pictured three young men traditionally dressed in <u>lavalavas</u>. They hail from the Polynesian-speaking outer island of Kapingamarangi, Pohnpei State, FSM. The young men stand proud, dignified, alert, and clean-cut.[591] On the next page (68), are four "modern" male youths, at the district centre petrol station, Kolonia, Pohnpei State, FSM. All four youths are skulking, furtive, dirty-looking no-goods.[592] The third, and most stark of the Carlos Viti photos examined, shows two teenaged boys "hangin' out" in front of a pool hall (or billiards room) in a village on Babeldoab, Palau. The picture of the slouching, sullen-looking, frizzy-haired <u>moke</u> punks says it all—as anyone who has ever lived (or even been) in Honolulu or any large U.S. Mainland city well knows.[593] The hot, angry eyes and hostile body-language of the "Westernized" Pacific Island adolescents are all too frighteningly familiar to North Americans and, increasingly, Australians and New Zealanders.

"Nearly everywhere in Micronesia, women appear to have held at least four major roles, all of which were critical to the functioning of their society. They were guardians of the land, keepers of the peace, counselors of family and community matters, and producers of cultural valuables."[594]
--Francis X. Hezel, S.J., <u>The New Shape of Old Island Cultures</u>, 2001

The Big Guns in the Pacific history and anthropology departments at the University of Hawai'i all sing the same, sad, Swan Song about: (1) The always-declining health standards, and (2) the decline of extended "family values" throughout Micronesia; (3) coterminous with the ever-ascending rise in population and lack of employment. On all the latter there is general concord among the academicians—who, in turn, are in agreement with travel writers and journalists. "Westernization" has been the common bugaboo, but, as we have seen, there has been sharp discord as to what degree and on which facet. The old U.S. Marine, Earl Hinz, has pointed out "the ills of western civilization", while U. Hawai'i Pacific historian, the eminent Robert C. Kiste, has fingered "Western education" as the cultural culprit. <u>In Tides of History: The Pacific Islands in the Twentieth Century</u>,[595] Robert C. Kiste's contributed chapter, 'Towards decolonisation', turned out to be the only essay in the entire collection with even a hint of "conservative" tilt. (The fine works published by U.H. Press are no exception to the unwritten pc rule!) Professor Kiste's list of reasons for Micronesia's on-going degradation include:

> "Urbanisation, wage labour, Western education [a-ha!], and the incongruities between federal programmes and traditional cultures have all served to undermine traditional forms of social control. The importance of extended kin groupings has declined, and traditional leaders have lost much of their authority. Teenage delinquency and alcohol abuse are common facts of urban life. Suicide rates for young Micronesian males are among the highest in the world. Rapid increases in population also pose a serious problem for the near future. The number of Micronesians has increased more than threefold since World War Two, and in some instances, such as the Marshalls, the population has quadrupled. Family planning is virtually non-existent. Employment opportunities are far short of the number of school graduates or dropouts, and unemployment contributes further to problems of delinquency and substance abuse."[596]

Dr. Kiste, who helped edit Tides of History (1994), five years later co-edited (with Mac Marshall) a superb and definitive anthology of anthropological essays on Micronesia—with contributions by Pacific historians David Hanlon, Francis X. Hezel, S.J. and Robert C. Kiste; himself an historian (rather than an anthropologist). Indeed, American Anthropology in Micronesia: An Assessment[597] became my main bibliographical textbook for this Addendum. Professor Kiste has greatly helped me in bridging the nexus between Pacific history and Micronesian anthropology. And without the works of Doctor Hanlon (1988, 1998) and Father Hezel (1994, ed., 1995)... I would know virtually nothing of the rich histories of the Carolines (FSM, Palau) or the Marshalls (RMI).

Before getting to Father Hezel's chapter in the anthology, I must make mention of an item in Karen L. Nero's 'American Anthropological Studies of Micronesian Arts.' Ms. Nero has reported that, in [dubious?] honour of young U.S. Peace Corps volunteers, one of the more popular (Micronesian!) T-shirts sold at the Yap Co-op during the late 1960s... "depicted a line of loincloth-clad dancers, one of whom was [W]hite."[598] Did the Yapese T-shirt portray an inspiring example of a young Peace Corps volunteer earnestly trying to "go Native"? Or was the T-shirt a subtle Micronesian instance of poking silent fun at the clumsy attempts of a clueless outsider trying to fit in? (Answered either way, the scenario could be considered absurd.)

Francis X. Hezel, S.J. has been referenced throughout this Chapter on Micronesia. His writings have been both educational and pleasureable reading for me. And Father Hezel, as a Roman Catholic Jesuit priest and Island high-school teacher, has that extra-special humane touch which collegiate academicians seem to rarely possess. (And if they do, lose it over time with the proliferation of academic letters after their name.)[599] After living, teaching, and writing on Chuuk for many years, Father Hezel now heads the Micronesian Seminar at Kolonia on Pohnpei. He penned a pivotal essay in American Anthropology in Micronesia: An Assessment (1999), and then went on to produce his own modest but significant work, The

New Shape of Old Island Cultures: A Half Century of Social Change in Micronesia (2001). I received this slender volume just in time to utilise as an important resource for my Addendum. (I am truly grateful to U.H. Press for getting it to me so quickly.)

For 'Social Problems Research in Micronesia', in the American Anthropology in Micronesia collection, Father Hezel has cited a 1985 study by Dennis T.P. Keene of young female runaways in the Marshall Islands. This highly visible sub-group in Majuro, RMI—known as kokan ("traders")—had left home and lived wherever they could find shelter. These young women...

"...ranging in age from fifteen to their early twenties, drank alcohol, smoked, ran around with men, and often supported themselves by prostitution. Many were newcomers to Majuro. Most had dropped out of school and had experienced tension with their families that often culminated in the parents cutting off their daughters' hair to humiliate them or throwing them out of the house altogether. Keene (1992) suggested that the flagrant promiscuity of these girls could be viewed as a defiant response to their problems with their families."[600]

Father Hezel would expand his examination of Micronesia's sociological problems, two years later, in The New Shape of Old Island Cultures. If Fr. Hezel high-lighted bad behaviour (coming from poor choices), in his earlier work, as occurring outside the Micronesian home... he has spot-lighted, in his later work, a far subtler sociocultural danger afflicting the mind and soul of Micronesia. The danger, of course, is the "mainstream" cultural fluff and garbage emanating from the septic tanks of Manhattan and Hollywood, U.S.A.—steadily and inexorably piped into the Pacific Basin through the twin sewer-lines of satellite television and video cassettes. And here, on this point, the nay-sayers to "Westernization of the Pacific" are 100% dead right. For although a number of "violent content" (i.e., martial arts) films are produced in East Asia (China, Hong Kong, Taiwan), the overwhelming majority are from America. And just about all "adult" releases (i.e., porno movies), with the exception of some from Europe, are made in the good ol' U.S.A. (I'll venture to say that American "mainstream"/whitebread culture has little to offer Islanders except NASCAR imbecility, "Hooters" restaurants, cheerleaders, condomania, "pot", neo-paganism, the Playboy Philosophy, and really lame popular music.) As expected and anticipated, Fr. Hezel vividly demonstrates the American visual assault on the Micronesian psyche:

"Video rentals, especially, open the family up to the world of R-rated movies—nothing of great importance if young men were watching by themselves, but a shocking departure from custom when the assembled family, males and females together, watch a love scene in an R-rated movie.... In some families someone might grab the remote control and fast-forward through the steamy parts of the video to spare everyone embarrassment, but in others this does not happen. Instead, the family may sit frozen when sex scenes begin, responding only with nervous laughter or perhaps awkward silence. Yet, even such a reaction is a gauge of the enormous encroachments that recent change has made on the customs that once protected a key value in the Micronesian constellation: cross-gender respect in the family."[601]

There is more, much more, on the sensitive topic of cross-gender relationships in The New Shape of Old Island Cultures. In fact, Fr. Hezel has written an entire sub-section on changing sexual mores in Micronesia; titled appropriately 'The Sanctuary of the Family'.[602] My own sub-section in this Addendum, 'Micronesia in America', relies almost entirely on information provided by Fr. Hezel in his absorbing study.

To lightly finish off this first part of the Addendum, I will cite P.F. Kluge, a generally cheerful but also downbeat commentator on Micronesia. A former Peace Corps volunteer, PF. Kluge knows Micronesia well. He was personally acquainted with Lazarus Sali'i, the late president of Palau. Indeed, Sali'i's tragic death opens Kluge's very revealing book, The Edge of Paradise: America in Micronesia (1993).[603] But Kluge's description of restaurant dining in the Islands, however patron-resistant to the potential tourist, is pure hilarity. It is also eerily reminiscent of that long-forgotten 1950s U.S. anthropologist's depiction of a Palauan dinner served in an American Quonset hut (see note 578):

"I'd forgotten how disappointing it can be, traveling through the islands, to be confronted with chopped steak, french fires, white bread, instant coffee. There are exceptions: the Village [on Pohnpei] does a decent job, especially with mangrove crab; some Japanese restaurants serve local fish. But there's nothing like that look on a tourist's face, sitting on a tropical island, confronted with some canned pineapple from the Philippines and Thailand. I've seen that look a hundred times, astonishment that contains the beginnings of wisdom. You picture beaches, you get mangrove swamps. You imagine native markets teeming with fruits and vegetables, you get rows of dusty canned goods, freezers full of rock-hard mastodon meat that could have been carved out of a glacier...."[604]

Back in a more serious tone, but still very funny, is PF. Kluge's clear understanding of the high hopes and grandiose dreams of new nation-building—no different, really, in Micronesia than anywhere else in the so-called "Third World":

"Meanwhile, Kolonia [Pohnpei, FSM] is a government town, with offices behind gas stations, above grocery stores and dress shops, along the main street, and down by the docks and in an old hospital across from the Spanish Wall. The place has the hasty, improvised appearance of a government in exile, on the lam after the fall of Washington or Richmond. Soon there will be a new capitol that a Korean construction firm is close to finishing, an $11 million campus that Washington paid for, outside town at a place called Palikir. It's a gorgeous, hopeful site, with rain-forested peaks on one side, hills rolling into mangrove swamps on the other; a great sense of new beginnings, compared to sad, soggy Kolonia. There's only one drawback that I can see, and even it may turn out to the good. Near the new capitol is a conical little hill, a volcanic plug that's known in Ponapean as pewshin malek. These days, what with the new capitol so nearby, you have to press before people tell you what it means in English: chicken shit."[605]

As Palikir is the new political capital of the Federated States of Micronesia, thus an important place in its own right, Kluge's overtly cavalier attitude may be excused by his obvious love for the Islands and the Carolinian people. (P.F. Kluge was not optimistic about the FSM's economic future, either.)[606] More's the reason: As an American citizen (too), I'd hate to see the United States—the alien nation that has done most to nurture and help Micronesia—destroy the Islands while destroying herself by committing sociocultural suicide.

Micronesia in America

"Today there are so many Micronesian communities scattered (like islands) throughout Mainland U.S.A., that an entire doctoral dissertation was researched and written for the University of Iowa in 1997 by a Ms. Linda A. Allen titled 'Enid"Atoll": A Marshallese migrant community in the midwestern United States.'."[607]
--C.D. Bay-Hansen, 2001

Has the great, harsh continent of Eldollarado already become a vast Third World colony? And if so, where do Micronesians fit in with American "multiculturalism"? With Cubans in Vermont and Laotians in Wisconsin, can Palauans really be acclimated to Denver, Colorado? Or Marshallese in Enid, Oklahoma? In other words, Can Islanders be acclimated to Middle America so well as in diverse California, Hawai'i, or Guam? The answer is both "Yes" and "No.". The cultural hodge-podge and ethnic mish-mosh of confused, crazily-quilted California represents one thing; the strong, vibrant all-American Heartland of Colorado or Oklahoma means something else entirely. So both "Yes" and "No" can be proven right. Giff Johnson, writing for Pacific Magazine in July 2001, reported:

"Walk into Kojo's, a popular Japanese restaurant in Kansas City, Missouri, and you'll find that most of the waitresses, busboys and cooks are from Pohnpei. Check out the Tyson chicken packing plant in Springdale, Arkansas where you can't move 10 feet down the assembly line without bumping into a Marshall Islander. And that's not to mention Sea World in Florida, which resonates with workers from the Federated States of Micronesia. Ditto the Kyoto Restaurant chain on the southeastern seaboard, while nursing homes for the elderly throughout the [Mainland] States find scores of FSM and Marshalls citizens handling nine-to-five jobs."[608]

Giff Johnson does remind his Pacific-oriented readers of the disquieting fact that the U.S. government, in its Immigration and Naturalization Service (INS) form, fully intends to start limiting the free entry of citizens from the Freely Associated States (FAS—the formerly American TTPI) to the U.S.A. There are no census figures on Micronesians currently living on Guam, the CNMI, Hawai'i, or the U.S. Mainland, but Giff Johnson guesses FAS citizens to number in the "tens of thousands".[609] In defense of Island in-migration to the U.S.A., Johnson has referred to the reaction of a number of Micronesian chiefs and politicians:

"Some island leaders wonder why the U.S. government is so exercised about the issue. Pohnpei Senator Peter Christian, the lead Compact negotiator for the Federated States of Micronesia, says in actuality, the majority of FAS islanders are working and paying taxes or going to school. Leaders in the Marshalls wonder why the U.S. is focusing so much attention on the FAS—which are loyal U.S. allies—when legal FAS migration to the U.S. in no way compares to the waves of other third country nationals attempting to get into the U.S."[610]

Some of the screening and crackdown is well and good; like the denying (since December 2000) entry to the U.S.A. of Marshallese convicted of felonious offenses in the RMI. But, Johnson clinches:

"The number of FAS citizens streaming to America is likely to increase the next two years particularly if islanders feel they need to get to the U.S. before more stringent regulations are implemented. But even so, just like the tiny islands they come from, the numbers represent just a drop in a huge ocean."[611]

Here I am in total agreement with Giff Johnson—and his key phrases are, "legal [my emphasis] FAS migration" and "the numbers represent just a drop in a huge ocean". Not only are the populations of Micronesia's tiny islands relatively small, but the actual figures of illegal aliens who have literally pushed their way into the U.S.A. are really huge. These "illegals" (numbering in the many millions) have foregone waiting in the long immigration queues outside the U.S. consulates of the world. After skipping the lines the illegals unlawfully break into America, and then expect—often demand—housing, education, and medical care: all paid for by legal immigrants and the long-suffering American taxpayer. Once in America, many illegals continue to transgress the laws of the land... many ending up as semi-permanent wards of the State, in prison. Micronesians who are FAS citizens, as former American colonials, have essentially more right to a U.S. passport than I do, a former subject of the king of Norway.[612] It is my considered opinion that law-abiding Micronesian citizens of the Freely Associated States should continue to enjoy free and unrestricted access to the U.S.A. and all her possessions.

"On Memorial Day weekend, I walked into a gym in Costa Mesa in which hundreds of Marshallese were watching an all-Marshallese basketball tournament. Teams from distant places—Hawai'i, Arizona, and Oregon, among others—represented the other Marshallese communities invited to celebrate the weekend in a renewal of old cultural bonds in their new home."[613]
--Francis X. Hezel, S.J., The New Shape of Old Island Cultures, 2001

Is the scene above, related from Father Hezel's absorbing new book, an example of Marshallese acclimation to U.S. multiculturalism in their new American home? Again, the question could be answered "Yes" or "No": "Yes" in the sense that Micronesians are blending into the vast American [multi-]cultural landscape; "No" in the way which Micronesians are retaining their Island heritage and keeping their communities intact in themselves. Fr. Hezel has offered several examples of these geographically dispersed but nonetheless cohesive communities:

"In the past few years, dozens of Pohnpeians were hired to do fruit picking in Oregon, while others were recruited to become physical therapists in a home for the aged in South Carolina. There is a large contingent of Micronesians working at Sea World In Orlando, Florida. Hundreds of Marshallese are working in Arkansas on a gigantic poultry farm. The recruiters continue to target the islands even now as a source of cheap and reliable untrained labor for large businesses in the United States....

"....In more recent years, instant Micronesian migrant communities have been springing up as large numbers of islanders were recruited for a single business. The scores of Marshallese brought into Arkansas to work on the poultry farms around the state is one of the most striking examples."[614]

For years, reports Fr. Hezel, Costa Mesa, California, was the "official" Marshallese overseas community. (Costa Mesa is a "blue collar" town in otherwise wealthy, "white collar" Orange County.) Two decades prior to the Compact of Free Association (1986)—when Marshallese seeking work streamed [mostly] to Hawai'i and the Mainland—Island youths had been attending Southern California College; a private institution administered by the (Pentacostal) Assemblies of God. After the Compact went into effect, the California Marshallese population grew quickly—from 300 in 1991, to 400 in 1995, to an eventual estimated 800 Islanders by 1999. Parents brought their children to Costa Mesa from the RMI to receive a better education, and other "homesick" Marshallese, residing in different areas of the Mainland, came to the California community to join their landsmenn.[615] By the early 1990s, the migrant Island households in Costa Mesa were "almost mirror images of those in the Marshalls in their size and composition."[616] Fr. Hezel has told us that other Micronesian communities have expanded following a similar pattern:

"....Often, there is a community college that Micronesian students once attended in significant numbers or a business recruiting Micronesian labor that served as the original magnet for the migrant population. Corsicana, Texas, a small town not far from Dallas in which 500 or 600 Micronesians are living, is the home of Navarro College, a favorite destination for Chuukese and Yapese students several years ago. A small community eventually gathered around the students who chose to stay after finishing school. Kansas City, which houses Park College, another favorite of the Micronesian college-bound in past years, now has several hundred Micronesian residents, including perhaps 300 Pohpeians."[617]

Once settled in their communities on Guam, Hawai'i, and the U.S. Mainland, Islanders keep the two-way Micronesia/America traffic flowing with goods as well as people. Fr. Hezel has written that Chuukese would send fresh fish and pounded breadfruit in ice-chests to relatives in Guam. A few days later, those same ice-chests would be returned to family in Chuuk—filled with frozen chicken and U.S. goodies, purchased cheaply on Guam.[618] Chuukese residents of Guam would also send back boxes of second-hand clothing, bought for nearly nothing at the Salvation Army or Catholic Charities. Marshallese settled in California and Hawai'i also send home food and clothing—but include such luxury items as perfume, keyboards, computer games and other electronic equipment.[619]

And a Micronesian from the Freely Associated States (FAS) really can go home again! The present Ibedul, the highest-ranked chief of Koror, Palau (Oreor, Belau), had been serving in the U.S. Army for several years when he was recalled to the Islands to assume his rightful title.[620] Modern Americans might laugh at lingering notions of title, chief, and clan, but Micronesians still regard seriously the ideal of tribe and extended family. Fr. Hezel has briefly and clearly spelled out—anthropologically no less—the central concept of Micronesian society:

> "In the tension that invariably exists between the individual and the social group, traditional Pacific societies have always come down strongly on the side of subordinating the individual's interests to those of the community, even at the cost of suppressing the creative impulses of the individual. Micronesian societies have repeatedly eschewed creativity at the expense of chaos; they have continually chosen conformity for the sake of consensus. Individualism, although capable of releasing powerful innovative energies, can threaten the fragile unity of the small island population. Yet, a source of livelihood independent of one's kin is the first step in the release of the individual from the powerful ties that bound him or her to the extended family."[621]

Independence and individualism—with all their freedoms from the family, both good and bad—can lead to social detachment and cultural disconnection for the Micronesian in America. Thus Pacific Islanders (generally) residing on the U.S. Mainland prefer living in "multigenerational households" (as extended families); indeed, this life-style is actually proliferating in the new century.[622] According to the U.S. Census Bureau, families of three, or more, generations dwell under the same roof in close to 4 million American homes. Multigenerational households are often seen among grandparents helping care for a daughter's children, or grown children looking after aging parents. But the most common multigenerational households, an Associated Press story has reported, are found among new immigrant groups:

> "By state, Hawaii had the highest proportion of homes with multigenerational families—8 percent of the state's 403,000 households, according to the 2000 census report released Friday. California was next, at 6 percent.
> "That is not surprising because those two states have higher concentrations of Asian[s] and Pacific Islanders, and recent immigrant families in general, said demographer Paul Ong of the University of California, Los Angeles.
> "New immigrants tend to settle in areas where they have family members already, for financial reasons and because of a need for some connection to home, Ong said."[623]

Now that large (proportional) numbers of Micronesian communities have been planted—and growing—on the U.S. Mainland, what's next? Will there be continued "Bosnification"[624] (scattered but culturally-intact ethnic communities) for Islanders in America, or "Brazilianisation"[625] (<u>real</u> ethnic-cleansing in the Great Tan Melting Flesh-Pot)? As FAS Micronesians are already <u>de facto</u> American, national loyalty poses no problem. And as FAS Micronesians are (still) free to travel from the Islands to the U.S. Mainland and back… cultural contact and Micronesian continuity remain assured. (Note: A sociocultural advantage and political status other immigrants to the U.S.A. don't enjoy.) But with the relentless passage of time and space, the generations of Islanders will merge into the American Mainstream sociocultural polity through enculturation and assimilation; facilitated by intermarriage with racial and ethnic "others". What then?

The Bosnification of Micronesians in America will transmogrify into Brazilianisation a-pace: "Pacific Islanders" on the U.S. Mainland will eventually disappear, as ethnic and cultural entities (they are so few in number!), into the Great Tan Melting Flesh-Pot. Young "bad-dude" Pacific Islanders (led by Polynesian Samoans, Melanesian Malaitans, Micronesian Chuukese?) shall have to physically patrol their neighbourhood, culturally protect their turf from "jive-ass" bands of marauding African-Americans or Hispanic/Latino gangs. To socioculturally survive the failed schools and mean streets of Honolulu or California, Pacific Islanders ought to (naturally) ally themselves with Asian groups—Korean, Chinese, or Filipino. (Micronesians from Guam and the Northern Marianas live next door to, and culturally resemble most, the Filipinos.) After all, "Asians" and "Pacific Islanders" are lumped together by the U.S. Census Bureau! So meanwhile?

Meanwhile, Micronesians must try to be <u>the best</u> Americans while retaining their Island identity in a totally debased and ruinously immoral cultural environment. Being a super-good American means conforming to—and honouring—the highest standards of Judaeo-Christian values, which have been part and parcel of the Anglo-American heritage[626] since the Republic's inception in 1776. But spiritual and cultural conditions have changed drastically in the American mainstream since about 1965. At the dawn of the third millennium, the U.S.A. is post-Modern, post-Feminist, post-Christian. Today, the impressionable children of often deeply-devout Micronesians can watch <u>shlock</u> "entertainers" Jenny Jones, Ricki Lake, and Jerry Springer on American <u>daytime</u> T.V.[627] Besides utilising cheap vulgarity and sexual innuendo to hike their ratings, these bottom-feeders take full advantage of racial tensions to shamelessly exploit reverse-race baiting. (Setting up one race against another was the political trick of choice practised by liberals during the Clinton administration—1993-2001—with devastating effect.) A favourite ploy, especially of Springer's, is "playing the race card:" Leeringly presenting mixed-couples as invariably "Minority" men/White women.[628] (White men are sexually-impotent, hateful bigots is the inferred message…except of course, for Jerry himself.) How do these racial resentments and blanket stereotyping, presented daily over the multi-media, affect Pacific Islanders? Therefore, besides having to cope with the ugly pornographic or violent videos shown back home, migrant Island parents have to contend with the whole soul-sucking burden of U.S. Mainland/mainstream popular culture.

America has surely devolved into an un-promising land of junk-food, junk-science, junk-culture. From the mansions of Malibu, California, to the humblest home in Heartland America, human existence seems to have degenerated into an insatiable four-way appetite for sex, food, drugs and alcohol. So what would I recommend (as a course of action) for the Micronesian family (extended or not) living on the Mainland in Century 21? (The same agenda I'd advise for Norwegian-Americans). 1) Work hard now and save money for the future; 2) obey the laws of the land and provide your keiki with a sound education; and 3) partake of the good things in American life and be a dutiful U.S. citizen, while neither forgetting your Micronesian roots nor forsaking your Island heritage. For you—and your extended family/community—might have to return home someday; perhaps sooner than you think, as the United States ship of state is sinking and sinking fast. Patriotic Virginia author David Baldacci (Absolute Power, Total Control et al.) has described an imaginary America's End of Days scenario to perfection, in what I'll call "Danny Buchanan's Nightmare". In Saving Faith, reformed D.C. lobbyist "Danny Buchanan" has envisioned:

"'I have this recurring dream, you see. In my dream America keeps getting richer and richer, fatter and fatter. Where an athlete gets a hundred million dollars to bounce a ball, a movie star earns twenty million to act in trash and a model gets ten million to walk around in her underwear. Where a nineteen-year-old can make a billion dollars in stock options by using the Internet to sell us more things we don't need faster than ever.... And where a ["Beltway"] lobbyist can earn enough to buy his own plane.... We keep hoarding the wealth of the world. Anybody gets in the way, we crush them, in a hundred different ways, while selling them the message of America the Beautiful. The world's remaining superpower, right?"[629]

"'Then, little by little, the rest of the world wakes up and sees us for what we are: a fraud. And they start coming for us. In log boats and propeller planes and God knows how else. First by the thousands, then by the millions and then by the billions. And they wipe us out. Stuff us all down some pipe and flush us for good. You, me, the ballplayers, the movie stars, the supermodels, Wall Street, Hollywood and Washington. The true land of make believe.'"[630]

Afterword: At Long Last

Both Future Fish 2000 and Future Fish 2001 were researched and written (October 1991—October 2001) solely on the author's own time, effort, and money. No foundation funds or academic grants were ever loaned or donated to this writer during the production of either book. But there are six individuals to thank personally for sending me invaluable information pertaining to Chapter 4, "A Modest Proposal" for Micronesia. I will list them (below) in order of appearance:

Joseph D. McDermott, FAS Desk Off., U.S. Dept. of the Interior; Lynne D. Pangelinan, Protocol Asst., Embassy of the U.S.A., Kolonia, Pohnpei, FSM: Michaela Saimon and Len Isotoff, Adm. Asst./Secty., National Fisheries Corp., Pohnpei, FSM; Eugene Pangelinan, Deputy Dir., Micronesian Maritime Authority, Palikir, Pohnpei, FSM; and Josie Tamate, Project Econ., Forum Fisheries Agency, Honiara, Solomon Islands. (Mr. Pangelinan of the MMA kindly sent me a whole packet of FSM fishery materials—crucial for my research—and requested a copy of my completed report on Micronesia for the MMA library at Palikir. I am extremely proud to comply. Thank you all so much!)

Finally getting the opportunity to write on the fisheries of the Central-Western Pacific (after all these years!) has reminded me of four Australians to whom I am still deeply indebted since my journey "Down Under" in March 1982. Dr. Ian Campbell, now teaching Pacific history at Canterbury University, Christchurch, NZ (whom I knew from U. Hawai'i at Mānoa), shepherded me around and about Adelaide, S.A., and the university there. Among all the Evelyn Waugh-type thatched hair-style pomposities at U. Adelaide, Ian C. Campbell stood out as a "fair-dinkum bloke" (a regular fellow) and an "ocker strine" (a real Aussie/"Ozzie"). "Good on yer and thanks, Mate!" My second and third Australians were the late Ivar Saether (1989) and his wife Evy, both natives (like my wife, myself, and elder son) of southeastern Norway. I was hospitably received on the very last evening of my stay in South Australia—and ended up corresponding with the Saether family for sixteen years! "Mange takk, Evy!"

My last—and least- Australian to whom I am eternally grateful is the absolute dastard (!) who served as Australian chief consul at Honolulu, Hawai'i, in 1982. His offhanded and dismissive manner, his arrogant and suspicious attitude regarding my visa application... convinced me that "The Lucky Country" was not the least bit interested in attracting qualified new immigrants, able to support themselves, to her shores. Or was it just me? I recall wondering. Anyway, within an hour, that [un]worthy had thoroughly convinced me that Australia ought NOT be my long-range projected destination, after all (fifteen years of youthful fantasies about emigrating to The Lucky Country). I had sighed, shrugged philosophically, and enjoyed my trip to the Land of Oz. But I have often speculated since then: How many other hopeful aspirants had that chief consul dissuaded from attempting to start a new life in Australia?

I am very glad today for that rude and abrupt Australian chief consul, as he unwittingly did me a great favour in 1982. I ended up remaining in North America, and presently live happily in the Evergreen State. Judging from personal letters and newspaper clippings sent from Down Under, along with travel programmes and documentary films shown on television, The Lucky Country is not so lucky any more. Gangs of youths, sporting "dreadlocks" and "mohawk" hair-do's, with nose, lip, ear, and tongue studs very much in evidence as fashion accoutrements, roam the streets of Sydney and Melbourne... as they do the streets of Los Angeles and San Francisco or Liverpool and London. If fashion may be regarded as "statement", the Antipodean youths-referred to above—display all the symbolic symptoms of sociocultural blight of the 21st century Western world. As both Australia (the new "Big Easy") and New Zealand ("Babeland" II) are condom cultures, Sydney (especially since the Y2K Olympiade) has become the "Queen City" of the South Pacific.

So why should I worry? As a Pacific historian and evangelical Christian, I perceive Australia and New Zealand—where many Pacific Islanders work and reside—acting in tandem to disseminate junk culture ("all the ills of western civilization") throughout Melanesia and Polynesia. (This junk culture emanates from the same "Westernized" source-cesspool from which the post-War United States has liberally piped "dying waters"[631] to a thirsty Micronesia.) But <u>that</u> is a spiritual problem for Australians and New Zealanders of Faith to address. <u>Soli Gloria Deo!</u>

C.D. B-H.
Port Angeles, Wash.
October 2001

Notes to Addendum/Afterword

566. Originally planned as Appendix J

567. Allan Bloom, <u>The Closing of the American Mind</u> (New York, N.Y.: Simon & Schuster, Inc., 1987), p.362.

568. <u>Ibid</u>., p.368.

569. See 'People', <u>Pacific Magazine</u>, May/June 1999, p.50.

570. Author's note: I attended Pacific history classes with (the then Mr.) David Hanlon at U. Hawai'i during 1979-1981 (those of Dr. Tim McNaught and Dr. John J. Stephan). Having seen the 'People' blurb (see above note) in <u>Pacific Magazine</u> (May/June 1999), I sent off a letter to (the now Dr.) Hanlon on 4 June 1999, but received (surprise!) no answer. Either my letter had not reached the good Doctor, or my correspondence had quickly (more likely) found its billet in Hanlon's File 13 (wastepaper basket). Aging academicians get uppity, and David Hanlon was so inclined twenty years ago!

571. David Hanlon, <u>Upon a Stone Altar: A History of the Island of Pohnpei to 1890</u> (Honolulu: University of Hawai'i Press, 1988.)

572. <u>Ibid</u>., p. 172.

573. <u>Ibid</u>., p. 105.

574. <u>Ibid</u>. p. 208.

575. See David Hanlon, <u>Remaking Micronesia: Discourses over Development in a Pacific Territory, 1944-1982</u> (Honolulu: University of Hawai'i Press, 1998), p.13, <u>passim</u>.

576. David Hanlon, 'American Anthropology's History in Micronesia.' In Robert C. Kiste and Mac Marshall, eds, <u>American Anthropology in Micronesia: An Assessment</u> (Honolulu: University of Hawai'i Press, 1999), p.76. (My brackets)

577. Margaret Meade, ed., <u>Cultural Patterns and Technical Change</u> (New York: New American Library, 1955), p.133. Reprinted by arrangement with UNESCO. (My brackets)

578. <u>Ibid</u>., p.129. (From 'Studies of Whole Cultures')

579. Robert Spegal, 'Eating Themselves to Death', <u>Pacific Magazine</u>, Vol. 26, No. 9, Iss. 53, Sept. 2001, p.9.

580. Op. cit., Robert Spegal, p.9.

581. <u>Ibid</u>., p.10. (My emphases)

582. Robert Keith-Reid, 'Starving Amidst Plenty' <u>ibid</u>., p.5.

583. David Stanley, <u>Micronesia Handbook: Guide to the Caroline, Gilbert, Mariana, and Marshall Islands</u> (Chico, Calif.: Moon Publications, Inc., 1992 ed.), p.38. Author's note: I strongly suspect, however, that boomer/liberal Stanley detested the hometown "village elders" of his own (probably misspent) youth!

584. <u>Ibid</u>., p.177

Notes to Addendum/Afterword (cont'd.)

585. Ibid., p.177.

586. Ibid., p.70.

587. Op. cit., David Stanley, pp.265-266. (Stanley's parenthesis)

588. Earl Hinz, Pacific Island Battlegrounds of World War II: Then and Now (Honolulu: The Bess Press, Inc., 1995), p.36.

589. Ibid., p.36. (My brackets and emphases)

590. Ibid., p.78. NB: Those "most determined" being mainly SCUBA-divers who make the trip to explore, underwater, "the ghost fleet of the Truk Lagoon". (Question: What's the Chuukese word for udel/pakalolo?)

591. Reilly Ridgell, Pacific Nations and Territories: The Islands of Micronesia, Melanesia, and Polynesia (Honolulu: The Bess Press, 1995 rev. 3rd ed.), p.67.

592. Ibid., p.68.

593. Ibid., p.77.

594. Francis X. Hezel, S.J., The New Shape of Old Island Cultures: A Half Century of Social Change in Micronesia (Honolulu: University of Hawai'i, 2001), p.57.

595. Honolulu: University of Hawai'i Press, 1994. K.R. Howe, Robert C. Kiste, Brij V. Lal, editors.

596. Op. cit., Robert C. Kiste, p.236.

597. Honolulu: University of Hawai'i Press, 1999. Edited by Robert C. Kiste and Mac Marshall.

598. Karen L. Nero, 'American Anthropological Studies of Micronesian Arts', ibid., p.274. Author's note: Yes, Ms. Nero, T-shirt design, like "handicrafts", is art too!

599. See Ch. 10, ibid., Donald H. Rubinstein 'Medical Anthropology, Health, and Medical Services'. In describing neurological disorders among southern Guamanian Chamorros, i.e., suffering from [para] "lytico-bodig", Dr. Rubinstein never once cites the very fine medical travelogue on—and thorough diagnosis of—that very disease (and colour blindness) in Micronesia by the brilliant Oliver Sacks (The Island of the Color Blind, and Cycad Island. Alfred A. Knopf, New York, 1997). Author's note: Don't these intellectual big-wigs ever talk to each other? Or are they "too smart by half"?!

600. Francis X. Hezel, S.J., 'Social Problems in Micronesia', ibid., p.320.

601. Op. cit., The New Shape of Old Island Cultures, pp.116-117.

602. Ibid.; see also pp.113-114, ff.

603. Honolulu: University of Hawai'i Press, 1993. Author's note: Starting my research on Micronesia in 1996, P.F. Kluge's book served as my entrée to the Islands.

604. Ibid., p.84. (My brackets)

Notes to Addendum/Afterword (cont'd.)

605. Ibid., p.68.

606. Kluge has commented: "Island fantasies go glimmering. Big ideas too. I've seen investors come in with plantations, mines, hotels, fishing fleets on their minds, only to find land was parceled and entangled beyond belief... that labor was spoiled and scarce.... Economic development, island style, meant marrying outside money—Filipino, Korean, or Chinese. The local role in all this was that of agent and landlord, something like the minister, organist, and witness at Las Vegas wedding chapels who'll marry almost anybody, no questions asked, for a fee." (Kluge, p.84.)

607. Robert C. Kiste, 'A Half Century in Retrospect', American Anthropology in Micronesia, p.465. Cf. also p.517.

608. Giff Johnson, 'Pursuing the Micronesian Dream', Pacific Magazine, Vol. 26, No.7, Iss. 151, July 2001, p.16. (My brackets)

609. Ibid., p.16.

610. Op. cit., p.17. NB: Surely Giff Johnson means Third World country nationals?!

611. Op. cit., p.17.

612. Author's note: Of my family, only my daughter born in Tacoma, Washington (1976), is a true-blue, real-live Yankee Doodle Dandy. I am a naturalised U.S. citizen, my wife is a legal alien with "green card" status, and both my sons are also naturalised U.S. citizens. (The elder was born in Norway, and the younger in Canada.)

613. Francis X. Hezel, S.J., The New Shape of Old Island Cultures, p.151. (Cf. Giff Johnson, note 608.)

614. Ibid., pp.145, 150.

615. Author's note: I can relate to that. We moved to Ballard in 1976 for the same reason—to be, live and work with fellow Norwegian-Americans.

616. Op. cit., F.X. Hezel, P.150.

617. Ibid., p.150.

618. Ibid., p.153.

619. Ibid., p.153.

620. Ibid., p.154. Author's note: As opposed to the Reklai of Melekeok! (See any history of Palau.)

621. Op. cit., p.158.

622. Genaro C. Armas, 'Families opening homes', Peninsula Daily News, 7 September 2000, p.C-1.

623. Ibid., p.C-1.

Notes to Addendum/Afterword (cont'd.)

624. From the Bosnia-Herzegovina region/republic in the former Yugoslavia (capital: Sarajevo). There, the differences are religious rather than ethnic. Aside from the few Jews, Gypsies, and [Muslim} Albanians, the population of Bosnia-Herzegovina is overwhelmingly South (Jugo) Slav. But Serbian Orthodox, Croation Catholic, and Bosnian Muslim communities are—by choice—separate and distinct.

625. From Brasíl in South America. I spent several weeks during 1963 anchored off Rio de Janeiro (state of Guanabara), Brazíl. Even though there were—and probably still are—intact communities of Germans, Italians, and Japanese scattered throughout the countryside (not to mention the vast Tupi-Guaraní native areas of the interior), Brazilians are very ethnically and racially mixed. "Cariocas", the light-hearted and indolent denizens of Rio, vary in skin-tone from pure white to very black to all shades in between. The blended "mainstream" Portuguese-idiomatic culture (Lusitanic rather than Hispanic?!) might be described as "Africanized Latino", due to the overwhelming predominance of African music, dance, and macumba (vodoun culture) from Brazíl's impoverished northeast. On the face of it, Cariocas (all Brasileiros?!) appear to exist solely for carnaval (party-time). The stereotypical Hollywood Pepper-and-Salt fun-couple, Brazilian version, has been immortalised by the deeply black Pele and the Clairol-blonde Xuxa. As for Brazíl's native Indio population—are they (and their aboriginal culture) destined merely to add the colour Red to the Great Tan Melting Flesh-Pot? Does a similar fate await Micronesians in an "Africanized"Anglo-America?

626. Author's note: By "Anglo-American" I mean English-speaking (Anglophone) America; the English language as the common (cultural) and binding tongue of the Republic. "Anglo-European" is a loose ethnodemographic term I have used for all English-speaking U.S. "White folks" (with origins from Iceland to Armenia; also Christians and Jews from West Asia, the old "Near East") who identify with the cultural traditions (and Judeo-Christian values) of Western Civilisation.

627. Initiated by toxic T.V. shlockmeister Geraldo Rivera, who still "exposes" himself daily on C-NBC. Needless to relate, Rivera was one of the more rabid Bill Clinton defenders.

628. Question: Have the bleached-blonde, beach-bunny bimboes, cavorting on the prurient, globa lly-viewed Baywatch , contributed to a kind of sexual "racial-profiling"? Author's note: I vividly remember shopping at Aina Haina centre (outside Honolulu on the Kalanianaole Highway) sometime between 1978-1982, and beholding an attractive young haole woman wearing nothing but a tanga and a smirk. All the "Locals" were staring at her— the men glaring in lust; the women glancing in disgust. I have never, ever, felt so ashamed at being a White Guy. Was I wrong for feeling that way?

629. David Baldacci, Saving Faith (New York, N.Y.: Warner Books, Inc., 1999), p.160. (My brackets) Author's note: The final copy for my book was completed in the shocking aftermath of the 11 September 2001, "9/11/01", triple terrorist skyjacking/airliner-bombing of the World Trade Center in New York, N.Y., and the Pentagon in Washington D.C. A fourth terrorist bombing-by-plane was thwarted by some stout-hearted American men who gave their lives in diverting their hijacked airliner to crash near Pittsburgh, Penna. But [the now Man-of-the Hour] "Hizzoner" Sir Rudy Giuliani, chief mourner/celebrant Sir Reggie Dwight (a/k/a "Elton John"), and all those blubbering, snuffling "sunshine patriots" and talking-heads who keep telling us, from safety over (and over) the air-waves, how much they "love this country" and "God Bless America".... These are not the people who will put Humpty Dumpty ("Fun City") together again! If that is to be ever accomplished, the job will be done by the real American heroes—folks like the firemen (NYFD)—especially—and policemen (NYPD) of New York City.

Notes to Addendum/Afterword (concl'd.)

630. <u>Ibid.</u>, p.161.

631. In direct contrast to the "living waters" described in the Holy Bible.

Bibliography for Addendum

32. Bloom, Allan. <u>The Closing of the American Mind.</u> Simon & Schuster, Inc., New York, N.Y., 1987.

33. Hezel, Francis X., S.J. <u>The New Shape of Old Island Cultures: A Half Century of Social Change in Micronesia</u>. University of Hawai'i Press, Honolulu, 2001.

34. Hunter, James Davison. <u>Culture Wars: The Struggle To Define America</u>. HarperCollins Publishers, Inc., New York, N.Y., 1991.

35. Kiste, Robert C., Mac Marshall, eds. <u>American Anthropology in Micronesia: An Assessment</u>. University of Hawai'i Press, Honolulu, 1999.

Supplemental Micronesia/United States Bibliography

1. Anderson, Kevin. The X-Files: Ground Zero. HarperPrism, New York, N.Y., 1995.

2. Clarke, Thurston. Equator: A Journey. Avon Books, New York, 1990.

3. Congdon, Don, ed. Combat: The War with Japan. Dell Publishing Co., Inc., New York, N.Y., 1962.

4. Däniken, Erich von. The Gold of the Gods. Bantam Books, New York, 1974. Translated by Michael Heron.

5. _____. Pathway to the Gods: The Stones of Kiribati. Berkley Books, New York, 1984.

6. Delgado, James P. Ghost Fleet: The Sunken Ships of Bikini Atoll. University of Hawai'i Press, Honolulu, 1996.

7. Dixon, Roland B., Ph. D. Part IV, "Micronesia', The Mythology of All Races, Vol. IX. Cooper Square Publishers, Inc., New York, 1964.

8. Fahey, James J. Pacific War Diary: 1942-1945. Zebra Books, Kensington Publishing Corp., New York, N.Y., 1963.

9. Fisher, Jon. Uninhabited Ocean Islands. Loompanics Unlimited, Port Townsend, Wash., 1991.

10. Gailey, Harry A. The War in the Pacific: From Pearl Harbor to Tokyo Bay. Presidio Press, Novato, Calif., 1995.

11. Grimble, Arthur Francis and H.E. Maude, ed. Tungaru Traditions: Writings on the Atoll Culture of the Gilbert Islands. University of Hawai'i Press, Honolulu, 1989. (The Grimble Papers: 1916-1926)

12. Groves, General Leslie R. Now It Can Be Told: The Story of the Manhattan Project. Reprinted by Da Capo Press, with new introduction by Dr. Edward Teller, 1983. (Orig. published in 1962 by Harper, New York.)

13. Hinz, Earl. Pacific Island Battlegrounds of World War II: Then and Now. The Bess Press, Honolulu, 1995.

14. Kellerman, Jonathan. The Web. Bantam Books, New York, 1996.

15. Manchester, William. Goodbye Darkness: A Memoir of the Pacific War. Dell Publishing Co., New York, 1987 ed.

16. Page, Thomas. Sigmet Active. Times Books, New York, N.Y., 1978.

17. Ridgell, Reilly. Pacific Nations and Territories: The Islands of Micronesia, Melanesia, and Polynesia. The Bess Press, Honolulu, 1995.

18. Sacks, Oliver. The Island of the Colorblind, and Cycad Island. Alfred A. Knopf, New York, 1997.

19. Taylor, Theodore. The Bomb. Avon Books, New York, N.Y., 1995.

Supplemental Micronesia/United States Bibliography (concl'd.)

20. White, Geoffrey M. and Lamont Lindstrom, eds. The Pacific Theater: Island Representations of World War II. University of Hawai'i Press, Honolulu, 1989.

Frontispiece Title:

Swift, Jonathan. A Modest Proposal and Other Satirical Works. Dover Publications, Inc., New York, 1996. Reprinted from The Works of Jonathan Swift (London: Henry G. Bohn, 1851). Orig. title essay written in 1729.

Appendix **A**

Changes in the Japanese Economy [mid-1970s]

	Before the Oil Crisis	After the Oil Crisis
General characteristic		
Era	"Mass" economy	"Quality" economy
Goal	High-speed growth	Information
	Mass production	Perfection of quality (services, softening)
Industrial structure		
Special characteristic	Heaviness	Lightness
Strongest industry	Iron and steel	Electronics
	Automobiles	Communications
Attitude toward trade	Emphasis on exports	Emphasis on imports
Finance		
Character of government	Large government	Small government
Economic system	Keynes	Monetarism
		Supply-side economics
International environment		
Economic sphere	Era of the West	Era of the Pacific

Source: Shōtarō Ishinomori, Japan Inc. (Berkeley, Calif.:University of California Press, 1988), p.159.

Appendix B

The Paracel and Spratly Islands

Paracel Islands: In Chinese, Hsi-sha ch'ün-tao (Pinyin: Xisha Qundao)
                In Vietnamese, Quan Dao Huang Sa
                130 small coral islands in South China Sea
250 mi. east of Viet-Nam, 220 mi. southeast of Hainan island (China)
                Occupied by China during the 1980s-
                Claimants: China, Taiwan (ROC), Viet-Nam
                1974: China drives out Vietnamese.

Spratly Islands: Midway between Viet-Nam and Philippines,
                 inhabited by seabirds and turtles --
    1933-1939: Held by France
    Post World War II: 1951 - Japan renounced all claims
                Claimed by China, Taiwan, Viet-Nam
                    1955 -- Philippine claim
    1970s: Oil prospects
    1973: S. Viet-Nam (RVN) claimed Spratly Islands as part of Phuoc Toy province --
to forestall Chinese occupation (as in Paracels), S. Viet-Nam occupies 3 of
Spratlys - Taiwan moved troops onto Itu Aba Island
    1976: Philippines moved forces onto 7 of islands;
built air-strip on Pagasa Island and announced plans to construct fishing
harbour and fish refrigeration plant --
    1980s: Spratlys claimed by China, Taiwan, Viet-Nam and Malaysia
  June 1983: Malaysia occupied Turumbu Layang-Layang reef -- Except for China, all claimants
had garrisons on various islands.

Source: 'Micropaedia' (Chicago, Ill.:Encyclopaedia Brittanica, Inc., 1990).

Appendix C

Knut Hamsun, "Artist of Scepticism"[1]

Knut Hamsun (né Knut Pederson) was born 1859 in Lom, central Norway, and died 1952 near Grimstad in southern Norway. He grew up in northern Norway, on the Lofoten Islands and in Bodø. As a young man, Hamsun visited the United States twice during the 1880s. The first time in America, Hamsun laboured at Elroy, Wisconsin, and in a lumberyard at Madelia, Minnesota. On his second sojourn, Hamsun worked as a streetcar conductor in Chicago and as a farm-hand in North Dakota.

A novelist, dramatist, and poet, Knut Hamsun won the Nobel Prize for Literature in 1920. Works written in his swift and pungent prose included Hunger ("Sult") in 1890, Mysteries ("Mysterier") in 1892, Pan in 1894, Victoria in 1898, and Growth of the Soil ("Markens Grøde") in 1917. Hamsun led the neo-Romantic revolt at the turn of the century, and rescued the novel from excessive naturalism. He furiously attacked such cultural icons as Henrik Ibsen and Leo Tolstoy. Hamsun's impulsive, lyrical style had an "electrifying effect on European writers",[2] and his own fierce individualism was influenced by August Strindberg and Friedrich Nietzsche. Like theirs, Hamsun's works were replete with asocial anti-heroes.

But time, and his deep antipathy towards traditional and modern Western Civilisation, put Hamsun at odds with his own countrymen and on the wrong side of history. For when Nazi Germany invaded and occupied neutral Norway (1940-1945), Knut Hamsun -- now eighty years old -- collaborated with the enemy. The burnt-out firebrand was imprisoned but then released on account of his advanced age by the kindly Norwegians. Hamsun has been admired and hailed by such unlikely figures as Maxim Gorky, Thomas Mann, and Isaac Bashevis Singer (go figure).

Knut Hamsun's literary legacy to Norway, America, and the world, appears to embody a state-sponsored individual liberty (i.e., libertinism), along with a lack of individual responsibility. Today's Socialist Norway and Libertarian America seem to personify much bread (to assuage hunger, want) and many circuses (to make life amusing, meaningful), together with a state-sponsored personal freedom (i.e., utter selfishness unencumbered by individual accountability).

Notes

1. From Isaac Bashevis Singer, 'Knut Hamsun, Artist of Scepticism', Introduction to <u>Hunger</u>, p.V. See source below.

2. <u>The New Encyclopaedia Britannica</u>, 'Micropaedia', Vol.5 (Chicago: Encyclopaedia Britannica, Inc., 1990 ed.), p.671.

Bibliography

Hamsun, Knut. <u>Hunger ("Sult")</u>. The Noonday Press; Farrar, Straus and Giroux, New York, 1996 ed. Translated by Robert Bly. (Orig. published in 1890).

Author's note: There is now <u>Hamsun</u> the movie! Jan Troell's film, released in 1997, stars Max von Sydow as Knut Hamsun. It has been critically acclaimed by James Bowman, U.S. editor of the <u>Times Literary Supplement.</u> (The American Spectator, December 1997)

Appendix D

"17 May In Our Hearts"
(Syttende mai i våre hjerter)

One of my earliest memories is of 17 May 1948 in Oslo, Norway. I was standing somewhere behind, and to the side, of King Haakon VII and the entire Norwegian royal family, on the palace balcony. It was three years to the month since Norway had been liberated from a five-year Nazi occupation, and the rippling red, white, and blue flags of freedom flew everywhere. Smaller Norwegian flags were waved by the crowds of joyous citizenry trooping by the royal palace to honour their King. Haakon VII, the former Prince Carl of Denmark, stood lean and straight, every inch the King of Norway, acknowledging his subjects' loyalty and veneration.

Why was I there? As a Norwegian war-orphan, I had been adopted several months previously by Charles Ulrick Bay, the first American ambassador to Norway after World War II. But even at age four years, I sensed that this was a special day for Norway, Norwegians, and their King.

Since that 17 May, I have perhaps taken a too-cavalier attitude towards Norway's national day. After all, when Norway received her Constitution at Eidsvoll on 17 May 1814, it wasn't as if Norway had been liberated from a cruel and oppressive Danish régime. Danish cavalry did not ride through the streets of Christiania rounding up, or beating up, Norwegian dissidents. That was hardly the Danish--or the later Swedish--way! And the Nazi occupation of Norway surely didn't match in ferocity the brutal suppression of Poland...but it was, nonetheless, a sudden and shocking loss of sovereignty, that served as a rude wake-up call for Norway to reconsider what it really meant to be free and Norwegian.

Recent threats to Norwegian independence have been the neighbouring Soviet Union and an expansive European Union. "Well," as President Ronald Reagan would have said, the "Evil Empire" collapsed at the end of the 1980s, and in the mid-1990s Norway wisely voted to stay out of the clutches of the Evil Eurocrats at Brussels. They only existing threat to Norwegian freedom--in my opinion--is an impacted, leftist leadership in Oslo which is all too willing to genuflect to ideas of global governance.

"17 May In Our Hearts" (cont).
(Syttende mai i våre hjerter)

Today, the 17th of May is being celebrated all over Norway and also in Bay Ridge, Brooklyn; Ballard, North Seattle; and in Burnaby, B.C. But a somber 17 May is being observed in those recently-flooded areas of North Dakota, Minnesota, and Manitoba where so many Norwegians (and other Scandinavians) in North America have chosen to live. Together we can wish them all well, cheering: "Hei hurra for den syttende mai!"

--Remarks made at Scandia Hall, S/N Olympic Lodge 2-037
Port Angeles, Wash. 17 May 1997.

Appendix **E**

Why I Have Resigned from Sons of Norway

Commencing in January 1999, the Sons of Norway <u>Viking</u> editorship promised their readers that most democratic of institutions -- a public forum. At long last, I told myself, it's about time; but it would prove too little, too late. A little less than a decade ago, I wrote two letters to (then) S/N International President Thorleif Bryn, protesting the much-ballyhooed 1989 S/N "Norwegian of the Year." For although a privileged track-star athlete living in Colorado, the Sons of Norway appointee had been caught cheating on her U.S. alien residency ("green card") status. My letters to the S/N International President elicited no response; I guessed that Mr. Bryn had been too busily engaged in his yearly custom-built Volvo exchange. A response from President Bryn would have meant a whole lot to this [then] new and lowly S/N member.

And the Sons of Norway <u>Viking</u> didn't have to field or reply to [what ought to have been an avalanche of] probing letters concerning the Mike "Lotsa Mozza" Colozza insurance scam-débâcle of several years ago. Consequently, Sons of Norway took its good old time in <u>finally</u> issuing a self-excusing, euphemised, explicatory statement. One could hear the foot-dragging from away out there in Minneapolis, Minnesota, to all the way back here in Port Angeles, Washington.

Last but not least, on their very first Letters-to-the-Editor page, the <u>Viking</u> featured [not one but] two letters (one each) from Alf Lunder Knudsen and his daughter, Kathleen. It has been the Knudsens, <u>père et fille</u>, who have dual-handedly destroyed the <u>Western Viking</u> newspaper; devolving <u>WV</u> from a friendly neighbourhood weekly to a politically-correct tabloid. Under the aegis of the Knudsens, no disturbing political or sociological trend in Norway was ever challenged; no critical letter to the editor was ever thoughtfully answered, or even treated with the respect it merited. I didn't renew my off-and-on eighteen-year-old subscription to <u>WV</u> at the end of 1994, and I know of at least one Seattle-area attorney, plus an[other] Everett-area engineer, who have quietly done the same.

With the publication of the <u>Viking</u>'s August 1998 puff-piece on <u>Western Viking</u> and the (like-minded) <u>Norway Times</u>, augmented by the self-laudatory letters from both Knudsens in the January 1999 issue, I realised that the Norwegian-American culture war was lost before ever being waged. All three periodicals will continue to sell a synthetic, sanitised, and saccharine Norway to their [mostly] elderly, [sadly] clueless readers. In this manner, without <u>any</u> reality checks, Norway can be safely sold (and profitably by a humungous travel/tourism-support industry) as a culturally condomised and mythological Fantasyland -- complete with all those cute trolls, elves (<u>nisser</u>), and sprites (<u>vetter</u>), which are so dearly beloved by ageing North American New Agers.

Ms. Liv Dahl, Sons of Norway Heritage Programs Manager and Foundation Administrative Director, has encapsulated the Order's vision of our Norwegian inheritance in a June 1997 <u>Viking</u> article, 'Living in a Fairy Tale':

"....[W]hen my grandson Espen comes to visit [Minnesota, U.S.A.] this year, I shall dust off the Asbjørnsen <u>og</u> (and) Moe stories of my childhood and prepare him for a fairy-tale life.
"More and more I find that I take time to ponder things. Occasionally I become aware of ideas of great value, and when I act on them, I am always amply rewarded. I nurture the fire in the fireplace, and the flickering shadows hint at trolls and princes and gleaming castles in a fairy-tale land.
"Now I know what my parents knew, and I know what they did for me was good and lasting. They told me the deepest Norwegian secret -- masked in metaphor and poetry -- that life is a fairy tale of your own creation."

Oh, <u>wow</u>! How utterly wet and weedy! But [seriously, <u>folkens</u>...] life in modern Norway is <u>not</u> a fairy tale, and Ms. Dahl's disingenuous drivel does a terrible disservice to all S/N members. A slow learner, I nonetheless wrote Ms. Dahl my sentiments which (again) have received no response. After fighting a fruitless and lonely <u>kulturkamp</u> in our local Lodge (Olympic No. 2-037) for a [round] dozen years, I accept the fact that I must go. I will greatly miss those Lodge members with whom I have worked and laughed, dined and talked, taught and visited; several whose memorial services I have attended.

In <u>the Rituals/Order of Business/Ceremonies for Sons of Norway</u> (Minneapolis: International Headquarters, 1978 ed.), there is a special section on the initiation of new S/N members. In the ceremony the Lodge president states:

"...Our aim is to preserve, maintain, and promote interest in everything that is good and noble in the Norwegian national heritage"....(p.28)

Perhaps the above was at one time an avowed goal, but it is no longer, in my conviction, the case today.

CDB-H   May 1999

Appendix F

Canada and the United States: Animist Beaver versus the Spread-Eagle

"We [Americans], who high-hat the British for their Empire, have already taken the empire of a hundred whole nations of [R]ed men, together with assorted chunks of sod we lifted from Spain, Canada, France and Mexico."
--Philip Wylie, Generation of Vipers, 1942

"Canada either is an idea or does not exist. It is either an intellectual undertaking or it is little more than a resource--rich vacuum lying in the buffer zone just to the north of the[G]reat [E]mpire."
--John Ralston Saul, Reflections of a Siamese Twin, 1997

Ever since the buzz-phrase "sleeping with an elephant"--likening Canada's precarious relationship to the United States--became part of popular parlance throughout Canada during the late 1960s/early 1970s, a minor host of Canadian historians have achieved small fame and mini-fortune on U.S.A.-bashing.[1] At first glance regarding the Canada/United States relationship, the average American (who knows where Canada is) would regard academic Canada's attitude as mostly unfair, totally uncharitable, and definitely ungrateful. After all, thousands and thousands of Canadians, from Dan Aykroyd to John Kenneth Galbraith, have acted as good mahouts; riding the useful American elephant to attain material goals scarcely achievable in Canada. The indignant American can demand, What better neighbour could the Canadians have?

But this American, who has lived in both Québec and British Columbia, will concede that--historically speaking--academic Canada has a point. The young American eagle was spreading her trans-continental talons with the annexation of the Texas Republic in 1845, and that dreaded phrase, "manifest destiny", was coined during the same year. By 1850 California was admitted as the 31st state of the American Union, after the entire U.S. Southwest had abeen wrested from Mexico. (The other territorial dominoes would fall later.) After a bitterly fought Civil War (1861-1865), the victorious Grand Army of the United States--battle-hardened and almost a million man-strong--still stood in uniform six months after cessation of North/South hostilities.[2] RetiredCanadian Major-General Richard Rohmer has vividly depicted the fear and loathing in late 1866 north of the Can-Am border:

"On a cold day in December 1866, a lonely man arrives in London, England, on the eve of the greatest enterprise of his already eventful life: uniting four very different colonies into a conferation--Canada.[3]

"John Alexander Macdonald[4] faces opposition at every turn. He is the target of Fenian assassins.[5] His colleagues at the London Conference are by no means unanimous in their enthusiasm for [con]federation. Meanwhile, American expansionism, backed by the enormous post-Civil War U.S. Army, threatens to absorb the British colonies.

"When Macdonald hears diplomatic rumours of a possible U.S. purchasse of Russian North America (Alaska), he conceives a daring plan in order to prevent the subsequent annexation by the United States of far-off, vulnerable British Columbia..."[6]

Most of my readers are well aware of the American purchase of Alaska in 1867, engineered by U.S. Secretary of State William H. Seward ("Seward's Folly")[7], and the semi-final results of "Fifty-four forty or fight"[8] with the Canadian incorporation of British Columbia in 1871. I write "semi-final" for the Alaska/B.C./Yukon border would not be diplomatically settled until 1903, with the precise Can-Am boundary delimited during 1905. Briefly stated, gold was discovered at Bonanza Creek in Canada's Yukon Territory in 1897; the fifty thousand prospectors, prostitutes, gamblers, and gunmen rushing to the Klondike were overwhelmingly of U.S. Origin. An alarmed Ottawa quickly established a Yukon territorial government and, as swiftly, dispatched a 200-man "Yukon Field Force" of regular troops--sent overland from British Columbia. The Yukon Field Force was organised to (1) assist the local North West Mounted Police in maintaining order, and (2) to assert Canadian sovereignty.[9]

In May 1898, an Anglo-Canadian-American conference at Washington, D. C., set up a trilateral Joint High Commission to resolve all border questions, but all fell apart in less than a year with no agreement on Alaska/B.C./Yukon arbitration. And so it went, dragging on until the Hay-Pauncefote Treaty of 1901, wherein Great Britain lost her right to participate in the Panama Canal, thus forfeiting a crucial bargaining chip on the Alaska/B.C./Yukon boundary.[10] Also during 1901 and worse still for Canada, U.S. President William McKinley fell to an assassin's bullet at Buffalo, New York, and was replaced by Vice-President Theodore Roosevelt (1858-1919)--the truculent bully-boy and conspicuous recent Spanish-American War hero of San Juan Hill. As president, the aggressive Roosevelt lost no time in refusing third-party arbitration; choosing instead to call for a six-man commission composed of three participants from each side. Roosevelt's "impartial" appointees included the U.S. Secretary of War Elihu Root, Senator Henry Cabot Lodge of Massachusetts, and former G.O.P. senator from Washington State, George Turner(!). The Anglo-Canadian side was made up of the lieutenant-governor of Québec, a prominent Toronto lawyer, and British representative Lord Alverstone, chief justice of England. Teddy Roosevelt in his belligerent, blustering way privately warned the British that, should the Commission fail to sustain the American position, he, the President, would deploy the U.S. Armed Forces to secure the American position.[11]

North American historians John Herd Thompson(U.S.A.) and Stephen J. Randall (Canada) have commented extensively on the lop-sided 1903 Alaska/B.C./Yukon boundary dispute, and its lasting (malign) influence on subsequent Can-Am relations:

> "The outcome of the [C]ommission's inquiry is infamous in Canadian historical mythology. Lord Alverstone, this version runs, sold out Canada in the interest of Anglo-American accord, and the two Canadians on the panel accordingly refused to sign the decision. The [C]ommission awarded...the vital coastal strip north from fifty-six degrees forty minutes north latitude to the United States.[12] There were minor concessions to Canada...of little consequence...
> ..."The Alaskan boundary dispute was a decisive moment in Anglo-Canadian relations with the United States. Its resolution removed what Theodore Roosevelt referred to as the last serious point of friction between the United States and Great Britain. The appointment of a majority of Canadians [Sir Louis Jetté and Allan B. Aylesworth] to the boundary commission proved useless in guarding Canada's interests, but its failure was part of the emergence of an independent Canadian tradition in foreign policy. For Canadians, the boundary dispute was a spur to Canadian nationalism, anti-Americanism, and more than a tinge of resentment against British authority, a sentiment that tended to give English and French Canadians common cause..."[13]

At the dawn of the 20th century, the United States had gained political control over the Philippines, Guam, American Samoa, Hawai'i, Panama, Cuba, and Puerto Rico. Canada had been given a brief taste of bully jingoism by Great Britain in her involvement during the Boer War (1899-1902), but the Spanish-American War (1898) had crowned the U.S.A. as a world power/imperial player. Thompson and Randall have wryly assessed that [Canadian] "Neighbors to the north could only watch with their usual combination of envy and tense expectation."[14]

Reflections on John Ralston Saul's "Siamese Twin"

"What could it possibly mean to say that Canada is an animist country?...Animism is the opposite of a goal-oriented idea of social order."[15] ..."What then about the impossible idea that all people belong to all communities? It could be said that this mythology is Canada's contribution to the ongoing western debate over the nature of nationalism"...[16]

In 1997, Viking (Canada) published John Ralston Saul's <u>Reflections of a Siamese Twin: Canada at the End of the Twentieth Century</u>. Since release, John Ralston Saul's book has received the acclamation of the

Canadian "fourth estate", and the adoring (reading) public has made Reflections... #1 Bestseller in Canada. All right (i.e., left)-thinking Canadians, tout the reviewers, should read John Ralston Saul's Reflections. Saul's book was highly recommended to me by an earnest young woman at Chapters bookstore in Victoria, B.C., who assured me that Reflections was the defining last word on Canada at century's end. I bought the ponderous (500+ pp.) volume with some trepidation, as I badly needed a political update on Canada (since Referendum 1995) for Future Fish 2001.

My initial reservations were well founded. For Mr. John Ralston Saul turned out to be one of those social élites who always knows what's best for the public sector; vid., "the people." Although constantly referring to the "élites"--indeed he has devoted an entire chapter in Reflections to élites "--Mr. Saul is hardly a member of the Great Unwashed ("the people") himself. The "Siamese Twin" in Saul's title is Anglophone/Francophone Canada, not Canada/U.S.A. In Reflections, the U.S.A. is summarily dismissed as the Evil Empire to the South--far, far away spiritually (the farther the better) from Canada; True North. How else can one explain such absurd observations as:

(1) "[T]he dominant anti-intellectual populism--real and false--which has been central to American history",[17] or (2) "[T]he American ideal--from its very beginning--contained, indeed survived upon, slavery"[18]? And again, "As the century wears on, so a very different American view penetrates ever deeper into our society, thanks to their domination--direct and indirect--of every imaginable means of communication. By their [American] definition, individualism is limited to equality of opportunity, the result being real inequality."[19] Huh? Say what?

Mr. John Saul Ralston's Reflections on Canada, this former Canadian resident avers, are equally nonsensical and plainly asinine. He even disparges the widely-held perception of Canada's possessing "European" qualities. In fact, John Ralston Saul seems to condemn everything in Canadian culture that smacks of anything European, Christian, bourgeois. Saul decries "corporate" globalism, but embraces planetary multiculturalism. To this writer, surely the two go hand-in hand? (www.disney.com?) No sector, public or private, is so prissily precious as the CBC/PBS Can-Am cultural élite. (There's that word again!) Saul loudly sings the praises of such jet-set Canadians as creepy-crawly neo-Darwinist David Suzuki, pompous eco-windbag Maurice Strong, matrifocal gender-feminist Margaret Atwood, and post-Christian Jungian Robertson Davies.[20] Yike! Those are the very Canadians this American has for years regarded with fear and loathing!

(However, John Ralston Saul can not understand why Lucien Bouchard's mother, who lives at Lac Saint-Jean, seems so...French provincial!)[21]

But Mr. John Ralston Saul surely knows Canada best; after all, his grand polemic has sold widely among Canada's governing and clerisy classes...and Reflections obviously poses no threat to those who really rule Canada. In two revealing passages, for example, Saul demonstrates his deeply-held commitment to Leftist solidarity:

(1) "No Canadian government has ever been defeated in a general election by a party running to its right. In other words, Canadians have never consciously voted for the choice of the right,"[22] and (2) "[T]he Health Ministry and Medicare have become symbols for the humanist idea of society which Canadians keep insisting they want."[23]

Interlarded throughout his contentious Reflections are Saul's bugbears: The Château Clique, "corporatism", Family Compact, Montréal School, "neo-conservatives", the Orange Order, Ultramontane movement; and bogeymen: William Aberhart, Lucien Bouchard, Bishop Bourget, Maurice Duplessis, Abbé Groulx, Mike Harris, Ralph Klein, Bernard Landry, both Ernest and Preston Manning, D'Alton McCarthy, Honoré Mercier, Brian Mulroney, Jacques Parizeau. (There are so many Canadian institutions and Canadian leaders which John Ralston Saul intensely dislikes, that I have listed them in alphabetical order.) A basic premise of Saul's has been that centralised government by liberal élites in Canada must never ever be threatened by American-type "false populism" or American-type economic "corporatism".

Gadzooks! It took a Herculean effort for me to finally get through Reflections of a Siamese Twin. Whatever happened to those pragmatic, practical Canadians I have encountered in greater Montréal or on Vancouver Island? I seriously doubt that Reflections was a bestseller among the denizens of Chicoutimi, Québec, or Sudbury, Ontario. (Or those of Prince Rupert, B.C., either.) My final reflection on John Ralston Saul's Siamese Twin is that now I might believe the apocryphal tale told of two hockey teams, one Canadian and one American. When the Canadian hockey team plays away games south of the border, it is cheered. But when the American hockey team plays away games north of the border, it is jeered. Or is this only the small, meaner Canada projected by John Ralston Saul?

NOTES

1. A notable example of this has been Pierre Berton of Pierre Berton Enterprizes Ltd.

2. Richard Rohmer, <u>John A.'s Crusade: A Novel</u> (Toronto: Stoddart Publishing Co., Limited, 1995, p.34.

3. Upper Canada (Ontario), Lower Canada (Québec), New Brunswick, and Nova Scotia.

4. Known affectionately as Sir John A and The Old Chieftain (1815-1891). P.M. of Canada 1867-1873, 1878-1891.

5. Richard Rohmer, pp. 36, 91. Author's note: The Irish Republican Brotherhood--the Fenians--was composed of several thousand Irish-Americans who had fought for the Union in the Civil War. The Fenian volunteers launched a series of violent forays into British North America between 1866 and 1871, somehow hoping to precipitate Anglo-American hostilities which would (somehow) result in an independent Ireland (Éire). There was no official U.S. complicity in the Fenian raids and, though bloody, the incursions ended in failure. But the Fenian invasions helped galvanise the British North American colonies to confererate into the Dominion of Canada, 1867. (Manitoba was incorporated in 1870 and British Columbia in 1871.) See Thompson and Randall (1994), p. 38.

6. Op.cit., Richard Rohmer, inside covers. NB: General Rohmer is also the author of <u>Ultimatum</u> (1973), about a U.S. takeover of Canada after heroic resistance (!).

7. William H. Seward served as U.S. Secretary of State from 1861-1869. British prime minister Lord Palmerston described Seward as a " `vapouring, blustering, ignorant man' ". (In Thompson and Randall, p. 35).

8. A U.S. political slogan used in 1846, especially popular among expansionist Democrats, which expressed the desire for American ownership of the <u>entire</u> Oregon country--all the way up to 54° 40` North. This included that part of British North America claimed north of 49° North.

9. John Herd Thompson and Steven J. Randall, <u>Canada and the United States: Ambivalent Allies</u> (Athens and London: The University of Georgia Press, 1994), p. 66.

10. <u>Ibid</u>., p. 67.

11. <u>Ibid</u>., p. 68.

12. NB: Today's Southeast Alaska, U.S.A.

13. Op.cit. Thompson and Randall, pp. 68-69. NB: All brackets and contents this writer's. See also J.R. Saul (1997), pp. 371-372.

14. <u>Ibid</u>., pp. 66, 65. Author's note: Thompson and Randall are the only historians I know of who have referred to Chiang Kai-Shek in his pinyin-transliterated name; thus "Jiang Jieshi" (p. 186).

15. John Ralston Saul, <u>Reflections of a Siamese Twin: Canada at the End of the Twentieth Century</u> (Toronto: Penguin Books Canada Ltd., 1998), p. 185.

16. <u>Ibid</u>., p 438.

Notes (con't)

17. Ibid., 76-77.

18. Ibid., p.123.

19. Ibid., pp. 505-506. NB: Brackets and emphasis mine.

20. Ibid., pp. 99-100; passim.

21. Ibid., pp. 448.

22. Ibid., p. 66. Author's note: Does Saul really believe that? Or is it just more Saul Canadian myth?

23. Ibid., p.496.

Canada Bibliography

1. Rohmer, Richard. Sir John A's Crusade: A Novel. Stoddart Publishing Co., Limited, Toronto, 1995.

2. Saul, John Ralston. Reflections of a Siamese Twin: Canada at the End of the Twentieth Century. Penguin Books Canada Ltd., Toronto, 1998.

3. Thompson, John Herd and Stephen J. Randall. Canada and the United States: Ambivalent Allies. The University of Georgia Press, Athens and London, 1994.

Appendix G

Revisionist Winston Churchill Bibliography

1. Asprey, Robert B. War in the Shadows. Garden City: Doubleday, 1975.

2. Bohlen, Charles E. Witness to History: 1929-1969. New York: Norton, 1973.

3. Ciechanowski, Jan. Defeat in Victory. New York: Doubleday, 1947.

4. Colby, Benjamin. 'Twas a Famous Victory. New Rochelle: Arlington House, 1974.

5. Fair, Charles. From the Jaws of Victory. New York: Simon and Schuster, 1971.

6. Flynn, John T. As We Go Marching. New York: Free Life, 1944.

7. Gilbert, Martin, ed. Churchill. Englewood Cliffs: Prentice-Hall, 1967.

8. Halle, Kay, ed. Winston Churchill on America and Britain. New York: Walker, 1970.

9. James, Robert Rhodes. Churchill: A Study in Failure, 1900-1939. New York: World, 1970.

10. Mee, Charles L., Jr. Meeting at Potsdam. New York: M. Evans, 1975.

11. Payne, Robert. The Great Man. New York: Coward, McCann & Geoghegan, 1974.

12. Simpson, Colin. The Lusitania. Boston: Little, Brown, 1972.

13. Taylor, A.J.P. et al. Churchill Revised. New York: Dial Press, 1969.

14. Thompson, R.W. Generalissimo Churchill. New York: Scribner's, 1973.

15. _____. Winston Churchill: The Yankee Marlborough. Garden City: Doubleday, 1963.

16. Veale, F.J.P. Advance to Barbarism. New York: Devin-Adair, rev. 1968 ed.

Source: Bibliography gleaned from William P. Hoar, Architects of Conspiracy: An Intriguing History (Belmont, Mass.: Western Islands Publishers, 1984), passim.

Appendix H

<u>The Late, Great United States: [In]Famous in [Pop]Song, [Wild] History, and [Urban] Legend</u>

"Ignorance is not bliss—it is oblivion. Determined ignorance is the hastiest kind of oblivion. Yours [America's] is the most inexplicably determined in all the swing of the centuries. So your oblivion will be the greatest."
--Philip Wylie, <u>Generation of Vipers</u>, 1942[1]

" '[American] [t]radition...sometimes brings down truth that history has let slip, but it is oftener the wild babble of the time, such as was formerly spoken of at the fireside and now congeals in newspapers...' "
--Nathaniel Hawthorne, <u>The House of the Seven Gables</u>, 1851[2]

So far, halfway through Year Zero in the new millennium, my favorite ad on US. television has been one for Ameritrade. The commercial shows a stressed middle-aged businessman, "Phil" (probably an Italian-American), who is tired, fed-up, and has just entered a "personal grooming" salon. While having one of his hairy hands manicured by an Asian-American woman, Phil is startled to hear that other salon employees (women of varying shades of colour) are busy trading on-line... and...right then and there—are actually <u>receiving</u> e-mailed information on the results of their purchases! Phil is so impressed by what he sees Ameritrade doing for its subscribers, that he leans back in his chair and calls for "Bobby"—the salon's single male (white, gay?) employee—to give him the full tonsorial works. All is well, everyone is happy, and life is good in the U.S. urban heartland. People of all ages, races and sexual proclivities get along (is the message), and multiculturalism is working in America.

The writer finds this particular advertisement at once pleasing and saddening. Although depicting an overall warm and fuzzy image of the cut-out characters concerned, the briefly moving picture is a false portrayal of racial relations in End-Time America. And therein lies the rub: What <u>is</u> to be believed in America at the very start of the new century? Are rainforests disappearing (and at what rate)? Is there global warming and is there an ozone depletion? Are both caused by human activity? World crisis mongering and rumours of environmental devastation have been jump-started by the Lyin' King White House since it's inception in 1993. And there are unanswered allegations, implications, and inferences put out there by our supremely evil arbiters of pop-culture: <u>Is</u> Promise Keepers a "hate" group? <u>Can</u> the

Confederate States of America (1861-1865) be equated to National Socialist Germany (1933-1945) or Imperial Japan (1895-1945)? One wonders as all this dross is fast becoming part and parcel of American current wisdom (cw).

Indeed, mainstream current wisdom in America is fast becoming indistinguishable from "urban legend." According to Jan Harold Brunvand of the U. of Utah, American urban (i.e., modern) legends are usually… "highly captivating and plausible, but mainly fictional, oral narratives that are widely told as true stories…[and] even a new story partly disseminated by the mass media soon becomes folklore if it passes into oral tradition and develops variations."[3] Although the phrase "urban legend" has only been in popular usage since the late 1970s/early 1980s in America, iconoclastic writer Phil Wylie wrote of "bastard legends" in 1942:

"In order to perpetrate upon ourselves the monstrous half-consciousness by which we have been living, we have, perforce, developed bastard legends. These legends suit half of a human personality, most admirably. Their development, as such, was unconscious—for we chose unconsciousness by our act of spiritual self-aggrandizement and the repudiation of the dark half of our heritage. The phenomenon, to wiser and more detached observers than most of us, in itself supplies an amusing description of our fallacy. Like the worship of any false god, it reveals its counterfeit nature by what it omits—even though the worshiper chooses to regard it as whole."[4]

The difference being that, during the 1990's the development of "bastard legends" has been a conscious endeavour by those same evil arbiters of American pop-culture and pop-history. How else might one explain the theory (i.e., new urban legend) that White (viz., Jewish) doctors have purposefully injected (infected) Black America with the HIV? How else can one explicate another bastard legend that—with extreme prejudice—crack-cocaine was spread throughout [Black] South Central L.A. by the (evil White) C.I.A.? Millions of Black Americans today truly believe both cruel lies. The very same folks that disseminated these destructive urban (inner city) legends are those identical evil arbiters of American pop-history who have distributed apocryphal cable TV offerings like "The Tuskegee Airmen" (1995) and "Buffalo Soldiers" (1999) in which Black Americans are shown that they—and they alone—(1) won the Great War, and (2) conquered the Wild West. Movies and myths are powerful cultural forces, and they are being effectively utilized by those arbiters to set up ("blaxploitation") the Black American population for a nefarious ("divide and conquer") socio-political agenda. It is the cultural equivalent of crisis/fear mongering in the worst way…and it's all bastard/urban legend!

To research for Appendix H, I have had the pleasure of reading several "revisionist/alternative" history books. Philip Wylie would have called this an examination of "the dark half of our heritage."[5] To do this, I had to dig up so-called conspiratorialist histories of the U.S.A., from both the Far Left and Far Right. (Philip Wylie, albeit briefly, does a good job himself. Imagine criticising Mom, Coca-Cola and the American Way during World War II!) As sources, I used two revisionist works by members of the John Birch Society, and an alternative work by Black comedian Dick Gregory.[6] The reader might think that Black activist Gregory and White "Bircher" A. Ralph Epperson would have no historical opinions in common—but the reader would be wrong! Far Left and Far Right revisionist/alternative U.S. histories are eerily alike...such as fraternal twins. Indeed, Gregory and Epperson's works share three commonalities: (1) Harbouring a virulent hatred for the leading American capitalist families of Rockefeller, Morgan, and DuPont; (2) making mention of the natural cancer-cure found in apricot seeds; and (3) repeated references to Edward Gibbon's Decline and Fall of the Roman Empire (1776-1788) and America's parallel history. (In 1952, Wylie commented that..."The Romans traded bread and circuses for their tentative tyrannical powers."[8] Sounds like the present [Billy Liar] administration's m.o. to me!)

The value in reading revisionist/alternative history is found in the questions which are asked; questions such as: Did Franklin D. Roosevelt know of the impending attack on Pearl Harbor (7 December 1941) beforehand? Or, did President Harry S. Truman know of Japanese attempts to communicate their intentions of surrendering before dropping "Little Boy" (the uranium gun-type bomb) on Hiroshima (6 August 1945) and "Fat Man" (the plutonium implosion-type bomb) on Nagaskai (9 August 1945)? U.S. historians—of all persuasions—are still studying both of these disturbing questions more than a half-century later. Any answers other (i.e., "bastard legends") than the absolute truth would be a gross insult to those who perished at Pearl Harbor, Hawai'i, and at Hiroshima and Nagasaki, Japan.

In A.D. 2000, revisionist/alternative history is equated with bastard/urban legend by mainstream academia—as it has been in America since at least World War I (1914-1918). But even thirty years ago, there was that strange similarity between Far Left and Far Right history. In his 1970 screed Do It!, Jerry Rubin (he of the notorious Chicago Seven) wrote "The right wing is usually right too—The John Birch Society understands the world we live in better than fools like [liberal historian] Arthur Schlesinger Jr. and [liberal columnist] Max Lerner who don't know what—is happening!"[9]

So what—or whom—to believe in America 2000? Not much--or many—for sure. The entire year of 1999 was dominated by the looming Y2K collapse, a technological melt-down, after which life in America might never be the same again. Simultaneously, while Rome burned, the recently impeached Nero-like President Clinton played his saxophone and raised yet more Democrat mega-bucks; aided and abetted by public offender, Gollywood's Robin Williams. Meanwhile, at the millennium's end, the presidential consort spoke darkly of a "vast right-wing conspiracy" seeking to unseat her fraudulent husband from power. Were the First Lady's words indicative of a paranoid personality, or merely the epigraph of yet another new urban legend?

The average American will never know, and he is rightfully confused—tsedrayt, tsetummelt, and tsedoodelt (all three Yiddish words meaning "confused")[10] –and I go back to watching old movies, The X-Files, and "Scream T.V."

NOTES

1. Pocket Cardinal, New York, N.Y., 1955 ed. NB: Brackets and contents mine.

2. Quoted by Jan Harold Brunvand, <u>The Choking Doberman and Other "New" Urban Legends</u> (New York: W. W. Norton & Company, 1984), p. 150.

3. Op.cit., Jan Harold Brunvand, pp. IX, X. NB: Emphasis mine.

4. Op.cit., Philip Wylie, <u>ibid</u>., Ch. V, 'A Specimen American Myth', p. 43. NB: Emphasis mine.

5. <u>Ibid</u>., p. 43.

6. Author's note: In Gregory's book (see Bibliography) there were the expected mini-hagiographies of the usual Great Black Men (Turner, Vesey, Douglass, King et al.), but nary a mention of Booker T. Washington. Booker T. was not merely a Great Black Man, he was one of the greatest <u>Americans</u> who ever lived! Did Gregory consider Washington just another ol' "Uncle Tom"?

7. Author's note: Indeed, A. Ralph Epperson devotes an entire chapter to the healing powers in, and unavailability of, (in the USA), laetrile.

8. Op.cit., Philip Wylie, p. 212.

9. Jerry Rubin, <u>Do It</u>! (New York: Ballantine Books, 1970), p. 148. Cited by A. Ralph Epperson, p. 428.

10. Thank you, Leo Rosten, for <u>The Joys of Yiddish</u> (New York, N.Y.: Simon & Schuster, 1991 ed.).

An Alternative U.S. Bibliography

1. Brunvand, Jan Harold. The Choking Doberman and Other "New" Urban Legends. W. W. Norton & Company, New York, 1984.

2. Epperson, A. Ralph. The Unseen Hand: An Introduction to the Conspiratorial View of History. Publius Press, Tucson, Arizona, 1985.

3. Gregory, Richard Claxton (Dick). No More Lies: The Myth and Reality of American History. Harper & Row, Publishers, New York, 1972 ed.

4. Heard, Alex. Apocalypse Pretty Soon: Travels in End-Time America. Doubleday, New York, 1999.

5. Hoar, William P. Architects of Conspiracy: An Intriguing History. Western Island Publishers, Belmont, Mass., 1984.

6. Wylie, Philip. Generation of Vipers. Pocket Books, Inc., New York, rev. 1955 ed. (Orig. published in 1942)

Appendix I

A Pacific Fish and Shellfish Gazetteer[1,2,3]

	Common Name	Scientific Taxonomy
Crabs:	Blue king crab	Paralithodes platypus
	Brown/golden king crab	Lithodes aequspina
	Centollas = Chilean king crabs	
	langostino colorado	Pleuroncodes monodon
	langostino amarillo	Cervimunida johni
	Japanese snow crab	Chinoecetes japonicus
Mussel:	Pacific Coast blue mussel	Mytilus trossilus
Clam:	Geoduck clam	Panope abrupta (alternate taxonomy)
Scallops:	Chilean farmed scallop	Argopecten purpuratus
	Sea scallop	Placopecten magellanicus
	Bay scallop	Argopecten irradians
	Calico scallop	Argopecten gibbus
	Weathervane scallop	Patinopecten caurinus
	Chinese (native) bay scallop	Chlamys farreri
Shrimps:	Black tiger shrimp	Penaeus monodon
	Green (grooved) tiger shrimp	P. semisculatus
	Brown tiger shrimp	P. esculentus
	Banana shrimp	P. merguiensis
	Western white shrimp	P. vannamei
	Japanese kuruma ebi	P. japonicus
	Yellow shrimp	Metapenaeus brevicomis
Squids:	Big Peruvian squid	Dosidicus gigas
	Seven star flying squid	Martialia hyadesi
Oysters:	Kumamoto oyster	Crassostrea sikamea
	Suminoe oyster	C. rivularis
	Eastern oyster	C. virginica
	European (Belon) "flats"	Ostrea edulis
	Chiloe oyster	O. chilensis
	Olympia oyster	O. lurida
Finfish:	Yellow-fin sole	Limanda aspera
	Dover ("slime") sole	Microstomus pacificus
	California halibut	Paralichthys californicus
	California white sea bass	Cynoscion nobilis
	Rainbow trout	Oncorhynchus mykiss
		(new taxonomy)
	Goosefish/monkfish	
	(Japanese ankō)	Lophius setigerus
	Swordfish	Xiphias gladius
	Sickle pomfret	Taratichthys steindachneri
	Lustrous pomfret	Eumegistus illustris
	(Hawaiian monchong)	

A Pacific Fish and Shellfish Gazetteer (cont'd.)

Antarctic whiting, hake	Merluccius australis
Silver hake, whiting	M. bilnearis
(Spanish merluza)	
Kingklip cusk-eel, NZ ling	Genypterus blacodes
(Chilean congrio)	G. chilensis

NEW ZEALAND FISHES[4]

Hoki	Macruronus novaezelandiae
Orange roughy	Hoplostethus atlanticus
Smooth oreo dory	Pseudocyttus maculatus
Black oreo dory	Allocyttus sp.
John Dory	Zeus faber
Blue mackerel	Scomber australasicus
Southern blue whiting	Micromesistius australis
Seabream (NZ snapper)	Pagrus auratus
Red seabream	Chrysophrys auratus
Blue nose	Hyperoglyche antarctica
School shark	Galeorhinus australis
Hapuka (NZ grouper)	Polyprion oxygeneios
Paua (NZ abalone)	Haliotis iris
Arrow squid	Nototodarus sloani
Rock lobster	Jasus edwardsii
NZ greenshell mussel	Perna canaliculus
Littleneck clam	Austrovenus stutchburyi
Sea scallop	Pecten novaezealandiae
Queen scallop	Chlamys Delicatula (sic)

HAWAIIAN FISHES

	Local Name
Bigeye/yellowfin tuna	ahi
skipjack tuna	aku
bonito	auhopo
Shortbill spearfish	hebi
Blue marlin	kajiki
Hawaiian grouper	kapu'upu'u
Striped marlin	nairagi
Hwn. snapper	onaga
Wahoo	ono
Hwn. snapper	opakapaka
Albacore	tombo
Hwn. grey snapper	uku

Sources:

1. <u>Alaska Fisherman's Journal</u>, Seattle, Wash.
2. <u>Seafood Leader</u> magazine, Seattle, Wash.
3. <u>The Register-Guard</u> newspaper, Eugene, Ore.
4. New Zealand Fishing Industry Board, Wellington, N.Z.

A Pacific Fish and Shellfish Gazetteer (cont'd.)

FISHES OF JAPAN[5]

abalone (small)	tokabushi
anchovy	katakuchi-iwashi, shiko-iwashi
angler, goosefish	ankō
Atka mackerel	hokke
blowfish, puffer	fugu
bonito, skipjack tuna	katsuo
capelin	shishamo
charr	iwana
cuttlefish	mongo-ika
fan mussel	taira-gai
flying fish	tobiuo
flounder (right-eye)	ohyō
Greenland halibut	karasugarei
Japanese bluefish	mutsu
Japanese scallop	hotate-gai
	(Pecten yessoensis)
little neck clam	asari
Pacific barracuda	kamasu
perch	akauo
pilchard	ma-iwashi
pollock	suketōdara
saury	sanma
sea bass	suzuki, fukko, seigo
sea bream	amadai
sea bream (black)	kurodai
sea bream (red)	madai
sea cucumber	namako
sea mussel	igai
shark	fuka
smelt	kisu
sockeye (red) salmon	beni-zake
Spanish mackerel	sawara
sturgeon	chōzame
swordfish	makajiki, shutome, tachiuo
trout (rainbow)	nijimasu
tuna (bluefin)	kuro-maguro
tuna (yellow fin)	kihada-maguro
yellowtail, jack	buri, hamachi, inada, warasa, wakashi

Sources:

5. Species above were those not listed in Fisheries of the Pacific Northwest Coast Vol. 1 (New York: Vantage Press, Inc., 1991).

6. Camy Condon and Sumiko Ashizawa, The Japanese Guide to Fish Cooking (Tokyo: Shufunotomo Co., Ltd., 1990).

456

A Pacific Fish and Shellfish Gazetteer (cont'd.)

THE TUNAS: TRIBE TUNNINI, FAMILY SCOMBRIDAE[7,8]

Name:

	English	Scientific	Japanese	Hawaiian
Species:	Albacore	(Thunnus alalunga)	tombo	ahipalaha
	Bigeye tuna	(Thunnus obesus)	mebachi/shibi	ahipo'o-nui
	Black skipjack	(Euthynnus lineatus)	suma-zoku	
	Blackfin tuna	(Thunnus atlanticus)	taiseiyōmaguro	
	Bluefin tuna	(Thunnus thynnus)	kuromaguro	
	Bullet mackerel	(Auxis rochei)		
	Frigate mackerel	(Auxis thazard)		
	Kawakawa	(Euthynnus affinis)	yaito	
	Little tunny	(Euthynnus alletteratus)	sama-rui	
	Longtail tuna	(Thunnus tonggol)	koshinaga	
	Skipjack tuna	(Katsuwonus pelamis)	katsuo	aku
	Southern bluefin	(Thunnus maccoyii)	minamimaguro	
	Yellowfin tuna	(Thunnus albacares)	kihada	ahimalailena

Commercial Fish Species in the Federated States of Micronesia[9]

Marketing name	Scientific name	Japanese name
Yellowfin tuna	(Thunnus albacares)	kihada
Bigeye tuna	(Thunnus obesus)	mebachi
Skipjack tuna	(Katsuwonus pelamis)	katsuo
Black jack	(Caranx lugubris)	kappore
Amberjack	(Seriola rivoliana)	hirenaga-kanpachi
Red snapper	(Lutjanus bohar)	barafuedai
Rock cod	(Plectropoma laevis)	sujiara
Onaga	(Etelis coruscans)	hamadai
Ehu	(Etelis carbunculus)	hachijyo-akamatsu
Opakapaka	(Pristipomoides flavipinnis)	kimme fuedai
Lehi	(Aphareus rutilans)	ohguchi-ishichibiki
Uku	(Aprion virescens)	aochibiki
Mahi mahi	(Coryphaena hippurus)	shiira
Wahoo (Ono)	(Acanthocybium solandri)	kamasusawara (okikamasu)

Sources:

7. <u>1982 Word Record Game Fishes</u>. IGFA, Fort Lauderdale, Fla., 1982.

8. "Hawaii Seafood". State of Hawaii, Ocean Resources Branch; Honolulu, Hawaii, 1993.

9. National Fisheries Corporation, Kolonia, Pohnpei, FSM. NB: In English, the Hawaiian marketing name of <u>ehu</u> is squirrelfish snapper, <u>lehi</u> is silverjaw jobfish, <u>onaga</u> is ruby/longtail snapper, <u>opakapaka</u> is yelloweye snapper, and <u>uku</u> is gray jobfish/snapper. These are all reef-fishes. Thanks to Hawaiian TV programme, "Let's Go Fishing", with Hari Kojima, which was popular during my time in the Islands (1978-1982).

A Pacific Fish and Shellfish Gazetteer (concl'd.)

FISHES OF PALAU[10]

Name

Palauan	English	Scientific
Species: Cherapruk 1	spiny lobster	Panulirus sp.
Chersuuch	mahimahi (dolphinfish)	Coryphaena hippurus
Choas	sea cucumber	Holothuria atra
Dub	fish poison made from	Derris root
Katsuo (from Japanese)	skipjack tuna	Katsuwonus pelamis
Keskas/mersad	wahoo	Acanthocybium solandri
Manguro (from Japanese)	yellowfin tuna	Thunnus albacares
Mokorkor/ Biturchturch	shark/double line mackerel	Grammatorcynus bicarinatus
Ngelngal	Spanish mackerel	Scomberomorus commersoni
Soda/katsuo (from Japanese)	mackerel tuna	Euthynnus affinis
Tkuu	yellowfin tuna	Thunnus albacares
Wii	golden jack	Gnathanodon speciosus

Giant Clams of Kiribati (TRIDACNIDAE)[11]

Name

Gilbertese	Scientific
Te kima	T. gigas
Te were	T. maxima
Te were matai	T. crocea
Te neitoro	H. hippopus

Source:

10. R.E. Johannes, Words of the Lagoon: Fishing and Marine Lore in the Palau District of Micronesia (Los Angeles: University of California Press, 1992 ed.).

11. Daniel Knop, Giant Clams; A Comprehensive Guide to the Identification and Care of Tridacnid Clams (Ettlingen, Germany: Dähne Verlag, GmbH, 1996).

Acknowledgements -- Cons and Pros

I received a rude shock when I tried to enlist the help of fellow fisheriographers within the Pacific Northwest industry. Brad M. of <u>National Fisherman</u>, Steve S. of <u>Pacific Fishing</u>, Wayne L. of <u>Seafood Leader</u>, Joe U. of <u>Alaska Fisherman's Journal</u>, and even local-yokel Harriet U. F. of Port Angeles, were all beseeched for help -- either directly or indirectly -- and chose to respond either marginally, evasively, or not at all. A literary pox on all their houses!

But my most sincere thanks go to Dr. Finnvald Hedin of Everett, Washington, for exemplary inspiration for all that is admirable in Norwegian - American Culture. And I must gratefully mention Mr. Francis E. Caldwell of Port Angeles, Washington (and Ketchikan, Alaska), for tacit reminders in keeping <u>Future Fish</u> on track as fisheriography. And, finally, without Ms. ("Miz") Ellie McCormick, owner/manager of Angeles Temporary Services, there would have been no <u>FutureFish</u> at all! Thanks to <u>you</u> all.

```
                                            C.D.B-H.
                                            Port Angeles, Wash.
                                            July 1998
```

ISBN 1553692934